Giampiero Palatucci, Tuomo Kuusi (Eds.)
Recent developments in Nonlocal Theory

Giampiero Palatucci, Tuomo Kuusi (Eds.)

Recent developments in Nonlocal Theory

Edited by
Managing Editor: Agnieszka Bednarczyk-Drąg
Series Editor: Gianluca Crippa
Language Editor: Adam Tod Leverton

DE GRUYTER

ISBN 978-3-11-057155-4
e-ISBN (PDF) 978-3-11-057156-1
e-ISBN (EPUB) 978-3-11-057203-2

Library of Congress Cataloging-in-Publication Data
A CIP catalog record for this book has been applied for at the Library of Congress.

Bibliographic information published by the Deutsche Nationalbibliothek
The Deutsche Nationalbibliothek lists this publication in the Deutsche Nationalbibliografie; detailed bibliographic data are available on the Internet at http://dnb.dnb.de.

♾ Printed on acid-free paper
Printed in Germany

www.degruyter.com

Contents

Preface

This book is devoted to recent advances in the theory and applications of Partial Differential Equations and energy functionals related to the fractional Laplacian operator $(-\Delta)^s$ and to more general integro-differential operators with singular kernel of fractional differentiability order $0 < s < 1$.

After being investigated firstly in Potential Theory and Harmonic Analysis, fractional operators defined via singular integral are currently attracting great attention in different research fields related to Partial Differential Equations with nonlocal terms, since they naturally arise in many different contexts. The literature is really too wide to attempt any reasonable account here, and the progress achieved in the last few years has been very important.

For this, we proposed the leading experts in the field to present their community recent results together with strategy, methods, sketches of the proofs, and related open problems.

The contributions to this book are the following,

Chapter 1: Heat kernel for nonsymmetric nonlocal operators

Z.-Q. CHEN and X. ZHANG present a survey on the recent progress in the study of heat kernels for a wide class of nonsymmetric nonlocal operators, by focusing on the existence and some sharp estimates of the heat kernels and their corresponding connection to jump diffusions.

Chapter 2: Fractional harmonic maps

F. DA LIO gives an overview of the recent results on the regularity and the compactness of fractional harmonic maps, by mainly focusing on the so-called *horizontal* $1/2$-*harmonic maps*, which arise from several geometric problems such as for instance in the study of free boundary manifolds. The author describes the techniques that have been introduced in a series of very recent important papers in order to investigate the regularity of these maps. Some natural applications to geometric problems are also mentioned.

Chapter 3: Obstacle problems involving the fractional Laplacian

D. DANIELLI and S. SALSA investigate fractional obstacle problems, by firstly presenting the very important results concerning the analysis of the solution and the free boundary of the obstacle for the fractional Laplacian, mainly based on the extension method. Then, the authors consider the two time-dependent models which can be seen as the parabolic counterparts of the stationary fractional obstacle problem as well as the Signorini problem in the cylinder, by discussing some regularity properties of the solutions and the free boundary.

https://doi.org/10.1515/9783110571561-001

Chapter 4: Nonlocal minimal surfaces: interior regularity, quantitative estimates and boundary stickiness

S. Dipierro and E. Valdinoci present various important results related to the surfaces which minimize the nonlocal perimeter functional. The authors discuss the interior regularity and some rigidity properties (in both a quantitative and a qualitative way) of these nonlocal minimal surfaces, together with their quite surprising boundary behavior.

Chapter 5: Eigenvalue bounds for the fractional Laplacian: a review

R. L. Frank reviews some recent developments concerning the eigenvalues of the fractional Laplacian and fractional Schrödinger operators. In particular, the author focuses his attention on Lieb–Thirring inequalities and their generalizations, as well as semi-classical asymptotics.

Chapter 6: A survey on the conformal fractional Laplacian and some geometric applications

M. d. M. Gonzalez reports on recent developments on the conformal fractional Laplacian, both from the analytic and geometric points of view, with a special sight towards the Partial Differential Equations community. Among other investigations, the author explains the construction of the conformal fractional Laplacian from a purely analytic point of view, by relating its original definition coming from Scattering Theory to a Dirichlet-to-Neumann operator for a related elliptic extension problem, thus allowing for an analytic treatment of Yamabe-type problems in the nonlocal framework. Several examples and related opens problems are presented.

Chapter 7: Jump processes, nonlocal operators and regularity

M. Kassmann reviews some basic concepts of Probability Theory, by focusing on the jump processes and their connection to nonlocal operators. Then, the author explains how to use jump processes for proving regularity results for a very general class of integro-differential equations.

Chapter 8: Regularity issues involving the fractional p-Laplacian

T. Kuusi, G. Mingione, and Y. Sire deal with a general class of nonlinear integro-differential equations involving measure data, mainly focusing on zero order potential estimates. The nonlocal elliptic operators considered are possibly degenerate or singular and cover the case of the fractional p-Laplacian operator with measurable coefficients. The authors report recent related existence and regularity results by providing different, more streamlined proofs.

Chapter 9: Boundary regularity, Pohozaev identities, and nonexistence results

X. Ros-Oton surveys some recent results on nonlocal Dirichlet problems driven by a class of integro-differential operators, whose model case is the fractional Laplacian. The author discusses in detail the fine boundary regularity of the solutions, by sketching the main proofs and the involved blow-up techniques. Related Pohozaev identities strongly based on the aforementioned boundary regularity results are also presented, by showing how they can be used in order to deduce nonexistence and unique continuation properties.

Chapter 10: Variational and topological methods for nonlocal fractional periodic equations

G. Molica Bisci reports on recent existence and multiplicity results for nonlocal fractional problems under periodic boundary conditions. The abstract approach is based on variational and topological methods. More precisely, for subcritical equations, mountain pass and linking-type nontrivial solutions are obtained, as well as solutions for parametric problems, followed by equations at resonance and the obtention of multiple solutions using pseudo-index theory. Finally, in order to overcome the difficulties related to the lack of compactness in the critical case, the author performs truncation arguments and the Moser iteration scheme in the fractional Sobolev framework. Some related open problems are briefly presented.

Chapter 11: Change of scales for crystal dislocation dynamics

S. Patrizi presents various results for a class of evolutionary equations driven by fractional operators, naturally arising in Crystallography, whose corresponding solution has the physical meaning of the atom dislocation inside a crystal structure. Since different scales come into play in such a description, different models have been adopted in order to deal with phenomena at atomic, microscopic, mesoscopic and macroscopic scale. By looking at the asymptotic states of the solutions of equations modeling the dynamics of dislocations at a given scale, it is shown in particular that one can deduce the model for the motion of dislocations at a larger scale.

For the sake of the reader, these contributions are preceded by an introduction, **Essentials of nonlocal operators**, redacted by C. Bucur, which aims at providing some basic knowledge of nonlocal operators. Must-know notions on the fractional Laplacian and on more general nonlocal operators are addressed. The expert users may completely skip this preliminary chapter.

Finally, we would like to thank all the authors who kindly accepted to write their contributions for this book. We appreciated very much both their effort in ensuring a large accessibility of their own chapters to a wide audience, and the fact that each contribution does bring as well new perspectives and proposals, by stimulating the expert in the field. We would also like to thank all the referees who have contributed with their constructive reviews on the improvement of the whole book. Special thanks are lastly due to Mauro Palatucci who created the cover image, and to Agnieszka Bednarczyk-Drag who assisted us in preparing this book.

We hope that our readers enjoy the inspiring insights into the variety of the recent research topics within the nonlocal theory presented in this book.

Parma, 2017, May 4 Tuomo Kuusi, Giampiero Palatucci

Claudia Bucur

Essentials of Nonlocal Operators

Abstract: This preliminary chapter aims at providing some basic knowledge on nonlocal operators. Notions which are necessary to know about the fractional Laplacian and about more general nonlocal operators will be addressed. The expert users may skip this introduction.

The goal of this preliminary chapter is to bring the non-expert reader closer to the beautiful world of nonlocal operators. By no means exhaustive, this introduction gives a glance at some basic definitions, notations and well known results related to a few aspects of some nonlocal operators. With these premises, we take a look at fractional Sobolev spaces, at the fractional Laplacian and at a more general class of nonlocal operators (of which the fractional p-Laplacian is the typical representative).

0.1 Fractional Sobolev Spaces

Fractional Sobolev spaces are a classical argument in harmonic and functional analysis (see for instance [17, 23]). The last decades have seen a revival of interest in fractional Sobolev spaces, both for their mathematical importance and for their use in the study of nonlocal operators and nonlocal equations. In this section, we give an introduction to the topic and state some preliminary results, following the approach in [10] (the interested reader should check this very nice paper for the detailed argument).

To begin with, we recall the definition of a $C^{k,\alpha}$ domain. Let $k \in \boldsymbol{N}$, $\alpha \in (0,1]$ and let $\Omega \subseteq \mathbb{R}^n$ be an open bounded set. We define

$$\begin{aligned} Q &:= \{x = (x', x_n) \in \mathbb{R}^{n-1} \times \mathbb{R} \quad \text{s.t. } |x'| < 1, |x_n| < 1\}, \\ Q_+ &:= \{x = (x', x_n) \in \mathbb{R}^{n-1} \times \mathbb{R} \quad \text{s.t. } |x'| < 1, 0 < x_n < 1\}, \\ Q_0 &:= \{x \in Q \quad \text{s.t. } x_n = 1\}. \end{aligned}$$

We say the domain Ω is of class $C^{k,\alpha}$ if there exists $M > 0$ such that for any $x \in \partial\Omega$ there exists a ball $B = B_r(x)$ for $r > 0$ and a isomorphism $T\colon Q \to B$ such that

$$T \in C^{k,\alpha}(\overline{Q}), \qquad T^{-1} \in C^{k,\alpha}(\overline{B}), \qquad T(Q_+) = B \cap \Omega, \qquad T(Q_0) = B \cap \partial\Omega \quad \text{and}$$
$$\|T\|_{C^{k,\alpha}(\overline{Q})} + \|T^{-1}\|_{C^{k,\alpha}(\overline{B})} \leqslant M.$$

We fix the fractional exponent $s \in (0,1)$ and the summability coefficient $p \in [1,\infty)$. Let $\Omega \subseteq \mathbb{R}^n$ be an open, possibly non-smooth domain. We define the fractional

Claudia Bucur, School of Mathematics and Statistics, University of Melbourne, 813 Swanston Street, Parkville VIC 3010, Australia, E-mail: c.bucur@unimelb.edu.au

https://doi.org/10.1515/9783110571561-002

Sobolev space $W^{s,p}(\Omega)$ as

$$W^{s,p}(\Omega) := \left\{ u \in L^p(\Omega) \text{ s.t. } \frac{|u(x) - u(y)|}{|x-y|^{\frac{n}{p}+s}} \in L^p(\Omega \times \Omega) \right\}. \tag{0.1.1}$$

This space is naturally endowed with the norm

$$\|u\|_{W^{s,p}(\Omega)} := \left(\int_\Omega |u|^p \, dx \right)^{\frac{1}{p}} + \left(\iint_{\Omega \times \Omega} \frac{|u(x) - u(y)|^p}{|x-y|^{n+sp}} \, dx \, dy \right)^{\frac{1}{p}}, \tag{0.1.2}$$

where the second term on the right hand side

$$[u]_{W^{s,p}(\Omega)} := \left(\iint_{\Omega \times \Omega} \frac{|u(x) - u(y)|^p}{|x-y|^{n+sp}} \, dx \, dy \right)^{\frac{1}{p}} \tag{0.1.3}$$

is the so-called *Gagliardo semi-norm.*

We define $W_0^{s,p}(\Omega)$ as the closure of $C_0^\infty(\Omega)$ in norm $\| \cdot \|_{W^{s,p}(\Omega)}$. Moreover

$$W_0^{s,p}(\mathbb{R}^n) = W^{s,p}(\mathbb{R}^n),$$

as stated in Theorem 2.4 in [10]. In other words, the space $C_0^\infty(\mathbb{R}^n)$ of smooth functions with compact support is dense in $W^{s,p}(\mathbb{R}^n)$ (actually this happens for any $s > 0$).

We point out for $p = 2$ the particular Hilbert spaces

$$H^s(\Omega) := W^{s,2}(\Omega)$$

and

$$H_0^s(\Omega) := W_0^{s,2}(\Omega),$$

that are related to the fractional Laplacian, that we introduce in the upcoming Section 0.2.

Fractional Sobolev spaces satisfy some of the classical embeddings properties (see Chapters 2 and 5 in [10] for the proofs and more details on this argument). Let $u \colon \Omega \subseteq \mathbb{R}^n \to \mathbb{R}$ be a measurable function. Then we have the following.

Proposition 0.1. *Let $0 < s \leqslant s' < 1$ and let $\Omega \subseteq \mathbb{R}^n$ be an open set. Then*

$$\|u\|_{W^{s,p}(\Omega)} \leqslant C \|u\|_{W^{s',p}(\Omega)}$$

for a suitable positive constant $C = C(n, s, p) \geqslant 1$. In other words we have the continuous embedding

$$W^{s',p}(\Omega) \subseteq W^{s,p}(\Omega).$$

One may wonder what happens at the limit case, when $s' = 1$. If the open set Ω is smooth with bounded boundary, then the embedding is true, as stated in the next proposition.

Proposition 0.2. *Let $\Omega \subseteq \mathbb{R}^n$ be an open set of class $C^{0,1}$ with bounded boundary. Then*

$$\|u\|_{W^{s,p}(\Omega)} \leqslant C\|u\|_{W^{1,p}(\Omega)}$$

for a suitable positive constant $C = C(n, s, p) \geqslant 1$. In other words we have the continuous embedding

$$W^{1,p}(\Omega) \subseteq W^{s,p}(\Omega).$$

Fractional Sobolev spaces enjoy also quite a number of fractional inequalities: the Sobolev inequality is one of these. Indeed, for $p \in [1, \infty)$ and $n \geqslant sp$ we introduce the fractional Sobolev critical exponent

$$p^\star = \begin{cases} \dfrac{np}{n-sp} & \text{for } sp < n, \\ \infty & \text{for } sp = n. \end{cases}$$

Then we have the fractional counterpart of the Sobolev inequality:

Theorem 0.3. *For any $s \in (0, 1)$, $p \in (1, n/s)$ and $u \in C_0^\infty(\mathbb{R}^n)$ it holds that*

$$\|u\|_{L^{p^\star}(\mathbb{R}^n)} \leqslant C[u]_{W^{s,p}(\mathbb{R}^n)}.$$

Consequently, we have the continuous embedding

$$W^{s,p}(\mathbb{R}^n) \subseteq L^q(\mathbb{R}^n) \quad \textit{for any} \quad q \in [p, p^\star].$$

Proof. We give here a short proof, that can be found in [21] (or in [5], Theorem 3.2.1). We have that

$$|u(x)| \leqslant |u(x) - u(y)| + |u(y)|.$$

For a fixed R (that will be given later on), we integrate over the ball $B_R(x)$ and have that

$$|B_R(x)||u(x)| \leqslant \int_{B_R(x)} |u(x) - u(y)|\, dy + \int_{B_R(x)} |u(y)|\, dy = I_1 + I_2. \tag{0.1.4}$$

We apply the Hölder inequality for the exponents p and $p/(p-1)$ in the first integral and obtain that

$$\begin{aligned} I_1 &= \int_{B_R(x)} \frac{|u(x) - u(y)|}{|x-y|^{\frac{n+sp}{p}}} |x-y|^{\frac{n+sp}{p}}\, dy \\ &\leqslant R^{\frac{n+sp}{p}} \left(\int_{B_R(x)} \frac{|u(x) - u(y)|^p}{|x-y|^{n+sp}}\, dy \right)^{\frac{1}{p}} \left(\int_{B_R(x)} dy \right)^{\frac{p-1}{p}} \\ &\leqslant CR^{n+s} \left(\int_{\mathbb{R}^n} \frac{|u(x) - u(y)|^p}{|x-y|^{n+sp}}\, dy \right)^{\frac{1}{p}}. \end{aligned}$$

The Hölder inequality for $\frac{np}{n-sp}$ and $\frac{np}{n(p-1)+sp}$ gives in the second integral

$$\begin{aligned} I_2 &\leqslant \left(\int_{B_R(x)} |u(y)|^{\frac{np}{n-sp}}\,dy\right)^{\frac{n-sp}{np}} \left(\int_{B_R(x)} dy\right)^{\frac{n(p-1)+sp}{np}} \\ &\leqslant CR^{\frac{n(p-1)+sp}{p}} \left(\int_{\mathbb{R}^n} |u(y)|^{\frac{np}{n-sp}}\,dy\right)^{\frac{n-sp}{np}}. \end{aligned}$$

Dividing by R^n in (0.1.4) and renaiming the constants, it follows that

$$|u(x)| \leqslant CR^s \left[\left(\int_{\mathbb{R}^n} \frac{|u(x)-u(y)|^p}{|x-y|^{n+sp}}\,dy\right)^{\frac{1}{p}} + R^{-\frac{n}{p}} \left(\int_{\mathbb{R}^n} |u(y)|^{\frac{np}{n-sp}}\,dy\right)^{\frac{n-sp}{np}}\right].$$

We take now R such that

$$\left(\int_{\mathbb{R}^n} \frac{|u(x)-u(y)|^p}{|x-y|^{n+sp}}\,dy\right)^{\frac{1}{p}} = R^{-\frac{n}{p}} \left(\int_{\mathbb{R}^n} |u(y)|^{\frac{np}{n-sp}}\,dy\right)^{\frac{n-sp}{np}}$$

and we obtain

$$|u(x)| \leqslant C\left(\int_{\mathbb{R}^n} \frac{|u(x)-u(y)|^p}{|x-y|^{n+sp}}\,dy\right)^{\frac{n-sp}{np}} \left(\int_{\mathbb{R}^n} |u(y)|^{\frac{np}{n-sp}}\,dy\right)^{\frac{s(n-sp)}{n^2}}.$$

Raising to the power $\frac{np}{n-sp}$ and integrating over $\mathbb{R}^n$, we get that

$$\int_{\mathbb{R}^n} |u(x)|^{\frac{np}{n-sp}}\,dx \leqslant C \iint\limits_{\mathbb{R}^n\times\mathbb{R}^n} \frac{|u(x)-u(y)|^p}{|x-y|^{n+sp}}\,dx\,dy \left(\int_{\mathbb{R}^n} |u(y)|^{\frac{np}{n-sp}}\,dy\right)^{\frac{ps}{n}}.$$

This leads to the conclusion, namely

$$\left(\int_{\mathbb{R}^n} |u(x)|^{\frac{np}{n-sp}}\,dx\right)^{\frac{n-sp}{np}} \leqslant C\left(\iint\limits_{\mathbb{R}^n\times\mathbb{R}^n} \frac{|u(x)-u(y)|^p}{|x-y|^{n+sp}}\,dx\,dy\right)^{\frac{1}{p}}.$$

□

Using this fractional Sobolev inequality, one can prove the embedding $W^{s,p}(\Omega) \subseteq L^q(\Omega)$ for any $q \in [p, p^\star]$, for particular domains Ω for which a $W^{s,p}(\Omega)$ function can be extended to the whole of $\mathbb{R}^n$. These are the extension domains, defined as follows.

Definition 0.4. *For any $s \in (0,1)$ and $p \in [1,\infty)$, we say that $\Omega \subseteq \mathbb{R}^n$ is an extension domain for $W^{s,p}$ if there exists a positive constant $C = C(n,s,p,\Omega)$ such that for any $u \in W^{s,p}(\Omega)$ there exists $\tilde{u} \in W^{s,p}(\mathbb{R}^n)$ such that $\tilde{u} = u$ in Ω and*

$$\|\tilde{u}\|_{W^{s,p}(\mathbb{R}^n)} \leqslant C\|u\|_{W^{s,p}(\Omega)}.$$

A nice example of an extension domain is any open set of class $C^{0,1}$ with bounded boundary.

We state this continuous embedding in the following theorem.

Theorem 0.5. *Let $s \in (0,1)$ and $p \in [1,\infty)$ such that $n \geqslant sp$. Let $\Omega \subseteq \mathbb{R}^n$ be an extension domain for $W^{s,p}$. Then there exists a positive constant $C = C(n,s,p,\Omega)$ such that for any $u \in W^{s,p}(\Omega)$ it holds*

$$\|u\|_{L^q(\Omega)} \leqslant C\|u\|_{W^{s,p}(\Omega)} \quad \textit{for any} \quad q \in [p, p^\star].$$

In other words, we have the continuous embedding

$$W^{s,p}(\Omega) \subseteq L^q(\Omega) \quad \textit{for any} \quad q \in [p, p^\star].$$

Moreover, if Ω is bounded, the embedding holds for any $q \in [1, p^\star]$.

In the case $n < sp$, we have the following embedding (see Theorem 8.2 in [10]) :

Theorem 0.6. *Let $\Omega \subseteq \mathbb{R}^n$ be an extension domain for $W^{s,p}$ with no external cups. Then for any $p \in [1,\infty)$, $s \in (0,1)$ such that $sp > n$ there exists a positive constant $C = C(n,s,p,\Omega)$ such that*

$$\|f\|_{C^{0,\alpha}(\Omega)} \leqslant C\left(\|f\|^p_{L^p(\Omega)} + [u]^p_{W^{s,p}(\Omega)}\right)^{\frac{1}{p}}$$

for any $u \in L^p(\Omega)$ with $\alpha := \frac{sp-n}{p}$.

0.2 The Fractional Laplacian

The fractional Laplace operator has a long history in mathematics, in particular it is well known in probability as an infinitesimal generator of Lévy processes (A detailed presentation of this aspect can be found in Chapter 7). Furthermore, this operator has numerous applications in real life models that describe a nonlocal behaviour, such as in phase transitions, anomalous diffusion, crystal dislocation, minimal surfaces, materials science, water waves and many more. As a matter of fact, Chapter 11 presents some very nice results on a nonlocal model related to crystal dislocation.

Hence, there is a rich literature on the mathematical models involving the fractional Laplacian, and different aspects of this operator can be studied. In this book, Chapters 3, 5, 6 present in a self-contained manner some very interesting aspects of the fractional Laplacian. This section gives some basic definitions and makes some preliminary observations on the fractional Laplacian. For more detailed information, the reader can see the above mentioned chapters, and i.e. [5, 22] and other references therein.

We introduce at first some useful notations. Let $n \in \boldsymbol{N}$, we denote by $\mathcal{S}$ the Schwartz space of rapidly decaying functions

$$\mathcal{S} := \left\{ f \in C^\infty(\mathbb{R}^n) \text{ s.t. for all } \alpha, \beta \in \boldsymbol{N}_0^n,\ \sup_{x\in\mathbb{R}^n} |x^\beta D^\alpha f(x)| < \infty \right\}.$$

Endowed with the family of semi-norms

$$[f]_{\mathcal{S}}^{\alpha,N} = \sup_{x\in\mathbb{R}^n}(1+|x|)^N \sum_{|\alpha|\leqslant N} |D^\alpha f(x)|,$$

where $N = 1, 2, \ldots$, the Schwartz space is a locally convex topological space. We denote the space of tempered distributions, namely the topological dual of $\mathcal{S}$, by $\mathcal{S}'$ and use $\langle \cdot, \cdot \rangle$ for the dual pairing between $\mathcal{S}$ and $\mathcal{S}'$.

Let $s \in (0,1)$. For any $u \in \mathcal{S}$ we define the fractional Laplacian as the singular integral

$$\begin{aligned}(-\Delta)^s u(x) &:= C(n,s)\text{P.V.} \int_{\mathbb{R}^n} \frac{u(x)-u(y)}{|x-y|^{n+2s}}\,dy \\ &= C(n,s) \lim_{\varepsilon\to 0} \int_{\mathbb{R}^n\setminus B_\varepsilon(x)} \frac{u(x)-u(y)}{|x-y|^{n+2s}}\,dy,\end{aligned} \tag{0.2.1}$$

where $C(n,s)$ is a dimensional constant. The P.V. stands for "in the principal value sense" and is defined as above. The integral needs to be considered in principle values since, for $s \in \left(\frac{1}{2}, 1\right)$ the kernel $\frac{1}{|x-y|^{n+2s}}$ is singular in a neighborhood of x and this singularity is not integrable near x.

With a change of variables, one can also write the fractional Laplacian as

$$(-\Delta)^s u(x) = C(n,s) \lim_{\varepsilon\to 0} \int_{\mathbb{R}^n\setminus B_\varepsilon(0)} \frac{u(x)-u(x-y)}{|y|^{n+2s}}\,dy. \tag{0.2.2}$$

By putting $\tilde{y} = -y$ we have that

$$(-\Delta)^s u(x) = C(n,s) \lim_{\varepsilon\to 0} \int_{\mathbb{R}^n\setminus B_\varepsilon(0)} \frac{u(x)-u(x+\tilde{y})}{|\tilde{y}|^{n+2s}}\,dy$$

and summing this with (0.2.2), we obtain the following equivalent representation

$$(-\Delta)^s u(x) = \frac{C(n,s)}{2} \int_{\mathbb{R}^n} \frac{2u(x)-u(x-y)-u(x+y)}{|y|^{n+2s}}\,dy. \tag{0.2.3}$$

Notice that this latter formula does not require the P.V. formulation, since for u smooth enough[0.1], taking a second order Taylor expansion near the origin, the first order term vanishes by symmetry, and we are left only with the second order reminder, that makes the kernel integrable. More precisely, we have that

$$\int_{B_1} \frac{|2u(x)-u(x-y)-u(x+y)|}{|y|^{n+2s}}\,dy \leqslant C\|D^2u\|_{L^\infty(\mathbb{R}^n)} \int_{B_1} |y|^{-n-2s+2}\,dy < \infty \quad \text{and}$$

$$\int_{\mathbb{R}^n\setminus B_1} \frac{|u(x)-u(x-y)-u(x+y)|}{|y|^{n+2s}}\,dy \leqslant C\|u\|_{L^\infty(\mathbb{R}^n)} \int_{\mathbb{R}^n\setminus B_1} |y|^{-n-2s}\,dy < \infty.$$

0.1 For instance, one can take $u \in L^\infty(\mathbb{R}^n)$ and locally C^2.

The fractional Laplacian is well defined for a wider class of functions. Indeed, as one may find in [22], it is enough to require that u belongs to a weighted L^1 space and is locally Lipschitz. More precisely, we define

$$L^1_s(\mathbb{R}^n) := \left\{ u \in L^1_{\text{loc}}(\mathbb{R}^n) \text{ s.t. } \int_{\mathbb{R}^n} \frac{|u(x)|}{1+|x|^{n+2s}}\, dx < \infty \right\}$$

(notice that that $L^q(\mathbb{R}^n) \subseteq L^1_s(\mathbb{R}^n)$ for any $q \in [1,\infty)$). Let $\varepsilon > 0$ be sufficiently small. Then, if u belongs to $L^1_s(\mathbb{R}^n)$ and to $C^{0,2s+\varepsilon}$ (or $C^{1,2s+\varepsilon-1}$ for $s \geqslant 1/2$) in a neighborhood of $x \in \mathbb{R}^n$, the fractional Laplacian is well defined in x as in (0.2.3). Indeed, while the fact that $u \in L^1_s(\mathbb{R}^n)$ assures that

$$\int_{\mathbb{R}^n \setminus B_1} \frac{|2u(x) - u(x+y) - u(x-y)|}{|y|^{n+2s}}\, dy < \infty,$$

if, taking for instance $s \in (0, 1/2)$ and u that is $C^{0,2s+\varepsilon}$ in a neighborhood of x, one has that

$$\int_{B_1} \frac{|2u(x) - u(x+y) - u(x-y)|}{|y|^{n+2s}}\, dy \leqslant 2\int_{B_1} |y|^{\varepsilon - n}\, dy \leqslant c(\varepsilon).$$

For $u \in \mathcal{S}$, the fractional Laplacian can be expressed as a pseudo-differential operator, as stated in the following identity:

$$(-\Delta)^s u(x) = \mathcal{F}^{-1}\left(|\xi|^{2s}\hat{u}(\xi)\right)(x). \tag{0.2.4}$$

Here, we set the usual notation for the Fourier transform and its inverse, using $x, \xi \in \mathbb{R}^n$ as the space, respectively as the frequency variable,

$$\mathcal{F}f(\xi) = \hat{f}(\xi) := \int_{\mathbb{R}} f(x) e^{-ix\xi}\, dx$$

and

$$\mathcal{F}^{-1}f(x) = \check{f}(x) := \int_{\mathbb{R}} f(\xi) e^{i\xi x}\, d\xi.$$

We point out that we do not account for the normalization constants in this definition. We notice here that this expression returns the classical Laplace operator for $s = 1$ (and the identity operator, for $s = 0$).

The expressions in (0.2.3) and (0.2.4) are equivalent (see [10], Proposition 3.3 for the proof of this statement). There, the dimensional constant $C(n,s)$ introduced in (0.2.1), is defined as

$$C(n,s) := \left(\int_{\mathbb{R}^n} \frac{1-\cos(\eta_1)}{|\eta|^{n+2s}}\, d\eta \right)^{-1},$$

where η_1 is the first component of $\eta \in \mathbb{R}^n$. The explicit value of $C(n,s)$ is given by

$$C(n,s) = \frac{2^{2s} s \Gamma\left(\frac{n}{2}+s\right)}{\pi^{\frac{n}{2}} \Gamma(1-s)},$$

as it is very nicely proved in the Appendix A of Chapter 11. The interested reader can also see formula (3.1.15) and Appendix B in [5] (and other references therein) for different approaches to the computation of the constant.

At this point, relating to Section 0.1, there is an alternative definition of the fractional Hilbert space $H^s(\mathbb{R}^n)$ via Fourier transform. Let

$$\widehat{H}^s(\mathbb{R}^n) := \left\{ u \in L^2(\mathbb{R}^n) \text{ s.t. } \int_{\mathbb{R}^n} (1 + |\xi|^{2s}) |\widehat{u}(\xi)|^2 \, d\xi < \infty \right\}.$$

Then (see Proposition 3.4 in [10]) the two spaces are equivalent, indeed

$$[u]^2_{H^s(\mathbb{R}^n)} = \frac{2}{C(n,s)} \int_{\mathbb{R}^n} |\xi|^{2s} |\widehat{u}(\xi)|^2 \, d\xi.$$

Moreover, the connection between the fractional Laplacian and the fractional Hilbert space is clarified in Proposition 3.6 in [10], as in the next identity

$$[u]^2_{H^s(\mathbb{R}^n)} = \frac{2}{C(n,s)} \|(-\Delta)^{\frac{s}{2}} u\|^2_{L^2(\mathbb{R}^n)}.$$

We point out that for $u \in \mathcal{S}$, the fractional Laplacian $(-\Delta)^s u$ belongs to $C^\infty(\mathbb{R}^n)$, but $(-\Delta)^s u \notin \mathcal{S}$ (it is not true that it decays faster than any power of x). In particular, we define the linear space

$$\mathcal{P}_s := \left\{ f \in C^\infty(\mathbb{R}^n) \text{ s. t. for all } \alpha \in \boldsymbol{N}_0^n, \ \sup_{x \in \mathbb{R}^n} (1 + |x|^{n+2s}) |D^\alpha f(x)| < +\infty \right\}, \tag{0.2.5}$$

which endowed with the family of semi-norms

$$[f]^\alpha_{\mathcal{P}_s} := \sup_{x \in \mathbb{R}^n} (1 + |x|^{n+2s}) |D^\alpha f(x)|,$$

where $\alpha \in \boldsymbol{N}_0^n$, is a locally convex topological space; we denote by $\mathcal{P}'_s$ its topological dual and by $\langle \cdot, \cdot \rangle_s$ their pairing. Then one has for $u \in \mathcal{S}$ that $(-\Delta)^s u \in \mathcal{P}_s$ (see for instance, the bound (1.10) in [3]). The symmetry of the operator $(-\Delta)^s$ allows to define the fractional Laplacian in a distributional sense: for any $u \in L^1_s(\mathbb{R}^n) \subset \mathcal{P}'_s$ one defines

$$\langle (-\Delta)^s u, \varphi \rangle := \langle u, (-\Delta)^s \varphi \rangle_s \quad \text{for any} \quad \varphi \in \mathcal{S}.$$

These spaces are used in the definition of distributional solutions. Indeed, we say that $u \in L^1_s(\mathbb{R}^n)$ is a distributional solution of

$$(-\Delta)^s u = f, \quad \text{for } f \in \mathcal{S}'$$

if

$$\langle u, (-\Delta)^s v \rangle_s = \langle f, v \rangle \quad \text{for any } v \in \mathcal{S}.$$

Other type of solutions are defined for more general kernels in Section 0.3.

0.2.1 The harmonic extension

The fractional Laplacian can be obtained from a local operator acting in a space with an extra-dimension, via an extension procedure. This extension procedure was first introduced by Molčanov and Ostrovskiĭ in [20], where symmetric stable-processes are seen as traces of degenerate diffusion processes. We will follow here the approach of Caffarelli and Silvestre (see [6]), that relies on considering a local Neumann to Dirichlet operator in the half-space $\mathbb{R}^{n+1}_+ := \mathbb{R}^n \times (0,\infty)$. Consider for any $s \in (0,1)$ the number

$$a := 1 - 2s,$$

the function $u\colon \mathbb{R}^n \to \mathbb{R}$ and the problem in the non-divergence form

$$\begin{cases} \Delta_x U + \dfrac{a}{y}\partial_y U + \partial^2_{yy} U = 0 & \text{in } \mathbb{R}^{n+1}_+ \\ U(x,0) = u(x) & \text{in } \mathbb{R}^n. \end{cases} \tag{0.2.6}$$

The problem (0.2.6) can equivalently be written in the divergence form as

$$\begin{cases} \operatorname{div}(y^a \nabla U) = 0 & \text{in } \mathbb{R}^{n+1}_+ \\ U(x,0) = u(x) & \text{in } \mathbb{R}^n. \end{cases} \tag{0.2.7}$$

Then one has for any $x \in \mathbb{R}^n$, up to constants, that

$$-\lim_{y\to 0+} y^a \partial_y U(x,y) = (-\Delta)^s u(x). \tag{0.2.8}$$

Also, by using the change of variables $z = \left(\frac{y}{2s}\right)^{2s}$ the problem (0.2.6) is equivalent to

$$\begin{cases} \Delta_z U + z^\alpha \partial_{zz} U = 0 & \text{in } \mathbb{R}^{n+1}_+ \\ U(x,0) = u(x) & \text{in } \mathbb{R}^n \end{cases} \tag{0.2.9}$$

for $\alpha = -2a/(1-a) = (2s-1)/s$. In this case also, for any $x \in \mathbb{R}^n$ and with the right choice of constants, one has that

$$-\partial_z U(x,0) = (-\Delta)^s u(x). \tag{0.2.10}$$

One way to prove (0.2.8) (see [6] for more details on this and for alternative proofs) is by means of the Poisson kernel

$$P(x,y) = k_s \frac{y^{1-a}}{\left(|x|^2 + y^2\right)^{\frac{n+1-a}{2}}},$$

that by convolution with u gives an explicit solution of the problem (0.2.6) as

$$U(x,y) = \int_{\mathbb{R}^n} P(x-\xi, y) u(\xi)\, d\xi.$$

Notice that k_s is chosen such that

$$\int_{\mathbb{R}^n} P(x,y)\,dx = 1.$$

One can compute then (up to constants)

$$\begin{aligned}
\lim_{y\to 0+} y^a \frac{U(x,y)-U(x,0)}{y} &= \lim_{y\to 0+} y^{a-1}\left[\int_{\mathbb{R}^n} P(x-\xi,y)u(\xi)\,d\xi - u(x)\right] \\
&= \lim_{y\to 0+} y^{a-1}\int_{\mathbb{R}^n} \frac{y^{1-a}}{\left(|x-\xi|^2+y^2\right)^{\frac{n+1-a}{2}}}\,(u(\xi)-u(x))\,d\xi \\
&= \lim_{y\to 0+}\int_{\mathbb{R}^n} \frac{u(\xi)-u(x)}{\left(|x-\xi|^2+y^2\right)^{\frac{n+1-a}{2}}}\,d\xi \\
&= \int_{\mathbb{R}^n} \frac{u(\xi)-u(x)}{|x-\xi|^{n+1-a}}\,d\xi \\
&= -(-\Delta)^{\frac{1-a}{2}} u(x),
\end{aligned}$$

for u smooth enough. Recalling that $s = \frac{1-a}{2}$, this proves formula (0.2.8).

This extension procedure is useful when one solves an equation with the fractional Laplacian on the whole $\mathbb{R}^n$: it overcomes the difficulty of dealing with a nonlocal operator, by replacing it with a local (possibly degenerate) one. For instance, a nonlinear problem of the type

$$(-\Delta)^s u(x) = f(u) \text{ in } \mathbb{R}^n$$

is translated into the system

$$\begin{cases} \operatorname{div}(y^a \nabla U) = 0 \\ -\lim\limits_{y\to 0} y^a \partial_y U = f(u), \end{cases} \tag{0.2.11}$$

where one identifies $u(x) = U(x,0)$ in a trace sense. At this point, one works with a local operator, which is of variational type. Indeed, the equation in (0.2.11) is the Euler-Lagrange equation of the functional

$$I(U) = \int_{\mathbb{R}^{n+1}_+} y^a |\nabla U|^2\,dX.$$

Here we denoted $X = (x,y) \in \mathbb{R}^{n+1}_+$. See, for instance [11, 4], where a nonlinear, nonlocal elliptic problem in the whole space $\mathbb{R}^n$ is dealt with using variational techniques related to the local extended operator.

0.2.2 Maximum Principle and Harnack inequality

In this subsection, we introduce some natural tools for the study of equations involving the fractional Laplacian: Maximum Principles and the Harnack inequality. We point out that these two type of instruments fail if one wants to apply them in the classical fashion. More precisely, we need to take into account the nonlocal character of the operator and have to require some global information on the function.

First of all, a function is s-harmonic in $x \in \mathbb{R}^n$ if $(-\Delta)^s u(x) = 0$. Of course, the class of s-harmonic functions is not trivial, one example is the one-dimensional function $u(x) = (x_+)^s = \max\{0, x\}^s$, that satisfies $(-\Delta)^s u(x) = 0$ on the half line $x > 0$ (see Theorem 3.4.1 in [5]). See also [12] for some other interesting examples of functions for which one can explicitly compute the fractional Laplacian.

We notice now that, if u has a global maximum at x_0, then by definition (0.2.3) it is easy to check that $(-\Delta)^s u(x_0) \geqslant 0$. On the other hand, this is no longer true if u merely has a local maximum at x_0. The Maximum Principle goes as follows:

Theorem 0.7. *If $(-\Delta)^s u \geqslant 0$ in B_R and $u \geqslant 0$ in $\mathbb{R}^n \setminus B_R$, then $u \geqslant 0$ in B_R. Furthermore, either $u > 0$ in B_R, or $u \equiv 0$ in $\mathbb{R}^n$.*

Proof. Suppose by contradiction that there exists $\overline{x} \in B_R$ such that $u(\overline{x}) < 0$ is a minimum in B_R. Since u is positive outside B_R, this is a global minimum. Hence for any $y \in B_{2R}$ we have that $2u(\overline{x}) - u(x - y) - u(x + y) \leqslant 0$, while for $y \in \mathbb{R}^n \setminus B_{2R}$, the inequality $|\overline{x} \pm y| \geqslant |y| - |\overline{x}| \geqslant R$ assures that $u(\overline{x} \pm y) \geqslant 0$. It yields that

$$\begin{aligned}
0 &\leqslant (-\Delta)^s u(\overline{x}) \\
&= \int_{B_{2R}} \frac{2u(\overline{x}) - u(\overline{x} - y) - u(\overline{x} + y)}{|y|^{n+2s}}\, dy \\
&\quad + \int_{\mathbb{R}^n \setminus B_{2R}} \frac{2u(\overline{x}) - u(\overline{x} - y) - u(\overline{x} + y)}{|y|^{n+2s}}\, dy \\
&\leqslant \int_{\mathbb{R}^n \setminus B_{2R}} \frac{2u(\overline{x})}{|y|^{n+2s}}\, dy \\
&= Cu(\overline{x}) R^{-2s} < 0.
\end{aligned}$$

This gives a contradiction, hence $u(\overline{x}) \geqslant 0$.

Now, suppose that u is not strictly positive in B_R and there exists $x_0 \in B_R$ such that $u(x_0) = 0$. Then

$$(-\Delta)^s u(x_0) = \int_{\mathbb{R}^n} \frac{-u(x_0 - y) - u(x_0 + y)}{|y|^{n+2s}}\, dy \leqslant 0,$$

hence $(-\Delta)^s u(x_0) = 0$. Since $u \geqslant 0$ in $\mathbb{R}^n$, this happens only if $u \equiv 0$ in $\mathbb{R}^n$, and this concludes the proof. □

As said before, if a function is s-harmonic and positive only on the ball, this does not assure that the infimum and supremum on the half-ball are comparable (see [15] for a counter-example of this type). One needs some global information on the function. One simple assumption is to take the function nonnegative on the whole of $\mathbb{R}^n$. Then the Harnack inequality holds, as stated in the next Theorem.

Theorem 0.8. *Let $u\colon \mathbb{R}^n \to \mathbb{R}$ be nonnegative in $\mathbb{R}^n$ such that $(-\Delta)^s u = 0$ in B_1. Then there exists a constant $C = C(n,s) > 0$ such that*

$$\sup_{B_{1/2}} u \leqslant C \inf_{B_{1/2}} u.$$

One way to prove this Theorem is to use the harmonic extension defined in the previous Subsection 0.2.1. Namely, this result follows as the trace inequality on $\mathbb{R}^n \times \{y = 0\}$ of the Harnack inequality holding for the extended local (weighted) operator. See [6] for all the details of this proof.

Another formulation that loses the strong assumption that u should be nonnegative in $\mathbb{R}^n$ is given in the following theorem (see Theorem 2.3 in [16]):

Theorem 0.9. *There exists a positive constant c such that for any function $u\colon \mathbb{R}^n \to \mathbb{R}$ which is s-harmonic function in B_1, the following bound holds for any $x, y \in B_{1/2}$*

$$u(x) \leqslant C\left(u(y) + \int_{\mathbb{R}^n \setminus B_1} \frac{u_-(z)}{(|z|^2-1)^s |z|^n}\, dz\right).$$

Moreover, if the function u is nonnegative in B_1, then one has

$$u(x) \leqslant C\left(u(y) + \int_{\mathbb{R}^n \setminus B_1} \frac{u_-(z)}{|z|^{n+2s}}\, dz\right).$$

Here, u_- is the negative part of u, i.e. $u_-(x) = \max\{-u(x), 0\}$.

A Harnack inequality for more generals kernels is also stated further on in Subsection 0.3.1.

0.3 More General Nonlocal Operators

It is natural to continue the study of nonlocal phenomena by introducing more general type of operators. In particular, one can introduce the fractional p-Laplacian

$$(-\Delta)_p^s u(x) := \text{P.V.} \int_{\mathbb{R}^n} \frac{|u(x)-u(y)|^{p-2}\,(u(x)-u(y))}{|x-y|^{n+sp}}\, dy$$

(notice that for $p = 2$, one gets the fractional Laplacian defined in (0.2.1)). As a further topic, on can generalise this formula by taking instead of $|x-y|^{-n-sp}$ a different kernel. So, in this section we introduce briefly nonlocal operators obtained by means of more general kernels and make some remarks on the well-posedness of the definition. Moreover, we shortly define weak solutions and viscosity solutions, and provide a few known results on these type of solutions.

In this book, Chapter 9 present in detail some arguments related to these nonlocal operators (references therein are of guidance for the interested reader).

We define a general nonlocal operator of fractional parameter $s \in (0,1)$ and summability coefficient $p \in (1,\infty)$. Let $K\colon \mathbb{R}^n \times \mathbb{R}^n \to [0,\infty)$ be a kernel that satisfies

(i) K is a measurable function

(ii) K is symmetric, i.e.

$$K(x,y) = K(y,x) \text{ for almost any } (x,y) \in \mathbb{R}^n \times \mathbb{R}^n;$$

(iii) there exists $\lambda, \Lambda \geqslant 1$ such that

$$\lambda \leqslant K(x,y)|x-y|^{n+sp} \leqslant \Lambda \text{ for almost any } (x,y) \in \mathbb{R}^n \times \mathbb{R}^n$$

for some $p > 1$.

(0.3.1)

Then formally one defines for any $x \in \mathbb{R}^n$

$$\begin{aligned} \mathcal{L}_K u(x) &:= \text{P.V.} \int_{\mathbb{R}^n} |u(x)-u(y)|^{p-2} (u(x)-u(y)) K(x,y)\, dy \\ &= \lim_{\varepsilon \to 0} \int_{\mathbb{R}^n \setminus B_\varepsilon(x)} |u(x)-u(y)|^{p-2} (u(x)-u(y)) K(x,y)\, dy \end{aligned} \tag{0.3.2}$$

where by P.V. we intend "in the principal value sense", as defined in the last line of (0.3.2).

Let us take as an example the case $p = 2$ and a general kernel K satisfying (0.3.1) and see when $\mathcal{L}_K u(x)$ is pointwise defined. If the kernel K satisfies an additional condition of weak translation invariance, i.e.

$$K(x, x+z) = K(x, x-z) \text{ for a.e. } (x,z) \in \mathbb{R}^n \times \mathbb{R}^n \tag{0.3.3}$$

and the function u for $\gamma > 0$ is locally $C^{0,2s+\gamma}$ (or $C^{1,2s+\gamma-1}$ if $s > 1/2$) and integrable at infinity respect to the kernel K, then $\mathcal{L}_K u(x)$ is well defined for any $x \in \mathbb{R}^n$. Indeed, for $r > 0$ and $\varepsilon \in (0,r)$ we have that

$$\int_{B_r(x)\setminus B_\varepsilon(x)} (u(x)-u(y))K(x,y)\, dy = \int_{B_r(0)\setminus B_\varepsilon(0)} (u(x)-u(x+z))K(x,x+z)\, dz.$$

By the symmetry of the domain of integration and the additional property (0.3.3), we obtain

$$\begin{aligned} &\int_{B_r(x)\setminus B_\varepsilon(x)} (u(x)-u(y))K(x,y)\, dy \\ &= \frac{1}{2} \int_{B_r(0)\setminus B_\varepsilon(0)} (u(x)-u(x+z))\, K(x,x+z)\, dz \\ &\quad + \frac{1}{2} \int_{B_r(0)\setminus B_\varepsilon(0)} (u(x)-u(x-z))\, K(x,x-z)\, dz \\ &= \frac{1}{2} \int_{B_r(0)\setminus B_\varepsilon(0)} (2u(x)-u(x+z)-u(x-z))\, K(x,x-z)\, dz. \end{aligned}$$

Now, if u is in $C^{1,2s+\gamma-1}(B_r(x))$, we have that

$$\begin{aligned}|2u(x)-u(x+z)-u(x-z)| &= \left|\int_0^1 (\nabla u(x+tz)-\nabla u(x-tz))\cdot z\,dt\right| \\ &\leqslant [u]_{C^{1,2s+\gamma-1}(B_r(x))}\,|z|^{2s+\gamma}\int_0^1 (2t)^{2s+\gamma-1}\,dt \\ &\leqslant \frac{2^{2s+\gamma-1}}{2s+\gamma}\,[u]_{C^{1,2s+\gamma-1}(B_r(x))}\,|z|^{2s+\gamma}.\end{aligned}$$

Therefore we obtain

$$\begin{aligned}&\int_{B_r(x)\setminus B_\varepsilon(x)} (u(x)-u(y))K(x,y)\,dy \\ &\leqslant \frac{2^{2s+\gamma-1}}{2s+\gamma}\,[u]_{C^{1,2s+\gamma-1}(B_r(x))}\,\Lambda\int_{B_r(0)\setminus B_\varepsilon(0)} |z|^{-n+\gamma}\,dz \leqslant \frac{c(n)}{\gamma}\,[u]_{C^{1,2s+\gamma-1}(B_r(x))}\,r^\gamma.\end{aligned}$$

Hence, for such $\gamma > 0$, the principal value exists and, moreover,

$$\left|\int_{B_r(x)} (u(x)-u(y))K(x,y)\,dy\right| \leqslant c\,[u]_{C^{1,2s+\gamma-1}(B_r(x))}\,r^\gamma.$$

0.3.1 Some remarks on weak and viscosity solutions

We give now an idea of different concepts of solutions and give some introductory properties on solutions of linear equations of the type

$$\begin{cases}\mathcal{L}_K u(x) = 0 & \text{in } \Omega \subseteq \mathbb{R}^n \\ u \text{ satisfies some "boundary condition"} & \text{in } \mathbb{R}^n\setminus\Omega,\end{cases} \tag{0.3.4}$$

where $\Omega \subset \mathbb{R}^n$ is an open bounded set. Notice at first that the boundary condition is given in the whole of the complement of Ω. This depends on the nonlocal character of the operator.

We have seen that, in the case $p = 2$, adding the weak translation invariance on K and proving sufficient regularity on u, then $\mathcal{L}_K u$ is pointwise defined. In this case, pointwise solutions of problem (0.3.4) can be considered.

The concept of pointwise solution is however reductive; in general, the boundary data is given in a trace sense or one can guarantee less regularity on the solution. We introduce two other concepts of solution, the weak and the viscosity notions.

We fix $s \in (0,1)$ and $p \in (1,\infty)$. We consider the following Dirichlet problem, with given boundary data $g \in W^{s,p}(\mathbb{R}^n)$

$$\begin{cases}\mathcal{L}_K u(x) = 0 & \text{in } \Omega \subseteq \mathbb{R}^n \\ u(x) = g(x) & \text{in } \mathbb{R}^n\setminus\Omega.\end{cases} \tag{0.3.5}$$

We recall the definition of $W^{s,p}(\mathbb{R}^n)$ as in (0.1.1)

$$W^{s,p}(\mathbb{R}^n) := \left\{ v \in L^p(\mathbb{R}^n) \text{ s.t. } \frac{|v(x) - v(y)|}{|x - y|^{n/p+s}} \in L^p(\mathbb{R}^n \times \mathbb{R}^n) \right\}$$

and we say that $v \in W_0^{s,p}(\Omega)$ if $v \in W^{s,p}(\mathbb{R}^n)$ and $v = 0$ almost everywhere in $\mathbb{R}^n \setminus \Omega$. In principle, this is a different way of defining the space $W_0^{s,p}(\Omega)$ when Ω is not a bounded Lipschitz open set (see for example the observations in Appendix B in [2]). We define the convex spaces

$$\mathcal{K}_g^{\pm} := \{ v \in W^{s,p}(\mathbb{R}^n) \text{ s.t. } (g - v)_{\pm} \in W_0^{s,p}(\Omega) \}$$

and

$$\mathcal{K}_g := \mathcal{K}_g^+ \cap \mathcal{K}_g^- = \{ v \in W^{s,p}(\mathbb{R}^n) \text{ s.t. } v - g \in W_0^{s,p}(\Omega) \}.$$

The problem has a variational structure, and we introduce a functional whose minimization leads to the solution of the problem (0.3.5). For $u \in \mathcal{K}_g$ we define the functional

$$\mathcal{E}_K(u) := \int_{\mathbb{R}^n} \int_{\mathbb{R}^n} |u(x) - u(y)|^p K(x, y)\, dx\, dy\,. \tag{0.3.6}$$

We have the following definition:

Definition 0.10. *Let Ω be an open set of $\mathbb{R}^n$. We say that u is a weak subsolution (supersolution) of the problem* (0.3.5) *if $u \in \mathcal{K}_g^-$ ($\mathcal{K}_g^+$) and it satisfies*

$$\int_{\mathbb{R}^n} \int_{\mathbb{R}^n} |u(x) - u(y)|^{p-2}(u(x) - u(y))(\varphi(x) - \varphi(y))K(x, y)\, dx\, dy \leqslant (\geqslant) 0$$

for every nonnegative $\varphi \in W_0^{s,p}(\Omega)$. Moreover, a function u is a weak solution if $u \in \mathcal{K}_g$ is both a super and a subsolution of the problem (0.3.5), *i.e. if*

$$\int_{\mathbb{R}^n} \int_{\mathbb{R}^n} |u(x) - u(y)|^{p-2}(u(x) - u(y))(\varphi(x) - \varphi(y))K(x, y)\, dx\, dy = 0$$

for every nonnegative $\varphi \in W_0^{s,p}(\Omega)$.

Using the notion of weak solution introduced in definition (0.10), we have the following existence theorem.

Theorem 0.11 (Existence). *Let $s \in (0, 1)$, $p \in (1, \infty)$ and $g \in W^{s,p}(\mathbb{R}^n)$. Then there exists a unique minimizer u of $\mathcal{E}_K$ over $\mathcal{K}_g$. Moreover, a function $u \in \mathcal{K}_g$ is a minimizer of $\mathcal{E}_K$ over $\mathcal{K}_g$ if and only if it is a weak solution to the problem* (0.3.5).

One can prove the existence of a unique minimizer by standard variational techniques (see Theorem 2.3 in [9] for details). We give here a sketch of the proof that a minimizer of the energy is a solution of the problem (0.3.5) and vice-versa.

Sketch of the proof. Let u be a minimizer of the functional $\mathcal{E}_k$. Consider $u + t\varphi$ to be a perturbation of u with $\varphi \in W_0^{s,p}(\Omega)$. We compute formally

$$\begin{aligned} 0 &= \frac{d}{dt}\mathcal{E}_k(u+t\varphi)\Big|_{t=0} \\ &= \frac{d}{dt}\int_{\mathbb{R}^n}\int_{\mathbb{R}^n} |u(x)+t\varphi(x)-u(y)-t\varphi(y)|^p K(x,y)\,dx\,dy\Big|_{t=0} \\ &= p\int_{\mathbb{R}^n}\int_{\mathbb{R}^n} |u(x)-u(y)|^{p-2}\,(u(x)-u(y))\,(\varphi(x)-\varphi(y))\,dx\,dy. \end{aligned}$$

This proves that u is a weak solution of (0.3.5), as introduced in Definition (0.10).

On the other hand, if u is a weak solution of (0.3.5), let $v \in \mathcal{K}_g$ ant let $\varphi = u - v \in W_0^{s,p}(\Omega)$. Then we have that

$$\begin{aligned} 0 &= \int_{\mathbb{R}^n}\int_{\mathbb{R}^n} |u(x)-u(y)|^{p-2}\,((u(x)-u(y))\,(\varphi(x)-\varphi(y))\,K(x,y)\,dx\,dy \\ &= \int_{\mathbb{R}^n}\int_{\mathbb{R}^n} |u(x)-u(y)|^p K(x,y)\,dx\,dy \\ &\quad - \int_{\mathbb{R}^n}\int_{\mathbb{R}^n} |u(x)-u(y)|^{p-2}(u(x)-u(y))(v(x)-v(y))K(x,y)\,dx\,dy. \end{aligned}$$

Using the Young inequality, we continue

$$\begin{aligned} 0 &\geqslant \frac{1}{p}\int_{\mathbb{R}^n}\int_{\mathbb{R}^n} |u(x)-u(y)|^p K(x,y)\,dx\,dy - \frac{1}{p}\int_{\mathbb{R}^n}\int_{\mathbb{R}^n} |v(x)-v(y)|^p K(x,y)\,dx\,dy \\ &= \mathcal{E}_k(u) - \mathcal{E}_k(v). \end{aligned}$$

Hence $\mathcal{E}_k(u) \leqslant \mathcal{E}_k(v)$ for any $v \in \mathcal{K}_g$ and therefore the weak solution $u \in \mathcal{K}_g$ minimizes the functional $\mathcal{E}_k$. □

In order to obtain some boundedness and regularity results, we introduce the important concept of nonlocal tail (given in [9]). The nonlocal tail takes into account the contribution of a function "coming from far", namely it allows to have a quantitative control of the "nonlocality" of the operator. The definition goes as follows:

$$\mathrm{Tail}(v; x_0, R) := \left[R^{sp} \int_{\mathbb{R}^n \setminus B_R(x_0)} \frac{|v(x)|^{p-1}}{|x-x_0|^{n+sp}} \right]^{\frac{1}{p-1}}. \tag{0.3.7}$$

Notice that this quantity is finite when $v \in L^q(\mathbb{R}^n)$, with $q \geqslant p-1$ and $R > 0$.

With this in hand, we have the following local boundedness result (see Theorem 1.1 in [9] for the proof and details).

Lemma 0.12 (Local boundedness). *Let $s \in (0,1)$, $p \in (1,\infty)$ and let $u \in W^{s,p}(\mathbb{R}^n)$ be a weak solution of the problem* (0.3.5). *Let $r > 0$ such that $B_r(x_0) \subseteq \Omega$. Then*

$$\sup_{B_{r/2}(x_0)} u \leqslant \delta \mathit{Tail}\left(u_+; x_0, \frac{r}{2}\right) + c\delta^{-\frac{(p-1)n}{sp^2}} \left(\fint_{B_r(x_0)} u_+^p\,dx \right)^{\frac{1}{p}},$$

where $u_+ = \max\{u, 0\}$ is the positive part of u and $c = c(n, p, s, \lambda, \Lambda)$.

Here, $\delta \in (0,1]$ behaves as an interpolation parameter between local and nonlocal terms.

Using the nonlocal tail, one can state also the Harnack inequality in this general case (see Theorem 1.1 in [8] for the proof of the statement).

Theorem 0.13. *Let $u \in W^{s,p}(\mathbb{R}^n)$ be a weak solution of (0.3.5) and $u \geqslant 0$ in $B_R(x_0) \subset \Omega$. Then for any $B_r := B_r(x_0) \subset B_{\frac{R}{2}}(x_0)$ we have that*

$$\sup_{B_r} u \leqslant C \inf_{B_r} u + C \left(\frac{r}{R}\right)^{\frac{sp}{p-1}} Tail(u_-; x_0, R),$$

where $u_- = \max\{-u, 0\}$ is the negative part of u and $C = C(n, s, p, \lambda, \Lambda)$.

We point out that a Harnack inequality for nonlocal general operators in the case $p = 2$ is obtained in [1].

Viscosity solutions take into account solutions which are only continuous. The idea is to "trap" the solution, which needs to be only continuous, between two functions which are C^2 (or at least $C^{1,\gamma}$). We introduce here the notion of viscosity solution for the problem (0.3.5), as given in [7].

Definition 0.14. *Let $u\colon \mathbb{R}^n \to \mathbb{R}$ be an upper (lower) semi-continuous function on $\overline{\Omega}$. The function u is said to be a subsolution (supersolution) of $\mathcal{L}_K u = 0$ and we write $\mathcal{L}_K u \leqslant 0$ ($\mathcal{L}_K u \geqslant 0$) if the following happens. If:*

- *x is any point in Ω*
- *$N := N(x) \subset \Omega$ is a neighborhood of x*
- *φ is some $C^2(\overline{N})$ function*
- *$\varphi(x) = u(x)$*
- *$\varphi(y) > u(y)$ for any $y \in N \setminus \{x\}$*

then, setting

$$v := \begin{cases} \varphi, & \text{in } N \\ u, & \text{in } \mathbb{R}^n \setminus N \end{cases}$$

we have that $\mathcal{L}_K v \leqslant 0$ ($\mathcal{L}_K v \geqslant 0$). Moreover, u is a viscosity solution if it is both a subsolution and a supersolution.

Existence and uniqueness of viscosity solutions of problems such as (0.3.5) are established in [14]. We introduce here a Hölder regularity result for viscosity solutions of the problem (0.3.5) (see [18] for more details and Theorem 1 therein for the proof).

Theorem 0.15. *Let $s \in (0,1)$ and $p \in (1,\infty)$ (in the case $p < 2$ we require additionally that $p > 1/(1-s)$). Assume that K satisfies $K(x,y) = K(x,-y)$ and there exist $\Lambda \geqslant \lambda > 0$, $M > 0$ and $\gamma > 0$ such that*

$$\frac{\lambda}{|y|^{n+sp}} \leqslant K(x,y) \leqslant \frac{\Lambda}{|y|^{n+sp}} \quad \textit{for } y \in B_2, x \in B_2$$

and

$$0 \leqslant K(x,y) \leqslant \frac{M}{|y|^{n+\gamma}} \quad \textit{for } y \in \mathbb{R}^n \backslash B_{1/4}, x \in B_2,$$

Let $u \in L^\infty(\mathbb{R}^n)$ be a viscosity solution of $\mathcal{L}_K u = 0$ in B_2. Then u is Hölder continuous in B_1 and in particular there exist $\alpha = \alpha(\lambda, \Lambda, M, \gamma, p, s)$ and $C = C(\lambda, \Lambda, M, \gamma, p, s)$ such that

$$\|u\|_{C^\alpha(B_1)} \leqslant C\|u\|_{L^\infty(\mathbb{R}^n)}.$$

Of course, much remains to be said about the arguments we presented in this introduction, and about the nonlocal setting in general. The fractional Laplace operator and operators of a more general type introduced here will be studied and beautifully presented in the following Chapters 3, 5, 6, 7, 8, 9, 11. Other very interesting topics are dealt with in upcoming chapters. In Chapter 1 some bounds on heat kernels for nonsymmetric nonlocal equations are obtained. Chapter 2 deals with fractional harmonic maps. In Chapter 4, nonlocal minimal surfaces are discussed. Furthermore, Chapter 10 deals with the existence of a weak solution of some fractional nonlinear problems with periodic boundary conditions.

Bibliography

[1] R. F. Bass and M. Kassmann. Harnack inequalities for non-local operators of variable order. *Transactions of the American Mathematical Society*, pages 837–850, 2005.

[2] L. Brasco, E. Lindgren, and E. Parini. The fractional Cheeger problem. *Interfaces Free Bound*, 16(3): 419–458, 2014.

[3] C. Bucur. Some observations on the Green function for the ball in the fractional Laplace framework. *Commun. Pure Appl. Anal.*, 15(2): 657–699, 2016.

[4] C. Bucur, M. Medina, A fractional elliptic problem in $\mathbb{R}^n$ with critical growth and convex nonlinearities, arXiv preprint arXiv:1609.01911, 2016.

[5] C. Bucur and E. Valdinoci. Nonlocal diffusion and applications. *Lecture Notes of the Unione Matematica Italiana*, 20: xii+155, 2016.

[6] L. Caffarelli and L. Silvestre. An extension problem related to the fractional Laplacian. *Comm. Partial Differential Equations*, 32(7-9):1245–1260, 2007.

[7] L. Caffarelli and L. Silvestre. Regularity theory for fully nonlinear integro-differential equations. *Communications on Pure and Applied Mathematics*, 62(5):597–638, 2009.

[8] A. Di Castro, T. Kuusi, and G. Palatucci. Nonlocal Harnack inequalities. *J. Funct. Anal.*, 267(6):1807–1836, 2014.

[9] A. Di Castro, T. Kuusi, and G. Palatucci. Local behavior of fractional p-minimizers. *Ann. Inst. H. Poincaré Anal. Non Linéaire*, 33:1279–1299, 2016.

[10] E. Di Nezza, G. Palatucci, and E. Valdinoci. Hitchhiker's guide to the fractional Sobolev spaces. *Bull. Sci. Math.*, 136(5):521–573, 2012.

[11] S. Dipierro, M. Medina, and E. Valdinoci. Fractional elliptic problems with critical growth in the whole of $\mathbb{R}^n$. *Lecture Notes (Scuola Normale Superiore)*, 2017.

[12] B. Dyda, A. Kuznetsov, and M. Kwaśnicki,Fractional Laplace operator and Meijer G-function, *Constr. Approx.*, 45(3):427–448, 2017.

[13] G. Franzina and G. Palatucci. Fractional p-eigenvalues. *Riv. Mat. Univ. Parma*, 5(2): 315–328, 2014.

[14] H. Ishii and G. Nakamura. A class of integral equations and approximation of p-laplace equations. *Calculus of Variations and Partial Differential Equations*, 37(3-4):485–522, 2010.

[15] M. Kassmann. *The classical Harnack inequality fails for non-local operators*. SFB 611, 2007.

[16] M. Kassmann. A new formulation of Harnack's inequality for nonlocal operators. *Comptes Rendus Mathematique*, 349(11):637–640, 2011.

[17] N. S. Landkof, *Foundations of modern potential theory*, [Translated from the Russian by A. P. Doohovskoy], Springer-Verlag, New York-Heidelberg, 1972.

[18] E. Lindgren. Hölder estimates for viscosity solutions of equations of fractional p-Laplace type. *NoDEA Nonlinear Differential Equations Appl.*, 23(5), Art. 55, 18, 2016.

[19] G. Mingione. The Calderón-Zygmund theory for elliptic problems with measure data. *Annali della Scuola Normale Superiore di Pisa–Classe di Scienze*, 6(2):195–261, 2007.

[20] S. A. Molčanov and E. Ostrovskiĭ. Symmetric stable processes as traces of degenerate diffusion processes. *Teor. Verojatnost. i Primenen.*, 14:127–130, 1969.

[21] A. C. Ponce. *Elliptic PDEs, Measures and Capacities. From the Poisson equation to Nonlinear Thomas-Fermi problems*, volume 23. European Mathematical Society (EMS), Zürich, 2015.

[22] L. Silvestre. Regularity of the obstacle problem for a fractional power of the Laplace operator. *Comm. Pure Appl. Math.*, 60(1):67–112, 2007.

[23] Elias M. Stein, Singular integrals and differentiability properties of functions, *Princeton Mathematical Series, No. 30, Princeton University Press, Princeton, N.J., 1970.*

Zhen-Qing Chen and Xicheng Zhang

Heat Kernels for Non-symmetric Non-local Operators

Abstract: We survey some recent progress in the study of heat kernels for a class of non-symmetric non-local operators. We focus on the existence and sharp two-sided estimates of the heat kernels and their connection to jump diffusions.

Keywords: Discontinuous Markov process, diffusion with jumps, non-local operator, pseudo-differential operator, heat kernel estimate, Lévy system

MSC: Primary 60J35, 47G30, 60J45; Secondary: 31C05, 31C25, 60J75

1.1 Introduction

Second order elliptic differential operators and diffusion processes take up, respectively, a central place in the theory of partial differential equations (PDE) and the theory of probability. There are close relationships between these two subjects. For a large class of second order elliptic differential operators $\mathcal{L}$ on $\mathbb{R}^d$, there is a diffusion process X on $\mathbb{R}^d$ associated with it so that $\mathcal{L}$ is the infinitesimal generator of X, and vice versa. The connection between $\mathcal{L}$ and X can also be seen as follows. The fundamental solution (also called heat kernel) for $\mathcal{L}$ is the transition density function of X. For example, when

$$\mathcal{L} = \frac{1}{2}\sum_{i,j=1}^{d} a_{ij}(x)\frac{\partial^2}{\partial x_i \partial x_j} + \sum_{i=1}^{d} b_i(x)\frac{\partial}{\partial x_i},$$

where $(a_{ij}(x))_{1\leqslant i,j\leqslant d}$ is a $d \times d$ symmetric matrix-valued continuous function on $\mathbb{R}^d$ that is uniformly elliptic and bounded, and $b(x) = (b_1(x), \ldots, b_d(x))$ is a bounded $\mathbb{R}^d$-valued function, there is a unique diffusion $X = \{X_t, t \geqslant 0; \mathbb{P}_x, x \in \mathbb{R}^d\}$ on $\mathbb{R}^d$ that solves the martingale problem for $(\mathcal{L}, C_c^2(\mathbb{R}^d))$. That is, for every $x \in \mathbb{R}^d$, there is a unique probability measure $\mathbb{P}_x$ on the space $C([0,\infty); \mathbb{R}^d)$ of continuous $\mathbb{R}^d$-valued functions on $[0,\infty)$ so that $\mathbb{P}_x(X_0 = x) = 1$ and for every $f \in C_c^2(\mathbb{R}^d)$,

$$M_t^f := f(X_t) - f(X_0) - \int_0^t \mathcal{L}f(X_s)ds$$

is a $\mathbb{P}_x$-martingale. Here $X_t(\omega) = \omega(t)$ is the coordinate map on $C([0,\infty); \mathbb{R}^d)$. It is also known that $(X, \mathbb{P}_x)$ is the unique weak solution to the following stochastic differential

Zhen-Qing Chen, Department of Mathematics, University of Washington, Seattle, WA 98195, USA, E-mail: zqchen@uw.edu
Xicheng Zhang, School of Mathematics and Statistics, Wuhan University, Hubei 430072, P. R. China, E-mail: XichengZhang@gmail.com

https://doi.org/10.1515/9783110571561-003

equation

$$dX_t = \sigma(X_t)dW_t + b(X_t)dt, \qquad X_0 = x,$$

where W_t is a d-dimensional Brownian motion and $\sigma(x) = a(x)^{1/2}$ is the symmetric square root matrix of $a(x) = (a_{ij}(x))_{1\leqslant i,j\leqslant d}$.

When a is Hölder continuous, it is known that $\mathcal{L}$ has a jointly continuous heat kernel $p(t,x,y)$ with respect to the Lebesgue measure on $\mathbb{R}^d$ that enjoys the following Aronson's estimate (see Theorem 1.8 below): there are constants $c_k > 0$, $k = 1, \cdots, 4$, so that

$$c_1 t^{-d/2}\exp(-c_2|x-y|^2/t) \leqslant p(t,x,y) \leqslant c_3 t^{-d/2}\exp(-c_4|x-y|^2/t) \tag{1.1.1}$$

for all $t > 0$ and $x, y \in \mathbb{R}^d$. The kernel $p(t,x,y)$ is the transition density function of the diffusion X.

As many physical and economic systems exhibit discontinuity or jumps, in-depth study on non-Gaussian jump processes are called for. See for example, [6, 31, 37, 43] and the references therein. The infinitesimal generator of a discontinuous Markov process in $\mathbb{R}^d$ is no longer a differential operator but rather a non-local (or, integro-differential) operator. For instance, the infinitesimal generator of an isotropically symmetric α-stable process in $\mathbb{R}^d$ with $\alpha \in (0,2)$ is up to a constant multiple a fractional Laplacian operator $\Delta^{\alpha/2} := -(-\Delta)^{\alpha/2}$. During the past several years there is also a lot of interest from the theory of PDE (such as singular obstacle problems) to study non-local operators; see, for example, [9, 45] and the references therein. A lot of progress has been made in the last fifteen years on the development of the De Giorgi-Nash-Moser-Aronson type theory for non-local operators. For example, Kolokoltsov [38] obtained two-sided heat kernel estimates for certain stable-like processes in $\mathbb{R}^d$, whose infinitesimal generators are a class of pseudo-differential operators having smooth symbols. Bass and Levin [4] used a completely different approach to obtain similar estimates for discrete time Markov chain on $\mathbb{Z}^d$, where the conductance between x and y is comparable to $|x-y|^{-n-\alpha}$ for $\alpha \in (0,2)$. In Chen and Kumagai [18], two-sided heat kernel estimates and a scale-invariant parabolic Harnack inequality (PHI in abbreviation) for symmetric α-stable-like processes on d-sets are obtained. Recently in [19], two-sided heat kernel estimates and PHI are established for symmetric non-local operators of variable order. The De Giorgi-Nash-Moser-Aronson type theory is studied very recently in [20] for symmetric diffusions with jumps. We refer the reader to the survey articles [11, 30] and the references therein on the study of heat kernels for symmetric non-local operators. However, for non-symmetric non-local operators, much less is known. In this article, we will survey the recent development in the study of heat kernels for non-symmetric non-local operators. We will concentrate on the recent progress made in [26, 27] and [14]. In Section 1.5 of this paper, we summarize some other recent work on heat kernels for non-symmetric non-local operators. We also take this opportunity to fill a gap in the proof of [26, (3.20)], which is (1.3.23) of this paper. The proof in [26] works for the case $|x| \geqslant t^{1/\alpha}$. In Section 3, a proof is supplied for the case $|x| \leqslant t^{1/\alpha}$. In fact, a slight modification of the original proof for [26, Theorem 2.5] gives a better estimate (1.3.20) than (1.3.23).

In this survey, we concentrate on heat kernel on the whole Euclidean spaces and on the work that the authors are involved. We will not discuss Dirichlet heat kernels in this article.

1.2 Lévy Process

A Lévy process on $\mathbb{R}^d$ is a right continuous process $X = \{X_t; t \geqslant 0\}$ having left limits that has independent stationary increments. It is uniquely characterized by its Lévy exponenent ψ:

$$\mathbb{E}_0 \exp(i\xi \cdot X_t) = \exp(-t\psi(\xi)), \quad \xi \in \mathbb{R}^d. \tag{1.2.1}$$

Here for $x \in \mathbb{R}^d$, the subscript x in the mathematical expectation $\mathbb{E}_x$ and the probability $\mathbb{P}_x$ means that the process X_t starts from x. The Lévy exponent ψ admits a unique decomposition:

$$\psi(\xi) = ib \cdot \xi + \sum_{i,j=1}^{d} a_{ij}\xi_i\xi_j + \int_{\mathbb{R}^d} \left(1 - e^{i\xi \cdot z} + i\xi \cdot z \mathbb{1}_{\{|z| \leqslant 1\}}\right) \Pi(dz), \tag{1.2.2}$$

where $b \in \mathbb{R}^d$ is a constant vector, (a_{ij}) is a non-negative definite symmetric constant matrix, and $\Pi(dz)$ is a positive measure on $\mathbb{R}^d \setminus \{0\}$ so that $\int_{\mathbb{R}^d} (1 \wedge |z|^2)\Pi(dz) < \infty$. The Lévy measure $\Pi(dz)$ has a strong probabilistic meaning. It describes the jumping intensity of X making a jump of size z. Denote by $\{P_t; t \geqslant 0\}$ the transition semigroup of X; that is, $P_t f(x) = \mathbb{E}_x f(X_t) = \mathbb{E}_0 f(x + X_t)$. For an integrable function f, its Fourier transform is defined to be $\hat{f}(\xi) = \int_{\mathbb{R}^d} e^{i\xi \cdot x} f(x)dx$. Then we have by (1.2.1) and Fubini's theorem,

$$\begin{aligned}
\widehat{P_t f}(\xi) &= \int_{\mathbb{R}^d} e^{i\xi \cdot x} \mathbb{E}_0 f(x + X_t)dx = \mathbb{E}_0 \left[e^{-i\xi \cdot X_t} \left(\int_{\mathbb{R}^d} e^{i\xi \cdot (x + X_t)} f(x + X_t)dx \right) \right] \\
&= e^{-t\psi(-\xi)} \hat{f}(\xi).
\end{aligned}$$

If we denote the infinitesimal generator of $\{P_t; t \geqslant 0\}$ (or X) by $\mathcal{L}$, then

$$\widehat{\mathcal{L}f}(\xi) = \frac{d}{dt}\Big|_{t=0} \widehat{P_t f}(\xi) = -\psi(-\xi)\hat{f}(\xi). \tag{1.2.3}$$

Hence $-\psi(-\xi)$ is the Fourier multiplier (or symbol) for the infinitesimal generator $\mathcal{L}$ of X. One can derive a more explicit expression for the generator $\mathcal{L}$: for $f \in C_c^2(\mathbb{R}^d)$,

$$\mathcal{L}f(x) = \sum_{i,j=1}^{d} a_{ij} \frac{\partial^2 f}{\partial x_i \partial x_j}(x) + b \cdot \nabla f(x) + \int_{\mathbb{R}^d} \left(f(x+z) - f(x) - \nabla f(x) \cdot z \mathbb{1}_{\{|z| \leqslant 1\}} \right) \Pi(dz). \tag{1.2.4}$$

When $b = 0$, $\Pi = 0$ and $(a_{ij}) = \mathbf{I}_{d\times d}$ the identity matrix, that is when $\psi(\xi) = |\xi|^2$, X is a Brownian motion in $\mathbb{R}^d$ with variance $2t$ and infinitesimal generator $\Delta := \sum_{i=1}^d \frac{\partial^2}{\partial x_i^2}$. When $b = 0$, $a_{ij} = 0$ for all $1 \leqslant i, j \leqslant d$ and $\Pi(dz) = \mathcal{A}(d, -\alpha)|z|^{-(d+\alpha)}dz$ for $0 < \alpha < 2$, where $\mathcal{A}(d, -\alpha)$ is a normalizing constant so that $\psi(\xi) = |\xi|^\alpha$, X is a rotationally symmetric α-stable process in $\mathbb{R}^d$, whose infinitesimal generator is the fractional Laplacian $\Delta^{\alpha/2} := -(-\Delta)^{\alpha/2}$.

Unlike in case of Brownian motion, explicit formula for the transition density function of symmetric α-stable processes is not known except for a very few cases. However we can get its two-sided estimates as follows. It follows from (1.2.1) that under $\mathbb{P}_0$, (i) AX_t has the same distribution as X_t for every $t > 0$ and rotation A (an orthogonal matrix); (ii) for every $\lambda > 0$, $X_{\lambda t}$ has the same distribution as $\lambda^{1/\alpha}X_t$. Let $p(t, x)$ be the density function of X_t under $\mathbb{P}_0$; that is,

$$p(t,x) = (2\pi)^{-d}\int_{\mathbb{R}^d} e^{-ix\cdot\xi}e^{-t|\xi|^\alpha}\,d\xi.$$

Then $p(t, x)$ is a function of t and $|x|$ and $p(t, x) = t^{-d/\alpha}p(1, t^{-1/\alpha}x)$. Using Fourier's inversion, one gets

$$\lim_{|x|\to\infty} |x|^{d+\alpha}p(1,x) = \alpha 2^{\alpha-1}\pi^{-(d/2+1)}\sin(\alpha\pi/2)\Gamma((d+\alpha)/2)\Gamma(\alpha/2).$$

(See Pólya [42] when $d = 1$ and Blumenthal-Getoor [7, Theorem 2.1] when $d \geqslant 2$.) It follows that $p(1, x) \asymp 1 \wedge \frac{1}{|x|^{d+\alpha}}$. Consequently,

$$p(t,x) \asymp t^{-d/\alpha} \wedge \frac{t}{|x|^{d+\alpha}} \asymp \frac{t}{(t^{1/\alpha}+|x|)^{d+\alpha}}. \tag{1.2.5}$$

Here for $a, b \in \mathbb{R}$, $a \wedge b := \min\{a, b\}$, and for two functions f, g, $f \asymp g$ means that f/g is bounded between two positive constants.

In real world, almost every media we encounter has impurities so we need to consider state-dependent stochastic processes and state-dependent local and non-local operators. Intuitively speaking, we need to consider processes and operators where $\psi(\xi)$ is dependent on x; that is, $\psi(x, \xi)$. If one uses Fourier multiplier approach (1.2.3), one gets pseudo differential operators. The connection between pseudo differential operators and Markov processes has been nicely exposited in N. Jacob [32]. In this survey, we take (1.2.4) as a starting point but with $a_{ij}(x)$, $b(x)$ and $\Pi(x, dz)$ being functions of $x \in \mathbb{R}^d$. That is,

$$\begin{aligned}\mathcal{L}f(x) &= \sum_{i,j=1}^d a_{ij}(x)\frac{\partial^2 f}{\partial x_i\partial x_j}(x) + b(x)\cdot\nabla f(x)\\ &\quad+\int_{\mathbb{R}^d}\Big(f(x+z)-f(x)-\nabla f(x)\cdot z\mathbb{1}_{\{|z|\leqslant 1\}}\Big)\,\Pi(x,dz).\end{aligned}$$

We will concentrate on the case where $\Pi(x, dz) = \frac{\kappa(x,z)}{|z|^{d+\alpha}}dz$ for some $\alpha \in (0, 2)$ and a measurable function $\kappa(x, z)$ on $\mathbb{R}^d \times \mathbb{R}^d$ satisfying for any $x, y, z \in \mathbb{R}^d$,

$$0 < \kappa_0 \leqslant k(x,z) \leqslant \kappa_1 < \infty, \qquad \kappa(x,z) = \kappa(x,-z), \tag{1.2.6}$$

and for some $\beta \in (0, 1)$,

$$|\kappa(x, z) - \kappa(y, z)| \leqslant \kappa_2 |x - y|^\beta. \tag{1.2.7}$$

1.3 Stable-Like Processes and their Heat Kernels

In this section, we consider the case where $a_{ij} = 0$, $b = 0$ and $\Pi(x, dz) = \frac{\kappa(x,z)}{|z|^{d+\alpha}} dz$; that is,

$$\mathcal{L}f(x) = \text{p.v.} \int_{\mathbb{R}^d} (f(x + z) - f(x)) \frac{\kappa(x, z)}{|z|^{d+\alpha}} dz, \tag{1.3.1}$$

where $\kappa(x, z)$ is a function on $\mathbb{R}^d \times \mathbb{R}^d$ satisfying (1.2.6) and (1.2.7). Here p.v. stands for the Cauchy principal value, that is,

$$\mathcal{L}f(x) = \lim_{\varepsilon \to 0} \int_{\{z \in \mathbb{R}^d : |x| \geqslant \varepsilon\}} (f(x + z) - f(x)) \frac{\kappa(x, z)}{|z|^{d+\alpha}} dz.$$

Since $\kappa(x, z)$ is symmetric in z, when $f \in C_b^2(\mathbb{R}^d)$, we can rewrite $\mathcal{L}f(x)$ as

$$\mathcal{L}f(x) = \int_{\mathbb{R}^d} \left(f(x + z) - f(x) - \nabla f(x) \cdot z \mathbb{1}_{\{|z| \leqslant 1\}}\right) \frac{\kappa(x, z)}{|z|^{d+\alpha}} dz. \tag{1.3.2}$$

The non-local operator $\mathcal{L}$ of (1.3.1) typically is not symmetric, as oppose to non-local operator given by

$$\widetilde{\mathcal{L}}f(x) := \lim_{\varepsilon \to 0} \int_{\{y \in \mathbb{R}^d : |y - x| \geqslant \varepsilon\}} (f(y) - f(x)) \frac{c(x, y)}{|x - y|^{d+\alpha}} dz \tag{1.3.3}$$

in the distributional sense. Here $c(x, y)$ is a symmetric function that is bounded between two positive constants. The operator $\widetilde{\mathcal{L}}$ is the infinitesimal generator of the symmetric α-stable-like process studied in Chen and Kumagai [18], where it is shown that $\widetilde{\mathcal{L}}$ has a jointly Hölder continuous heat kernel that admits two-sided estimates in the same form as (1.2.5).

The following result is recently established in [26].

Theorem 1.1. ([26, Theorem 1.1]) *Under* (1.2.6) *and* (1.2.7), *there exists a unique non-negative jointly continuous function* $p(t, x, y)$ *in* $(t, x, y) \in (0, 1] \times \mathbb{R}^d \times \mathbb{R}^d$ *solving*

$$\frac{\partial}{\partial t} p(t, x, y) = \mathcal{L}p(t, \cdot, y)(x), \; x \neq y, \tag{1.3.4}$$

and enjoying the following four properties:

(i) *(Upper bound) There is a constant* $c_1 > 0$ *so that for all* $t \in (0, 1]$ *and* $x, y \in \mathbb{R}^d$,

$$p(t, x, y) \leqslant c_1 t (t^{1/\alpha} + |x - y|)^{-d-\alpha}. \tag{1.3.5}$$

(ii) *(Hölder's estimate) For every $\gamma \in (0, \alpha \wedge 1)$, there is a constant $c_2 > 0$ so that for every $t \in (0, 1]$ and $x, y, z \in \mathbb{R}^d$,*

$$|p(t,x,z) - p(t,y,z)| \leqslant c_2|x-y|^{\gamma} t^{1-(\gamma/\alpha)} \left(t^{1/\alpha} + |x-z| \wedge |y-z|\right)^{-d-\alpha}. \tag{1.3.6}$$

(iii) *(Fractional derivative estimate) For all $x \neq y \in \mathbb{R}^d$, the mapping $t \mapsto \mathcal{L}p(t, \cdot, y)(x)$ is continuous on $(0, 1]$, and*

$$|\mathcal{L}p(t,\cdot,y)(x)| \leqslant c_3(t^{1/\alpha} + |x-y|)^{-d-\alpha}. \tag{1.3.7}$$

(iv) *(Continuity) For any bounded and uniformly continuous function $f : \mathbb{R}^d \to \mathbb{R}$,*

$$\lim_{t \downarrow 0} \sup_{x \in \mathbb{R}^d} \left| \int_{\mathbb{R}^d} p(t,x,y) f(y) dy - f(x) \right| = 0. \tag{1.3.8}$$

Moreover, we have the following conclusions.

(v) *The constants c_1, c_2 and c_3 in* (i)-(iii) *above can be chosen so that they depend only on $(d, \alpha, \beta, \kappa_0, \kappa_1, \kappa_2)$, $(d, \alpha, \beta, \gamma, \kappa_0, \kappa_1, \kappa_2)$, and $(d, \alpha, \beta, \kappa_0, \kappa_1, \kappa_2)$, respectively.*

(vi) *(Conservativeness) For all $(t, x, y) \in (0, 1] \times \mathbb{R}^d \times \mathbb{R}^d$, $p(t, x, y) \geqslant 0$ and*

$$\int_{\mathbb{R}^d} p(t,x,y) dy = 1. \tag{1.3.9}$$

(vii) *(C-K equation) For all $s, t \in (0, 1]$ with $t+s \in (0, 1]$ and $x, y \in \mathbb{R}^d$, the following Chapman-Kolmogorov equation holds:*

$$\int_{\mathbb{R}^d} p(t,x,z) p(s,z,y) dz = p(t+s,x,y). \tag{1.3.10}$$

(viii) *(Lower bound) There exists $c_4 = c_4(d, \alpha, \beta, \kappa_0, \kappa_1, \kappa_2) > 0$ so that for all $t \in (0, 1]$ and $x, y \in \mathbb{R}^d$,*

$$p(t,x,y) \geqslant c_4 t (t^{1/\alpha} + |x-y|)^{-d-\alpha}. \tag{1.3.11}$$

(ix) *(Gradient estimate) For $\alpha \in [1, 2)$, there exists $c_5 = c_5(d, \alpha, \beta, \kappa_0, \kappa_1, \kappa_2) > 0$ so that for all $x \neq y$ in $\mathbb{R}^d$ and $t \in (0, 1]$,*

$$|\nabla_x \log p(t,x,y)| \leqslant c_5 t^{-1/\alpha}. \tag{1.3.12}$$

1.3.1 Approach

We now sketch the main idea behind the proof of Theorem 1.1.

To emphasize the dependence of $\mathcal{L}$ in (1.3.1) on κ, we write it as $\mathcal{L}^{\kappa}$. For each fixed $y \in \mathbb{R}^d$, we consider a symmetric Lévy process (starting from 0) with Lévy measure

$\Pi_y(dz) = \frac{\kappa(y,z)}{|z|^{d+\alpha}}dz$, and denote its marginal probability density function and infinitesimal generator by $p_y(t,x)$ and $\mathcal{L}^{\kappa(y)}$, respectively. Then we have

$$\frac{\partial}{\partial t}p_y(t,x) = \mathcal{L}^{\kappa(y)}p_y(t,x). \tag{1.3.13}$$

We use Levi's idea and search for heat kernel $p(t,x,y)$ for $\mathcal{L}^\kappa$ with the following form:

$$p(t,x,y) = p_y(t,x-y) + \int_0^t \int_{\mathbb{R}^d} p_z(t-s,x-z)q(s,z,y)dzdy \tag{1.3.14}$$

with function $q(s,z,y)$ to be determined below. We want

$$\frac{\partial}{\partial t}p(t,x,y) = \mathcal{L}^\kappa p(t,\cdot,y)(x) = \mathcal{L}^{\kappa(x)}p(t,\cdot,y)(x).$$

Formally,

$$\begin{aligned}\frac{\partial}{\partial t}p(t,x,y) =& \mathcal{L}^{\kappa(y)}p_y(t,x-y) + q(t,x,y) + \int_0^t \int_{\mathbb{R}^d} \partial_t p_z(t-s,x-z)q(s,z,y)dzds\\ =& \mathcal{L}^{\kappa(y)}p_y(t,x-y) + q(t,x,y) + \int_0^t \int_{\mathbb{R}^d} \mathcal{L}^{\kappa(z)}p_z(t-s,x-z)q(s,z,y)dzds,\end{aligned}$$

while

$$\begin{aligned}\mathcal{L}^{\kappa(x)}p(t,x,y) =& \mathcal{L}^{\kappa(x)}p_y(t,x-y) + \int_0^t \int_{\mathbb{R}^d} \mathcal{L}^{\kappa(x)}p_z(t-s,x-z)q(s,z,y)dzds\\ =& \mathcal{L}^{\kappa(x)}p_y(t,x-y) + \int_0^t \int_{\mathbb{R}^d} q_0(t-s,x,z)q(s,z,y)dzds\\ &+ \int_{\mathbb{R}^d} \mathcal{L}^{\kappa(z)}p_z(t-s,x-z)q(s,z,y)dzds,\end{aligned}$$

where

$$q_0(t,x,z) = (\mathcal{L}^{\kappa(x)} - \mathcal{L}^{\kappa(z)})p_z(t,x-z).$$

It follows from (1.3.13) that $q(t,x,y)$ should satisfy

$$q(t,x,y) = q_0(t,x,y) + \int_0^t \int_{\mathbb{R}^d} q_0(t-s,x,z)q(s,z,y)dzds. \tag{1.3.15}$$

Thus for the construction and the upper bound heat kernel estimates of $p(t,x,y)$, the main task is to solve $q(t,x,y)$, and to make the above argument rigorous. We use Picard's iteration to solve (1.3.15). For $n \geqslant 1$, define

$$q_n(t,x,y) = \int_0^t \int_{\mathbb{R}^d} q_0(t-s,x,z)q_{n-1}(s,z,y)dzds. \tag{1.3.16}$$

Then it can be shown that

$$q(t,x,y) := \sum_{n=0}^\infty q_n(t,x,y) \tag{1.3.17}$$

converges absolutely and locally uniformly on $(0, 1] \times \mathbb{R}^d \times \mathbb{R}^d$. Moreover, $q(t, x, y)$ is jointly continuous in (t, x, y) and has the following upper bound estimate

$$|q(t, x, y)| \leqslant C\left(\varrho_0^\beta + \varrho_\beta^0\right)(t, x - y),$$

where

$$\varrho_\gamma^\beta(t, x) := \frac{t^{\gamma/\alpha}(|x|^\beta \wedge 1)}{(t^{1/\alpha} + |x|)^{d+\alpha}}.$$

We then need to address the following issues.

(i) Show that $p(t, x, y)$ constructed through (1.3.14) and (1.3.17) is non-negative, has the property $\int_{\mathbb{R}^d} p(t, x, y)dy = 1$ and satisfies the Chapman-Kolmogorov equation.
(ii) The kernel $p(t, x, y)$ has the claimed two-sided estimates, and derivative estimates.
(iii) Uniqueness of $p(t, x, y)$.

This requires detailed studies on the kernel $p_\alpha^\kappa(t, x - y)$ for the symmetric Lévy process with Lévy measure $\frac{\kappa(z)}{|z|^{d+\alpha}}dz$, including its fractional derivative estimates, and its continuous dependence on $\kappa(z)$, which will be outlined in the next two subsections.

1.3.2 Upper bound estimates

Key observation: For any symmetric function $\kappa(z)$ with $\kappa_0 \leqslant \kappa(z) \leqslant \kappa_1$, let $\widehat{\kappa}(z) := \kappa(z) - \frac{\kappa_0}{2}$. Since the Lévy process with Lévy measure $\frac{\kappa(z)}{|z|^{d+\alpha}}dz$ can be decomposed as the independent sum of Lévy processes having respectively Lévy measures $\frac{\widehat{\kappa}(z)}{|z|^{d+\alpha}}dz$ and $\frac{\kappa_0/2}{|z|^{d+\alpha}}dz$, we have

$$p_\alpha^{\kappa(z)}(t, x) = \int_{\mathbb{R}^d} p_\alpha^{\kappa_0/2}(t, x - y)p_\alpha^{\widehat{\kappa}(z)}(t, y)dy.$$

Thus the gradient and fractional derivative estimates on $p_\alpha^{\kappa(z)}(t, x)$ can be obtained from those on $p_\alpha^{\kappa_0/2}(t, x)$. On the other hand, it follows from [18] that there is a constant $c = c(d, \kappa_0, \kappa_1) \geqslant 1$ so that

$$c^{-1}\varrho_\alpha^0(t, x) \leqslant p_\alpha^{\kappa(z)}(t, x) \leqslant c\varrho_\alpha^0(t, x) \quad \text{for all } t > 0 \text{ and } x \in \mathbb{R}^d. \tag{1.3.18}$$

First one can establish that for $\gamma_1, \gamma_2, \beta_1, \beta_2 \geqslant 0$ with $\beta_1 + \gamma_1 > 0$ and $\beta_2 + \gamma_2 > 0$,

$$\begin{aligned} &\int_0^t \int_{\mathbb{R}^d} \varrho_{\gamma_1}^{\beta_1}(t - s, x - z)\varrho_{\gamma_2}^{\beta_2}(s, z)dzds \\ \leqslant\ & \mathcal{B}\left(\tfrac{\gamma_1+\beta_1}{\alpha}, \tfrac{\gamma_2+\beta_2}{\alpha}\right)\left(\varrho^0_{\gamma_1+\gamma_2+\beta_1+\beta_2} + \varrho^{\beta_1}_{\gamma_1+\gamma_2+\beta_2} + \varrho^{\beta_2}_{\gamma_1+\gamma_2+\beta_1}\right)(t, x), \end{aligned} \tag{1.3.19}$$

where $\mathcal{B}$ denotes the usual β-function.

Next we establish the continuous dependence of $p_\alpha^{\kappa(z)}(t,y)$ on the symmetric function $\kappa(z)$. Let $\kappa(z)$ and $\widetilde{\kappa}(z)$ be two symmetric functions that are bounded between κ_0 and κ_1. Then for every $0<\gamma<\alpha/4$, there is a constant $c>0$ so that the following estimates hold for all $t\in(0,1]$ and $x\in\mathbb{R}^d$,

$$
\begin{aligned}
|p_\alpha^{\kappa(z)}(t,x)-p_\alpha^{\widetilde{\kappa}(z)}(t,x)| &\leqslant c\|\kappa-\widetilde{\kappa}\|_\infty\,\varrho_\alpha^0(t,x), &(1.3.20)\\
|\nabla_x p_\alpha^{\kappa(z)}(t,x)-\nabla_x p_\alpha^{\widetilde{\kappa}(z)}(t,x)| &\leqslant c\|\kappa-\widetilde{\kappa}\|_\infty t^{-1/\alpha}\,\varrho_\alpha^0(t,x), &(1.3.21)\\
\int_{\mathbb{R}^d}|\delta_{p_\alpha^\kappa}(t,x;z)-\delta_{p_\alpha^{\widetilde{\kappa}}}(t,x;z)|\frac{dz}{|z|^{d+\alpha}} &\leqslant c\|\kappa-\widetilde{\kappa}\|_\infty\,\varrho_0^0(t,x). &(1.3.22)
\end{aligned}
$$

Here $\|\kappa-\widetilde{\kappa}\|_\infty:=\sup_{x\in\mathbb{R}^d}|\kappa(z)-\widetilde{\kappa}(z)|$ and $\delta_f(t,x;z):=f(t,x+z)+f(t,x-z)-2f(t,x)$. The above estimates are established in [26, Theorem 2.5], but with an extra term on their right hand sides. For example, (1.3.20) corresponds to [26, (2.30)] where the estimate is

$$|p_\alpha^{\kappa(z)}(t,x)-p_\alpha^{\widetilde{\kappa}(z)}(t,x)|\leqslant c\|\kappa-\widetilde{\kappa}\|_\infty\left(\varrho_\alpha^0+\varrho_{\alpha-\gamma}^\gamma\right)(t,x). \tag{1.3.23}$$

We take this opportunity to fill a gap in the proof of [26, (3.20)]. The proof there works only for $|x|\geqslant t^{1/\alpha}$ and $t\in(0,1]$, as in this case, by [26, (2.2)],

$$
\begin{aligned}
\int_0^t\int_{\mathbb{R}^d}\varrho_0^\gamma(t-s,x-y)\varrho_{\alpha-\gamma}^0(s,x)dyds &\leqslant c_1\varrho_{\alpha-\gamma}^0(t,x)\int_0^t\int_{\mathbb{R}^d}\varrho_0^\gamma(s,y)dyds\\
&\leqslant c_2\varrho_{\alpha-\gamma}^0(t,x)t^{\gamma/\alpha}=c_2\varrho_\alpha^0(t,x),
\end{aligned}
$$

which gives (1.3.23) by line 8 on p.284 of [26]. On the other hand, one deduces by the inverse Fourier transform that

$$\sup_{y\in\mathbb{R}^d}\left|p_\alpha^{\kappa(z)}(t,y)-p_\alpha^{\widetilde{\kappa}(z)}(t,y)\right|\leqslant(2\pi)^d\int_{\mathbb{R}^d}|e^{-t\psi_\kappa(\xi)}-e^{-t\psi_{\widetilde{\kappa}}(\xi)}|d\xi\leqslant c_3\|\kappa-\widetilde{\kappa}\|_\infty t^{-d/\alpha}.$$

Thus when $|x|\leqslant t^{1/\alpha}$,

$$\left|p_\alpha^{\kappa(z)}(t,x)-p_\alpha^{\widetilde{\kappa}(z)}(t,x)\right|\leqslant c_3\|\kappa-\widetilde{\kappa}\|_\infty t^{-d/\alpha}\leqslant c_4\|\kappa-\widetilde{\kappa}\|_\infty\varrho_\alpha^0(t,x).$$

In fact, by a slight modification of the original proof given in [26] for (1.3.23), we can get estimate (1.3.20). Indeed, by the symmetry of $\mathcal{L}^{\kappa(z)}$ and $\mathcal{L}^{\widetilde{\kappa}(z)}$,

$$
\begin{aligned}
&p_\alpha^{\kappa(z)}(t,x)-p_\alpha^{\widetilde{\kappa}(z)}(t,x)\\
&=\int_0^t\frac{d}{ds}\left(\int_{\mathbb{R}^d}p_\alpha^{\kappa(z)}(s,y)p_\alpha^{\widetilde{\kappa}(z)}(t-s,x-y)dy\right)ds\\
&=\int_0^t\left(\int_{\mathbb{R}^d}\left(\mathcal{L}_\alpha^{\kappa(z)}p_\alpha^{\kappa(z)}(s,\cdot)(y)p_\alpha^{\widetilde{\kappa}(z)}(t-s,x-y)\right.\right.\\
&\qquad\left.\left.-p_\alpha^{\kappa(z)}(s,y)\mathcal{L}_\alpha^{\widetilde{\kappa}(z)}p_\alpha^{\widetilde{\kappa}(z)}(t-s,\cdot)(x-y)\right)dy\right)ds\\
&=\int_0^{t/2}\left(\int_{\mathbb{R}^d}p_\alpha^{\kappa(z)}(s,y)\left(\mathcal{L}_\alpha^{\kappa(z)}-\mathcal{L}_\alpha^{\widetilde{\kappa}(z)}\right)p_\alpha^{\widetilde{\kappa}(z)}(t-s,\cdot)(x-y)dy\right)ds
\end{aligned}
$$

$$+\int_{t/2}^{t}\left(\int_{\mathbb{R}^d} p_\alpha^{\tilde\kappa(z)}(t-s,x-y)\left(\mathcal{L}_\alpha^{\kappa(z)}-\mathcal{L}_\alpha^{\tilde\kappa(z)}\right)p_\alpha^{\kappa(z)}(s,\cdot)(y)dy\right)ds.$$

Hence by (1.3.18) and [26, (2.28)],

$$\begin{aligned}
|p_\alpha^{\kappa(z)}(t,x)-p_\alpha^{\tilde\kappa(z)}(t,x)| &\leqslant c\|\kappa-\tilde\kappa\|_\infty \int_0^{t/2}\int_{\mathbb{R}^d} \varrho_\alpha^0(s,y)\varrho_0^0(t-s,x-y)dyds \\
&\quad + c\|\kappa-\tilde\kappa\|_\infty \int_{t/2}^{t}\int_{\mathbb{R}^d} \varrho_0^0(s,y)\varrho_\alpha^0(t-s,x-y)dyds \\
&\leqslant \frac{c\|\kappa-\tilde\kappa\|_\infty}{t}\int_0^t\int_{\mathbb{R}^d} \varrho_\alpha^0(s,y)\varrho_\alpha^0(t-s,x-y)dyds \\
&\leqslant c\|\kappa-\tilde\kappa\|_\infty\, \varrho_\alpha^0(t,x).
\end{aligned}$$

The same proof as that for [26, Theorem 2.5] but using (1.3.20) instead of (1.3.23) then gives (1.3.21)-(1.3.22).

Since

$$\mathcal{L}^{\kappa(z)}f(x)=\text{p.v.}\int_{\mathbb{R}^d}(f(x+z)-f(x))\frac{\kappa(z)}{|z|^{d+\alpha}}dz=\frac{1}{2}\int_{\mathbb{R}^d}\delta_f(x;z)\frac{\kappa(z)}{|z|^{d+\alpha}}dz,$$

estimate (1.3.22) implies that

$$|\mathcal{L}^{\kappa(z)}p_\alpha^{\kappa(z)}(t,x)-\mathcal{L}^{\tilde\kappa(z)}p_\alpha^{\tilde\kappa(z)}(t,x)|\leqslant c\|\kappa-\tilde\kappa\|_\infty\,\varrho_\alpha^0(t,x).$$

From these estimates, one can establish the first part ((i)-(iv)) of the Theorem 1.1 as well as

$$p(t,x,y)\geqslant ct^{-d/\alpha}\qquad \text{for } t\in(0,1] \text{ and } |x-y|\leqslant 3t^{1/\alpha}. \tag{1.3.24}$$

1.3.3 Lower bound estimates

The upper bound estimates in Theorem 1.1 are established using analytic method, while the lower bound estimate in Theorem 1.1 are obtained mainly by probabilistic argument.

From (i)-(iv) of Theorem 1.1, we see that $P_tf(x):=\int_{\mathbb{R}^d}p(t,x,y)f(y)dy$ is a Feller semigroup. Hence, it determines a Feller process $(\Omega,\mathcal{F},(\mathbb{P}_x)_{x\in\mathbb{R}^d},(X_t)_{t\geqslant 0})$ having strong Feller property on $\mathbb{R}^d$.

We first claim the following.

Theorem 1.2. *Let $\mathcal{F}_t:=\sigma\{X_s, s\leqslant t\}$. Then for each $x\in\mathbb{R}^d$ and every $f\in C_b^2(\mathbb{R}^d)$, under $\mathbb{P}_x$,*

$$M_t^f:=f(X_t)-f(X_0)-\int_0^t\mathcal{L}f(X_s)ds \ \text{ is an } \mathcal{F}_t\text{-martingale.} \tag{1.3.25}$$

In other words, $\mathbb{P}_x$ solves the martingale problem for $(\mathcal{L},C_b^2(\mathbb{R}^d))$. Thus $\mathbb{P}_x$ in particular solves the martingale problem for $(\mathcal{L},C_c^\infty(\mathbb{R}^d))$.

Sketch of Proof. For $f \in C_b^2(\mathbb{R}^d)$, define $u(t,x) = f(x) + \int_0^t P_s\mathcal{L}f(x)ds$. Then we have by (1.3.4) in Theorem 1.1 that

$$\mathcal{L}u(t,x) = \mathcal{L}f(x) + \int_0^t \mathcal{L}P_s\mathcal{L}f(x)ds = \mathcal{L}f(x) + \int_0^t \partial_s(P_s\mathcal{L}f)(x) = P_t\mathcal{L}f(x) = \partial_t u(t,x).$$

Since P_tf also satisfies the equation $\partial_t P_t f = \mathcal{L}(P_tf)$ with $P_0f = f$, we have

$$P_tf(x) = f(x) + \int_0^t P_s\mathcal{L}f(x)ds. \tag{1.3.26}$$

The desired property (1.3.25) now follows from (1.3.26) and the Markov property of X. □

Theorem 1.2 allows us to derive a Lévy system of X by following an approach from [16]. It is easy to see from (1.3.25) that $X_t = (X_t^1, \dots, X_t^d)$ is a semi-martingale. For any $f \in C_c^\infty(\mathbb{R}^d)$, we have by Itô's formula that

$$f(X_t) - f(X_0) = \sum_{i=1}^d \int_0^t \partial_i f(X_{s-})dX_s^i + \sum_{s\leqslant t} \eta_s(f) + \frac{1}{2}\gamma_t(f), \tag{1.3.27}$$

where

$$\eta_s(f) = f(X_s) - f(X_{s-}) - \sum_{i=1}^d \partial_i f(X_{s-})(X_s^i - X_{s-}^i) \tag{1.3.28}$$

and

$$\gamma_t(f) = \sum_{i,j=1}^d \int_0^t \partial_i\partial_j f(X_{s-})d\langle X^{i,c}, X^{j,c}\rangle_s. \tag{1.3.29}$$

Here $X^{i,c}$ is the continuous local martingale part of the semimartingale X^i and $\langle X^{i,c}, X^{j,c}\rangle$ is the covariational process of $X^{i,c}$ and $X^{j,c}$.

Now suppose that A and B are two bounded closed subsets of $\mathbb{R}^d$ having a positive distance from each other. Let $f \in C_c^\infty(\mathbb{R}^d)$ with $f = 0$ on A and $f = 1$ on B. Let M^f be defined as in (1.3.25). Clearly, $N_t^f := \int_0^t \mathbb{1}_A(X_{s-})dM_s^f$ is a martingale. Define

$$J(x,y) = k(x, y-x)/|y-x|^{d+\alpha}, \tag{1.3.30}$$

so $\mathcal{L}$ can be rewritten as

$$\mathcal{L}f(x) = \lim_{\varepsilon\to 0}\int_{\{|y-x|>\varepsilon\}} (f(y) - f(x))J(x,y)dy. \tag{1.3.31}$$

We get by (1.3.25)–(1.3.29) and (1.3.31),

$$\begin{aligned} N_t^f &= \sum_{s\leqslant t} \mathbb{1}_A(X_{s-})(f(X_s) - f(X_{s-})) - \int_0^t \mathbb{1}_A(X_s)\mathcal{L}f(X_s)ds \\ &= \sum_{s\leqslant t} \mathbb{1}_A(X_{s-})f(X_s) - \int_0^t \mathbb{1}_A(X_s)\int_{\mathbb{R}^d} f(y)J(X_s,y)dyds. \end{aligned}$$

By taking a sequence of functions $f_n \in C_c^\infty(\mathbb{R}^d)$ with $f_n = 0$ on A, $f_n = 1$ on B and $f_n \downarrow \mathbb{1}_B$, we get that, for any $x \in \mathbb{R}^d$,

$$\sum_{s\leqslant t} \mathbb{1}_A(X_{s-})\mathbb{1}_B(X_s) - \int_0^t \mathbf{1}_A(X_s)\int_B J(X_s, y)dyds$$

is a martingale with respect to $\mathbb{P}_x$. Thus,

$$\mathbb{E}_x\left[\sum_{s\leqslant t} \mathbb{1}_A(X_{s-})\mathbb{1}_B(X_s)\right] = \mathbb{E}_x\left[\int_0^t\int_{\mathbb{R}^d} \mathbb{1}_A(X_s)\mathbb{1}_B(y)J(X_s, y)dyds\right].$$

Using this and a routine measure theoretic argument, we get

$$\mathbb{E}_x\left[\sum_{s\leqslant t} f(X_{s-}, X_s)\right] = \mathbb{E}_x\left[\int_0^t\int_{\mathbb{R}^d} f(X_s, y)J(X_s, y)dyds\right]$$

for any non-negative measurable function f on $\mathbb{R}^d \times \mathbb{R}^d$ vanishing on $\{(x, y) \in \mathbb{R}^d \times \mathbb{R}^d : x = y\}$. Finally, following the same arguments as in [18, Lemma 4.7] and [19, Appendix A], we get

Theorem 1.3. *X has a Lévy system (J, t) with J given by* (1.3.3)*; that is, for any $x \in \mathbb{R}^d$ and any non-negative measurable function f on $\mathbb{R}_+ \times \mathbb{R}^d \times \mathbb{R}^d$ vanishing on $\{(s, x, y) \in \mathbb{R}_+ \times \mathbb{R}^d \times \mathbb{R}^d : x = y\}$ and $(\mathcal{F}_t)$-stopping time T,*

$$\mathbb{E}_x\left[\sum_{s\leqslant T} f(s, X_{s-}, X_s)\right] = \mathbb{E}_x\left[\int_0^T\left(\int_{\mathbb{R}^d} f(s, X_s, y)J(X_s, y)dy\right)ds\right]. \tag{1.3.32}$$

For a set $K \subset \mathbb{R}^d$, denote

$$\sigma_K := \inf\{t \geqslant 0 : X_t \in K\}, \quad \tau_K := \inf\{t \geqslant 0 : X_t \notin K\}.$$

Denote by $B(x, r)$ the open ball with radius r and center x. We need the following lemma (see [3, 18]).

Lemma 1.4. *For each $\gamma \in (0, 1)$, there exists $R_0 > 0$ such that for every $R > R_0$ and $r \in (0, 1)$,*

$$\mathbb{P}_x(\tau_{B(x,Rr)} \leqslant r^\alpha) \leqslant \gamma. \tag{1.3.33}$$

Proof. Without loss of generality, we assume that $x = 0$. Given $f \in C_b^2(\mathbb{R}^d)$ with $f(0) = 0$ and $f(x) = 1$ for $|x| \geqslant 1$, we set

$$f_r(x) := f(x/r), \quad r > 0.$$

By the definition of f_r, we have

$$\mathbb{P}_0(\tau_{B(0,Rr)} \leqslant r^\alpha) \leqslant \mathbb{E}_0\left[f_{Rr}(X_{\tau_{B(0,Rr)}\wedge r^\alpha})\right] \overset{(1.3.25)}{=} \mathbb{E}_0\left(\int_0^{\tau_{B(0,Rr)}\wedge r^\alpha} \mathcal{L}f_{Rr}(X_s)ds\right). \tag{1.3.34}$$

On the other hand, by the definition of $\mathcal{L}$, we have for $\lambda > 0$,

$$\begin{aligned}
|\mathcal{L}f_{Rr}(x)| &= \frac{1}{2}\left|\int_{\mathbb{R}^d}(f_{Rr}(x+z)+f_{Rr}(x-z)-2f_{Rr}(x))\kappa(x,z)|z|^{-d-\alpha}dz\right| \\
&\leqslant \frac{\kappa_1\|\nabla^2 f_{Rr}\|_\infty}{2}\int_{|z|\leqslant\lambda r}|z|^{2-d-\alpha}dz + 2\kappa_1\|f_{Rr}\|_\infty\int_{|z|\geqslant\lambda r}|z|^{-d-\alpha}dz \\
&= \kappa_1\frac{\|\nabla^2 f\|_\infty}{(Rr)^2}\frac{(\lambda r)^{2-\alpha}}{2(2-\alpha)}s_1 + 2\kappa_1\|f\|_\infty\frac{(\lambda r)^{-\alpha}}{\alpha}s_1 \\
&= \kappa_1 s_1\left(\frac{\|\nabla^2 f\|_\infty}{R^2}\frac{\lambda^{2-\alpha}}{2(2-\alpha)} + 2\|f\|_\infty\frac{\lambda^{-\alpha}}{\alpha}\right)r^{-\alpha},
\end{aligned}$$

where s_1 is the sphere area of the unit ball. Substituting this into (1.3.34), we get

$$\mathbb{P}_0(\tau_{B(0,Rr)}\leqslant r^\alpha)\leqslant\kappa_1 s_1\left(\frac{\|\nabla^2 f\|_\infty}{R^2}\frac{\lambda^{2(2-\alpha)}}{2-\alpha}+2\|f\|_\infty\frac{\lambda^{-\alpha}}{\alpha}\right).$$

Choosing first λ large enough and then R large enough yield the desired estimate. □

We can now proceed to establish the lower bound heat kernel estimate (1.3.11). By Lemma 1.4, there is a constant $\lambda\in(0,\frac{1}{2})$ such that for all $t\in(0,1)$,

$$\mathbb{P}_x(\tau_{B(x,t^{1/\alpha}/2)}>\lambda t)\geqslant\tfrac{1}{2}. \tag{1.3.35}$$

In view of the estimate (1.3.24), it remains to consider the case that $|x-y|>3t^{1/\alpha}$. Using (1.3.35) and the Lévy system of X,

$$\begin{aligned}
&\mathbb{P}_x(X_{\lambda t}\in B(y,t^{1/\alpha})) \\
\geqslant\ &\mathbb{P}_x\Big(X \text{ hits } B(y,t^{1/\alpha}/2) \text{ before } \lambda t \text{ and then travels less than} \\
&\qquad\qquad \text{distance } t^{1/\alpha}/2 \text{ for at least } \lambda t \text{ units of time}\Big) \\
\geqslant\ &\mathbb{P}_x(\sigma_{B(y,t^{1/\alpha}/2)}<\lambda t)\inf_{z\in B(y,t^{1/\alpha}/2)}\mathbb{P}_z(\tau_{B(z,t^{1/\alpha}/2)}>\lambda t) \\
\geqslant\ &c_1\mathbb{P}_x(X_{(\lambda t)\wedge\tau_{B(x,t^{1/\alpha})}}\in B(y,t^{1/\alpha}/2)) \\
=\ &\mathbb{E}_x\int_0^{(\lambda t)\wedge\tau_{B(x,t^{1/\alpha})}}\int_{B(y,t^{1/\alpha}/2)}J(X_s,u)\,du\,ds \\
\geqslant\ &c_2\mathbb{E}_x\left[(\lambda t)\wedge\tau_{B(x,t^{1/\alpha})}\right]\int_{B(y,t^{1/\alpha}/2)}\frac{1}{|x-y|^{d+\alpha}}du \\
\geqslant\ &c_3\frac{t^{(d+\alpha)/\alpha}}{|x-y|^{d+\alpha}}.
\end{aligned}$$

Thus

$$p(t,x,y)\ \geqslant\ \int_{B(y,t^{1/\alpha})}p(\lambda t,x,z)p((1-\lambda)t,z,y)\,dz$$

$$\begin{aligned} &\geq \mathbb{P}_x(X_{\lambda t} \in B(y, t^{1/\alpha})) \inf_{z \in B(y,t^{1/\alpha})} p((1-\lambda t, z, y) \\ &\geq c_4 t^{-d/\alpha} t^{(d+\alpha)/\alpha} \frac{1}{|x-y|^{d+\alpha}} \\ &= \frac{c_4 t}{|x-y|^{d+\alpha}}. \end{aligned}$$

This proves that

$$p(t, x, y) \geq c \left(t^{-d/\alpha} \wedge \frac{t}{|x-y|^{d+\alpha}} \right) \quad \text{for every } x, y \in \mathbb{R}^d \text{ and } t \leq 1.$$

1.3.4 Strong stability

In real world applications and modeling, state-dependent parameter $\kappa(x, z)$ of (1.3.1) is an approximation of real data. So a natural question is how reliable the conclusion is when using such an approximation. The following strong stability result is recently obtained in [27].

Theorem 1.5. *Suppose $\beta \in (0, \alpha/4]$, and κ and $\widetilde{\kappa}$ are two functions satisfying* (1.2.6) *and* (1.2.7). *Denote the corresponding fundamental solution of $\mathcal{L}^\kappa$ and $\mathcal{L}^{\widetilde{\kappa}}$ by $p^\kappa(t, x, y)$ and $p^{\widetilde{\kappa}}(t, x, y)$, respectively. Then for every $\gamma \in (0, \beta)$ and $\eta \in (0, 1)$, there exists a constant $C = C(d, \alpha, \beta, \kappa_0, \kappa_1, \kappa_2, \gamma, \eta) > 0$ so that for all $t \in (0, 1]$ and $x, y \in \mathbb{R}^d$,*

$$|p^\kappa_\alpha(t, x, y) - p^{\widetilde{\kappa}}_\alpha(t, x, y)| \leq C \|\kappa - \widetilde{\kappa}\|_\infty^{1-\eta} \left(1 + t^{-\gamma/\alpha}(|x-y|^\gamma \wedge 1)\right) \frac{t}{(t^{1/\alpha} + |x-y|)^{d+\alpha}}. \tag{1.3.36}$$

Here $\|\kappa - \widetilde{\kappa}\|_\infty := \sup_{x,z \in \mathbb{R}^d} |\kappa(x, z) - \widetilde{\kappa}(x, z)|$.

Observe that by (1.3.5) and (1.3.11), the term $\frac{t}{(t^{1/\alpha}+|x-y|)^{d+\alpha}}$ in (1.3.36) is comparable to $p^\kappa_\alpha(t, x, y)$ and to $p^{\widetilde{\kappa}}_\alpha(t, x, y)$. So the error bound (1.3.36) is also a relative error bound, which is good even in the region when $|x - y|$ is large.

Let $\{P^\kappa_t; t \geq 0\}$ and $\{P^{\widetilde{\kappa}}_t; t \geq 0\}$ be the semigroups generated by $\mathcal{L}^\kappa$ and $\mathcal{L}^{\widetilde{\kappa}}$, respectively. For $p \geq 1$, denote by $\|P^\kappa_t - P^{\widetilde{\kappa}}_t\|_{p,p}$ the operator norm of $P^\kappa_t - P^{\widetilde{\kappa}}_t$ in Banach space $L^p(\mathbb{R}^d; dx)$.

Corrolary 1.1. *Suppose $\beta \in (0, \alpha/4]$, and κ and $\widetilde{\kappa}$ are two functions satisfying* (1.2.6) *and* (1.2.7). *Then for every $\gamma \in (0, \beta)$ and $\eta \in (0, 1)$, there exists a constant $C = C(d, \alpha, \beta, \kappa_0, \kappa_1, \kappa_2, \gamma, \eta) > 0$ so that for every $p \geq 1$ and $t \in (0, 1]$,*

$$\|P^\kappa_t - P^{\widetilde{\kappa}}_t\|_{p,p} \leq C t^{-\gamma/\alpha} \|\kappa - \widetilde{\kappa}\|_\infty^{1-\eta}. \tag{1.3.37}$$

Theorem 1.5 is derived by estimating each $|q^\kappa_n(t, x, y) - q^{\widetilde{\kappa}}_n(t, x, y)|$ for $q^\kappa_n(t, x, y)$ and $q^{\widetilde{\kappa}}_n(t, x, y)$ of (1.3.16). Corollary 1.1 is a direct consequence of Theorem 1.5.

For uniformly elliptic divergence form operators $\mathcal{L}$ and $\widetilde{\mathcal{L}}$ on $\mathbb{R}^d$, pointwise estimate on $|p(t,x,y) - \widetilde{p}(t,x,y)|$ and the L^p-operator norm estimates on $P_t - \widetilde{P}_t$ are obtained in Chen, Hu, Qian and Zheng [15] in terms of the local L^2-distance between the diffusion matrix of $\mathcal{L}$ and $\widetilde{\mathcal{L}}$. Recently, Bass and Ren [5] obtained strong stability result for symmetric α-stable-like non-local operators of (1.3.3), with error bound expressed in terms of the L^q-norm on the function $c(x) := \sup_{y\in\mathbb{R}^d} |c(x,y) - \widetilde{c}(x,y)|$.

1.3.5 Applications to SDE driven by stable processes

Suppose that $\sigma(x) = (\sigma_{ij}(x))_{1\leqslant i,j\leqslant d}$ is a bounded continuous $d \times d$-matrix-valued function on $\mathbb{R}^d$ that is non-degenerate at every $x \in \mathbb{R}^d$, and Y is a (rotationally) symmetric α-stable process on $\mathbb{R}^d$ for some $0 < \alpha < 2$. It is shown in Bass and Chen [2, Theorem 7.1] that for every $x \in \mathbb{R}^d$, SDE

$$dX_t = \sigma(X_{t-})dY_t, \qquad X_0 = x, \tag{1.3.38}$$

has a unique weak solution. (Although in [2] it is assumed $d \geqslant 2$, the results there are valid for $d = 1$ as well.) The family of these weak solutions forms a strong Markov process $\{X, \mathbb{P}_x, x \in \mathbb{R}^d\}$. Using Itô's formula, one deduces (see the display above (7.2) in [2]) that X has generator

$$\mathcal{L}f(x) = \text{p.v.} \int_{\mathbb{R}^d} (f(x + \sigma(x)y) - f(x)) \frac{\mathcal{A}(d,-\alpha)}{|y|^{d+\alpha}} dy. \tag{1.3.39}$$

A change of variable formula $z = \sigma(x)y$ yields

$$\mathcal{L}f(x) = \text{p.v.} \int_{\mathbb{R}^d} (f(x+z) - f(x)) \frac{\kappa(x,z)}{|z|^{d+\alpha}} dz, \tag{1.3.40}$$

where

$$\kappa(x,z) = \frac{\mathcal{A}(d,-\alpha)}{|\det\sigma(x)|} \left(\frac{|z|}{|\sigma(x)^{-1}z|} \right)^{d+\alpha}. \tag{1.3.41}$$

Here $\det(\sigma(x))$ is the determinant of the matrix $\sigma(x)$ and $\sigma(x)^{-1}$ is the inverse of $\sigma(x)$. As an application of Theorem 1.1, we have

Corrolary 1.2. ([2, Corollary 1.2]) *Suppose that $\sigma(x) = (\sigma_{ij}(x))$ is a $d \times d$ matrix-valued function on $\mathbb{R}^d$ such that there are positive constants $\lambda_0, \lambda_1, \lambda_2$ and $\beta \in (0,1)$ so that*

$$\lambda_0 \mathbf{I}_{d\times d} \leqslant \sigma(x) \leqslant \lambda_1 \mathbf{I}_{d\times d} \quad \textit{for every } x \in \mathbb{R}^d, \tag{1.3.42}$$

and

$$|\sigma_{ij}(x) - \sigma_{ij}(y)| \leqslant \lambda_2 |x-y|^\beta \quad \textit{for } 1 \leqslant i,j \leqslant d. \tag{1.3.43}$$

Then the strong Markov process X formed by the unique weak solution to SDE (1.3.38) *has a jointly continuous transition density function $p(t,x,y)$ with respect to the Lebesgue* *measure on $\mathbb{R}^d$, and there is a constant $C \geqslant 1$ that depends only on $(d, \alpha, \beta, \lambda_0, \lambda_1)$ so that*

$$C^{-1} \frac{t}{(t^{1/\alpha} + |x-y|)^{d+\alpha}} \leqslant p(t,x,y) \leqslant C \frac{t}{(t^{1/\alpha} + |x-y|)^{d+\alpha}}$$

for every $t \in (0,1]$ and $x, y \in \mathbb{R}^d$. Moreover, $p(t,x,y)$ enjoys all the properties stated in the conclusions of Theorem 1.1 with $\kappa_0 = \mathcal{A}(d,-\alpha)\lambda_0^{d+\alpha}\lambda_1^{-d}$, $\kappa_1 = \mathcal{A}(d,-\alpha)\lambda_0^{-d}\lambda_1^{d+\alpha}$ and $\kappa_2 = \kappa_2(d, \lambda_0, \lambda_1, \lambda_2)$.

The following strong stability result for SDE (1.3.38) is a direct consequence of Corollary 1.1 and (1.3.41).

Corrolary 1.3. *Suppose that $\sigma(x) = (\sigma_{ij}(x))$ and $\tilde{\sigma}(x) = (\tilde{\sigma}_{ij}(x))$ are $d \times d$ matrix-valued functions on $\mathbb{R}^d$ satisfying conditions (1.3.42) and (1.3.43). Let $p(t,x,y)$ and $\tilde{p}(t,x,y)$ be the transition density functions of the corresponding strong Markov processes X and $\tilde{X}$ that solve SDE (1.3.38), respectively. Then for every $\gamma \in (0,\beta)$ and $\eta \in (0,1)$, there exists a constant $C = C(d,\alpha,\beta,\lambda_0,\lambda_1,\lambda_2,\gamma,\eta) > 0$ so that for all $t \in (0,1]$ and $x, y \in \mathbb{R}^d$,*

$$|p(t,x,y) - \tilde{p}(t,x,y)| \leqslant C\|\sigma - \tilde{\sigma}\|_\infty^{1-\eta} \left(1 + t^{-\gamma/\alpha}(|x-y|^\gamma \wedge 1)\right) \frac{t}{(t^{1/\alpha} + |x-y|)^{d+\alpha}}, \tag{1.3.44}$$

where $\|\sigma - \tilde{\sigma}\|_\infty := \sum_{i,j=1}^d \sup_{x,y\in\mathbb{R}^d} |\sigma_{ij}(x) - \tilde{\sigma}_{ij}(x)|$.

1.4 Diffusion with Jumps

In this section, we consider non-local operators that have both elliptic differential operator part and pure non-local part:

$$\mathcal{L}f(x) := \mathcal{L}^a f(x) + b \cdot \nabla f(x) + \mathcal{L}^\kappa f(x), \tag{1.4.1}$$

where

$$\mathcal{L}^a f(x) := \frac{1}{2}\sum_{i,j=1}^d a_{ij}(x)\partial_{ij}^2 f(x), \qquad b \cdot \nabla f(x) := \sum_{i=1}^d b_i(x)\partial_i f(x),$$

$$\mathcal{L}^\kappa f(x) := \int_{\mathbb{R}^d} \left(f(x+z) - f(x) - \mathbb{1}_{\{|z|\leqslant 1\}} z \cdot \nabla f(x)\right) \frac{\kappa(x,z)}{|z|^{d+\alpha}} dz.$$

Here $a(x) := (a_{ij}(x))_{1\leqslant i,j\leqslant d}$ is a $d \times d$-symmetric matrix-valued measurable function on $\mathbb{R}^d$, $b(x) : \mathbb{R}^d \to \mathbb{R}^d$ and $\kappa(x,z) : \mathbb{R}^d \times \mathbb{R}^d \to \mathbb{R}$ are measurable functions, and $\alpha \in (0,2)$.

For convenience, we assume $d \geqslant 2$. Throughout this section, we impose the following assumptions on a and κ:

($\mathbf{H}^a$) There are $c_1 > 0$ and $\beta \in (0,1)$ such that for any $x, y \in \mathbb{R}^d$,

$$|a(x) - a(y)| \leqslant c_1|x-y|^\beta, \tag{1.4.2}$$

and for some $c_2 \geqslant 1$,

$$c_2^{-1}\mathbf{I}_{d\times d} \leqslant a(x) \leqslant c_2\mathbf{I}_{d\times d}. \tag{1.4.3}$$

($\mathbf{H}^\kappa$) $\kappa(x,z)$ is a bounded measurable function and if $\alpha = 1$, we require

$$\int_{r<|z|\leqslant R} \kappa(x,z)|z|^{-d-1}dz = 0 \quad \text{for any } 0<r<R<\infty. \tag{1.4.4}$$

Note that when $\kappa(x,z)$ is a positive constant function,

$$\mathcal{L} = \mathcal{L}^a + b\cdot\nabla + c\Delta^{\alpha/2}$$

for some constant $c > 0$. A function f defined on $\mathbb{R}^d$ is said to be in Kato class $\mathbb{K}_2$ if $f \in L^1_{loc}(\mathbb{R}^d)$ and

$$\lim_{\delta\to0}\sup_{x\in\mathbb{R}^d}\int_0^\delta\int_{\mathbb{R}^d}\frac{t^{1/2}|f(y)|}{(t^{1/2}+|x-y|)^{d+2}}dydt = 0. \tag{1.4.5}$$

Let $q(t,x,y)$ be the fundamental solution of $\mathcal{L}^a$; see Theorem 1.8 below for more information. Since $\mathcal{L}$ can be viewed as a perturbation of $\mathcal{L}^a$ by $\mathcal{L}^{b,\kappa} := b\cdot\nabla + \mathcal{L}^\kappa$, heuristically the fundamental solution (or heat kernel) $p(t,x,y)$ of $\mathcal{L}$ should satisfy the following Duhamel's formula: for all $t > 0$ and $x, y \in \mathbb{R}^d$,

$$p(t,x,y) = q(t,x,y) + \int_0^t\int_{\mathbb{R}^d} p(r,x,z)\mathcal{L}^{b,\kappa}q(t-r,\cdot,y)(z)dzdr \tag{1.4.6}$$

or

$$p(t,x,y) = q(t,x,y) + \int_0^t\int_{\mathbb{R}^d} q(r,x,z)\mathcal{L}^{b,\kappa}p(t-r,\cdot,y)(z)dzdr. \tag{1.4.7}$$

The following is a special case of the main results in [14], where the corresponding results are also obtained for time-inhomogeneous operators.

Theorem 1.6. ([14, Theorem 1.1]) *Let $\alpha \in (0,2)$. Under* **($\mathbf{H}^a$)**, **($\mathbf{H}^\kappa$)** *and $b \in \mathbb{K}_2$, there is a unique continuous function $p(t,x;y)$ that satisfies* (1.4.6), *and*

(i) *(Upper-bound estimate) For any $T > 0$, there exist constants $C_0, \lambda_0 > 0$ such that for $t \in (0,T]$ and $x, y \in \mathbb{R}^d$,*

$$|p(t,x,y)| \leqslant C_0\left(t^{-d/2}e^{-\lambda_0|x-y|^2/t} + \frac{\|\kappa\|_\infty t}{(t^{1/2}+|x-y|)^{d+\alpha}}\right). \tag{1.4.8}$$

(ii) *(C-K equation) For all $s, t > 0$ and $x, y \in \mathbb{R}^d$, we have*

$$\int_{\mathbb{R}^d} p(s,x,z)p(t,z,y)dz = p(s+t,x,y). \tag{1.4.9}$$

(iii) *(Gradient estimate) For any $T > 0$, there exist constants $C_1, \lambda_1 > 0$ such that for $t \in (0, T]$ and $x, y \in \mathbb{R}^d$,*

$$|\nabla_x p(t, x, y)| \leqslant C_1 t^{-1/2} \left(t^{-d/2} e^{-\lambda_1 |x-y|^2/t} + \frac{\|\kappa\|_\infty t}{(t^{1/2} + |x-y|)^{d+\alpha}} \right). \tag{1.4.10}$$

(iv) *(Conservativeness) For any $t > 0$ and $x \in \mathbb{R}^d$, $\int_{\mathbb{R}^d} p(t, x, y) dy = 1$.*

(v) *(Generator) Define $P_t f(x) = \int_{\mathbb{R}^d} p(t, x, y) f(y) dy$. Then for any $f \in C_b^2(\mathbb{R}^d)$, we have*

$$P_t f(x) - f(x) = \int_0^t P_s \mathcal{L} f(x) ds. \tag{1.4.11}$$

(vi) *(Continuity) For any bounded and uniformly continuous function f, $\lim_{t \to 0} \|P_t f - f\|_\infty = 0$.*

Define $m_\kappa = \inf_{x \in \mathbb{R}^d} \operatorname{essinf}_{z \in \mathbb{R}^d} k(x, z)$.

Theorem 1.7. ([14, Theorem 1.3]) *If κ is a bounded function satisfying $(\mathbf{H}^\kappa)$ and that for each $x \in \mathbb{R}^d$,*

$$\kappa(x, z) \geqslant 0 \quad \text{for a.e. } z \in \mathbb{R}^d, \tag{1.4.12}$$

then $p(t, x, y) \geqslant 0$. Furthermore, if $m_\kappa > 0$, then for any $T > 0$, there are constants $C_1, \lambda_2 > 0$ such that for any $t \in (0, T]$ and $x, y \in \mathbb{R}^d$,

$$p(t, x, y) \geqslant C_1 \left(t^{-d/2} e^{-\lambda_2 |x-y|^2/t} + \frac{m_\kappa t}{(t^{1/2} + |x-y|)^{d+\alpha}} \right). \tag{1.4.13}$$

We have by Theorems 1.6 and 1.7 that when $\kappa \geqslant 0$, then there is a conservative Feller process $X = \{X_t, t \geqslant 0; \mathbb{P}_x, x \in \mathbb{R}^d\}$ having $p(t, x, y)$ as its transition density function with respect to the Lebesgue measure. It follows from (1.4.11) that X is a solution to the martingale problem for $(\mathcal{L}, C_b^2(\mathbb{R}^d))$.

When a is the identity matrix, $b = 0$ and $\kappa(x, z)$ is a positive constant, $\mathcal{L} = \Delta + c\Delta^{\alpha/2}$ for some positive constant $c > 0$. In this case, the corresponding Markov process X is a symmetric Lévy process that is the sum of a Brownian motion W and an independent rotationally symmetric α-stable process Y. Thus the heat kernel $p(t, x, y)$ for $\mathcal{L}$ is the convolution of the transition density function of W and Y. In this case, its two-sided bounds can be obtained through a direct calculation. Indeed such a computation is carried out in Song and Vondraček [46] by dividing into four cases with different expressions for each cases. The two-sided estimates (1.4.8) and (1.4.13) first appeared in Chen and Kumagai [20] for symmetric diffusions with jumps, including the symmetry Lévy process case.

Symmetric diffusions with jumps corresponding to symmetric non-local operators on $\mathbb{R}^d$ with variable coefficients of the the following form have been studied in [20]:

$$\mathcal{L} f(x) = \sum_{i,j=1}^d \frac{\partial}{\partial x_i} \left(a_{ij}(x) \frac{\partial f(x)}{\partial x_i} \right) + \lim_{\varepsilon \to 0} \int_{|x-y| > \varepsilon} (f(y) - f(x)) \frac{c(x, y)}{|x-y|^{d+\alpha}} dy, \tag{1.4.14}$$

where $a(x) := (a_{ij}(x))_{1\leqslant i,j\leqslant d}$ is a $d \times d$-symmetric matrix-valued measurable function on $\mathbb{R}^d$, $c(x, y)$ is a symmetric measurable function on $\mathbb{R}^d \times \mathbb{R}^d$ that is bounded between two positive constants, and $\alpha \in (0, 2)$. Clearly, when $a(x)$ is the identity matrix and $c(x, y)$ is a positive constant, the above non-local operator is $\Delta + c_0\Delta^{\alpha/2}$ for some $c_0 > 0$. Among other things, it is established in Chen and Kumagai [20] that the symmetric non-local operator $\mathcal{L}$ of (1.4.14) has a jointly Hölder continuous heat kernel $p(t, x, y)$ and there are positive constants c_i, $1 \leqslant i \leqslant 4$ so that

$$\begin{aligned} &c_1 \left(t^{-d/2} \wedge t^{-d/\alpha}\right) \wedge \left(t^{-d/2} e^{-c_2|x-y|^2/t} + t^{-d/\alpha} \wedge \frac{t}{|x-y|^{d+\alpha}}\right) \\ &\leqslant p(t, x, y) \leqslant c_3 \left(t^{-d/2} \wedge t^{-d/\alpha}\right) \wedge \left(t^{-d/2} e^{-c_4|x-y|^2/t} + t^{-d/\alpha} \wedge \frac{t}{|x-y|^{d+\alpha}}\right) \end{aligned} \tag{1.4.15}$$

for all $t > 0$ and $x, y \in \mathbb{R}^d$. It is easy to see that for each fixed $T > 0$, the two-sided estimates (1.4.15) on $(0, T] \times \mathbb{R}^d \times \mathbb{R}^d$ is equivalent to

$$\begin{aligned} &\tilde{c}_1 \left(t^{-d/2} e^{-c_2|x-y|^2/t} + \frac{t}{(t^{1/2} + |x-y|)^{d+\alpha}}\right) \leqslant p(t, x, y) \\ &\leqslant \tilde{c}_3 \left(t^{-d/2} e^{-c_4|x-y|^2/t} + \frac{t}{(t^{1/2} + |x-y|)^{d+\alpha}}\right). \end{aligned}$$

When a is the identity matrix and $b = 0$, the results in Theorems 1.6 and 1.7 have been obtained recently in [50] for $\kappa(x, z)$ that is symmetric in z.

1.4.1 Approach

The approach in [14] is to treat $\mathcal{L}$ as $\mathcal{L}^a$ under lower order perturbation $b \cdot \nabla + \mathcal{L}^\kappa$, and thus one can construct the fundamental solution for $\mathcal{L}$ from that of $\mathcal{L}^a$ through Duhamel's formula.

The following result is essentially known in literature; see [29] (see also [14, Theorem 2.3]).

Theorem 1.8. *Under* **(H^a)**, *there exists a nonnegative continuous function $q(t, x, y)$, called the fundamental solution or heat kernel of $\mathcal{L}^a$, with the following properties:*

(i) *(Two-sided estimates) For any $T > 0$, there exist constants $C, \lambda > 0$ such that for $t \in (0, T]$ and $x, y \in \mathbb{R}^d$,*

$$C^{-1} t^{-d/2} e^{-\lambda^{-1}|x-y|^2/t} \leqslant q(t, x, y) \leqslant C t^{-d/2} e^{-\lambda|x-y|^2/t}. \tag{1.4.16}$$

(ii) *(Gradient estimate) For $j = 1, 2$ and $T > 0$, there exist constants $C, \lambda > 0$ such that for $t \in (0, T]$ and $x, y \in \mathbb{R}^d$,*

$$|\nabla_x^j q(t, x, y)| \leqslant C t^{-(d+j)/2} e^{-\lambda|x-y|^2/t}. \tag{1.4.17}$$

(iii) *(Hölder estimate in y) For $j = 0, 1$, $\eta \in (0, \beta)$ and $T > 0$, there exist constants $C, \lambda > 0$ such that for $t \in (0, T]$, $x, y, z \in \mathbb{R}^d$,*

$$|\nabla_x^j q(t,x,y) - \nabla_x^j q(t,x,z)| \leqslant C|y-z|^\eta t^{-(d+j+\eta)/2} \left(e^{-\lambda|x-y|^2/t} + e^{-\lambda|x-z|^2/t}\right). \tag{1.4.18}$$

Moreover, for bounded measurable $f : \mathbb{R}^d \to \mathbb{R}$, let $Q_t f(x) := \int_{\mathbb{R}^d} q(t,x,y)f(y)dy$. We have

(iv) *(Continuity) For any bounded and uniformly continuous function f, $\lim_{t\to 0} \|Q_t f - f\|_\infty = 0$.*

(v) *(C-K equation) For all $0 \leqslant t < r < s < \infty$, $Q_t Q_s = Q_{t+s}$.*

(vi) *(Conservativeness) For all $0 \leqslant t < s < \infty$, $Q_t 1 = 1$.*

(vii) *(Generator) For any $f \in C_b^2(\mathbb{R}^d)$, we have*

$$Q_t f(x) - f(x) = \int_0^t Q_s \mathcal{L}^a f(x) dr = \int_0^t \mathcal{L}^a Q_s f(x) ds.$$

As mentioned earlier, it is expected that the fundamental solution $p(t,x,y)$ of $\mathcal{L}$ should satisfy Duhamel's formula (1.4.6). We construct $p(t,x,y)$ recursively. Let $p_0(t,x,y) = q(t,x,y)$, and define for $n \geqslant 1$,

$$p_n(t,x,y) := \int_0^t \int_{\mathbb{R}^d} p_{n-1}(t-s,x,z)\mathcal{L}^{b,\kappa} q(s,\cdot,y)(z)dzds.$$

Using Theorem 1.8, one can show that $p_n(t,x,y)$ is well defined and that $\sum_{n=0}^\infty p_n(t,x,y)$ converges locally uniformly to some function $p(t,x,y)$, and that $p(t,x,y)$ is the unique solution stated in Theorem 1.6. The positivity (1.4.12) of Theorem 1.7 can be established by using Hille-Yosida-Ray theorem and Courrége's first theorem.

The Gaussian part in the lower bound estimate on $p(t,x,y)$ in Theorem 1.7 is obtained from the near diagonal lower bound estimate on $p(t,x,y)$ and a chaining ball argument, while the pure jump part in the lower bound estimate on $p(t,x,y)$ is obtained by using a probabilistic argument through the Lévy system, similar to that in Section 3.

1.4.2 Application to SDE

Let $\sigma(x)$ be a $d \times d$-matrix valued function on $\mathbb{R}^d$ that is uniformly elliptic and bounded, and each entry σ_{ij} is β-Hölder continuous on $\mathbb{R}^d$, $b \in \mathbb{K}_2$ and $\widetilde{\sigma}$ a bounded $d \times d$-matrix valued measurable function on $\mathbb{R}^d$. Suppose X solves the following stochastic differential equation:

$$dX_t = \sigma(X_t)dB_t + b(X_t)dt + \widetilde{\sigma}(X_{t-})dY_t,$$

where B is a Brownian motion on $\mathbb{R}^d$ and Y is a rotationally symmetric α-stable process on $\mathbb{R}^d$. By Itô's formula, the infinitesimal generator $\mathcal{L}$ of X is of the form $\mathcal{L}^a + b\cdot\nabla + \mathcal{L}^\kappa$

with $a(x) = \sigma(x)\sigma(x)^*$ and

$$\kappa(x,z) = \frac{\mathcal{A}(d,-\alpha)}{|\det\widetilde{\sigma}(x)|}\left(\frac{|z|}{|\widetilde{\sigma}(x)^{-1}z|}\right)^{d+\alpha}.$$

So by Theorems 1.6 and 1.7, X has a transition density function $p(t,x,y)$ satisfying the properties there. If in addition, $\widetilde{\sigma}$ is uniformly elliptic, then for any $T > 0$,

$$\begin{aligned} c_1\left(t^{-d/2}e^{-\lambda_1|x-y|^2/t} + \frac{t}{(t^{1/2}+|x-y|)^{d+\alpha}}\right) &\leqslant p(t,x,y) \\ \leqslant c_2\left(t^{-d/2}e^{-\lambda_2|x-y|^2/t} + \frac{t}{(t^{1/2}+|x-y|)^{d+\alpha}}\right) & \end{aligned}$$

for $t \in (0,T]$ and $x, y \in \mathbb{R}^d$.

1.5 Other Related Work

In this section, we briefly mention some other recent work on heat kernels of non-symmetric non-local operators.

Using a perturbation argument, Bogdan and Jakubowski [8] constructed *a particular* heat kernel (also called fundamental solution) $q^b(t,x,y)$ for operator $\mathcal{L}^b := \Delta^{\alpha/2} + b\cdot\nabla$ on $\mathbb{R}^d$, where $d \geqslant 1$, $\alpha \in (1,2)$ and b is a function on $\mathbb{R}^d$ that is in a suitable Kato class. It is based on the following heuristics: $q^b(t,x,y)$ of $\mathcal{L}^b$ can be related to the fundamental solution $p(t,x,y)$ of $\mathcal{L}^0 = \Delta^{\alpha/2}$, which is the transition density of the rotationally symmetric α-stable process Y, by the following Duhamel's formula:

$$q^b(t,x,y) = p(t,x,y) + \int_0^t\int_{\mathbb{R}^d} q^b(s,x,z)\,b(z)\cdot\nabla_z p(t-s,z,y)dzds. \tag{1.5.1}$$

Applying the above formula recursively, one expects that

$$q^b(t,x,y) := \sum_{k=0}^{\infty} q_k^b(t,x,y) \tag{1.5.2}$$

is a fundamental solution for $\mathcal{L}^b$, where $q_0^b(t,x,y) := p(t,x,y)$ and for $k \geqslant 1$,

$$q_k^b(t,x,y) := \int_0^t\int_{\mathbb{R}^d} q_{k-1}^b(s,x,z)\,b(z)\cdot\nabla_z p(t-s,z,y)dz.$$

It is shown in [8] that the series in (1.5.2) converges absolutely and, for every $T > 0$, such defined $q^b(t,x,y)$ is a conservative transition density function and is comparable to $p(t,x,y)$ on $(0,T]\times\mathbb{R}^d\times\mathbb{R}^d$. Recall that $p(t,x,y)$ has two-sided estimate (1.2.5). In [23], Chen and Wang showed that the Markov process X_t having $q^b(t,x,y)$ as its transition density function is the unique solution to the martingale problem

$(\mathcal{L}^b, C_b^2(\mathbb{R}^d))$; moreover, it is the unique weak solution to the following stochastic differential equation:

$$dX_t = dY_t + b(X_t)dt,$$

where Y_t is the rotationally symmetric α-stable process on $\mathbb{R}^d$. Dirichlet heat kernel estimate for $\mathcal{L}^b$ in a bounded $C^{1,1}$ open set has been obtained in [16]. In [34, 35], Kim and Song extended results in [1, 8, 17] to $\Delta^{\alpha/2} + \mu \cdot \nabla$, where $\mu = (\mu_1, \dots, \mu_d)$ are signed measures in suitable Kato class. These can be regarded as heat kernels for fractional Laplacian under gradient perturbation. Heat kernel estimates for relativistic stable processes and for mixed Brownian motions and stable processes with drifts have recently been studied in [24] and [13], respectively. See [12] for drift perturbation of subordinate Brownian motion of pure jump type and its heat kernel estimate. While in [52], Xie and Zhang considered the critical operator $\mathcal{L}^b := a\Delta^{1/2} + b \cdot \nabla$, where for some $0 < c_0 < c_1$, $a : \mathbb{R}^d \to [c_0, c_1]$ and $b : \mathbb{R}^d \to \mathbb{R}^d$ are two Hölder continuous functions. They established two-sided estimates for the heat kernel of $\mathcal{L}^b$ by using Levi's method as described in Subsection 3.1.

In the same spirit, Wang and Zhang in [48] considered more general fractional diffusion operators over a complete Riemannian manifold perturbed by a time-dependent gradient term, and showed two-sided estimates and gradient estimate of the heat kernel. More precisely, let M be a d-dimensional connected complete Riemannian manifold with Riemannian distance ρ. Let Δ^M be the Laplace-Beltrami operator. Suppose that the heat kernel $p(t, x, y)$ of Δ^M with respect to the Riemannian volume dx exists and has the following two-sided estimates:

$$c_1 t^{-d/2} e^{-c_2 \rho(x,y)^2/t} \leqslant p(t, x, y) \leqslant c_3 t^{-d/2} e^{-c_4 \rho(x,y)^2/t}, \quad t > 0, x, y \in M, \tag{1.5.3}$$

and gradient estimate

$$|\nabla_x p(t, x, y)| \leqslant c_5 t^{-(d+1)/2} e^{-c_4 \rho(x,y)^2/t}, \tag{1.5.4}$$

where ∇_x denotes the covariant derivative. Let P_t be the corresponding semigroup, that is,

$$P_t f(x) := \int_M p(t, x, y) f(y) dx, \quad f \in C_b(M).$$

For $0 < \alpha < 2$, consider the $(\alpha/2)$-stable subordination of P_t

$$P_t^{(\alpha)} := \int_0^\infty P_s \, \mu_t^{(\alpha/2)}(ds), \quad t \geqslant 0,$$

where $\mu_t^{(\alpha/2)}$ is a probability measure on $[0, \infty)$ with Laplace transform

$$\int_0^\infty e^{-\lambda s} \mu_t^{(\alpha/2)}(ds) = e^{-t\lambda^{\alpha/2}}, \quad \lambda \geqslant 0.$$

Then $P_t^{(\alpha)}$ is a C_0-contraction semigroup on $C_b(M)$. Let $\mathcal{L}^{(\alpha)}$ be the infinitesimal generator of $P_t^{(\alpha)}$. In [48], Wang and Zhang considered the following operator

$$\mathcal{L}_{b,c}^{(\alpha)} f(t, x) := \mathcal{L}^{(\alpha)} f(x) + \langle b(t, x), \nabla_x f(x) \rangle + c(t, x) f(x), \ f \in C_b^2(M),$$

where $b : \mathbb{R}_+ \times M \to TM$ and $c : \mathbb{R}_+ \times M \to \mathbb{R}$ are measurable. For $\alpha \in (0, 2)$, one says that a measurable function $f : \mathbb{R}_+ \times M \to \mathbb{R}$ belongs to Kato's class $\mathbb{K}_\alpha$ if

$$\lim_{\varepsilon\to 0} \sup_{(t,x)\in[0,\infty)\times M} \varepsilon^{1/\alpha} \int_0^\varepsilon\int_M \frac{s^{1-1/\alpha}(\varepsilon - s)^{-1/\alpha}|f(t \pm s, y)|}{(s^{1/\alpha} + \rho(x, y))^{d+\alpha}} dyds = 0.$$

Notice that when $\alpha = 2$ and f is time-independent, $\mathbb{K}_\alpha$ is the same as in (1.4.5).

The following result is shown in [48].

Theorem 1.9. *Assume* (1.5.3), (1.5.4) *and* $\alpha \in (1, 2)$. *If* $|b|, c \in \mathbb{K}_\alpha$, *then there is a unique continuous function* $p^{(\alpha)}_{b,c}(t, x; s, y)$ *having the following properties:*

(i) *(Two-sided estimates) There is a constant* $c_1 > 0$ *such that for all* $t - s \in (0, 1], x, y \in M$,

$$c_1^{-1} \frac{t - s}{((t - s)^{1/\alpha} + \rho(x, y))^{d+\alpha}} \leqslant p^{(\alpha)}_{b,c}(t, x; s, y) \leqslant c_1 \frac{t - s}{((t - s)^{1/\alpha} + \rho(x, y))^{d+\alpha}}.$$

(ii) *(Gradient estimate) There is a constant* $c_2 > 0$ *such that for all* $t - s \in (0, 1], x, y \in M$,

$$|\nabla_x p^{(\alpha)}_{b,c}(t, x; s, y)| \leqslant c_2 \frac{(t - s)^{1-1/\alpha}}{((t - s)^{1/\alpha} + \rho(x, y))^{d+\alpha}}.$$

(iii) *(C-K equation) For any* $0 \leqslant s < r < t$ *and* $x, y \in M$,

$$p^{(\alpha)}_{b,c}(t, x; s, y) = \int_M p^{(\alpha)}_{b,c}(t, x; r, z)p^{(\alpha)}_{b,c}(r, z; s, y)dz.$$

(iv) *(Generator) If* $b \in C([0, \infty); L^1_{loc}(M, dx; TM))$ *and* $c \in C([0, \infty); L^1_{loc}(M, dx; \mathbb{R}))$, *then for any* $\varphi, \psi \in C^2_0(M)$,

$$\lim_{t\downarrow s} \frac{1}{t - s} \int_M \psi(P^{b,c}_{t,s}\varphi - \varphi)dx = \int_M \psi\mathcal{L}^{(\alpha)}_{b,c}(s, \cdot)\varphi dx, \quad s \geqslant 0,$$

where $P^{b,c}_{t,s}\varphi := \int_M p^{(\alpha)}_{b,c}(t, \cdot; s, y)\varphi(y)dy$.

The above results indicate that, under suitable Kato class condition, heat kernel estimates are stable under gradient perturbation.

In [21], Chen and Wang studied heat kernels for fractional Laplacian under nonlocal perturbation of high intensity; that is, heat kernels for

$$\mathcal{L}^\kappa f(x) = \Delta^{\alpha/2} f(x) + \mathcal{S}^\kappa f(x), \quad f \in C^2_b(\mathbb{R}^d), \tag{1.5.5}$$

where

$$\mathcal{S}^\kappa f(x) := \mathcal{A}(d, -\beta) \int_{\mathbb{R}^d} \left(f(x + z) - f(x) - \nabla f(x) \cdot z\mathbb{1}_{\{|z|\leqslant 1\}}\right) \frac{\kappa(x, z)}{|z|^{d+\beta}} dz \tag{1.5.6}$$

for some $0 < \beta < \alpha < 2$ and a real-valued bounded function $\kappa(x,z)$ on $\mathbb{R}^d \times \mathbb{R}^d$ satisfying

$$\kappa(x,z) = \kappa(x,-z) \qquad \text{for every } x, z \in \mathbb{R}^d.$$

Uniqueness and existence of fundamental solution $q^\kappa(t,x,y)$ is established in [21]. The approach is also a perturbation argument by viewing $\mathcal{L}^\kappa = \Delta^{\alpha/2} + \mathcal{S}^\kappa$ as a lower order perturbation of $\mathcal{L}^0 = \Delta^{\alpha/2}$ by $\mathcal{S}^\kappa$. So heuristically, the fundamental solution (or heat kernel) $q^\kappa(t,x,y)$ of $\mathcal{L}^\kappa$ should satisfy the following Duhamel's formula:

$$q^\kappa(t,x,y) = p(t,x,y) + \int_0^t \int_{\mathbb{R}^d} q^\kappa(t-s,x,z)\mathcal{S}_z^\kappa p(s,z,y)dzds \tag{1.5.7}$$

for $t > 0$ and $x, y \in \mathbb{R}^d$. Here the notation $S_z^\kappa p(s,z,y)$ means that the non-local operator S^κ is applied to the function $z \mapsto p(s,z,y)$. Similar notation will also be used for other operators, for example, $\Delta_z^{\alpha/2}$. Applying (1.5.7) recursively, it is reasonable to conjecture that $\sum_{n=0}^\infty q_n^\kappa(t,x,y)$, if convergent, is a solution to (1.5.7), where $q_0^\kappa(t,x,y) := p(t,x,y)$ and

$$q_n^\kappa(t,x,y) := \int_0^t \int_{\mathbb{R}^d} q_{n-1}^\kappa(t-s,x,z)\mathcal{S}_z^\kappa p(s,z,y)dzds \quad \text{for } n \geqslant 1. \tag{1.5.8}$$

The hard part is the estimates on $\mathcal{S}_z^\kappa p(s,z,y)$ and on each $q_n^\kappa(t,x,y)$. In contrast to the gradient perturbation case, the fundamental solution to the non-local perturbation (1.5.5) does not need to be positive, and when the kernel is positive, it does not need to be comparable to $p(t,x,y)$. One can rewrite $\mathcal{L}^\kappa$ of (1.5.5) as follows:

$$\mathcal{L}^\kappa f(x) = \int_{\mathbb{R}^d} \left(f(x+z) - f(x) - \langle \nabla f(x), z\rangle \mathbb{1}_{\{|z|\leqslant 1\}}\right) j^\kappa(x,z)dz,$$

where

$$j^\kappa(x,z) = \frac{\mathcal{A}(d,-\alpha)}{|z|^{d+\alpha}} \left(1 + \frac{\mathcal{A}(d,-\beta)}{\mathcal{A}(d,-\alpha)} \kappa(x,z)\, |z|^{\alpha-\beta}\right). \tag{1.5.9}$$

It is shown in [21] that the fundamental solution $q^\kappa \geqslant 0$ if $j^\kappa(x,z) \geqslant 0$; that is, if

$$\kappa(x,z) \geqslant -\frac{\mathcal{A}(d,-\alpha)}{\mathcal{A}(d,-\beta)} |z|^{\beta-\alpha} \quad \text{for a.e. } z \in \mathbb{R}^d. \tag{1.5.10}$$

When $\kappa(x,z)$ is continuous in x, the above condition is also necessary for the non-negativity of $q^\kappa(t,x,y)$. Under condition (1.5.10), various sharp heat kernel estimates have been obtained in [21]. In particular, it is shown in [21] that if there are constants $0 < c_1 \leqslant c_2$ so that

$$\frac{c_1}{|z|^{d+\alpha}} \leqslant j^\kappa(x,z) \leqslant \frac{c_2}{|z|^{d+\alpha}} \quad \text{for } x, z \in \mathbb{R}^d,$$

then for every $T > 0$, $q^\kappa(t,x,y) \asymp p(t,x,y)$ on $(0,T] \times \mathbb{R}^d \times \mathbb{R}^d$. Dirichlet heat kernel estimates for $\mathcal{L}^\kappa$ of (1.5.5) has recently been studied in Chen and Yang [25].

In a subsequent work [50], Wang studied fundamental solution for $\Delta + \mathcal{S}^\kappa$ and its two-sided heat kernel estimates. In [17], Chen, Kim and Song established stability of heat kernel estimates under (local and non-local) Feynman-Kac transforms for a class of jump processes; see also C. Wang [47] on a related work. Very recently, stability of heat kernel estimates for diffusions with jumps (both symmetric and non-symmetric) under Feynman-Kac transform has been studied in Chen and Wang [22]. On the other hand, by employing the strategy and road map from Chen and Zhang [26] as outlined in Section 3 of this paper, Kim, Song and Vondracek [36] has extended Theorem 1.1 to more general non-local operator $\mathcal{L}$ of (1.3.1) with $\frac{1}{|z|^{d+\alpha}}$ being replaced by the density of Lévy measure of certain subordinate Brownian motions. In a recent work [10], X. Chen, Z.-Q. Chen and J. Wang have used Levi's freezing coefficient method to obtain upper and lower bound estimates for heat kernels of the following type of non-local operators of variable order:

$$\mathcal{L}f(x) := \int_{\mathbb{R}^d} \left(f(x+z) - f(x) - \nabla f(x) \cdot z \mathbb{1}_{\{|z| \leqslant 1\}}\right) \frac{\kappa(x,z)}{|z|^{d+\alpha(x)}}\, dz, \quad f \in C_c^2(\mathbb{R}^d),$$

where $\alpha(x)$ is a Hölder continuous function on $\mathbb{R}^d$ such that

$$0 < \alpha_1 \leqslant \alpha(x) \leqslant \alpha_2 < 2 \quad \text{for all } x \in \mathbb{R}^d,$$

and $\kappa(x, z)$ satisfies conditions (1.2.6)-(1.2.7).

Very recently Chen and Zhang [28] have improved the results of Theorem 1.3.1 by dropping the symmetry assumption on $k(x, z)$ in z from (1.2.6). See also [33].

In this survey, we mainly concentrate on the quantitive estimates of the heat kernels of non-symmetric nonlocal operators. For derivative formula of the heat kernel associated with stochastic differential equations with jumps, we refer the interested reader to [53, 51, 49]. For other results on the existence and smoothness of heat kernels or fundamental solutions for non-symmetric jump processes or non-local operators under Hörmander's type conditions, see [44, 40] for the studies of linear Ornstein-Uhlenbeck processes with jumps, and [54, 55, 56] and the references therein for the studies of general stochastic differential equations with jumps. We do not survey these results in this article since the arguments in the above references are mainly based on the Malliavin calculus and thus belong to another topic.

Acknowledgment: We thank the referee for helpful comments, in particular for pointing out a gap in the proof of (2.30) in [26].

Research partially supported by NSF Grant DMS-1206276 and by NNSFC grant of China (Nos. 11731009).

Bibliography

[1] R. F. Bass and Z.-Q. Chen, Brownian motion with singular drift. *Ann. Probab.* **31** (2003), 791–817.

[2] R. F. Bass and Z.-Q. Chen, Systems of equations driven by stable processes. *Probab. Theory Relat. Fields* **134** (2006), 175–214.

[3] R. F. Bass and D. A. Levin, Harnack inequalities for jump processes. *Potential Anal.* **17** (2002), 375–388.

[4] R. F. Bass and D. A. Levin, Transition probabilities for symmetric jump processes. *Trans. Amer. Math. Soc.* **354** (2002), 2933–2953.

[5] R. F. Bass and H. Ren, Meyers inequality and strong stability for stable-like operators. *J. Funct. Anal.* **265** (2013), 28-48.

[6] J. Bertoin. *Lévy Processes*. Cambridge University Press, 1996.

[7] R. M. Blumenthal and R. K. Getoor, Some theorems on stable processes. *Trans. Amer. Math. Soc.* **95** (1960), 263-273.

[8] K. Bogdan and T. Jakubowski, Estimates of heat kernel of fractional Laplacian perturbed by gradient operator. *Commun. Math. Phys.* **271**, (2007)179-198.

[9] L. A. Caffarelli, S. Salsa and Luis Silvestre. Regularity estimates for the solution and the free boundary to the obstacle problem for the fractional Laplacian. *Invent. Math.* **171(1)** (2008) 425–461.

[10] X. Chen, Z.-Q. Chen and J. Wang, Heat kernel for non-local operators with variable order. preprint.

[11] Z.-Q. Chen, Symmetric jump processes and their heat kernel estimates. *Sci. China Ser. A.* **52** (2009), 1423-1445.

[12] Z.-Q. Chen and X. Dou, Drift perturbation of subordinate Brownian motions with Gaussian component. *Sci. China Math.* **59** (2016), 239-260.

[13] Z.-Q. Chen and E. Hu, Heat kernel estimates for $\Delta + \Delta^{\alpha/2}$ under gradient perturbation. *Stochastic Process. Appl.* **125** (2015), 2603-2642.

[14] Z.-Q. Chen, E. Hu, L. Xie and X. Zhang, Heat kernels for non-symmetric diffusions operators with jumps. *J. Differential Equations* **263** (2017) 6576–6634.

[15] Z.-Q. Chen, Y. Hu, Z. Qian and W. Zheng, Stability and approximations of symmetric diffusion semigroups and kernels. *J. Funct. Anal.* **152** (1998), 255-280.

[16] Z.-Q. Chen, P. Kim and R. Song, Dirichlet heat kernel estimates for fractional Laplacian with gradient perturbation. *Ann. Probab.* **40** (2012), 2483-2538.

[17] Z.-Q. Chen, P. Kim and R. Song, Stability of Dirichlet heat kernel estimates for non-local operators under Feynman-Kac perturbation. *Trans. Amer. Math. Soc.* **367** (2015), 5237-5270.

[18] Z.-Q. Chen and T. Kumagai, Heat kernel estimates for stable-like processes on d-sets. *Stochastic Process. Appl.* **108** (2003), 27-62.

[19] Z.-Q. Chen and T. Kumagai, Heat kernel estimates for jump processes of mixed types on metric measure spaces. *Probab. Theory Related Fields* **140** (2008), 277-317.

[20] Z.-Q. Chen and T. Kumagai, A priori Hölder estimate, parabolic Harnack inequal-

ity and heat kernel estimates for diffusions with jumps. *Revista Mathemática Iberoamericana* **26** (2010), 551-589.

[21] Z.-Q. Chen and J.-M. Wang, Perturbation by non-local operators. To appear in *Ann. Inst. Henri Poincaré Probab. Statist.*

[22] Z.-Q. Chen and Lidan Wang, Stability of heat kernel estimates for diffusions with jumps under non-local Feynman-Kac perturbations. arXiv:1702.04489 [math.PR]

[23] Z.-Q. Chen and Longmin Wang, Uniqueness of stable processes with drifts. *Proc. Amer. Math. Soc.* **144** (2016), 2661-2675.

[24] Z.-Q. Chen and Longmin Wang, Heat kernel estimates for relativistic stable processes with singular drifts. Preprint.

[25] Z.-Q. Chen and T. Yang, Dirichlet heat kernel estimates for fractional Laplacian under non-local perturbation. Preprint.

[26] Z.-Q. Chen and X. Zhang, Heat kernels and analyticity of non-symmetric jump diffusion semigroups. *Probab. Theory Relat. Fields* **165** (2016), 267-312.

[27] Z.-Q. Chen and X. Zhang, Strong stability of heat kernels of non-symmetric stable-like operators. In *Stochastic Analysis and Related Topics, A Festschrift in honor of Rodrigo Banuelos*, p. 57–65. Birkhauser, 2017.

[28] Z.-Q. Chen and X. Zhang, Heat kernels for time-dependent non-symmetric stable-like operators. arXiv:1709.04614. To appear in *J. Math. Anal. Appl.*

[29] A. Friedman, *Partial Differantial Equations of Parabolic Type*. Prentice-Hall, Englewood Cliffs, N.J., 1975.

[30] A. Grigor'yan, J. Hu and K.-S. Lau, Heat kernels on metric measure spaces. *Geometry and analysis of fractals*, 147-207, Springer Proc. Math. Stat., 88, Springer, Heidelberg, 2014.

[31] A. Janicki and A. Weron, *Simulation and Chaotic Behavior of α-Stable Processes*. Dekker, 1994.

[32] N. Jacob, *Pseudo Differential Operators and Markov Processes*. Vo. I-III. Imperial College Press, London, 2001/2002/2005.

[33] P. Jin, Heat kernel estimates for non-symmetric stable-like processes. arXiv: arXiv:1709.02836

[34] P. Kim and R. Song, Stable process with singular drift. *Stochastic Process Appl.* **124** (2014), 2479–2516.

[35] P. Kim and R. Song, Dirichlet heat kernel estimates for stable processes with singular drift in unbounded $C^{1,1}$ open sets. *Potential Anal.* **41** (2014), 555-581.

[36] P. Kim, R. Song and Z. Vondracek, Heat kernels of non-symmetric jump processes: beyond the stable case. arXiv:1606.02005v2 [math.PR]

[37] J. Klafter, M. F. Shlesinger and G. Zumofen, Beyond Brownian motion. *Physics Today,* **49** (1996), 33–39.

[38] V. Kolokoltsov, Symmetric stable laws and stable-like jump-diffusions. *Proc. London Math. Soc.* **80** (2000), 725–768.

[39] T. Komatsu, On the martingale problem for generators of stable processes with perturbations. *Osaka J. Math.* **21** (1984), 113-132.

[40] A. Kulik: Conditions for existence and smoothness of the distribution density for Ornstein-Uhlenbeck processes with Lévy noises. *Theory Probab. Math. Statist.* no.79(2009), 23-38.

[41] A. Matacz, Financial modeling and option theory with the truncated Lévy process. *Int. J. Theor. Appl. Finance* **3(1)** (2000), 143–160.
[42] G. Pólya, On the zeros of an integral function represented by Fourier's integral. *Messenger of Math.* **52** (1923), 185-188.
[43] G. Samorodnitsky and M. S. Taqqu, *Stable Non-Gaussian Random Processes*. Chapman & Hall, New York-London, 1994.
[44] E. Priola and J. Zabczyk, Densities for Ornstein-Uhlenbeck processes with jumps. *Bull. Lond. Math. Soc.* **41** (2009), 41-50.
[45] L. Silvestre, Regularity of the obstacle problem for a fractional power of the Laplace operator. *Comm. Pure Appl. Math.* **60** (2007), 67-112.
[46] R. Song and Z. Vondraček, Parabolic Harnack inequality for the mixture of Brownian motion and stable process. *Tohoku Math. J.* **59** (2007), 1–19.
[47] C. Wang, On estimates of the density of Feynman-Kac semigroups of α-stable-like processes. *J. Math. Anal. Appl.* **348** (2008), 938–970.
[48] F.-Y. Wang and X. Zhang, Heat kernel for fractional diffusion operators with perturbations. *Forum Mathematicum* **27** (2015), 973-994.
[49] F.-Y. Wang, L. Xu and X. Zhang X., Gradient estimates for SDEs Driven by Multiplicative Lévy Noise. *J. Funct. Anal.* **269** (2015), 3195-3219.
[50] J.-M. Wang, Laplacian perturbed by non-local operators. *Math. Z.* **279** (2015), 521-556.
[51] L. Wang, L. Xie and X. Zhang, Derivative formulae for SDEs driven by multiplicative α-stable-like processes, *Stochastic Process Appl.* **125** (2015), 867-885.
[52] L. Xie and X. Zhang, Heat kernel estimates for critical fractional diffusion operator. *Studia Math.* **224** (2014), 221-263.
[53] X. Zhang, Derivative formula and gradient estimate for SDEs driven by α-stable processes. *Stochastic Process Appl.* **123** (2013), 1213-1228.
[54] X. Zhang, Fundamental solution of kinetic Fokker-Planck operator with anisotropic nonlocal dissipativity. *SIAM J. Math. Anal.* **46**, No. 3 (2014), 2254-2280.
[55] X. Zhang, Densities for SDEs driven by degenerate α-stable processes. *Ann. Probab.* **42** (2014), 1885-1910.
[56] X. Zhang, Fundamental solutions of nonlocal Hörmander's operators. *Commun. Math. Stat.* **4** (2016), 359-402.
[57] X. Zhang, Fundamental solutions of nonlocal Hörmander's operators II. *Ann. Probab.*, **45** (2017), 1799–1841.

Francesca Da Lio

Fractional Harmonic Maps

Abstract: The theory of α-harmonic maps has been initiated some years ago by the author and Tristan Rivière in [8]. These maps are critical points of the following nonlocal energy

$$\mathcal{L}^{\alpha}(u) = \int_{\mathbb{R}^k} |(-\Delta)^{\frac{\alpha}{2}} u(x)|^2 dx^k\,, \tag{2.0.1}$$

where $u \in \dot{H}^{\alpha}(\mathbb{R}^k, \mathcal{N})$, $\mathcal{N} \subset \mathbb{R}^m$ is an at least C^2 closed (compact without boundary) n-dimensional smooth manifold. In a recent paper [10] we also introduce the notion of *horizontal α-harmonic maps*. Precisely, given a C^1 plane distribution P_T on all $\mathbb{R}^m$, these are maps $u \in \dot{H}^{\alpha}(\mathbb{R}^k, \mathbb{R}^m)$, $\alpha \geqslant 1/2$, satisfying

$$\begin{cases} P_T(u)\nabla u = \nabla u & \text{in } \mathcal{D}'(\mathbb{R}^k) \\ P_T(u)(-\Delta)^{\alpha} u = 0 & \text{in } \mathcal{D}'(\mathbb{R}^k). \end{cases}$$

If the distribution of planes is integrable, then we recover the case of α-harmonic maps with values into a manifold. We will concentrate here to the case $\alpha = 1/2$ and $k = 1$ which corresponds to a critical situation. Such maps arise from several geometric problems such as for instance in the study of free boundary manifolds. After giving an overview of the recent results on the regularity and the compactness of horizontal 1/2-harmonic maps, we will describe the techniques that have been introduced in [8, 9] to investigate the regularity of such maps and mention some relevant applications to geometric problems.

2.1 Overview

Since the early 50's the analysis of critical points to conformal invariant Lagrangians has raised a special interest, due to the important role they play in physics and geometry.

The most elementary example of a 2-dimensional conformal invariant Lagrangian is the Dirichlet Energy

$$\mathcal{L}(u) = \int_D |\nabla u(x,y)|^2 dxdy\,, \tag{2.1.1}$$

where $D \subseteq R^2$ is an open set, $u\colon D \to \mathbb{R}^m$ and ∇u is the gradient of u.

We can define the Lagrangian (2.1.1) in the set of maps taking values in an at least C^2 closed n- dimensional submanifold $\mathcal{N} \subseteq \mathbb{R}^m$. In this case critical points $u \in W^{1,2}(D, \mathcal{N})$ of $\mathcal{L}$ satisfy in a weak sense the equation

Francesca Da Lio, Department of Mathematics, ETH Zürich, Rämistrasse 101, 8092 Zürich, Switzerland, E-mail: francesca.dalio@math.ethz.ch

https://doi.org/10.1515/9783110571561-004

$$-\Delta u \perp T_u\mathcal{N}\,, \tag{2.1.2}$$

where $T_\xi\mathcal{N}$ is the tangent plane a $\mathcal{N}$ at the point $\xi \in \mathcal{N}$, or in a equivalent way

$$-\Delta u = A(u)(\nabla u, \nabla u) := A(u)(\partial_x u, \partial_x u) + A(u)(\partial_y u, \partial_y u), \tag{2.1.3}$$

where $A(\xi)$ is the second fundamental form at the point $\xi \in \mathcal{N}$ (see for instance [17]). The equation (2.1.3) is called the *harmonic map equation* into $\mathcal{N}$.

In the case when $\mathcal{N}$ is an oriented hypersurface of $\mathbb{R}^m$ the harmonic map equation reads as

$$-\Delta u = \nu(u)\langle\nabla\nu(u), \nabla u\rangle\,, \tag{2.1.4}$$

where ν is the unit normal vector field to $\mathcal{N}$.

The key point to get the regularity of the harmonic maps with values into the sphere S^{m-1} was to rewrite the r.h.s of the equations as a sum of a Jacobians. More precisely Hélein in [17] wrote the equation (2.1.4) in the form

$$-\Delta u = \nabla^\perp B \cdot \nabla u, \tag{2.1.5}$$

where $\nabla^\perp B = (\nabla^\perp B_{ij})$ with $\nabla^\perp B_{ij} = u_i\nabla u_j - u_j\nabla u_i$, (for every vector field $v\colon \mathbb{R}^2 \to \mathbb{R}^m$, $\nabla^\perp v$ denotes the $\pi/2$ rotation of the gradient ∇v, namely $\nabla^\perp v = (-\partial_y v, \partial_x v))$.

The r.h.s of (2.1.5) can be written actually as a sum of Jacobians:

$$\nabla^\perp B_{ij}\nabla u_j = \partial_x u_j \partial_y B_{ij} - \partial_y u_j \partial_x B_{ij}\,.$$

This particular structure permitted to apply to the equation (2.1.5) the following result

Theorem 2.1. *[28] Let D be a smooth bounded domain of $\mathbb{R}^2$. Let a and b be two measurable functions in D whose gradients are in $L^2(D)$. Then there exists a unique solution $\varphi \in W^{1,2}(D)$ to*

$$\begin{cases} -\Delta\varphi = \dfrac{\partial a}{\partial x}\dfrac{\partial b}{\partial y} - \dfrac{\partial a}{\partial y}\dfrac{\partial b}{\partial x}, & \text{in } D \\ \varphi = 0 & \text{on } \partial D\,. \end{cases} \tag{2.1.6}$$

Moreover there exists a constant $C > 0$ independent of a and b such that

$$||\varphi||_\infty + ||\nabla\varphi||_{L^2} \leqslant C||\nabla a||_{L^2}||\nabla b||_{L^2}\,.$$

In particular φ is a continuous in D.

In the case of an oriented hypersurface $\mathcal{N}$ of $\mathbb{R}^m$ by using the fact that ∇u is orthogonal to $\nu(u)$ the equation (2.1.4) can be reformulated as follows

$$-\Delta u^i = \sum_{j=1}^m \left(\nu(u)^i\,\nabla(\nu(u))_j - \nu(u)_j\,\nabla(\nu(u))^i\right)\cdot\nabla u^j \quad . \tag{2.1.7}$$

Unlike the sphere case there is no reason for which the vector field

$$\nu(u)^i\,\nabla(\nu(u))_j - \nu(u)_j\,\nabla(\nu(u))^i$$

is divergence-free. What remains true is the anti-symmetry of the matrix

$$\Omega := \left(\nu(u)^i\,\nabla(\nu(u))_j - \nu(u)_j\,\nabla(\nu(u))^i\right)_{i,j=1\cdots m} \quad . \tag{2.1.8}$$

Actually Rivière in [20] identified the anti-symmetry of the 1-form in (2.1.8) as the essential structure of equation (2.1.4) and he succeeded in writing the harmonic map system in the form of a conservation law whose constituents satisfy elliptic equations with a Jacobian structure to which Wente's regularity result (Theorem 2.1) could be applied.

Let us now introduce $P_T(z)$, $P_N(z)$ the orthogonal projections respectively to the tangent space $T_z\mathcal{N}$ and to the normal space $(T_z\mathcal{N})^\perp$. Then the equation (2.1.2) can be re-formulated as follows

$$P_T(u)\Delta u = 0, \quad \text{in } \mathcal{D}'(D). \tag{2.1.9}$$

We are going to release the assumption that the field of orthogonal projections is associated to a sub-manifold $\mathcal{N}$ and to consider the equation (2.1.9) for a general field of orthogonal projections P_T and for horizontal maps u satisfying

$$P_N(u)\nabla u = 0, \quad \text{in } \mathcal{D}'(D). \tag{2.1.10}$$

We will assume that $P_T \in C^1(\mathbb{R}^m, \mathcal{M}_m(\mathbb{R}))$ and $P_N \in C^1(\mathbb{R}^m, \mathcal{M}_m(\mathbb{R}))$ satisfy

$$\begin{cases} P_T \circ P_T = P_T \quad P_N \circ P_N = P_N \\ P_T + P_N = I_m \\ \forall\, z \in \mathbb{R}^m \quad \forall\, U, V \in T_z\mathbb{R}^m \quad < P_T(z)U, P_N(z)V >= 0 \\ \|\partial_z P_T\|_{L^\infty(\mathbb{R}^m)} < +\infty \end{cases} \tag{2.1.11}$$

where $< \cdot, \cdot >$ denotes the standard scalar product in $\mathbb{R}^m$. In other words P_T is a C^1 map into the orthogonal projections of $\mathbb{R}^m$. For such a distribution of projections P_T we denote by

$$n := \text{rank}(P_T).$$

Such a distribution identifies naturally with the distribution of n-planes given by the images of P_T (or the Kernel of P_T) and conversely, any C^1 distribution of n-dimensional planes defines uniquely P_T satisfying (2.1.11).

For any $\alpha \geqslant 1/2$ and for $k \geqslant 1$ we define the space of *H^α-Sobolev horizontal maps*

$$\mathfrak{H}^\alpha(\mathbb{R}^k) := \left\{u \in H^\alpha(\mathbb{R}^k, \mathbb{R}^m); \quad P_N(u)\nabla u = 0 \quad \text{in } \mathcal{D}'(\mathbb{R}^k)\right\}.$$

Observe that this definition makes sense since we have respectively $P_N \circ u \in H^\alpha(\mathbb{R}^k, \mathcal{M}_m(\mathbb{R}))$ and $\nabla u \in H^{\alpha-1}(\mathbb{R}^k, \mathbb{R}^m)$. Next we give the following definition.

Definition 2.2. *Given a C^1 plane distribution P_T in $\mathbb{R}^m$ satisfying (2.1.11), a map u in the space $\mathfrak{H}^\alpha(\mathbb{R}^k)$ is called* **horizontal α-harmonic** *with respect to P_T if*

$$\forall\, i = 1 \cdots m \qquad \sum_{j=1}^{m} P_T^{ij}(u)(-\Delta)^\alpha u_j = 0 \qquad \text{in } \mathcal{D}'(\mathbb{R}^k) \tag{2.1.12}$$

and we shall use the following notation

$$P_T(u)\,(-\Delta)^\alpha u = 0 \qquad \text{in } \mathcal{D}'(\mathbb{R}^k).$$

When the plane distribution P_T is *integrable* that is to say when

$$\forall\, X, Y \in C^1(\mathbb{R}^m, \mathbb{R}^m) \quad P_N[P_T X, P_T Y] \equiv 0, \tag{2.1.13}$$

where $[\cdot,\cdot]$ denotes the Lie Bracket of vector-fields, using Frobenius theorem the plane distribution corresponds to the tangent plane distribution of a $n-$dimensional *foliation* $\mathcal{F}$. A smooth map u in $\mathfrak{H}^\alpha(\mathbb{R}^m)$ takes values everywhere into a *leaf* of $\mathcal{F}$ that we denote $\mathcal{N}^n$ and we are back to the classical theory of harmonic maps into manifolds. Observe that our definition includes the case of α-harmonic maps with values into a sub-manifold of the euclidean space and horizontal with respect to a plane distribution in this sub-manifold. Indeed it is sufficient to add to such a distribution the projection to the sub-manifold and extend the all to a tubular neighborhood of the sub-manifold.

In [10] we have proved the following result

Theorem 2.3 (Theorem 2.1, [10]). *Let P_T be a C^1 distribution of planes (or projections) satisfying (2.1.11). Any map $u \in \mathfrak{H}^1(D)$*

$$P_T(u)\,\Delta u = 0 \quad \text{in } \mathcal{D}'(D) \tag{2.1.14}$$

is in $\cap_{\delta<1} C^{0,\delta}_{loc}(D)$.

The main idea to prove Theorem 2.3 is to show that u satisfies an elliptic Schrödinger type system with an antisymmetric potential $\Omega \in L^2(\mathbb{R}^k, \mathbb{R}^k \otimes so(m))$ (whose construction depends on P_T) of the form

$$-\Delta u = \Omega \cdot \nabla u. \tag{2.1.15}$$

Hence, following the analysis in [20] the authors deduced in dimension 2 the local existence on a disk D of $A \in L^\infty \cap W^{1,2}(D, Gl_m(\mathbb{R}))$ and $B \in W^{1,2}(D, \mathcal{M}_m(\mathbb{R}))$, depending both on $P_T(u)$, such that

$$\operatorname{div}(A\,\nabla u) = \nabla^\perp B \cdot \nabla u \tag{2.1.16}$$

from which the regularity of u can be deduced using Wente's Theorem 2.1.[2.1]

2.1 We denote by $so(m)$ the space of antisymmetric matrices of order m and by GL_m the space of invertible matrices of order m.

Now we turn our attention to an analogous *fractional* problem in dimension 1. We consider the following Lagrangian that we will call *L–energy* (*L* stands for "Line")

$$\mathcal{L}^{1/2}(u) := \int_{\mathbb{R}} |(-\Delta)^{1/4}u|^2\, dx \tag{2.1.17}$$

within

$$\dot{H}^{1/2}(\mathbb{R}, \mathcal{N}) := \left\{ u \in \dot{H}^{1/2}(\mathbb{R}, \mathbb{R}^m) \; ; \; u(x) \in \mathcal{N} \text{ for a. e. } x \in \mathbb{R} \right\}.$$

The operator $(-\Delta)^\alpha$ on $\mathbb{R}$ is defined by means of the Fourier transform as follows

$$\widehat{(-\Delta)^\alpha u} = |\xi|^{2\alpha}\hat{u}\,,$$

(given a function f, both $\hat{f}$ and $\mathcal{F}[f]$ denote the Fourier transform of f).

The Lagrangian (2.1.17) is invariant with respect to the Möbius group and it satisfies the following identity

$$\int_{\mathbb{R}} |(-\Delta)^{1/4}u(x)|^2 dx = \inf\left\{ \int_{\mathbb{R}^2_+} |\nabla\tilde{u}|^2 dx : \tilde{u} \in W^{1,2}(\mathbb{R}^2_+, \mathbb{R}^m), \;\; \text{trace } \tilde{u} = u \right\}.$$

In [8] we introduced the following Definition:

Definition 2.4. *A map $u \in \dot{H}^{1/2}(\mathbb{R}, \mathcal{N})$ is called a weak 1/2-harmonic map into $\mathcal{N}$ if for any $\varphi \in \dot{H}^{1/2}(\mathbb{R}, \mathbb{R}^m) \cap L^\infty(\mathbb{R}, \mathbb{R}^m)$ there holds*

$$\frac{d}{dt}\mathcal{L}^{1/2}(\pi_{\mathcal{N}}(u + t\varphi))_{|_{t=0}} = 0,$$

where $\Pi_{\mathcal{N}}$ is the orthogonal projection on $\mathcal{N}$.

In short we say that *a weak 1/2-harmonic map is a critical point of $\mathcal{L}^{1/2}$ in $\dot{H}^{1/2}(\mathbb{R}, \mathcal{N})$ for perturbations in the target.*

Weak 1/2-harmonic maps satisfy the Euler-Lagrange equation

$$\nu(u) \wedge (-\Delta)^{1/2}u = 0 \qquad \text{in } \mathcal{D}'(\mathbb{R}). \tag{2.1.18}$$

Let $\Pi_{-i}\colon S^1\backslash\{-i\} \to \mathbb{R}$, $\Pi_{-i}(\xi + i\eta) = \frac{\xi}{1+\eta}$ be the stereographic projection from the south pole, then the following relation between the 1/2 Laplacian in $\mathbb{R}$ and in S^1 holds:

Proposition 2.5 (Proposition 4.1, [7]). *Given $u : \mathbb{R} \to \mathbb{R}^m$, we set $v := u \circ \Pi_{-i} : S^1 \to \mathbb{R}^m$. Then $u \in L_{\frac{1}{2}}(\mathbb{R})$[2.2] if and only if $v \in L^1(S^1)$. In this case*

$$(-\Delta)^{\frac{1}{2}}_{S^1} v(e^{i\vartheta}) = \frac{((-\Delta)^{\frac{1}{2}}_{\mathbb{R}}u)(\Pi_{-i}(e^{i\vartheta}))}{1 + \sin\vartheta} \quad \text{in } \mathcal{D}'(S^1\backslash\{-i\}), \tag{2.1.19}$$

2.2 We recall that $L_{\frac{1}{2}}(\mathbb{R}) := \left\{ u \in L^1_{loc}(\mathbb{R}) : \int_{\mathbb{R}} \frac{|u(x)|}{1+x^2} dx < \infty \right\}$

Observe that $(1+\sin(\vartheta))^{-1} = |\Pi'_{-i}(\vartheta)|$, *and hence we have*

$$\int_{S^1} (-\Delta)^{\frac{1}{2}} v(e^{i\vartheta})\,\varphi(e^{i\vartheta})\,d\vartheta = \int_{\mathbb{R}} (-\Delta)^{\frac{1}{2}} u(x)\,\varphi(\Pi_{-i}^{-1}(x))\,dx \quad \textit{for every } \varphi \in C_0^\infty(S^1\backslash\{-i\}).$$

From (2.1.19) and the invariance of the Lagrangian (2.1.17) with respect to the trace of conformal maps in $\mathbb{C}$ it follows that a map $u \in \dot{H}^{1/2}(\mathbb{R}, \mathcal{N})$ is weak $1/2$-harmonic in $\mathbb{R}$ if and only if $v = u \circ \Pi_{-i} \in \dot{H}^{1/2}(S^1, \mathcal{N})$ is weak $1/2$-harmonic in S^1.

Indeed $v \in \dot{H}^{1/2}(S^1, \mathcal{N})$ satisfies

$$\nu(v) \wedge (-\Delta)^{1/2} v = 0 \qquad \text{in } \mathcal{D}'(S^1\backslash\{-i\}). \tag{2.1.20}$$

Consider now the stereographic projection from the north pole $\Pi_i \colon S^1\backslash\{i\} \to \mathbb{R}$, $\Pi_i(\xi + i\eta) = \frac{\xi}{1-\eta}$ and $\tilde{u} = v \circ \Pi_i^{-1} = u \circ \frac{1}{z}$. Since $\frac{1}{z} \colon \mathbb{C}\backslash\{0\} \to \mathbb{C}\backslash\{0\}$ is a conformal map, $\tilde{u} \in \dot{H}^{1/2}(\mathbb{R}, \mathcal{N})$ is weak $1/2$-harmonic in $\mathbb{R}\backslash\{0\}$. By applying Proposition 2.2 in [5] (a singular point removability type result on $\mathbb{R}$) we deduce that $\tilde{u}$ is weak $1/2$-harmonic in $\mathbb{R}$ and in particular continuous in $\mathbb{R}$. Therefore not only v is weak $1/2$-harmonic in S^1 but we deduce that

$$\lim_{x \to +\infty} u(x) = \lim_{x \to +\infty} u(x) \ \text{and} \ \lim_{\substack{z \to -i^+ \\ z \in S^1}} v(z) = \lim_{\substack{z \to -i^- \\ z \in S^1}} v(z).$$

Fractional harmonic maps appear in several geometric problems and we mention some of them below.

1. The first application is the connection between weak $1/2$-harmonic maps and free boundary minimal disks. The following characterization of weak $1/2$-harmonic maps of S^1 into sub-manifolds of $\mathbb{R}^n$ holds, (see [7] and [18]).

Theorem 2.6. *Let* $u \in \dot{H}^{1/2}(S^1, \mathcal{N})$, *where* $\mathcal{N}$ *is a n-dimensional closed smooth sub-manifold of* $\mathbb{R}^m$. *If u is a nontrivial weak* $1/2$-*harmonic map, then its harmonic extension* $\tilde{u} \in W^{1,2}(D, \mathbb{R}^m)$ *is conformal and*

$$\nu(u) \wedge \frac{\partial \tilde{u}}{\partial r} = 0 \quad \textit{in } \mathcal{D}'(S^1). \tag{2.1.21}$$

From Theorem 2.6 it follows that $\tilde{u}$ is a minimal disk whose boundary lies in $\mathcal{N}$ and meets $\mathcal{N}$ orthogonally, namely its outward normal vector $\frac{\partial \tilde{u}}{\partial r}$ is othogonal to $\mathcal{N}$ at each point of $\tilde{u}(S^1)$. Moreover we can deduce the following two characterizations of $1/2$-harmonic maps in the case where $\mathcal{N} = S^1$ and $\mathcal{N} = S^2$.

Theorem 2.7. *i) Weak* $1/2$-*harmonic maps* $u \colon S^1 \to S^1$ *with* $deg(u) = 1$ *coincide with the trace of Möbius transformations of the disk* $B^2(0,1) \subseteq \mathbb{R}^2$.

ii) If $u \colon S^1 \to S^2$ *is a weak* $1/2$-*harmonic map then* $u(S^1)$ *is an equatorial plane and it is the composition of weak* $1/2$-*harmonic map* $u \colon S^1 \to S^1$ *with an isometry* $\tau \colon S^2 \to S^2$.

2. Another geometrical application concerns the so-called Steklov eigenvalue problem that is the first eigenvalue σ_1 of the Dirichlet-to-Neumann map on some Riemannian surfaces (M, g) with boundary ∂M. In [14] the authors show the following

Theorem 2.8 (Proposition 2.8, [14]). *If M is a surface with boundary, and g_0 is a metric on M with*

$$\sigma_1(g_0)L_{g_0}(\partial M) = \max_g \sigma_1(g)L_g(\partial M),$$

where $L_g(\partial M)$ is the lenght of ∂M, the max is over all smooth metrics on M in the conformal class of g_0. Then there exist independent eigenfunctions $u_1, \dots, u_n$ corresponding to the eigenvalue $\sigma_1(g_0)$ which give a conformal minimal immersion $u = (u_1, \dots, u_n)$ of M into the unit ball B^n and $u(M)$ is a free boundary solution. That is, $u\colon (M, \partial M) \to (B^n, \partial B^n)$ is a harmonic map such that $u(\partial M)$ meets ∂B^n orthogonally. Hence $u|_{\partial M}$ is $1/2$-harmonic.

3. $1/2$-harmonic maps appear in the asymptotics of fractional Ginzburg-Landau equation, (see [18]) and in connections with regularity of critical knots of Möbius energy (see [2]).

The theory of weak $1/2$ harmonic maps with values into a closed n-dimensional sub-manifold $\mathcal{N}$ has been initiated some years ago by the author and Tristan Rivière in [8]. Since then several extensions have been considered (see [4, 12, 9]). The main novelty in the regularity of $1/2$-harmonic was the re-formulation of the Euler-Lagrange equation in terms of special algebraic quantities called *3-terms commutators* which are roughly speaking bilinear pseudo-differential operators satisfying some *integrability by compensation* properties.

As in the local case we can consider a plane distribution P_T satisfying (2.1.11) and solutions of

$$P_T(u)\,(-\Delta)^{1/2}u = 0 \quad \text{in } \mathcal{D}'(\mathbb{R}) \tag{2.1.22}$$

under the constraint $P_N(u)\,\nabla u = 0$ in $\mathcal{D}'(\mathbb{R})$. Maps $u \in \mathfrak{H}^{1/2}(\mathbb{R})$ satisfying (2.1.22) are called *horizontal $1/2$-harmonic maps*. One of the main result in [10] is the following Theorem.

Theorem 2.9. *Let P_T be a C^1 distribution of planes satisfying (2.1.11). Any map $u \in \mathfrak{H}^{1/2}(\mathbb{R})$*

$$P_T(u)\,(-\Delta)^{1/2}u = 0 \quad \text{in } \mathcal{D}'(\mathbb{R}) \tag{2.1.23}$$

is in $\cap_{\delta<1} C^{0,\delta}_{loc}(\mathbb{R})$.

In [10] conservation laws corresponding to *horizontal $1/2$-harmonic maps* have been discovered: locally, modulo some smoother terms coming from the application of non-local operators on cut-off functions, we construct out of $P_T(u)$ $A \in L^\infty \cap$

$\dot{H}^{1/2}(\mathbb{R}, Gl_m(\mathbb{R}))$ and $B \in \dot{H}^{1/2}(\mathbb{R}, \mathcal{M}_m(\mathbb{R}))$ such that

$$(-\Delta)^{1/4}(A\,v) = \mathfrak{J}(B, v) + \text{cut-off}, \tag{2.1.24}$$

where $v := (P_T(-\Delta)^{1/4}u, \mathcal{R}(P_N(-\Delta)^{1/4}u))^t$, $\mathcal{R}$ denotes the Riesz operator defined by $\widehat{\mathcal{R}f}(\xi) = i\frac{\xi}{|\xi|}\hat{f}$ and $\mathfrak{J}$ is a bilinear pseudo-differential operator satisfying

$$\|\mathfrak{J}(B, v)\|_{\dot{H}^{-1/2}(\mathbb{R})} \leqslant C\,\|(-\Delta)^{1/4}B\|_{L^2(\mathbb{R})}\,\|v\|_{L^2(\mathbb{R})}. \tag{2.1.25}$$

As we will see later, the conservation law (2.1.24) will be crucial in the quantization analysis of sequences of horizontal $1/2$-harmonic maps.

By assuming that $P_T \in C^2(\mathbb{R}^m)$ and by bootstrapping arguments one gets that every horizontal $1/2$-harmonic map $u \in \mathfrak{H}^{1/2}(\mathbb{R})$ is $C^{1,\alpha}_{loc}(\mathbb{R})$, for every $\alpha < 1$ (see [11]).

We would like to mention that in the non-integrable case it seems not feasible to get the regularity of the horizontal $1/2$-harmonic maps by the techniques in [23] or [18] which consist in transforming the a-priori non-local PDE (2.1.18) into a local one and in performing *ad-hoc* extensions and reflections.

Also in the nonintegrable case the following geometric characterization holds.

Proposition 2.10. *An element in $\mathfrak{H}^{1/2}$ satisfying (2.1.22) has a harmonic extension $\tilde{u}$ in $B^2(0,1)$ which is conformal and hence it is the boundary of a minimal disk whose exterior normal derivative $\partial_r\tilde{u}$ is orthogonal to the plane distribution given by P_T.*

Example : We consider the following field of non-integrable projections in $\mathbb{C}^2\setminus\{0\}$.

$$P_T(z)\,Z := Z - |z|^{-2}\,[Z\cdot(z_1, z_2)\,(z_1, z_2) + Z\cdot(iz_1, iz_2)\,(iz_1, iz_2)]\,. \tag{2.1.26}$$

An example of u satisfying (2.1.23) is given by solutions to the system

$$\begin{cases} \dfrac{\partial\tilde{u}}{\partial r}\wedge u\wedge iu = 0 \quad \text{in } \mathcal{D}'(S^1) \\[2ex] u\cdot\dfrac{\partial u}{\partial\vartheta} = 0 \quad \text{in } \mathcal{D}'(S^1) \qquad\qquad \text{an at least} \\[2ex] iu\cdot\dfrac{\partial u}{\partial\vartheta} = 0 \quad \text{in } \mathcal{D}'(S^1) \end{cases} \tag{2.1.27}$$

where $\tilde{u}$ denotes the harmonic extension of u which happens to be conformal due to Proposition 2.10 and define a minimal disk. An example of such maps is given by

$$u(\vartheta) := \frac{1}{\sqrt{2}}(e^{i\vartheta}, e^{-i\vartheta}) \quad \text{where} \quad \tilde{u}(z,\overline{z}) = \frac{1}{\sqrt{2}}(z, \overline{z}). \tag{2.1.28}$$

Observe that the solution in (2.1.28) is also a $1/2$-harmonic map into S^3 and it would be interesting to investigate whether this is the unique solution.

From a geometrical point of view to find a solution to (2.1.23) means to find a minimal disk whose boundary is horizontal and the normal direction is vertical.

One natural question is to see if this problem is variational. A priori if $\tilde{u}$ is a critical point of the Dirichlet energy whose boundary is horizontal, then its exterior normal derivative $\partial_r \tilde{u}$ does not belong necessarily to $\mathrm{Im}(P_N)$. Despite the geometric relevance of equations (2.1.12) in the non-integrable case, it is however a-priori not the *Euler-Lagrange* equation of the variational problem consisting in finding the critical points of $\|(-\Delta)^{\alpha/2}u\|^2_{L^2}$ within $\mathfrak{H}^\alpha$ when P_T is not satisfying (2.1.13). This can be seen in the particular case where $\alpha = 1$ where the critical points to the *Dirichlet Energy* have been extensively studied in relation with the computation of *normal geodesics* in sub-riemannian geometry. We then introduce the following definition:

Definition 2.11. *A map u in $\mathfrak{H}^\alpha$ is called* **variational $\alpha-$harmonic** *into the plane distribution P_T if it is a critical point of the $\|(-\Delta)^{\alpha/2}u\|^2_{L^2}$ within variations in $\mathfrak{H}^\alpha$ i.e. for any $u_t \in C^1((-1,1),\mathfrak{H}^\alpha)$ we have*

$$\frac{d}{dt}\|(-\Delta)^{\alpha/2}u_t\|^2_{L^2}\Big|_{t=0} = 0. \tag{2.1.29}$$

Example of variational harmonic maps from S^1 into a plane distribution is given by the sub-riemannian geodesics.

A priori the equation (2.1.22) is not the Euler-Lagrange equation associated to (2.1.29). The main difficulty is that we have not a pointwise constraint but a constraint on the gradient. In order to study critical points of (2.1.29) we use a convexification of the above variational problem following the spirit of the approach introduced by Strichartz in [27] for *normal geodesics* in sub-riemannian geometry. We prove in particular for the case $\alpha = 1/2$ that the smooth critical points of

$$\begin{aligned}\mathcal{L}^{1/2}(u,\xi) := &\int_{S^1} \frac{|(-\Delta)_0^{-1/4}(P_T(u)\xi)|^2}{2}\, d\vartheta \\ &- \int_{S^1}\left\langle (-\Delta)_0^{-1/4}(P_T(u)\xi), (-\Delta)_0^{-1/4}\left(P_T(u)\frac{du}{d\vartheta}\right)\right\rangle d\vartheta \\ &- \int_{S^1}\left\langle (-\Delta)_0^{-1/4}(P_N(u)\xi), (-\Delta)_0^{-1/4}\left(P_N(u)\frac{du}{d\vartheta}\right)\right\rangle d\vartheta\end{aligned} \tag{2.1.30}$$

in the co-dimension m <u>Hilbert subspace</u> of $\dot{H}^{1/2}(S^1,\mathbb{R}^m)\times \dot{H}^{-1/2}(S^1,\mathbb{R}^m)$ given by[2.3]

$$\mathfrak{E} := \left\{\begin{array}{c} (u,\xi)\in \dot{H}^{1/2}(S^1,\mathbb{R}^m)\times H^{-1/2}(S^1,\mathbb{R}^m) \quad \text{s.t.} \\ \left(P_N(u),\frac{du}{d\vartheta}\right)_{\dot{H}^{1/2},\dot{H}^{-1/2}} = 0 \\ (-\Delta)_0^{-1/4}(P_T(u)\xi)\in L^2(S^1) \quad \text{and} \quad (-\Delta)_0^{-1/4}\left(P_T(u)\frac{du}{d\vartheta}\right)\in L^2(S^1)\end{array}\right\}$$

2.3 Given $f \in \dot{H}^{1/2}$, $g \in \dot{H}^{-1/2}$ we denote by $(f,g)_{\dot{H}^{1/2},\dot{H}^{-1/2}}$ the duality between f and g.

at the point where the constraint $\left(P_N(u), \frac{du}{d\vartheta}\right)_{\dot H^{1/2},\dot H^{-1/2}} = 0$ is **non-degenerate** are "variational 1/2-harmonic" into the plane distribution P_T in the sense of definition 2.11. It remains open the regularity of critical points of (2.1.30) or even of the 1/2 energy (2.1.17) in $\mathfrak{H}^{1/2}$ in the case when the constraint $\left(P_N(u), \frac{du}{d\vartheta}\right)_{\dot H^{1/2},H^{-1/2}} = 0$ is degenerate.

In a joint paper with P. Laurain and T. Rivière we investigate compactness and quantization properties of sequences of horizontal 1/2 harmonic maps $u_k \in \mathfrak{H}^{1/2}(\mathbb{R})$ by extending the results obtained by the author in [5] in the case of 1/2-harmonic maps with values into a sphere. Our first main result is the following:

Theorem 2.12. [Theorem 1.2 in [6]] *Let $u_k \in \mathfrak{H}^{1/2}(\mathbb{R})$ be a sequence of horizontal 1/2-harmonic maps such that*

$$\|u_k\|_{\dot H^{1/2}} \leqslant C, \quad \|(-\Delta)^{1/2}u_k\|_{L^1} \leqslant C. \tag{2.1.31}$$

Then it holds:

1. *There exist $u_\infty \in \mathfrak{H}^{1/2}(\mathbb{R})$ and a possibly empty set $\{a_1,\dots,a_\ell\}$, $\ell \geqslant 1$, such that up to subsequence*

$$u_k \to u_\infty \quad \text{in } \dot W^{1/2,p}_{loc}(\mathbb{R}\setminus\{a_1,\dots,a_\ell\}),\ p \geqslant 2 \text{ as } k \to +\infty \tag{2.1.32}$$

   and

$$P_T(u_\infty)(-\Delta)^{1/2}u_\infty = 0, \quad \text{in } \mathcal{D}'(\mathbb{R}). \tag{2.1.33}$$

2. *There is a family $\tilde u^{i,j}_\infty \in \dot{\mathfrak{H}}^{1/2}(\mathbb{R})$ of horizontal 1/2-harmonic maps ($i \in \{1,\dots,\ell\}, j \in \{1,\dots,N_i\}$), such that up to subsequence*

$$\left\|(-\Delta)^{1/4}\left(u_k - u_\infty - \sum_{i,j} \tilde u^{i,j}_\infty((x - x^k_{i,j})/r^k_{i,j})\right)\right\|_{L^2_{loc}(\mathbb{R})} \to 0, \quad \text{as } k \to +\infty. \tag{2.1.34}$$

   for some sequences $r^k_{i,j} \to 0$ and $x^k_{i,j} \in \mathbb{R}$.

As we have already remarked in [6] the condition $\|(-\Delta)^{1/2}u_k\|_{L^1} \leqslant C$ is always satisfied in the case the maps u_k take values into a closed manifold of $\mathbb{R}^m$ (case of sequences of 1/2 harmonic maps) as soon as $\|u_k\|_{\dot H^{1/2}} \leqslant C$. This follows from the fact that if u is a 1/2-harmonic maps with values into a closed manifold of $\mathcal{N}$ of $\mathbb{R}^m$ then the following inequality holds (see Proposition 5.1 in [6])

$$\|(-\Delta)^{1/2}u\|_{L^1(\mathbb{R})} \leqslant C\|(-\Delta)^{1/4}u\|^2_{L^2(\mathbb{R})}. \tag{2.1.35}$$

Hence in the case of 1/2-harmonic maps defined in S^1 we have the following corollary.

Corollary 2.13. [Corollary 1.1 in [6]] *Let $\mathcal{N}$ be a closed C^2 submanifold of $\mathbb{R}^m$ and let $u_k \in H^{1/2}(S^1,\mathcal{N})$ be a sequence of 1/2-harmonic maps such that*

$$\|u_k\|_{\dot H^{1/2}(S^1)} \leqslant C \tag{2.1.36}$$

then the conclusions of Ttheseheorem 2.12 hold. In particular up to subsequence we have the following energy identity

$$\lim_{k\to+\infty}\int_{S^1}|(-\Delta)^{1/4}u_k|^2\,d\vartheta=\int_{S^1}|(-\Delta)^{1/4}u_\infty|^2\,d\vartheta+\sum_{i,j}\int_{S^1}|(-\Delta)^{1/4}\tilde{u}_\infty^{i,j}|^2\,d\vartheta \quad (2.1.37)$$

where $\tilde{u}_\infty^{i,j}$ are the **bubbles** *associated to the weak convergence.*

For the moment it remains open whether the bound (2.1.35) holds or not in the general case of horizontal 1/2-harmonic maps.

The compactness issue (first part of Theorem 2.12) is quite standard. The most delicate part is the quantization analysis consisting in verifying that there is no dissipation of the energy in the region between u_∞ and the *bubbles* $\tilde{u}_\infty^{i,j}$ and between the bubbles themselves (the so-called *neck-regions*). Such an analysis has been achieved in [6] by performing a precise asymptotic development of horizontal 1/2-harmonic maps in these neck-regions, that was possible thanks to the conservation law (2.1.24) and an application of new Pohozaev-type identities in 1-D discovered in [6]. We refer the reader to [6] for a complete description of compactness and quantization issues of horizontal 1/2-harmonic maps.

We conclude this section by mentioning that the partial regularity of 1/2-harmonic map in dimension $k \geqslant 2$ with values into a sphere has been been deduced in [18] from existing regularity results of harmonic maps with free boundary. Schikorra [25] has also studied the partial regularity of weak solutions to nonlocal linear systems with an antisymmetric potential in the supercritical case under a crucial monotonicity assumption on the solutions which allows us to reduce it to the critical case.

It still remains open a direct proof of the partial regularity without an ad-hoc monotonicity assumption.

2.2 3-Commutators Estimates

As we have already mentioned in the previous section, when the notion of 1/2-harmonic map was introduced in [8], one of the main novelty was the re-formulation of the Euler-Lagrange equation in terms of *three-terms-commutators* which have played a key role in all the results that have been obtained later.

In this section we will introduce such commutators and recall some important estimates and properties. Such properties will be crucial to get regularity results of 1/2-harmonic maps and to re-write the system (satisfied by a horizontal 1/2-harmonic map)

$$\begin{cases} P_T(u)(-\Delta)^{1/2}u=0\\[2ex] P_N(u)\nabla u=0\end{cases} \quad (2.2.1)$$

in term of a conservation law.

We first introduce some functional spaces.

$\mathcal{H}^1(\mathbb{R}^n)$ denotes the Hardy space which is the space of L^1 functions f on $\mathbb{R}^n$satisfying

$$\int_{\mathbb{R}^n} \sup_{t\in\mathbb{R}} |\varphi_t * f|(x)\, dx < +\infty \quad ,$$

where $\varphi_t(x) := t^{-n}\ \varphi(t^{-1}x)$ and where φ is some function in the Schwartz space $\mathcal{S}(\mathbb{R}^n)$ satisfying $\int_{\mathbb{R}^n} \varphi(x)\, dx = 1$. For more properties on the Hardy space $\mathcal{H}^1$ we refer to [15, 16, 26].

The $L^{2,\infty}(\mathbb{R})$ is the space of measurable functions f such that

$$\sup_{\lambda>0} \lambda |\{x \in \mathbb{R}\ :\ |f(x)| \geqslant \lambda\}|^{1/2} < +\infty\,.$$

$L^{2,1}(\mathbb{R})$ is the Lorentz space of measurable functions satisfying

$$\int_0^{+\infty} |\{x \in \mathbb{R}\ :\ |f(x)| \geqslant \lambda\}|^{1/2} d\lambda < +\infty\ .$$

In [8] the following two *three-terms commutators* have been introduced:

$$T(Q, v) := (-\Delta)^{1/4}(Qv) - Q(-\Delta)^{1/4}v + (-\Delta)^{1/4}Qv \tag{2.2.2}$$

and

$$S(Q, v) := (-\Delta)^{1/4}[Qv] - \overline{\mathcal{R}}(Q\mathcal{R}(-\Delta)^{1/4}v) + \overline{\mathcal{R}}((-\Delta)^{1/4}Q\mathcal{R}v), \tag{2.2.3}$$

where $\mathcal{R}$ is the Riesz operator.

In [8] the authors obtained the following estimates.

Theorem 2.14. *Let $v \in L^2(\mathbb{R})$, $Q \in \dot{H}^{1/2}(\mathbb{R})$. Then $T(Q, v), S(Q, v) \in H^{-1/2}(\mathbb{R})$ and*

$$\|T(Q, v)\|_{H^{-1/2}(\mathbb{R})} \leqslant C\, \|Q\|_{\dot{H}^{1/2}(\mathbb{R})} \|v\|_{L^{2,\infty}(\mathbb{R})}\ ; \tag{2.2.4}$$

$$\|S(Q, v)\|_{H^{-1/2}(\mathbb{R})} \leqslant C\, \|Q\|_{\dot{H}^{1/2}(\mathbb{R})} \|v\|_{L^{2,\infty}(\mathbb{R})}\ . \tag{2.2.5}$$

We observe that under our assumptions $u \in \dot{H}^{1/2}(\mathbb{R}, \mathbb{R}^m)$ and $Q \in \dot{H}^{1/2}(\mathbb{R}, \mathcal{M}_{\ell\times m}(\mathbb{R}))$ each term individually in T and S - like for instance $(-\Delta)^{1/4}(Q(-\Delta)^{1/4}u)$ or $Q(-\Delta)^{1/2}u$ ... - are not in $H^{-1/2}$ but the special linear combination of them constituting T and S are in $H^{-1/2}$. In a similar way, in dimension 2, $J(a, b) := \frac{\partial a}{\partial x}\frac{\partial b}{\partial y} - \frac{\partial a}{\partial y}\frac{\partial b}{\partial x}$ satisfies, as a direct consequence of Wente's theorem 2.1

$$\|J(a, b)\|_{\dot{H}^{-1}} \leqslant C\, \|a\|_{\dot{H}^1}\, \|b\|_{\dot{H}^1} \tag{2.2.6}$$

whereas, individually, the terms $\frac{\partial a}{\partial x}\frac{\partial b}{\partial y}$ and $\frac{\partial a}{\partial y}\frac{\partial b}{\partial x}$ are not in H^{-1}.

Actually in [5] we improve the estimates on the operators T, S.

Theorem 2.15. *Let $v \in L^2(\mathbb{R})$, $Q \in \dot{H}^{1/2}(\mathbb{R})$. Then $T(Q,v), S(Q,v) \in \mathcal{H}^1(\mathbb{R})$ and*

$$\|T(Q,v)\|_{\mathcal{H}^1(\mathbb{R})} \leqslant C\|Q\|_{\dot{H}^{1/2}(\mathbb{R})}\|v\|_{L^2(\mathbb{R})}\,. \tag{2.2.7}$$

$$\|S(Q,v)\|_{\mathcal{H}^1(\mathbb{R})} \leqslant C\|Q\|_{\dot{H}^{1/2}(\mathbb{R})}\|v\|_{L^2(\mathbb{R})}\,. \tag{2.2.8}$$

We refer the reader to [8] and [5] for the proof of respectively Theorem 2.14 and Theorem 2.15. We just mention that the above estimates is based on a well-known tool in harmonic analysis, the Littlewood-Paley dyadic decomposition of unity that we briefly recall here. Such a decomposition can be obtained as follows. Let $\varphi(\xi)$ be a radial Schwartz function supported in $\{\xi \in \mathbb{R}^n : |\xi| \leqslant 2\}$, which is equal to 1 in $\{\xi \in \mathbb{R}^n : |\xi| \leqslant 1\}$. Let $\psi(\xi)$ be the function given by

$$\psi(\xi) := \varphi(\xi) - \varphi(2\xi)$$

ψ is then a "bump function" supported in the annulus $\{\xi \in \mathbb{R}^n : 1/2 \leqslant |\xi| \leqslant 2\}$.

Let $\psi_0 = \varphi$, $\psi_j(\xi) = \psi(2^{-j}\xi)$ for $j \neq 0$. The functions ψ_j, for $j \in \mathbb{Z}$, are supported in $\{\xi \in \mathbb{R}^n : 2^{j-1} \leqslant |\xi| \leqslant 2^{j+1}\}$ and they realize a dyadic decomposition of the unity:

$$\sum_{j\in\mathbb{Z}} \psi_j(x) = 1\,.$$

We further denote

$$\varphi_j(\xi) := \sum_{k=-\infty}^{j} \psi_k(\xi)\,.$$

The function φ_j is supported on $\{\xi,\ |\xi| \leqslant 2^{j+1}\}$.

For every $j \in \mathbb{Z}$ and $f \in \mathcal{S}'(\mathbb{R})$ we define the Littlewood-Paley projection operators P_j and $P_{\leqslant j}$ by

$$\widehat{P_j f} = \psi_j \hat{f} \quad \widehat{P_{\leqslant j} f} = \varphi_j \hat{f}\,.$$

Informally P_j is a frequency projection to the annulus $\{2^{j-1} \leqslant |\xi| \leqslant 2^j\}$, while $P_{\leqslant j}$ is a frequency projection to the ball $\{|\xi| \leqslant 2^j\}$. We will set $f_j = P_j f$ and $f^j = P_{\leqslant j} f$.

We observe that $f^j = \sum_{k=-\infty}^{j} f_k$ and $f = \sum_{k=-\infty}^{+\infty} f_k$ (where the convergence is in $\mathcal{S}'(\mathbb{R})$).

Given $f, g \in \mathcal{S}'(\mathbb{R})$ we can split the product in the following way

$$fg = \Pi_1(f,g) + \Pi_2(f,g) + \Pi_3(f,g), \tag{2.2.9}$$

where

$$\begin{aligned} \Pi_1(f,g) &= \sum_{-\infty}^{+\infty} f_j \sum_{k\leqslant j-4} g_k = \sum_{-\infty}^{+\infty} f_j g^{j-4}\,; \\ \Pi_2(f,g) &= \sum_{-\infty}^{+\infty} f_j \sum_{k\geqslant j+4} g_k = \sum_{-\infty}^{+\infty} g_j f^{j-4}\,; \end{aligned}$$

$$\Pi_3(f,g) \quad = \quad \sum_{-\infty}^{+\infty} f_j \sum_{|k-j|<4} g_k\,.$$

We observe that for every j we have

$$\mathrm{supp}\mathcal{F}[f^{j-4}g_j] \subset \{2^{j-2} \leqslant |\xi| \leqslant 2^{j+2}\};$$

$$\mathrm{supp}\mathcal{F}[\textstyle\sum_{k=j-3}^{j+3} f_j g_k] \subset \{|\xi| \leqslant 2^{j+5}\}\,.$$

The three pieces of the decomposition (2.2.9) are examples of paraproducts. Informally the first paraproduct Π_1 is an operator which allows high frequences of f ($\sim 2^j$) multiplied by low frequences of g ($\ll 2^j$) to produce high frequences in the output. The second paraproduct Π_2 multiplies low fequences of f with high frequences of g to produce high fequences in the output. The third paraproduct Π_3 multiply high frequences of f with high frequences of g to produce comparable or lower frequences in the output. For a presentation of these paraproducts we refer to the reader for instance to the book [16].

The compensations of the 3 different terms in $T(Q,v)$ will be clear just from the Littlewood-Paley decomposition of the different products. With this regards to get for instance the estimate (2.2.7) we shall need the following groupings

– i) For $\Pi_1(T(Q,v))$ we proceed to the following decomposition

$$\Pi_1(T(Q,v)) = \underbrace{\Pi_1((-\Delta)^{1/4}(Qv))}+\underbrace{\Pi_1 Q(-\Delta)^{1/4}v + (-\Delta)^{1/4}Qv}\,.$$

– ii) For $\Pi_2(R(Q,u))$ we decompose as follows

$$\Pi_2(T(Q,v)) = \underbrace{\Pi_2((-\Delta)^{1/4}(Qv) - Q(-\Delta)^{1/4}v)}+\underbrace{\Pi_2((-\Delta)^{1/4}Qv)}\,.$$

– ii) Finally, for $\Pi_3(R(Q,u))$ we decompose as follows

$$\Pi_3(T(Q,v)) = \underbrace{\Pi_3((-\Delta)^{1/4}(Qv))}-\underbrace{\Pi_3(Q(-\Delta)^{1/4}v)}+\underbrace{\Pi_3((-\Delta)^{1/4}Qv)}\,.$$

The following *2-terms commutators* have also been used in [9, 10]:

$$F(Q,v) := \mathcal{R}[Q]\mathcal{R}[v] - Qv. \tag{2.2.10}$$

$$\Lambda(Q,v) := Qv + \mathcal{R}[Q\mathcal{R}[v]]. \tag{2.2.11}$$

Theorem 2.16. [Theorem 3.6 in [10]] *For $f, v \in L^2$ it holds*

$$\|F(f,v)\|_{H^{-1/2}(\mathbb{R})} \leqslant C\|f\|_{L^2(\mathbb{R})}\|v\|_{L^{2,\infty}(\mathbb{R})}, \tag{2.2.12}$$

and

$$\|F(f,v)\|_{\mathcal{H}^1(\mathbb{R})} \leqslant C\|f\|_{L^2(\mathbb{R})}\|v\|_{L^2(\mathbb{R})}\,. \square \tag{2.2.13}$$

Theorem 2.17. [Theorem 3.7 in [10]] *For $Q \in \dot{H}^{1/2}(\mathbb{R})$, $v \in L^2(\mathbb{R})$ it holds*

$$\|(-\Delta)^{1/4}(\Lambda(Q,v))\|_{\mathcal{H}^1(\mathbb{R})} \leqslant C\|Q\|_{H^{1/2}(\mathbb{R})}\|v\|_{L^2(\mathbb{R})}\,. \tag{2.2.14}$$

Actually the estimate (2.2.13) is a consequence of the Coifman-Rochberg-Weiss estimate [3].

From Theorem 2.17 we deduce that under the same assumptions it holds $\Lambda(Q,v) \in L^{2,1}(\mathbb{R})$ with

$$\|\Lambda(Q,v)\|_{L^{2,1}(\mathbb{R})} \leqslant C\|Q\|_{H^{1/2}(\mathbb{R})}\|v\|_{L^2(\mathbb{R})}.$$

We finally remark that we can simply write the operator S as follows:

$$S(Q,v) = \overline{\mathcal{R}}T(Q,\mathcal{R}v) - \overline{\mathcal{R}}(-\Delta)^{1/4}[\Lambda(Q,\mathcal{R}v)]. \tag{2.2.15}$$

Therefore the estimate (2.2.8) for S can be deduced from the estimate (2.2.8) for the operator T and Theorem 2.17.

In [10] we have proved a sort of stability of of the operators T, S with respect to the multiplication by a function $P \in H^{1/2}(\mathbb{R}) \cap L^\infty(\mathbb{R})$. Roughly speaking if we multiply $T(Q,v)$ or $S(Q,v)$ by a function $P \in H^{1/2}(\mathbb{R}) \cap L^\infty(\mathbb{R})$ we get a decomposition into the sum of a function in the Hardy Space and a term which is the product of function in $L^{2,1}$ by one in L^2.

Theorem 2.18. [Multiplication of T by $P \in H^{1/2}(\mathbb{R}) \cap L^\infty(\mathbb{R})$] *Let $P, Q \in H^{1/2}(\mathbb{R}) \cap L^\infty(\mathbb{R})$ and $v \in L^2(\mathbb{R})$. Then*

$$PT(Q,v) = J_T(P,Q,v) + \mathcal{A}_T(P,Q)v, \tag{2.2.16}$$

where

$$\mathcal{A}_T(P,Q) = P(-\Delta)^{1/4}[Q] + (-\Delta)^{1/4}[P]Q - (-\Delta)^{1/4}[PQ] \in L^{2,1}$$

with

$$\|\mathcal{A}_T(P,Q)\|_{L^{2,1}} \leqslant C\|(-\Delta)^{1/4}[P]\|_{L^2}\|(-\Delta)^{1/4}[Q]\|_{L^2}, \tag{2.2.17}$$

and

$$J_T(P,Q,v) := T(PQ,v) - T(P,Qv) \in \mathcal{H}^1(\mathbb{R})$$

with

$$\|J_T(P,Q,v)\|_{\mathcal{H}^1(\mathbb{R})} \leqslant C(\|P\|_{L^\infty} + \|Q\|_{L^\infty})\left(\|(-\Delta)^{1/4}[P]\|_{L^2} + \|(-\Delta)^{1/4}[Q]\|_{L^2}\right)\|v\|_{L^2}. \tag{2.2.18}$$

Proof of Theorem 2.18. We have

$$\begin{aligned} PT(Q,v) &= P(-\Delta)^{1/4}[Qv] - PQ(-\Delta)^{1/4}[v] + P(-\Delta)^{1/4}[Q]v \\ &= \{P(-\Delta)^{1/4}[Q] - (-\Delta)^{1/4}[PQ] + (-\Delta)^{1/4}[P]Q\}v \\ &+ (-\Delta)^{1/4}[PQv] - PQ(-\Delta)^{1/4}v + (-\Delta)^{1/4}[PQ]v \end{aligned}$$

$$
\begin{aligned}
&- \quad \left((-\Delta)^{1/4}[PQv] + P(-\Delta)^{1/4}(Qv) - (-\Delta)^{1/4}[P]Qv\right)\\
&= \quad [P(-\Delta)^{1/4}[Q] + (-\Delta)^{1/4}[P]Q - (-\Delta)^{1/4}[PQ]]v\\
&+ \quad T(PQ, v) - T(P, Qv).
\end{aligned}
$$

Finally the estimates (2.2.17), (11.5.4) follow from Theorems 3.2 and 3.3 in [10].

An analogous property holds for the operator $\mathcal{R}S$. We just state the Theorem and we refer for proof to Theorem 3.10 in [10].

Theorem 2.19. **[Multiplication of $\mathcal{R}S$ by a rotation $P \in H^{1/2}(\mathbb{R}) \cap L^\infty(\mathbb{R})$]** *Let $P, Q \in H^{1/2}(\mathbb{R}) \cap L^\infty(\mathbb{R})$ and $v \in L^2(\mathbb{R})$. Then*

$$
P\mathcal{R}[S(Q, v)] \quad = \quad \mathcal{A}_S(P, Q)v + J_S(P, Q, v) \tag{2.2.19}
$$

where $\mathcal{A}_S(P, Q) \in L^{2,1}$, $J_S(P, Q, v) \in \mathcal{H}^1(\mathbb{R})$ with

$$
\|\mathcal{A}_S(P, Q)\|_{L^{2,1}} \leqslant C\|(-\Delta)^{1/4}[P]\|_{L^2}\|(-\Delta)^{1/4}[Q]\|_{L^2},
$$

and

$$
\|J_S(P, Q, v)\|_{\mathcal{H}^1(\mathbb{R})} \leqslant C(\|P\|_{L^\infty} + \|Q\|_{L^\infty})\left(\|(-\Delta)^{1/4}[P]\|_{L^2} + \|(-\Delta)^{1/4}[Q]\|_{L^2}\right)\|v\|_{L^2}.
$$

We just mention that the operators $\mathcal{A}_S(P, Q), J_S(P, Q, v)$ and $\mathcal{A}_T(P, Q), J_T(P, Q, v)$ can be expressed in turn as a combinations of the operators F, T, S.

Remark 2.1. We remark without going into detail that in 2-D the Jacobian $J(a, b) = \nabla(a)\nabla^\perp(b)$ satisfies a stability property enjoyed by the operators (2.2.2), (2.2.3), (2.2.10) with respect to the multiplication by $P \in W^{1,2}(\mathbb{R}^2) \cap L^\infty(\mathbb{R}^2)$ as well. More precisely we may define the following two zero-order pseudo-differential operators: $\mathrm{Grad}(X) := \nabla\mathrm{div}(-\Delta)^{-1}(X)$, $\mathrm{Rot}(Y) = \nabla^\perp\mathrm{curl}(-\Delta)^{-1}(Y)$. If $a, b \in W^{1,2}(\mathbb{R}^2)$ and $P \in W^{1,2}(\mathbb{R}^2) \cap L^\infty(\mathbb{R}^2)$ then

$$
\begin{aligned}
J(a, b) &= \nabla(a)\nabla^\perp(b) \\
&= \mathrm{Grad}(\nabla(a))\,\mathrm{Rot}(\nabla^\perp(b)) - \mathrm{Rot}(\nabla(a))\,\mathrm{Grad}(\nabla^\perp(b));
\end{aligned} \tag{2.2.20}
$$

and

$$
\begin{aligned}
PJ(a, b) &= P\nabla(a)\nabla^\perp(b) \\
&= \underbrace{[P\mathrm{Grad}(\nabla(a)) - \mathrm{Grad}(P\nabla(a))]}_{\in L^{2,1}(\mathbb{R}^2)}\mathrm{Rot}(\nabla^\perp(b)) \\
&+ \underbrace{\mathrm{Grad}(P\nabla(a))\,\mathrm{Rot}(\nabla^\perp(b)) - \mathrm{Rot}(P\nabla(a))\,\mathrm{Grad}(\nabla^\perp(b))}_{\in\mathcal{H}^1(\mathbb{R}^2)}.
\end{aligned} \tag{2.2.21}
$$

2.3 Regularity of Horizontal 1/2-harmonic Maps and Applications

In this section we describe the regularity results we have obtained respectively in [8, 9, 10].

2.3.1 Case of 1/2-harmonic maps with values into a sphere

In [8] we started the investigation of weak 1/2-harmonic maps $u \in H^{1/2}(\mathbb{R}, S^{m-1})$ with values into the sphere S^{m-1} which are critical points of the Lagrangian

$$\mathcal{L}^{1/2}(u) = \int_{\mathbb{R}} |(-\Delta)^{1/4} u(x)|^2 dx. \tag{2.3.1}$$

The main novelty in [8] is the rewriting of the Euler-Lagrange equation. To this purpose we recall the following equivalent relations.

Theorem 2.20. *All weak 1/2-harmonic maps $u \in H^{1/2}(\mathbb{R}, S^{m-1})$ satisfy in a weak sense*

i) the equation

$$\int_{\mathbb{R}} (-\Delta)^{1/2} u \cdot v \, dx = 0, \tag{2.3.2}$$

for every $v \in H^{1/2}(\mathbb{R}, \mathbb{R}^m) \cap L^\infty(\mathbb{R}, \mathbb{R}^m)$ and $v \in T_{u(x)} S^{m-1}$ almost everywhere, or in a equivalent way

ii) the equation

$$(-\Delta)^{1/2} u \wedge u = 0 \ \text{in}\ \mathcal{D}', \tag{2.3.3}$$

or

iii) the equation

$$(-\Delta)^{1/4}(u \wedge (-\Delta)^{1/4} u) = T(Q, u) \ \text{in}\ \mathcal{D}', \tag{2.3.4}$$

with $Q = u \wedge \cdot$.

Proof of Theorem 2.20

i) The proof of (2.3.2) is analogous of Lemma 1.4.10 in [17].

Let $v \in H^{1/2}(\mathbb{R}, \mathbb{R}^m) \cap L^\infty(\mathbb{R}, \mathbb{R}^m)$ and $v \in T_{u(x)} S^{m-1}$. We have

$$\Pi_{S^{m-1}}(u + tv) = u + t w_t\,,$$

where $\Pi_{S^{m-1}}$ is the orthogonal projection onto S^{m-1} and

$$w_t = \int_0^1 \frac{\partial \Pi_{S^{m-1}}}{\partial y_j}(u + tsv) v^j ds\,.$$

Hence

$$\mathcal{L}^{1/2}(\Pi_{S^{m-1}}(u+tv)) = \int_{\mathbb{R}} |(-\Delta)^{1/4}u|^2 dx + 2t\int_{\mathbb{R}} (-\Delta)^{1/2}u\cdot w_t dx + o(t)\,,$$

as $t \to 0$.

Thus to be a critical point of (11.2.1) is equivalent to

$$\lim_{t\to 0}\int_{\mathbb{R}} (-\Delta)^{1/2}u\cdot w_t dx = 0\,.$$

Since $\Pi_{S^{m-1}}$ is smooth it follows that $w_t \to w_0 = d\Pi_{S^{m-1}}(u)(v)$ in $H^{1/2}(\mathbb{R},\mathbb{R}^m)\cap L^\infty(\mathbb{R},\mathbb{R}^m)$ and therefore

$$\int_{\mathbb{R}} (-\Delta)^{1/4}u\, d\Pi_{S^{m-1}}(u)(v)dx = 0\,.$$

Since $v \in T_{u(x)}S^{m-1}$ a.e., we have $d\Pi_{S^{m-1}}(u)(v) = v$ a.e. and thus equation (2.3.2) follows immediately.

ii) We prove (2.3.3). We take $\varphi \in C_0^\infty(\mathbb{R},\bigwedge_{m-2}(\mathbb{R}^m))$. The following holds

$$\int_{\mathbb{R}} \varphi\wedge u\wedge(-\Delta)^{1/2}u\,dx = \left(\int_{\mathbb{R}} *(\varphi\wedge u)\cdot(-\Delta)^{1/2}u\,dx\right) e_1\wedge\ldots\wedge e_m\,. \tag{2.3.5}$$

Claim : $v = *(\varphi\wedge u)\in \dot H^{1/2}(\mathbb{R},\mathbb{R}^m)$ [2.4] and $v(x)\in T_{u(x)}S^{m-1}$ a.e.

Proof of the claim.

The fact that $v\in H^{1/2}(\mathbb{R},\mathbb{R}^m)\cap L^\infty(\mathbb{R},\mathbb{R}^m)$ follows form the fact that its components are the product of two functions which are in $\dot H^{1/2}(\mathbb{R},\mathbb{R}^m)\cap L^\infty(\mathbb{R},\mathbb{R}^m)$, which is an algebra .

We have

$$v\cdot u = *(u\wedge\varphi)\cdot u = *(u\wedge\varphi\wedge u) = 0\,. \tag{2.3.6}$$

It follows from (2.3.2) and (2.3.5) that

$$\int_{\mathbb{R}} \varphi\wedge u\wedge(-\Delta)^{1/2}u\,dx = 0\,.$$

This shows that $(-\Delta)^{1/2}u\wedge u = 0$ in $\mathcal{D}'$, and we can conclude .

iii) As far as equation (2.3.4) is concerned it is enough to observe that $(-\Delta)^{1/2}u\wedge u = 0$ and $(-\Delta)^{1/4}u\wedge(-\Delta)^{1/4}u = 0$. □

The Euler Lagrange equation (2.3.4) will often be completed by the following "structure equation" which is a consequence of the fact that $u\in S^{m-1}$ almost everywhere:

2.4 the symbol $*$ we denote the Hodge-star operator, $*\colon \bigwedge_p(\mathbb{R}^m)\to\bigwedge_{m-p}(\mathbb{R}^m)$, defined by $*\beta = (e_1\wedge\ldots\wedge e_n)\bullet\beta$, the symbol $\bullet$ is the first order contraction between multivectors, for every $p = 1,\ldots,m$, $\bigwedge_p(\mathbb{R}^m)$ is the vector space of p-vectors.

Proposition 2.21. *All maps in $\dot{H}^{1/2}(\mathbb{R}, S^{m-1})$ satisfy the following identity*

$$(-\Delta)^{1/4}(u \cdot (-\Delta)^{1/4}u) = S(u\cdot, u) - \overline{\mathcal{R}}((-\Delta)^{1/4}u \cdot \mathcal{R}(-\Delta)^{1/4}u), \tag{2.3.7}$$

where, in general for an arbitrary integer n, for every $Q \in \dot{H}^{1/2}(\mathbb{R}^n, \mathcal{M}_{\ell\times m}(\mathbb{R}))$, $\ell \geqslant 1$ and $u \in \dot{H}^{1/2}(\mathbb{R}^n, \mathbb{R}^m)$, S is the operator defined by (2.2.3).

Proof of Proposition 2.21. We observe that if $u \in H^{1/2}(\mathbb{R}, \mathbb{R}^{m-1})$ then the Leibniz's rule holds. Thus

$$\nabla |u|^2 = 2u \cdot \nabla u \text{ in } \mathcal{D}'. \tag{2.3.8}$$

Indeed the equality (2.3.8) trivially holds if $u \in C_0^\infty(\mathbb{R}, \mathbb{R}^{m-1})$. Let $u \in H^{1/2}(\mathbb{R}, \mathbb{R}^{m-1})$ and $u_j \in C_0^\infty(\mathbb{R}, \mathbb{R}^m)$ be such that $u_j \to u$ as $j \to +\infty$ in $H^{1/2}(\mathbb{R}, \mathbb{R}^m)$. Then $\nabla u_j \to \nabla u$ as $j \to +\infty$ in $H^{-1/2}(\mathbb{R}, \mathbb{R}^{m-1})$. Thus $u_j \cdot \nabla u_j \to u \cdot \nabla u$ in $\mathcal{D}'$ and (2.3.8) follows.
If $u \in H^{1/2}(\mathbb{R}, S^{m-1})$, then $\nabla |u|^2 = 0$ and thus $u \cdot \nabla u = 0$ in $\mathcal{D}'$ as well. Thus u satisfies equation (2.3.7) and this conclude the proof. □

We remark that in the sphere case the term $\overline{\mathcal{R}}((-\Delta)^{1/4}u \cdot \mathcal{R}(-\Delta)^{1/4}u)$ is in the Hardy-Space $\mathcal{H}^1(\mathbb{R})$ as well (see Corollary 3.1 in [8]). The estimates (2.2.4) and (2.2.5) imply in particular that if $u \in \dot{H}^{1/2}(\mathbb{R}, \mathcal{S}^{m-1})$ is a 1/2-harmonic map then

$$\|(-\Delta)^{1/4}u\|_{L^2(\mathbb{R})} \leqslant C\|(-\Delta)^{1/4}u\|^2_{L^2(\mathbb{R})}\,. \tag{2.3.9}$$

where the constant C is independent of u.

From the inequality (2.3.9) it follows that if $\varepsilon_0 := \|(-\Delta)^{1/4}u\|_{L^2(\mathbb{R})}$ is small enough so that

$$C\varepsilon_0 < 1 \tag{2.3.10}$$

then the solution is constant. This the so-called *bootstrap test* and it is the key observation to prove Morrey-type estimates and to deduce Hölder regularity of 1/2-harmonic maps.

Indeed by combining Theorem 2.20, Proposition 2.21 and suitable localization estimates obtained in Section 4 in [8] we get the local Hölder regularity of weak 1/2-harmonic maps.

Theorem 2.22. [Theorem 5.2, [8]] *Let $u \in \dot{H}^{1/2}(\mathbb{R}, S^{m-1})$ be a weak 1/2-harmonic map. Then $u \in C^{0,\alpha}_{loc}(\mathbb{R}, S^{m-1})$, for all $\alpha \in (0, 1)$.*

Sketch of Proof of 2.22. The strategy of proof is to show some decrease energy estimates. From Proposition 4.1 and 4.2 in [8] by using the fact that $u \wedge (-\Delta)^{1/4}u$ and $u \cdot (-\Delta)^{1/4}u$ satisfy respectively (2.3.4) and (2.3.7) one deduces that there exist $C > 0$ depending on $\|(-\Delta)^{1/4}u\|_{L^2(\mathbb{R})}$, $\overline{k} \in \mathbb{Z}$ depending on ε_0 in (2.3.10), such that that for every $x_0 \in \mathbb{R}$, for all $k < \overline{k}$ the following estimate holds

$$||(-\Delta)^{1/4}u||^2_{L^2(B_{2^k})} \leqslant C\sum_{h=k}^{\infty}(2^{\frac{k-h}{2}})||(-\Delta)^{1/4}u||^2_{L^2(A_h)} \tag{2.3.11}$$

where $B_{2^k} = B(x_0, 2^k)$, $A_h = B_{2^{h+1}} \backslash B_{2^{h-1}}$. On the other hand one has

$$2^{-1} \sum_{h=-\infty}^{k-1} \|(-\Delta)^{1/4} u\|^2_{L^2(A_h)} \leqslant \|(-\Delta)^{1/4} u\|^2_{L^2(B_{2^k})} \leqslant \sum_{h=-\infty}^{k-1} \|(-\Delta)^{1/4} u\|^2_{L^2(A_h)} \tag{2.3.12}$$

By combining (2.3.11) and (2.3.12) we get

$$\sum_{h=-\infty}^{k-1} \|(-\Delta)^{1/4} u\|^2_{L^2(A_h)} \leqslant C \sum_{h=k}^{\infty} (2^{\frac{k-h}{2}}) \|(-\Delta)^{1/4} u\|^2_{L^2(A_h)}.$$

This implies by an iteration argument (see Proposition A.1 in [8], or Lemma A.1 in [24])

$$\sup_{\substack{x \in B(x_0, \rho) \\ 0<r<\rho/8}} r^{-\beta} \int_{B(x,r)} |(-\Delta)^{1/4} u|^2 dx \leqslant C, \tag{2.3.13}$$

for ρ small enough, for some $0 < \beta < 1$ independent on x_0 and $C > 0$ depending only on the dimension and on $\|(-\Delta)^{1/4} u\|^2_{L^2(\mathbb{R})}$.
Condition (2.3.13) yields that $u \in C^{0,\beta/2}_{loc}(\mathbb{R})$, (see for instance [1] or [11] for the details). By bootstrapping into the equations (2.3.4) and (2.3.7) we can deduce that $u \in C^{0,\alpha}_{loc}(\mathbb{R})$ for all $\alpha \in (0, 1)$. □

We mention that Schikorra in [24] and the author and Schikorra in [12] extended the local the Hölder continuity of respectively $k/2$-harmonic maps ($k > 1$ odd) and k/p-harmonic maps ($p \in (1, \infty)$, $k/p \in (0, k)$) from subsets of $\mathbb{R}^k$ into a sphere.

k/p-harmonic maps with values into a sphere are defined as critical points of the following nonlocal Lagrangian

$$\int_{\mathbb{R}^k} |(-\Delta)^{\frac{k}{2p}} u|^p \, dx^k,$$

where $u(x) \in S^{m-1}$, a.e. and $\int_{\mathbb{R}^k} |(-\Delta)^{\frac{k}{2p}} u|^p \, dx^k < +\infty$.

2.3.2 Case of 1/2-harmonic maps into a closed manifold

We consider the case of 1/2-harmonic maps with values into a closed C^2 n-dimensional manifold $\mathcal{N} \subset \mathbb{R}^m$. Let $\Pi_{\mathcal{N}}$ be the orthogonal projection on $\mathcal{N}$. We denote by P_T and P_N respectively the tangent and the normal projection to the manifold $\mathcal{N}$.

They verify the following properties: $(P_T)^t = P_T$, $(P_N)^t = P_N$ (namely they are symmetric operators), $(P_T)^2 = P_T$, $(P_N)^2 = P_N$, $P_T + P_N = Id$, $P_N P_T = P_T P_N = 0$.

In this case the Euler-Lagrange equation associated to the energy (11.2.1) and the structural equation can be expressed as follows:

$$\begin{cases} P_T(u)(-\Delta)^{1/2} u = 0 & \text{in } \mathcal{D}'(\mathbb{R}) \\ P_N \nabla u = 0 & \text{in } \mathcal{D}'(\mathbb{R}). \end{cases} \tag{2.3.14}$$

The second step is to reformulate the two equations in (2.3.14) by using the commutators introduced in the previous section. The Euler equation (2.3.4) and structural equation (2.3.7) become in this case respectively

$$(-\Delta)^{1/4}(P^T(-\Delta)^{1/4}u) = T(P^T, u) - \underbrace{((-\Delta)^{1/4}P^T)(-\Delta)^{1/4}u}_{(1)} . \tag{2.3.15}$$

and

$$(-\Delta)^{1/4}(\mathcal{R}(P^N(-\Delta)^{1/4}u)) = \mathcal{R}(S(P^N, u)) - \underbrace{((-\Delta)^{1/4}P^N)(\mathcal{R}(-\Delta)^{1/4}u)}_{(2)} . \tag{2.3.16}$$

Unlike the sphere case the term (1) in (2.3.15) is not zero and term (2) in (2.3.16) is not in the Hardy Space.

The main idea in Proposition 1.1 in [9] is the re-writing of the terms (1) and (2) and to show that $v = (P_T(-\Delta)^{1/4}u, \mathcal{R}P_N(-\Delta)^{1/4}u)^t$ satisfies a nonlocal Schrödinger type system with a antisymmetric potential. Precisely, we obtained the following result.

Proposition 2.23. [Proposition 1.1, [9]] *Let $u \in \dot{H}^{1/2}(\mathbb{R}, \mathcal{N})$ be a weak $1/2$-harmonic map. Then the following equation holds*

$$\begin{aligned}(-\Delta)^{1/4}v = (-\Delta)^{1/4}\begin{pmatrix} P_T(-\Delta)^{1/4}u \\ \mathcal{R}P_N(-\Delta)^{1/4}u \end{pmatrix} &= \tilde{\Omega} + \Omega_1 \begin{pmatrix} P_T(-\Delta)^{1/4}u \\ \mathcal{R}P_N(-\Delta)^{1/4}u \end{pmatrix} \\ &+ \Omega \begin{pmatrix} P_T(-\Delta)^{1/4}u \\ \mathcal{R}P_N(-\Delta)^{1/4}u \end{pmatrix},\end{aligned} \tag{2.3.17}$$

where $\Omega = \Omega \in L^2(\mathbb{R}, so(2m))$, $\Omega_1 = \Omega_1 \in L^{2,1}(\mathbb{R}, \mathcal{M}_{m\times m})$ with

$$\|\Omega\|_{L^2}, \|\Omega_1\|_{L^{2,1}} \leqslant C(\|P_T\|_{\dot{H}^{1/2}} + \|P_T\|^2_{\dot{H}^{1/2}}),$$

$$\tilde{\Omega} = \begin{pmatrix} -2F(\omega_1, (P_N\Delta^{1/4}u)) + T(P_T, (-\Delta)^{1/4}u) \\ -2F(\mathcal{R}((-\Delta)^{1/4}P_N), \mathcal{R}((-\Delta)^{1/4}u)) - 2F(\omega_2, P_N((-\Delta)^{1/4}u) + \mathcal{R}(S(P_N, (-\Delta)^{1/4}u)) \end{pmatrix}$$

$\omega_1, \omega_2 \in L^2(\mathbb{R}, \mathcal{M}_{m\times m})$ and

$$\|\omega_1\|_{L^2}, \|\omega_2\|_{L^2} \leqslant C(\|P_T\|_{\dot{H}^{1/2}} + \|P_T\|^2_{\dot{H}^{1/2}}).^{2.5}$$

We would like to make some comments on Proposition 2.23.

2.5 The matrices Ω, Ω_1, ω_1 and ω_2 are constructed out of the projection P_T.

In [20] and [21] the author proved the sub-criticality of local *a-priori* critical Schödinger systems of the form

$$\forall i = 1 \cdots m \qquad - \Delta u^i = \sum_{j=1}^{m} \Omega^i_j \cdot \nabla u^j\,, \tag{2.3.18}$$

where $u = (u^1, \cdots, u^m) \in W^{1,2}(D, \mathbb{R}^m)$ and $\Omega \in L^2(D, \mathbb{R}^2 \otimes so(m))$, or of the form

$$\forall i = 1 \cdots m \qquad - \Delta v^i = \sum_{j=1}^{m} \Omega^i_j \, v^j\,, \tag{2.3.19}$$

where $v \in L^{n/(n-2)}(B^n, \mathbb{R}^m)$ and $\Omega \in L^{n/2}(B^n, so(m))$. In each of these two situations the antisymmetry of Ω was responsible for the regularity of the solutions or for the stability of the system under weak convergence.

One of the main result in the paper [9] was to establish the sub-criticality of non-local Schrödinger systems of the form

$$(-\Delta)^{1/4} v = \Omega v + \Omega_1 v + \mathcal{Z}(Q, v) + g(x) \tag{2.3.20}$$

where $v \in L^2(\mathbb{R})$, $Q \in \dot{H}^{1/2}(\mathbb{R})$, $\mathcal{Z}\colon \dot{H}^{1/2}(\mathbb{R}) \times L^2(\mathbb{R}) \to \mathcal{H}^1(\mathbb{R})$ is a linear combination of the operators (2.2.10), (2.2.2) and (2.2.3) introduced in the previous section, $\Omega \in L^2(\mathbb{R}, so(m))$, $\Omega_1 \in L^{2,1}(\mathbb{R})$. Precisely we prove the following theorem which extends to a non-local setting the phenomena observed in [20] and [21] for the above local systems.

Theorem 2.24. [Theorem 1.1, [9]] *Let $v \in L^2(\mathbb{R})$ be a weak solution of (2.3.20). Then $v \in L^p_{loc}(\mathbb{R})$ for every $1 \leqslant p < +\infty$.*

From Theorem 2.24 it follows that $(-\Delta)^{1/4} u \in L^p_{loc}(\mathbb{R})$, for all $p \geqslant 1$ as well, (u as in Proposition 2.23). This implies that $u \in C^{0,\alpha}_{loc}$ for all $0 < \alpha < 1$, since $W^{1/2,p}_{loc}(\mathbb{R}) \hookrightarrow C^{0,\alpha}_{loc}(\mathbb{R})$ if $p > 2$ (see for instance [1]).

The main technique to prove Theorem 2.24 is to perform a *change of gauge* by rewriting the system after having multiplied v by a well chosen rotation valued map $P \in H^{1/2}(\mathbb{R}, SO(m))$. [2.6] In [20] the choice of P for systems of the form (2.3.18) was given by the geometrically relevant *Coulomb Gauge* satisfying

$$div\left[P^{-1}\nabla P + P^{-1}\Omega P\right] = 0\,. \tag{2.3.21}$$

In this context there is not hope to solve an equation of the form (2.3.21) with the operator ∇ replaced by $(-\Delta)^{1/4}$, since for $P \in SO(m)$ the matrix $P^{-1}(-\Delta)^{1/4}P$ is not in general antisymmetric. The novelty in [9] was to choose the gauge P satisfying the

2.6 $SO(m)$ is the space of $m \times m$ matrices R satisying $R^tR = RR^t = Id$ and $det(R) = +1$

following (maybe less geometrically relevant) equation which involves the antisymmetric part of $P^{-1}(-\Delta)^{1/4}P$[2.7]:

$$Asymm\left(P^{-1}(-\Delta)^{1/4}P\right) := 2^{-1}\left[P^{-1}(-\Delta)^{1/4}P - (-\Delta)^{1/4}P^{-1}P\right] = \Omega\,. \qquad (2.3.22)$$

The local existence of such P is given by the following theorem.

Theorem 2.25. *There exists $\varepsilon > 0$ and $C > 0$ such that for every $\Omega \in L^2(\mathbb{R}; so(m))$ satisfying $\int_{\mathbb{R}} |\Omega|^2 dx \leqslant \varepsilon$, there exists $P \in \dot{H}^{1/2}(\mathbb{R}, SO(m))$ such that*

$$\begin{cases} (i) \quad P^{-1}(-\Delta)^{1/4}P - (-\Delta)^{1/4}P^{-1}P = 2\,\Omega\,; \\ (ii) \quad \int_{\mathbb{R}} |(-\Delta)^{1/4}P|^2 dx \leqslant C \int_{\mathbb{R}} |\Omega|^2 dx\,. \end{cases} \qquad (2.3.23)$$

□

The proof of this theorem is established by following an approach introduced by K.Uhlenbeck in [29] to construct *Coulomb Gauges* for L^2 curvatures in 4 dimension. The construction does not provide the continuity of the map which to $\Omega \in L^2$ assigns $P \in \dot{H}^{1/2}$. This illustrates the difficulty of the proof of Theorem 10.4.5 which is not a direct consequence of an application of the local inversion theorem but requires more elaborated arguments.

Thus if the L^2 norm of Ω is small, Theorem 10.4.5 gives a P for which $w := Pv$ satisfies

$$\begin{aligned} (-\Delta)^{1/4}w &= -\left[P\Omega P^{-1} - (-\Delta)^{1/4}P\,P^{-1}\right] w + T(P, P^{-1}w) + P\Omega_1 P^{-1}w \\ &+ PZ(Q, P^{-1}w) = -Symm\left(((-\Delta)^{1/4}P)\,P^{-1}\right) w + T(P, P^{-1}w) \\ &+ P\Omega_1 P^{-1}w + PZ(Q, P^{-1}w)\,. \end{aligned} \qquad (2.3.24)$$

The matrix $Symm\left(((-\Delta)^{1/4}P)\,P^{-1}\right)$ belongs to $L^{2,1}(\mathbb{R})$ and this fact comes from the combination of the following lemma according to which

$$(-\Delta)^{1/4}(Symm\left(((-\Delta)^{1/4}P)\,P^{-1}\right)) \in \mathcal{H}^1(\mathbb{R})$$

and the sharp Sobolev embedding [2.8] which says that $f \in \mathcal{H}^1(\mathbb{R})$ implies that $(-\Delta)^{-1/4}f \in L^{2,1}$. Precisely we have

2.7 Given a $m \times m$ matrix M, we denote by $Asymm(M)$ and by $Symm(M)$ respectively the antisymmetric and the symmetric part of M, namely $Asymm(M) := \frac{M-M^t}{2}$ and $Symm(M) := \frac{M+M^t}{2}$, M^t is the transpose of M.

2.8 The fact that $v \in \mathcal{H}^1$ implies $(-\Delta)^{-1/4}v \in L^{2,1}$ is deduced by duality from the fact that $(-\Delta)^{1/4}v \in L^{2,\infty}$ implies that $v \in BMO(\mathbb{R})$. This last embedding has been proved by Adams in [1]

Lemma 2.26. *Let $P \in H^{1/2}(\mathbb{R}, SO(m))$ then $(-\Delta)^{1/4}(Symm\left((-\Delta)^{1/4}P\,P^{-1}\right))$ is in the Hardy space $\mathcal{H}^1(\mathbb{R})$ and the following estimates hold*

$$\|(-\Delta)^{1/4}[(-\Delta)^{1/4}P\,P^{-1} + P\,(-\Delta)^{1/4}P^{-1}]\|_{\mathcal{H}^1} \leqslant C\|P\|^2_{\dot{H}^{1/2}},$$

where $C > 0$ is a constant independent of P. This implies in particular that

$$\|Symm\left(((-\Delta)^{1/4}P)\,P^{-1}\right)\|_{L^{2,1}} \leqslant C\|P\|^2_{\dot{H}^{1/2}}. \tag{2.3.25}$$

The proof of Lemma 2.26 is a consequence of the Theorem 1.5 in [9].

By combining the different properties of the commutators (2.2.2), (2.2.3), (2.2.10) mentioned in section 2.2, in [10] we proved that the system (2.3.20) is “equivalent" to a conservation law.

Theorem 2.27. *Let $v \in L^2(\mathbb{R}, \mathbb{R}^m)$ be a solution of (2.3.20), where $\Omega \in L^2(\mathbb{R}, so(m))$, $\Omega_1 \in L^{2,1}(\mathbb{R})$, $\mathcal{Z}$ is a linear combination of the operators (2.2.10), (2.2.2) and (2.2.3), $\mathcal{Z}(Q, v) \in \mathcal{H}^1$ for every $Q \in \dot{H}^{1/2}$, $v \in L^2$ with*

$$\|\mathcal{Z}(Q, v))\|_{\mathcal{H}^1} \leqslant C\|Q\|_{\dot{H}^{1/2}}\|v\|_{L^2}.$$

There exists $\varepsilon_0 > 0$ such that if

$$(\|\Omega\|_{L^2} + \|\Omega_1\|_{L^{2,1}} + \|Q\|_{\dot{H}^{1/2}}) < \varepsilon_0,$$

then there exist $A \in \dot{H}^{1/2}(\mathbb{R}, GL_m(\mathbb{R})))$ and an operator $B \in \dot{H}^{1/2}(\mathbb{R})$ (both constructed out of (Ω, Ω_1, Q)) such that

$$\|A\|_{\dot{H}^{1/2}} + \|B\|_{\dot{H}^{1/2}} \leqslant C(\|\Omega\|_{L^2} + \|\Omega_1\|_{L^{2,1}} + \|Q\|_{\dot{H}^{1/2}}) \tag{2.3.26}$$

$$dist(\{A, A^{-1}\}, SO(m)) \leqslant C(\|\Omega\|_{L^2} + \|\Omega_1\|_{L^{2,1}} + \|Q\|_{\dot{H}^{1/2}}) \tag{2.3.27}$$

and

$$(-\Delta)^{1/4}[Av] = \mathcal{J}(B, v) + Ag, \tag{2.3.28}$$

where $\mathcal{J}$ is a linear operator in B, v, $\mathcal{J}(B, v) \in \mathcal{H}^1(\mathbb{R})$ and

$$\|\mathcal{J}(B, v)\|_{\mathcal{H}^1(\mathbb{R})} \leqslant C\|B\|_{\dot{H}^{1/2}}\|v\|_{L^2}. \tag{2.3.29}$$

We mention that the case of $k/2$-harmonic maps ($k \geqslant 3$ odd) with values into a closed manifold has been considered in [4].

2.3.3 Case of horizontal 1/2-harmonic maps

We release the assumption that the field of orthogonal projection P_T is *integrable* and associated to a sub-manifold $\mathcal{N}$ and to consider the equation (2.3.14) for a general field of orthogonal projections P_T defined on the whole of $\mathbb{R}^m$ and for *horizontal maps* u satisfying $P_T(u)\nabla u = \nabla u$.

Precisely we consider $P_T \in C^1(\mathbb{R}^m, \mathcal{M}_m(\mathbb{R}))$ and $P_N \in C^1(\mathbb{R}^m, \mathcal{M}_m(\mathbb{R}))$ such that

$$\begin{cases} P_T \circ P_T = P_T \quad P_N \circ P_N = P_N \\ P_T + P_N = I_m \\ \forall\, z \in \mathbb{R}^m \quad \forall\, U, V \in T_z(\mathbb{R}^m) \quad < P_T(z)U, P_N(z)V >= 0 \\ \|\partial_z P_T\|_{L^\infty(\mathbb{R}^m)} < +\infty \end{cases} \tag{2.3.30}$$

For such a distribution of projections P_T we denote by

$$n := \operatorname{rank}(P_T).$$

Such a distribution identifies naturally with the distribution of $n-$planes given by the images of P_T (or the Kernel of P_T) and conversely, any C^1 distribution of $n-$dimensional planes defines uniquely P_T satisfying (2.3.30).

We will present here the proof of the C^α_{loc} of horizontal $1/2$-harmonic maps which directly uses the conservation law (2.3.28) and which is a refinement of the arguments used in Theorem 2.24 (Theorem 1.1 in [9]). We premise the following result.

Theorem 2.28. *Let $m \in \boldsymbol{N}^*$, then there exists $\delta > 0$ such that for any $P_T, P_N \in \dot{H}^{1/2}(\mathbb{R}, \mathcal{M}_m)$ satisfying*

$$\begin{cases} P_T \circ P_T = P_T, \quad P_N = I_m - P_T \\ \forall\, X, Y \in \mathbb{R}^m, \text{ for a.e } x \in \mathbb{R} \quad < P_T(x)X, P_N(x)Y >= 0 \end{cases} \tag{2.3.31}$$

and

$$\int_{\mathbb{R}} |(-\Delta)^{1/4} P_T|^2 \, d\vartheta \leqslant \delta \tag{2.3.32}$$

then for any $f \in H^{-1/2}(\mathbb{R})$

$$(P_T + P_N \mathcal{R})\, f = 0 \quad \Longrightarrow \quad f = 0. \tag{2.3.33}$$

Proof of Theorem 2.28.
We first set $f := (-\Delta)^{1/2} u$. From (2.3.33) it follows that

$$\begin{cases} P_T(-\Delta)^{1/2} u = 0 \\ P_N \mathcal{R}(-\Delta)^{1/2} u = 0 \end{cases} \tag{2.3.34}$$

Then set $v = (P_T(-\Delta)^{1/4} u, \mathcal{R}(P_N(-\Delta)^{1/4} u))^t$. Therefore v satisfies a system of the form (2.3.20) with $\Omega \in L^2(\mathbb{R}, so(\mathbb{R}^m))$ $\Omega_1 \in L^{2,1}$, (Ω and Ω_1 depend on P_T), $\mathcal{Z}(P_T, v)$ is a linear operator in P_T, v, $\mathcal{Z}(P_T, v) \in \mathcal{H}^1$ with

$$\|\Omega\|_{L^2} \quad = \quad \|\Omega\|_{L^2} \leqslant C\|P_T\|_{\dot{H}^{1/2}}$$

$$\begin{aligned} \|\Omega_1\|_{L^{2,1}} &= \|\Omega_1\|_{L^{2,1}} \leqslant C\|P_T\|_{\dot H^{1/2}} \\ \|\mathcal{Z}(P_T, v))\|_{\mathcal{H}^1} &\leqslant C\|P_T\|_{\dot H^{1/2}}\|v\|_{L^2} \end{aligned}$$

From Theorem 2.27 it follows that if δ is small enough then there exist $A \in L^\infty \cap \dot H^{1/2}(\mathbb{R}, GL_m(\mathbb{R}))$ and $B \in \dot H^{1/2}(\mathbb{R}, \mathcal{M}_{m\times m}(\mathbb{R}))$ such that

$$(-\Delta)^{1/4}[Av] = \mathcal{J}(B, v) \tag{2.3.35}$$

and

$$\begin{aligned} \|A\|_{\dot H^{1/2}} + \|B\|_{\dot H^{1/2}} &\leqslant C\|P_T\|_{\dot H^{1/2}} \\ dist(\{A, A^{-1}\}, SO(m)) &\leqslant \leqslant C\|P_T\|_{\dot H^{1/2}} \\ \|\mathcal{J}(B, v)\|_{\mathcal{H}^1(\mathbb{R})} &\leqslant C\|B\|_{\dot H^{1/2}}\|v\|_{L^2}. \end{aligned} \tag{2.3.36}$$

From (2.3.35) and (2.3.36) it follows that

$$\begin{aligned} \|v\|_{L^2} &= \|A^{-1}Av\|_{L^2} \leqslant C\|A^{-1}\|_{L^\infty}\|Av\|_{L^2} \\ &\leqslant C\|(-\Delta)^{-1/4}\mathcal{J}(B, v)\|_{L^{2,1}} \leqslant C\|B\|_{\dot H^{1/2}}\|v\|_{L^2} \\ &\leqslant C\|P_T\|_{\dot H^{1/2}}\|v\|_{L^2} \leqslant C\delta\|v\|_{L^2}. \end{aligned} \tag{2.3.37}$$

Again if δ is small enough then (2.3.37) yields $v \equiv 0$ a.e. and therefore $f = 0$ a.e. as well. □

Proof of Theorem 2.9. The proof of Theorem 2.9 follows by combining Theorem 2.28 and localization arguments used in [9]. □

2.3.4 Applications

In this section we mention two geometric applications related to $1/2$-harmonic maps. We start by proving Theorem 2.7 .

Proof of Theorem 2.7 . 1) (see [5, 14, 18]). If $\mathcal{N} = S^1$, then its harmonic extension $\tilde u$, which is conformal thanks to Theorem 2.6, maps the unit disk $B^2(0, 1)$ into itsself because of the maximum principle. On the other hand it turns out that every conformal transformation with finite energy from $B^2(0, 1)$ into $B^2(0, 1)$ and sending S^1 into S^1 has to be a finite Blaschke product, namely there exist $d > 0, \vartheta_0 \in \mathbb{R}, a_1, \dots, a_d \in B^2(0, 1)$ such that

$$\tilde u(z) = \prod_{i=1}^{d} e^{i\vartheta_0}\frac{z - a_i}{1 - z\bar a_i}.$$

Since $deg(u) = 1$ then $d = 1$ and $\tilde u$ coincides with a Möbius transformation of the disk.

2) We are going to use the following result by Nitsche [19]: if Σ is a regular minimal immersion in $B^3(0, 1) \subset \mathbb{R}^3$ that meets $B^3(0, 1)$ orthogonally then $\partial\Sigma$ is a great circle.

Let $\tilde{u}\colon B^2(0,1) \to B^3(0,1)$ be the harmonic extension of u. In [11] it has been shown that $u \in C^{1,\alpha}(S^1)$, therefore $\tilde{u} \in C^{1,\alpha}(\overline{B}^2)$. Moreover $\tilde{u}$ is conformal in $\overline{B}^2(0,1)$ (see Proposition 2.29 below)[2.9] and by Maximum Principle $\tilde{u}$ takes values in $B^3(0,1)$. We set $h = |\tilde{u}|^2$. We have $-\Delta h \leqslant 0$, and $h = 1$ on S^2. By Hopf Boundary Lemma we have $\frac{\partial h}{\partial r} \neq 0$ on S^1. Since $\tilde{u}$ is conformal up to the boundary, this implies in particular $\nabla \tilde{u} \neq 0$ on S^1 and therefore $\tilde{u}$ is a minimal immersion up to the boundary. Since it meets $B^3(0,1)$ orthogonally then by Nitsche's result [19] $\tilde{u}(S^1) = u(S^1)$ is an equatorial circle. Let $T\colon S^2 \to S^2$ be an isometry,[2.10] $\sigma := \{az + by + cx = 0, a, b, c \in \mathbb{R}\}$ be a plane in $\mathbb{R}^3$ such that $u(S^1) = \sigma \cap S^2$. Define $\tau = T|_{\sigma\cap S^2}\colon \sigma \cap S^2 \to S^1$. Let $v := \tau \circ u\colon S^1 \to S^1$ and we show that it is $1/2$-harmonic in S^1.

$$\begin{cases} \Delta(\widetilde{\tau \circ u}) = 0 & \text{in } B^2 \\ \widetilde{\tau \circ u} = \tau \circ u & \text{in } \partial B^2 \end{cases} \tag{2.3.38}$$

Since τ can be identified with a rotation in $\mathbb{R}^3$, we have

$$\frac{\partial \widetilde{\tau \circ u}}{\partial \nu} = \tau \frac{\partial \tilde{u}}{\partial \nu}.$$

It follows that

$$\begin{aligned} (-\Delta)^{1/2}(\tau \circ u) &= \frac{\partial \widetilde{\tau \circ u}}{\partial \nu} = \tau \frac{\partial \tilde{u}}{\partial \nu} \\ &= \tau(-\Delta)^{1/2} u \parallel \tau \circ u\,. \end{aligned}$$

We can conclude the proof. □

Proposition 2.29. [Proposition 1.1, [10]] *An element in $\mathfrak{H}^{1/2}$ satisfying*

$$P_T(u)\,(-\Delta)^{1/2} u = 0 \quad \text{in } \mathcal{D}'(S^1) \tag{2.3.39}$$

has a harmonic extension $\tilde{u}$ in $B^2(0,1)$ which is conformal in $\overline{B}^2(0,1)$ and hence it is the boundary of a minimal disk whose exterior normal derivative $\partial_r \tilde{u}$ is orthogonal to the plane distribution given by P_T.

Proof of Proposition 2.29. We prove the result by assuming that $P_T \in C^2(\mathbb{R}^m)$. In that case we have that $u \in C^{1,\alpha}(S^1)$, (see [11]). Denote $\tilde{u}$ the harmonic extension of u. It is well known that the *Hopf differential* of $\tilde{u}$

$$|\partial_{x_1} \tilde{u}|^2 - |\partial_{x_2} \tilde{u}|^2 - 2i\,\langle \partial_{x_1} \tilde{u}, \partial_{x_2} \tilde{u}\rangle = f(z)$$

is holomorphic. Considering on $S^1 = \partial B^2$

$$2\,\langle \partial_r \tilde{u}, \partial_\vartheta \tilde{u}\rangle = -\sin 2\vartheta \left(|\partial_{x_1} \tilde{u}|^2 - |\partial_{x_2} \tilde{u}|^2\right) - \cos 2\vartheta\,(-2\,\langle \partial_{x_1} \tilde{u}, \partial_{x_2} \tilde{u}\rangle) = -\operatorname{Im}\left(z^2 f(z)\right).$$

2.9 We refer to the book [22] for an overview of the the regularity of minimal disks up to the boundary (solution of the Plateau problem)

2.10 The isometry group of the sphere S^2 is isomorphic to the group $SO(3)$ of orthogonal matrices.

Since $0 = P_T(u)\,(-\Delta)^{1/2}u = P_T(u)\,\partial_r\tilde{u}$ and $0 = P_N(u)\,\partial_\vartheta u = P_N(u)\,\partial_\vartheta\tilde{u}$ on S^1 we have that

$$\operatorname{Im}\left(z^2 f(z)\right) = 0 \quad \text{on } S^1.$$

Hence the holomorphic function $z^2 f(z)$ is equal to a real constant. Since $f(z)$ cannot have a pole at the origin we have that $z^2 f(z)$ is identically equal to zero and thus $\tilde{u}$ is conformal. □

Bibliography

[1] D.R. Adams *A note on Riesz potentials.* Duke Math. J. 42 (1975), no. 4, 765–778.

[2] S. Blatt, P. Reiter Schikorra, A. Schikorra, *Harmonic analysis meets critical knots. Critical points of the Möbius energy are smooth,* Trans. Amer. Math. Soc. 368 (2016), no. 9, 6391–6438.

[3] Coifman, R. R.; Rochberg, R.; Weiss, G. *Factorization theorems for Hardy spaces in several variables,* Ann. of Math. (2) 103 (1976), no. 3, 611-635.

[4] F. Da Lio, *Fractional harmonic maps into manifolds in odd dimension $n > 1$,* Calc. Var. Partial Differential Equations 48 (2013), no. 3-4, 421–445.

[5] F. Da Lio,*Compactness and bubble analysis for 1/2-harmonic maps,* Ann. Inst. H. Poincaré Anal. Non Linéaire 32 (2015), no. 1, 201–224.

[6] F. Da Lio, P. Laurain, T. Rivière, *A Pohozaev-type formula and Quantization of Half-Harmonic Maps,* arXiv:1607.05504.

[7] F. Da Lio, L. Martinazzi, T. Rivière, *Blow-up analysis of a nonlocal Liouville-type equations,* Analysis and PDE **8**, no. 7 (2015), 1757-1805.

[8] F. Da Lio, T. Rivière, *Three-term commutator estimates and the regularity of* $1/2$-*harmonic maps into spheres,* Anal. PDE 4 (2011), no. 1, 149–190.

[9] F. Da Lio, T. Rivière, *Sub-criticality of non-local Schrödinger systems with antisymmetric potentials and applications to half-harmonic maps,* Adv. Math. 227 (2011), no. 3, 1300–1348.

[10] F. Da Lio, T. Rivière, *Horizontal α-Harmonic Maps,* arXiv:1604.05461.

[11] F. Da Lio, T. Rivière, *Free Boundary Surfaces: A nonlocal Approach,* in preparation.

[12] F.Da Lio, A. Schikorra, *n/p-harmonic maps: regularity for the sphere case,* Adv. Calc. Var. 7 (2014), no. 1, 1–26.

[13] A. Fraser, R. Schoen, *Uniqueness theorems for free boundary minimal disks in space forms,* arXiv:1409.1632.

[14] A. Fraser, R. Schoen, *Minimal surfaces and eigenvalue problems,* Geometric analysis, mathematical relativity, and nonlinear partial differential equations, 105–121, Contemp. Math., 599, Amer. Math. Soc., Providence, RI, 2013.

[15] L. Grafakos, *Classical Fourier Analysis.* Graduate Texts in Mathematics 249, Springer (2009).

[16] L. Grafakos, *Modern Fourier Analysis.* Graduate Texts in Mathematics 250, Springer (2009).

[17] F. Hélein, *Harmonic maps, conservation laws and moving frames.* Cambridge Tracts in Mathematics, 150. Cambridge University Press, Cambridge, 2002.

[18] V. Millot, Y. Sire, *On a fractional Ginzburg-Landau equation and 1/2-harmonic maps into spheres,* Arch. Ration. Mech. Anal. 215 (2015), no. 1, 125–210.

[19] J. Nitsche, *Stationary partitioning of convex bodies*, Archive for Rational Mechanics and Analysis, 89, (1985), 1–19.

[20] T. Rivière, *Conservation laws for conformal invariant variational problems*, Invent. Math., 168 (2007), 1–22.

[21] T. Rivière, *Sub-criticality of Schrödinger Systems with Antisymmetric Potentials*, J. Math. Pures Appl. 95 (2011) 260–276.

[22] T. Rivière, *Conformally invariant Variational Problems*, book in preparation (2016).

[23] A. Scheven, *Partial regularity for stationary harmonic maps at a free boundary*, Math. Z. 253 (2006), , 135–157.

[24] A. Schikorra, *Regularity of n/2 harmonic maps into spheres*, Journal of Differential Equations 252 (2012) 1862–1911.

[25] A. Schikorra, *ε-regularity for systems involving non-local, antisymmetric operators,* Calc. Var. Partial Differential Equations 54 (2015), no. 4, 3531–3570.

[26] S. Semmes, *A primer on Hardy spaces, and some remarks on a theorem of Evans and Miller*, Comm. Partial Differential Equations 19 (1994), no. 1-2, 277–319.

[27] R. Strichartz, *Sub-Riemannian geometry*, J. Differential Geom. 24 (1986), no. 2, 221-263.

[28] H.C. Wente, *An existence theorem for surfaces of constant mean curvature*, J. Math. Anal. Appl. 26 1969 318–344.

[29] K. Uhlenbeck, *Connections with L^p bounds on curvature,* Comm. Math. Phys, 83, 31-42, 1982.

Donatella Danielli and Sandro Salsa

Obstacle Problems Involving the Fractional Laplacian

3.1 Introduction

Obstacle problems involving a fractional power of the Laplace operator appear in many contexts, such as in pricing of American options governed by assets evolving according to jump processes [26], or in the study of local minimizers of some nonlocal energies [24].

In the first part of this expository paper we are concerned with the stationary case, which can be stated in several ways. Given a smooth function $\varphi : \mathbb{R}^n \to \mathbb{R}$, $n > 1$, with bounded support (or at least rapidly decaying at infinity), we look for a continuous function u satisfying the following system:

$$\begin{cases} u \geqslant \varphi & \text{in } \mathbb{R}^n \\ (-\Delta)^s u \geqslant 0 & \text{in } \mathbb{R}^n \\ (-\Delta)^s u = 0 & \text{when } u > \varphi \\ u(x) \to 0 & \text{as } |x| \to +\infty. \end{cases} \tag{3.1.1}$$

Here we consider only the case $s \in (0, 1)$. The set $\Lambda(u) = \{u = \varphi\}$ is called *the contact or coincidence set*. The free boundary is the set

$$F(u) = \partial\Lambda(u).$$

The main theoretical issues in a constrained minimization problem are *optimal regularity of the solution* and the *analysis of the free boundary*.

If $s = 1$ and $\mathbb{R}^n$ is replaced by a bounded domain Ω our problem corresponds to the usual obstacle problem for the Laplace operator. The existence of a unique solution satisfying some given boundary condition $u = g$ can be obtained by minimizing the Dirichlet integral in $H^1(\Omega)$ under the constraint $u \geqslant \varphi$. The solution is the least superharmonic function greater or equal to φ in Ω, with $u \geqslant g$ on $\partial\Omega$, and inherits up to a certain level the regularity of φ ([33]). In fact, even if φ is smooth, u is only $C^{1,1}_{loc}$, which is the optimal regularity. A classical reference for the obstacle problem, including the regularity and the complete analysis of the free boundary is [18]. See also the recent book [56].

Donatella Danielli, Department of Mathematics, Purdue University, E-mail: danielli@math.purdue.edu
Sandro Salsa, Department of Mathematics, Politecnico di Milano, E-mail: sandro.salsa@polimi.it

https://doi.org/10.1515/9783110571561-005

Analogously, the existence of a solution u for problem (3.1.1) can be obtained by variational methods as the unique minimizer of the functional

$$J(v) = \int_{\mathbb{R}^n}\int_{\mathbb{R}^n} \frac{|v(x) - v(y)|^2}{|x-y|^{n+2s}}\,dxdy$$

over a suitable set of functions $v \geqslant \varphi$. We can also obtain u via a Perron type method, as the least supersolution of $(-\Delta)^s$ such that $u \geqslant \varphi$. By analogy (see also later the Signorini problem), when φ is smooth, we expect the optimal regularity for u to be $C^{1,s}$. This is indeed true, as it is shown by Silvestre in [63] when the contact set $\{u = \varphi\}$ is convex and by Caffarelli, Salsa, Silvestre in [22] in the general case.

The case $s = 1/2$ is strongly related to the so called *thin (or lower dimensional) obstacle problem* for the Laplace operator. To keep a connection with the obstacle problem for $(-\Delta)^s$, it is better to work in $\mathbb{R}^{n+1}$, writing $X = (x, y) \in \mathbb{R}^n \times \mathbb{R}$. The thin obstacle problem concerns the case in which the obstacle is not anymore $n+1$ dimensional, but supported instead on a smooth n–dimensional manifold $\mathcal{M}$ in $\mathbb{R}^{n+1}$. This problem and variations of it also arise in many applied contexts, such as flow through semi-permeable membranes, elasticity (known as the *Signorini problem*, see below), boundary control temperature or heat problems (see [29]).

More precisely, let Ω be a domain in $\mathbb{R}^{n+1}$ divided into two parts Ω^+ and Ω^- by $\mathcal{M}$. Let $\varphi : \mathcal{M} \to \mathbb{R}$ be the (thin) obstacle and g be a given function on $\partial\Omega$ satisfying $g > \varphi$ on $\mathcal{M} \cap \partial\Omega$.

The problem consists in the minimization of the Dirichlet integral

$$J(v) = \int_\Omega |\nabla v|^2$$

over the closed convex set

$$\mathbb{K} = \left\{ v \in H^1(\Omega) : v = g \text{ on } \partial\Omega \text{ and } v \geqslant \varphi \text{ on } \mathcal{M} \cap \partial\Omega \right\}.$$

Since we can perturb the solution u upwards and freely away from $\mathcal{M}$, it is apparent that u is superharmonic in Ω and harmonic in $\Omega \backslash \mathcal{M}$. One expects the continuity of the first derivatives along the directions tangential to $\mathcal{M}$, and the one sided continuity of normal derivatives ([33]). In fact (see [16]), on $\mathcal{M}$, u satisfies the following complementary conditions

$$u \geqslant \varphi,\ u_{\nu^+} + u_{\nu^-} \leqslant 0,\ (u - \varphi)(u_{\nu^+} + u_{\nu^-}) = 0$$

where $\nu^\pm$ are the interior unit normals to $\mathcal{M}$ from the $\Omega^\pm$ side. The free boundary here is given by the boundary of the set $\Omega \backslash \Lambda(u)$ in the relative topology of $\mathcal{M}$, and in general, we expect it is a $(n-1)$–dimensional manifold.

As mentioned above, a related problem is the *Signorini problem*[3.1] (or *boundary thin obstacle problem*), in which the manifold $\mathcal{M}$ is part of $\partial\Omega$ and one has to minimize

3.1 After Fichera, see ([31]), 1963.

the Dirichlet integral over the closed convex set

$$\mathbb{K} = \left\{ v \in H^1(\Omega) : v = g \text{ on } \partial\Omega \backslash \mathcal{M} \text{ and } v \geqslant \varphi \text{ on } \mathcal{M} \right\}.$$

In this case, u is harmonic in Ω and on $\mathcal{M}$ it satisfies the complementary conditions

$$u \geqslant \varphi,\ u_{\nu^+} \leqslant 0,\ (u - \varphi)\, u_{\nu^+} = 0.$$

If $\mathcal{M}$ is a hyperplane (say $\{y = 0\}$) and Ω is symmetric with respect to $\mathcal{M}$, then the thin obstacle in Ω and the boundary obstacle problems in Ω^+ or Ω^- are equivalent.

Let us see how these problems are related to the obstacle problem for the $(-\Delta)^{1/2}$. This is explained through the following remarks.

(*a*) *Reduction to a global problem.* Let $\Omega = B_1$ be the unit ball in $\mathbb{R}^{n+1}$ and $B_1' = B_1 \cap \{y = 0\}$. Let $\varphi : \mathbb{R}^n \to \mathbb{R}$ be a smooth obstacle, $\varphi < 0$ on $\partial B_1'$ and positive somewhere inside B_1'. Consider the following Signorini problem in $B_1^+ = B_1 \cap \{y > 0\}$:

$$\begin{cases} -\Delta u = 0 & \text{in } B_1^+ \\ u = 0 & \text{on } \partial B_1 \cap \{y > 0\} \\ u(x,0) \geqslant \varphi(x) & \text{in } \overline{B}_1' \\ u_y(x,0) \leqslant 0 & \text{in } \overline{B}_1' \\ u_y(x,0) = 0 & \text{when } u(x,0) > \varphi(x). \end{cases} \tag{3.1.2}$$

We want to convert the above problem in B_1 into a global one, that is in $\mathbb{R}^n \times (0, +\infty)$. To do this, let η be a radially symmetric cut-off function in B_1' such that

$$\{\varphi > 0\} \Subset \{\eta = 1\} \text{ and } \operatorname{supp}(\eta) \subset B_1'.$$

Extending ηu by zero outside B_1, we have $\eta(x)\, u(x,0) \geqslant \varphi(x)$ and also $(\eta u)_y(x,0) \leqslant 0$ for every $x \in \mathbb{R}^n$. Moreover, $(\eta u)_y(x,0) = 0$ if $\eta(x)\, u(x,0) > \varphi(x)$.

Let now v be the unique solution of the following Neumann problem in the upper half space, vanishing at infinity:

$$\begin{cases} \Delta v = \Delta(\eta u) & \text{in } \mathbb{R}^n \times \{y > 0\} \\ v_y(x,0) = 0 & \text{in } \mathbb{R}^n. \end{cases}$$

Then $w = \eta u - v$ is a solution of a global Signorini problem with $\varphi - v$ as the obstacle. Thus, the regularity of u in the local setting can be inferred from the regularity for the global problem.

The opposite statement is obvious.

(*b*) *Realization of* $(-\Delta)^{1/2}$ *as a Dirichlet-Neumann map.* Consider a smooth function $u_0 : \mathbb{R}^n \longrightarrow \mathbb{R}$ with rapid decay at infinity. Let $u : \mathbb{R}^n \times (0, +\infty) \longrightarrow \mathbb{R}$ be the unique solution of the Dirichlet problem

$$\begin{cases} \Delta u = 0 & \text{in } \mathbb{R}^n \times (0, +\infty) \\ u(x,0) = u_0(x) & \text{in } \mathbb{R}^n \end{cases}$$

vanishing at infinity. We call u the *harmonic extension of* u_0 to the upper half space.

Consider the *Dirichlet-Neumann map* $T : u_0(x) \longmapsto -u_y(x, 0)$. We have:

$$
\begin{aligned}
(Tu_0, u_0) &= \int_{\mathbb{R}^n} -u_y(x,0)\, u(x,0)\, dx \\
&= \int_{\mathbb{R}^n \times (0,+\infty)} \left\{ \Delta u(X)\, u(X) + |\nabla u(X)|^2 \right\} dX \\
&= \int_{\mathbb{R}^n \times (0,+\infty)} |\nabla u(X)|^2\, dX \geqslant 0
\end{aligned}
$$

so that T is a positive operator. Moreover, since u_0 is smooth and u_y is harmonic, we can write:

$$T \circ Tu_0 = -\partial_y(-\partial_y)u(x,0) = u_{yy}(x,0) = -\Delta u_0.$$

We conclude that

$$T = (-\Delta)^{1/2}.$$

As a consequence:

1. If $u = u(X)$ is a solution of the Signorini problem in $\mathbb{R}^n \times (0, +\infty)$, that is $\Delta u = 0$ in $\mathbb{R}^{n+1}$, $u(x, 0) \geqslant \varphi$, $u_y(x, 0) \leqslant 0$, and $(u - \varphi)\, u_y(x, 0) = 0$ in $\mathbb{R}^n$, then $u_0 = u(\cdot, 0)$ solves the obstacle problem for $(-\Delta)^{1/2}$.

2. If u_0 is a solution of the obstacle problem for $(-\Delta)^{1/2}$, then its harmonic extension to $\mathbb{R}^n \times (0, +\infty)$ solves the corresponding Signorini problem.

Therefore, the two problems are equivalent and any regularity result for one of them can be carried to the other one. More precisely, consider the optimal regularity for the solution u_0 of the obstacle problem for $(-\Delta)^{1/2}$, which is $C^{1,1/2}$. If we can prove a $C^{1,1/2}$ regularity of the solution u of the Signorini problem up to $y = 0$, then the same is true for u_0.

On the other hand, the $C^{1,\alpha}$ regularity of u_0 extends to u, via boundary estimates for the Neumann problem. Similarly, the analysis of the free boundary in the Signorini problem carries to the obstacle problem for $(-\Delta)^{1/2}$ as well and vice versa.

Although the two problem are equivalent, there is a clear advantage in favor of the Signorini type formulation. This is due to the possibility of avoiding the direct use of the non local pseudodifferential operator $(-\Delta)^{1/2}$, by localizing the problem and using local PDE methods, such as monotonicity formulas and classification of blow-up profiles.

At this point it is a natural question to ask whether there exists a PDE realization of $(-\Delta)^s$ for every $s \in (0, 1)$, $s \neq \frac{1}{2}$.

The answer is positive as it is shown by Caffarelli and Silvestre in [23]. Indeed in a weak sense we have that

$$(-\Delta)^s u_0(x) = -\kappa_a \lim_{y \to 0+} y^a u_y(x, y)$$

for a suitable constant κ_a, where u is the solution of the problem

$$\begin{cases} L_a u = \operatorname{div}\left(y^a \nabla u\right) = 0 & \text{in } \mathbb{R}^n \times (0, +\infty) \\ u\,(x, 0) = u_0\,(x) & \text{in } \mathbb{R}^n \end{cases}$$

vanishing at infinity, where $a = 2s - 1$. Coherently, we call u the *L_a−harmonic extension of u_0*.

Thus problem (3.1.1) is equivalent to the following Signorini problem for the operator $L_a = \operatorname{div}(y^a \nabla)$,

$$u\,(x, 0) \geqslant \varphi \qquad \text{in } \mathbb{R}^n \tag{3.1.3}$$

$$L_a u = \operatorname{div}\left(y^a \nabla u\right) = 0 \qquad \text{in } \mathbb{R}^n \times (0, +\infty) \tag{3.1.4}$$

$$\lim_{y \to 0+} y^a u_y\,(x, y) = 0 \qquad \text{when } u\,(x, 0) > \varphi\,(x) \tag{3.1.5}$$

$$\lim_{y \to 0+} y^a u_y\,(x, y) \leqslant 0 \qquad \text{in } \mathbb{R}^n. \tag{3.1.6}$$

For $y > 0$, u is smooth so that (3.1.4) is understood in the classical sense. The equations at the boundary (3.1.5) and (3.1.6) should be understood in a weak sense. Since in [63] it is shown that $u\,(x, 0) \in C^{1,\alpha}$ for every $\alpha < s$, for the range of values $2s - 1 < \alpha < s$, $\lim_{y \to 0+} y^a u_y\,(x, y)$ can be understood in the classical sense too.

The solution u of the above Signorini problem can be extended to the whole space by symmetrization, setting $u\,(x, -y) = u\,(x, y)$. Then, by the results in [23], condition (3.1.5) holds if and only if the extended u is a solution of $L_a u = 0$ across $y = 0$, where $u\,(x, 0) > \varphi\,(x)$. On the other hand, condition (3.1.6) is equivalent to $L_a u \leqslant 0$ in the sense of distributions. Thus, for the extended u, the Signorini problem translates into the following system:

$$\begin{cases} u\,(x, 0) \geqslant \varphi\,(x) & \text{in } \mathbb{R}^n \\ u\,(x, -y) = u\,(x, y) & \text{in } \mathbb{R}^{n+1} \\ L_a u = 0 & \text{in } \mathbb{R}^{n+1} \setminus \{(x, 0) : u\,(x, 0) = \varphi\,(x)\} \\ L_a u \leqslant 0 & \text{in } \mathbb{R}^{n+1}, \text{ in the sense of distributions.} \end{cases}$$

with u vanishing at infinity. Again, we can exploit the advantages to analyze the obstacle problem for a nonlocal operator in PDE form by considering a local version of it. Indeed, to study the optimal regularity properties of the solution we will focus on the following local version, where $\varphi : B' \longrightarrow \mathbb{R}$:

$$\begin{cases} u\,(x, 0) \geqslant \varphi\,(x) & \text{in } B_1' \\ u\,(x, -y) = u\,(x, y) & \text{in } B_1 \\ L_a u = 0 & \text{in } B_1 \setminus \{(x, 0) : u\,(x, 0) = \varphi\,(x)\} \\ L_a u \leqslant 0 & \text{in } B_1, \text{ in the sense of distributions.} \end{cases}$$

The above problem can be thought of as the minimization of the weighted Dirichlet integral

$$J_a\,(v) = \int_{B_1} |y|^a\, |\nabla u\,(X)|^2\, dX$$

over the set

$$\mathbb{K}_a = \left\{ v \in W^{1,2}\left(B_1, |y|^a\right) : u(x,0) \geqslant \varphi(x) \right\}.$$

In a certain sense, this corresponds to an obstacle problem, where the obstacle is defined in a set of codimension $1 + a$, where a is not necessarily an integer.

The operator L_a is degenerate elliptic, with a degeneracy given by the weight $|y|^a$. This weight belongs to the Muckenhoupt class $A_2\left(\mathbb{R}^{n+1}\right)$. We recall that a positive weight function $w = w(X)$ belongs to $A_2\left(\mathbb{R}^N\right)$ if

$$\left(\frac{1}{|B|}\int_B w\right)\left(\frac{1}{|B|}\int_B w^{-1}\right) \leqslant C$$

for every ball $B \subset \mathbb{R}^N$. For the class of degenerate elliptic operators of the form $Lu = \mathrm{div}(A(X)\nabla u)$, where

$$\lambda w(X)|\xi|^2 \leqslant A(X)\xi\cdot\xi \leqslant \Lambda w(X)|\xi|^2,$$

there is a well established potential theory for solutions in the weighted Sobolev space $W^{1,2}(\Omega, w)$ (Ω bounded domain in $\mathbb{R}^N$), defined as the closure of $C^\infty\left(\overline{\Omega}\right)$ in the norm

$$\left[\int_\Omega v^2 w + \int_\Omega |\nabla v|^2 w\right]^{1/2}$$

(see [30]). Since $w \in A_2$, the gradient of a function in $W^{1,2}(\Omega, w)$ is well defined in the sense of distributions and belongs to the weighted space $L^2(\Omega, w)$.

The outline of the first part. We intend here to give a brief review of the results concerning the analysis of the solution and the free boundary of the obstacle for the fractional Laplacian, mainly based on the extension method.

For the thin obstacle problem, Caffarelli in [16] proves that u is $C^{1,\alpha}$ up to $y = 0$, for some $\alpha \leqslant 1/2$. Subsequently Athanasopoulos and Caffarelli achieve the optimal regularity of the solution in [7]. In the case of zero obstacle, the regularity of the free boundary around a so called *nondegenerate* (or *stable or regular*) point is analyzed by Athanasopoulos, Caffarelli and Salsa in [10]. Indeed these last two papers opened the door to all subsequent developments.

In [37], Garofalo and Petrosyan start the analysis of $F(u)$ around non regular points (also for non zero obstacles). They obtain a stratification result for singular points, i.e. points of $F(u)$ of vanishing density for $\Lambda(u)$, in terms of homogeneity of suitable blow-ups of the solution.

The analysis of the obstacle problem for the operator $(-\Delta)^s$, $0 < s < 1$, starts with Silvestre in ([63]), which shows $C^{1,\alpha}$ estimates for the solution for any $\alpha < s$ and also $\alpha = s$ if the interior of the coincidence set is convex. Notably, Silvestre does not use any extension properties; his methods are purely nonlocal. A few years later, Caffarelli, Salsa and Silvestre ([22]) extend to the fractional Laplacian case the results in [10] on the optimal regularity and the analysis of the *regular part* of the free boundary.

Recently, in [12], Barrios, Figalli and Ros-Oton continued the work of [37], giving a complete picture of the free boundary under two basic assumptions. The first one is a strict concavity of the obstacle, the same assumption needed in the case of the classical obstacle problem. The second one prescribes zero boundary values of the solution and it turns out to be crucial.

Here, we shall focus mainly on the optimal regularity of the solution and on the analysis and structure of the free boundary, only mentioning, for brevity reasons, recent important results on higher regularity and extension to more general operators. In particular, we will give here an outline of the strategy used in the papers ([22]) and [12].

A few comments on the key concepts and tools that will repeatedly appear are in order.

Semi-convexity: it is a peculiarity of solutions of the obstacle problem. More precisely, semi-convexity along tangential directions τ (i.e. parallel to the plane $y = 0$) and semiconcavity along the $y-$direction consistently play a key role. It is noteworthy that, for global solution of the zero thin obstacle case, the tangential semi-convexity of u comes for free, since $u(X + h\tau)$ and $u(X - h\tau)$ are admissible nonnegative superharmonic functions, and therefore

$$\frac{1}{2}\left(u\left(X+h\tau\right)+u\left(X-h\tau\right)\right) \geqslant u\left(X\right).$$

Asymptotic profiles. From semi-convexity, one deduces that suitable *global asymptotic profiles* coming from blow-ups of u around a free boundary point (say, the origin) are tangentially convex and can be classified according to their homogeneity degree. From this it is an easy matter to deduce optimal regularity.

Frequency and monotonicity formulas. Frequency formulas of Almgren type, first introduced in the case $s = 1/2$ in ([10]) are key tools in carrying optimal regularity from global to local solutions. Other types of monotonicity formulas, such as Weiss or Monneau-types, first introduced in ([37]) for $s = 1/2$, play a crucial role in the analysis of non-regular points of the free boundary.

Carleson estimates and boundary Harnack principles are by now standard tools in the study of the optimal regularity of the free boundary, in our case around the so called regular points. Due to the non homogeneous right hand side in the equation, the Carleson estimate and boundary Harnack principle proved here are somewhat weaker than the usual ones. More recently, De Silva and Savin ([28]) have applied these principles to prove higher regularity of the free boundary.

We will always assume that the origin belongs to the free boundary.

The outline of the second part. In Section 3 we consider two time-dependent models, which can be thought of as parabolic counterparts of the systems (3.1.1) and (3.1.2). In the first part, Section 3.1, we discuss the regularity of solutions to the parabolic fractional obstacle problem

$$\begin{cases} \min\{-v_t + (-\Delta)^s v, v - \psi\} = 0 & \text{on } [0, T] \times \mathbb{R}^n \\ v(T) = \psi & \text{on } \mathbb{R}^n, \end{cases}$$

following [19]. In particular, under some assumptions on the obstacle ψ, the solution v is shown to be globally Lipschitz continuous in space-time. Moreover, v_t and $(-\Delta)^s v$ belong to suitable Hölder and logLipschitz spaces. The regularity in space is optimal, whereas the regularity in time is almost optimal in the cases $s = 1/2$ and $s \to 1^-$.

In Section 3.2 we give an overview of the parabolic Signorini problem

$$\begin{cases} \Delta v - \partial_t v = 0 & \text{in } \Omega_T := \Omega \times (0, T], \\ v \geqslant \varphi, \quad \partial_\nu v \geqslant 0, \quad (v - \varphi)\partial_\nu v = 0 & \text{on } \mathcal{M}_T := \mathcal{M} \times (0, T], \\ v = g & \text{on } \mathcal{S}_T := \mathcal{S} \times (0, T], \\ v(\cdot, 0) = \varphi_0 & \text{on } \Omega_0 := \Omega \times \{0\}. \end{cases}$$

Here $\mathcal{S} = \partial\Omega \backslash \mathcal{M}$. Similarly to the elliptic case, we are interested in the regularity properties of v, and the structure and regularity of the *free boundary* $\Gamma(v) = \partial_{\mathcal{M}_T}\{(x, t) \in \mathcal{M}_T \mid v(x, t) > \varphi(x, t)\}$, where $\partial_{\mathcal{M}_T}$ indicates the boundary in the relative topology of $\mathcal{M}_T$. Following [27], the analysis comprises the monotonicity of a generalized frequency function, the study of blow-ups and the ensuing regularity of solutions, the classification of free boundary points, and the regularity of the free boundary at so-called regular points.

3.2 The Obstacle Problem for the Fractional Laplacian

This section is devoted to the study of the fractional Laplacian obstacle problem that we recall below.

Given a smooth function $\varphi : \mathbb{R}^n \to \mathbb{R}$, with bounded support (or rapidly vanishing at infinity), we look for a continuous function u satisfying the following conditions:

- $u \geqslant \varphi$ in $\mathbb{R}^n$
- $(-\Delta)^s u \geqslant 0$ in $\mathbb{R}^n$
- $(-\Delta)^s u = 0$ when $u > \varphi$
- $u(x) \to 0$ as $|x| \to +\infty$.

We list below the main steps in the analysis of the problem that we are going to describe.

1. Construction of the solution and basic properties.

2. Lipschitz continuity, semiconvexity and $C^{1,\alpha}$ estimates.
3. Reduction to the thin obstacle for the operator L_a.
4. Optimal regularity for tangentially convex global solution.
5. Classification of asymptotic blow-up profiles around a free boundary point.
6. Optimal regularity of the solution.
7. Analysis of the free boundary at stable points: Lipschitz continuity.
8. Boundary Harnack Principles and $C^{1,\alpha}$ regularity of the free boundary at stable points.
9. Structure of the free boundary.

Steps 1 and 2 are covered in [63]. In particular from [63] we borrow the $C^{1,\alpha}$ estimates without proof, for brevity. The steps 3-8 follow basically the paper [22], except for step 2 and part of 5. In particular, the optimal regularity of global solutions follows a different approach, similar to the corresponding proof for the zero obstacle problem in [10]. Step 8 is taken from [12]. Finally, step 9 comes from [10] (Carleson estimate) and [22] (Boundary Harnack).

3.2.1 Construction of the solution and basic properties

We start by proving the existence of a solution. Observe that the proof fails for $n = 1$ and $s > 1/2$, because in this case it is impossible to have $(-\Delta)^s u \geqslant 0$ in $\mathbb{R}$ with u vanishing at infinity.

Let $\mathcal{S}$ be the set of rapidly decreasing C^∞ functions in $\mathbb{R}^n$. We denote by $\dot{H}^s$ the completion of $\mathcal{S}$ in the norm

$$\|f\|_{\dot{H}^s}^2 = \int_{\mathbb{R}^n}\int_{\mathbb{R}^n} \frac{|f(x) - f(y)|^2}{|x-y|^{n+2s}}\,dxdy \sim \int_{\mathbb{R}^n} |\xi|^{2s} \left|\hat{f}(\xi)\right|^2 d\xi.$$

Equipped with the inner product

$$\begin{aligned}(f,g)_{\dot{H}^s} &= \int_{\mathbb{R}^n}\int_{\mathbb{R}^n} \frac{(f(x)-f(y))(g(x)-g(y))}{|x-y|^{n+2s}}\,dxdy \\ &= 2\int_{\mathbb{R}^n} f(x)(-\Delta)^s g(x)\,dx \sim \int_{\mathbb{R}^n} |\xi|^{2s}\hat{f}(\xi)\overline{\hat{g}(\xi)}d\xi,\end{aligned}$$

$\dot{H}^s$ is a Hilbert space. Since we are considering $n \geqslant 2$ and $s < 1 \leqslant n/2$, it turns out that $\dot{H}^s$ coincides with the set of functions in $L^{2n/(n-2s)}$, for which the $\dot{H}^s$–norm is finite.

The solution u_0 of the obstacle problem is constructed as the unique minimizer of the strictly convex functional

$$J(v) = \|v\|_{\dot{H}^s}^2$$

over the closed, convex set $\mathbb{K}_s = \left\{v \in \dot{H}^s : v \geqslant \varphi\right\}$.

In the following proposition we gather some basic properties of u.

Proposition 3.1. *Let u_0 be the minimizer of the functional J over K_s. Then:*

(a) The function u_0 is a supersolution, that is $(-\Delta)^s u_0 \geqslant 0$ in $\mathbb{R}^n$ in the sense of measures. Thus, it is lower semicontinuous and, in particular, the set $\{u_0 > \varphi\}$ is open.
(b) u_0 is actually continuous in $\mathbb{R}^n$.
(c) If $u_0(x) > \varphi(x)$ in some ball B then $(-\Delta)^s u_0 = 0$ in B.

Proof. (a) Let $h \geqslant 0$ be any smooth function with compact support. If $t > 0$, $u_0 + th \geqslant \varphi$ so that

$$(u_0, u_0)_{\dot{H}^s} \leqslant (u_0 + th, u_0 + th)_{\dot{H}^s}$$

or

$$0 \leqslant 2t\,(u_0, h)_{\dot{H}^s} + t^2\,(h, h)_{\dot{H}^s} = \left((-\Delta)^s h\right)_{L^2} + t^2\,(h, h)_{\dot{H}^s}$$

from which

$$\left(u_0, (-\Delta)^s h\right)_{L^2} = \left((-\Delta)^s u_0, h\right)_{L^2} \geqslant 0.$$

Therefore $(-\Delta)^s u_0$ is a nonnegative measure and therefore is lower semicontinuous by Proposition A1. □

(b) The continuity follows from Proposition A2.

(c) For any test function $h \geqslant 0$, supported in B the proof in (a) holds also for $t < 0$. Therefore $(-\Delta)^s u_0 = 0$ *in* B. □

Corrolary 3.2. *The minimizer u_0* of the functional *J over K_s is a solution of the obstacle problem.* □

3.2.2 Lipschitz continuity and semiconvexity and $C^{1,\alpha}$ estimates

Following our strategy, we first show that, if φ is smooth enough, then the solution of our obstacle problem is Lipschitz continuous and semiconvex. We are mostly interested in the case $\varphi \in C^{1,1}$. When φ has weaker regularity, u_0 inherits corresponding weaker regularity (see [63]). We emphasize that the proof in this subsection depends only on the maximum principle and translation invariance.

Lemma 3.3. *The function u_0 is the least supersolution of $(-\Delta)^s$ such that $u_0 > \varphi$ and* $\liminf_{|x|\to\infty} u_0(x) \geqslant 0$.

Proof. Let v such that $(-\Delta)^s v \geqslant 0$, $v > \varphi$ and $\liminf_{|x|\to\infty} v(x) \geqslant 0$. Let $w = \min\{u_0, v\}$. Then w is lower-semicontinuous in $\mathbb{R}^n$ and is another supersolution above φ (by Propositions A1 and A4). We show that $w \geqslant u_0$.

Since $\varphi \leqslant w \leqslant u_0$, we have $w(x) = u_0(x)$ for every x in the contact set $\Lambda(u_0) = \{u_0 = \varphi\}$. In $\Omega = \{u_0 > \varphi\}$, u_0 solves $(-\Delta)^s u_0 = 0$ and w is a supersolution. By

Proposition 3.1 (b), u_0 is continuous. Then $w - u_0$ is lower-semicontinuous and $w \geqslant u_0$ from comparison. □

Corrolary 3.4. *The function* u_0 is bounded and $\sup u_0 \leqslant \sup \varphi$. *If the obstacle* φ *has a modulus of continuity* ω, *then* u_0 *has the same modulus of continuity. In particular, if* φ *is Lipschitz, then* u_0 *is Lipschitz and* $Lip(u_0) \leqslant Lip(\varphi)$.

Proof. By hypothesis $u_0 \geqslant \varphi$. The constant function $v(x) = \sup \varphi$ is a supersolution that is above φ. By Lemma 3.3, $u_0 \leqslant v$ in $\mathbb{R}^n$.

Moreover, since ω is a modulus of continuity for φ, for any $h \in \mathbb{R}^n$,

$$\varphi(x+h) + \omega(|h|) \geqslant \varphi(x)$$

for all $x \in \mathbb{R}^n$. Then, the function $u_0(x+h) + \omega(|h|)$ is a supersolution above $\varphi(x)$. Thus $u_0(x+h) + \omega(|h|) \geqslant u_0(x)$ for all $x, h \in \mathbb{R}^n$. Therefore u_0 has a modulus of continuity not larger than ω. □

Lemma 3.5. *Let* $\varphi \in C^{1,1}$ *and assume that* $\inf \partial_{\tau\tau}\varphi \geqslant -C$, *for any unit vector* τ. *Then* $\partial_{\tau\tau}u_0 \geqslant -C$ *too. In particular,* u_0 *is semiconvex.*

Proof. Since $\partial_{\tau\tau}\varphi \geqslant -C$, we have

$$\frac{\varphi(x+h\tau) + \varphi(x-h\tau)}{2} + Ch^2 \geqslant \varphi(x)$$

for every $x \in \mathbb{R}^n$ and $h > 0$. Therefore:

$$V(x) \equiv \frac{u_0(x+h\tau) + u_0(x-h\tau)}{2} + Ch^2 \geqslant \varphi(x)$$

and V is also a supersolution: $(-\Delta)^s V \geqslant 0$. Thus, by Lemma 3.3, $V \geqslant u$ so that:

$$\frac{u_0(x+h\tau) + u_0(x-h\tau)}{2} + Ch^2 \geqslant u_0(x)$$

for every $x \in \mathbb{R}^n$ and $h > 0$. This implies $\partial_{\tau\tau}u_0 \geqslant -C$. □

From the results in [63] we can prove a partial regularity result, under the hypothesis that φ is smooth.

Theorem 3.6. *Let* $\varphi \in C^2$. *Then* $u_0 \in C^{1,\alpha}$ *for every* $\alpha < s$ *and* $(-\Delta)^s u_0 \in C^\beta$ *for every* $\beta < 1 - s$.

The proof is long and very technical, so we refer to the original paper [63]. However, an idea of the proof in the case of the zero thin obstacle problem can be given without much effort. Indeed, from tangential semi-convexity we deduce (here u is harmonic outside $\Lambda(u)$)

$$u_{yy} \leqslant C \quad \text{in } B_1 \backslash \Lambda(u).$$

In particular, the function $u_y - Cy$ is monotone and bounded. Thus we are allowed to define in $B_1' : \sigma(x) = \lim_{y\to 0+} u_y(x,y)$. Since, in this case

$$0 \geqslant \Delta u = 2u_y \mathcal{H}^n_{|\Lambda(u)}$$

in the sense of measures, we have $\sigma(x) \leqslant 0$ in $\Lambda(u)$ and by symmetry, $\sigma(x) = 0$ in $B_1' \backslash \Lambda(u)$. We may summarize the properties of the solution of a zero thin obstacle problem in complementary form as follows:

$$\begin{cases} \Delta u \leqslant 0, \ u\Delta u = 0 & \text{in } B_1 \\ \Delta u = 0 & \text{in } B_1 \backslash \Lambda(u) \\ u(x,0) \geqslant 0, \ \sigma(x) \leqslant 0, \ u(x,0)\,\sigma(x) = 0 & \text{in } B_1' \\ \sigma(x) = 0 & \text{in } B_1' \backslash \Lambda(u) \end{cases}$$

Now, let u be a solution of the zero thin obstacle problem in B_1, normalized by $\|u\|_{L^2(B_1)} = 1$. To prove local $C^{1,\alpha}$ estimates, it is enough to show that $\sigma \in C^{0,\alpha}$ near the free boundary $F(u)$.

In fact, in the interior of $\Lambda(u)$, $u(x,0)$ is smooth and so is σ. On the other hand, on $\Omega' = B' \backslash \Lambda'(u)$, $\sigma = 0$. Thus, if we show that σ is $C^{0,\alpha}$ in a neighborhood of $F(u)$, then $u \in C^{1,\alpha}$ from both sides of the free boundary by standard estimates for the Neumann problem.

In particular it is enough to show uniform estimates around a free boundary point, say the origin. This can be achieved by a typical iteration procedure, in the De Giorgi style. We distinguish two steps:

Step 1: To show that near the free boundary we can locate large regions where $-\sigma$ grows at most linearly (estimates in measure of the oscillation of $-\sigma$).

Step 2: Using Poisson representation formula and semi-concavity, we convert the estimate in average of the oscillation of $-\sigma$, done in step 1, into pointwise estimates, suitable for a dyadic iteration of the type

$$u_y(X) \geqslant -\beta^k \quad \text{in } B'_{\gamma^k}(x_0) \times \left[0, \gamma^k\right] \qquad (x_0 \in F(u))$$

for some $0 < \gamma < 1$, $0 < \beta < 1$, and any $k \geqslant 0$.

The details can be found in the paper of Caffarelli [16] (see also the review paper [61]).

3.2.3 Thin obstacle for the operator L_a. Local $C^{1,\alpha}$ estimates

To achieve optimal regularity we now switch to the equivalent thin obstacle problem for the operator L_a as mentioned in the introduction and that we restate here:

$$\begin{cases} u(x,0) \geqslant \varphi(x) & \text{in } B_1' \\ u(x,-y) = u(x,y) & \text{in } B_1 \\ L_a u = 0 & \text{in } B_1 \backslash \Lambda(u) \\ L_a u \leqslant 0 & \text{in } B_1, \text{ in the sense of distributions.} \end{cases} \tag{3.2.1}$$

In the global setting (i.e. with B_1 replaced by $\mathbb{R}^n$), $u_0(x) = u(x,0)$ is the solution of the global obstacle problem for $(-\Delta)^s$ and

$$(-\Delta)^s u_0 = \kappa_a \lim_{y\to 0+} y^a u_y(x,y).$$

The estimates in Corollary 3.4 and Lemma 3.5 translate, after an appropriate localization argument and the use of boundary estimates for the operator L_a, into corresponding estimates for the solution of u. Namely:

Lemma 3.7. *Let $\varphi \in C^{2,1}(B_1')$ and u be the solution of (3.2.1). Then*

1. $\nabla_x u(X) \in C^\alpha(B_{1/2})$ *for every* $\alpha < s$;
2. $|y|^a u_y(X) \in C^\alpha(B_{1/2})$ *for every* $\alpha < 1 - s$;
3. $u_{\tau\tau}(X) \geqslant -C$ *in* $B_{1/2}$.

Proof. From Corollary 3.4 and Lemma 3.5 we have that the above estimates holds on $y = 0$. Since $\partial_{x_j} u$ and $u_{\tau\tau}$ also solve the equation $L_a w = 0$ in $B_1 \backslash \Lambda(u)$, the estimates 1 and 3 extend to the interior. On the other hand $w(x,y) = |y|^a u_y(X)$ solves the conjugate equation $\operatorname{div}(|y|^{-a}\nabla w(X))$ and we obtain 2. □

Remark 3.8. *Observe that u can only be $C^{1,\alpha}$ in both variables up to $y = 0$ only if $a \leqslant 0$. If $a > 0$, since $y^a u_y(X)$ has a non-zero limit for some x in the contact set, it follows that u_y cannot be bounded.*

We close this subsection with a compactness result, useful in dealing with blow-up sequences.

Lemma 3.9. *Let $\{v_j\}$ be a bounded sequence of functions in $W^{1,2}(B_1, |y|^a)$. Assume that there exists a constant C such that, in B_1:*

$$|\nabla_x v_j(X)| \leqslant C \quad \text{and} \quad |\partial_y v_j(X)| \leqslant C|y|^{-a} \tag{3.2.2}$$

and that, for each small $\delta > 0$, v_j is uniformly $C^{1,\alpha}$ in $B_{1-\delta} \cap \{|y| > \delta\}$.

Then, there exists a subsequence $\{v_{j_k}\}$ strongly convergent in $W^{1,2}\left(B_{1/2}, |y|^a\right)$.

Proof. From the results in [43], there is a subsequence, that we still call $\{v_j\}$, that converges strongly in $L^2\left(B_{1/2}, |y|^a\right)$. Since for each $\delta > 0$, v_j is uniformly bounded in $C^{1,\alpha}$ in the set $B_{1-\delta} \cap \{|y| > \delta\}$, we can extract a subsequence so that ∇v_j converges uniformly in $B_{1-\delta} \cap \{|y| > \delta\}$. Thus, ∇v_j converges pointwise in $B_1 \backslash \{y = 0\}$.

Now, from (3.2.2) and the fact that C and $|y|^{-a}$ both belong to $L^2\left(B_{1/2}, |y|^a\right)$, the convergence of each partial derivative of v_j in $L^2\left(B_{1/2}, |y|^a\right)$ follows from the dominated convergence theorem. □

3.2.4 Minimizers of the weighted Rayleigh quotient and a monotonicity formula

The next step towards optimal regularity is to consider tangentially convex global solutions.

Lemma 3.10. *Let ∇_ϑ denote the surface gradient on the unit sphere ∂B_1. Set, for $-1 < a < 1$,*

$$\lambda_{0,a} = \inf \left\{ \frac{\int_{\partial B_1^+} |\nabla_\vartheta w|^2 y^a dS}{\int_{\partial B_1^+} w^2 y^a dS} : w \in W^{1,2}\left(\partial B_1^+, y^a dS\right) : w = 0 \text{ on } \left(\partial B_1'\right)^+ \right\}$$

where $\left(\partial B_1'\right)^+ = \{(x', x_n) \in \partial B_1',\ x_n > 0\}$. Then the first eigenfunction, up to a multiplicative factor, is given by

$$w(x, y) = \left(\sqrt{x_n^2 + y^2} - x_n\right)^s \qquad s = (1 - a)/2$$

and [3.2]

$$\lambda_{0,a} = \frac{1 - a}{4}(2n + a - 1).$$

The following lemma gives a first monotonicity result.

Lemma 3.11. *Let w be continuous in B_1, $w(0) = 0$, $w(x, 0) \leqslant 0$, $w(x, 0) = 0$ on $\Lambda \subset \{y = 0\}$, $L_a w = 0$ in $B_1 \backslash \Lambda$. Assume that the set*

$$\left\{x \in B_r' : w(x, 0) < 0\right\}$$

is non empty and convex. Set

$$\beta(r) = \beta(r; w) = \frac{1}{r^{1-a}} \int_{B_r^+} \frac{y^a |\nabla w(X)|^2}{|X|^{n+a-1}} dX.$$

Then, $\beta(r)$ is bounded and increasing for $r \in (0, 1/2]$.

Proof. We have $L_a w^2 = 2w L_a w + 2y^a |\nabla w|^2 = 2y^a |\nabla w|^2$, so that

$$\beta(r) = \frac{1}{r^{1-a}} \int_{B_r^+} \frac{y^a |\nabla w(X)|^2}{|X|^{n+a-1}} dX = \frac{1}{2r^{1-a}} \int_{B_r^+} \frac{L_a\left(w^2\right)}{|X|^{n+a-1}} dX.$$

Now:

$$\beta'(r) = \frac{a - 1}{2r^{2-a}} \int_{B_r^+} \frac{L_a\left(w^2\right)}{|X|^{n+a-1}} dX + \frac{1}{r^n} \int_{\partial B_r^+} y^a |\nabla w|^2 dS.$$

3.2 Formally, the first eigenvalue can be obtained plugging $\alpha = s = (1 - a)/2$ and $n + a$ instead of n into the formula $\alpha(\alpha - 1) + n\alpha$.

Since $w(0,0) = 0$ and $y^a w_y(x,y)\, w(x,y) \to 0$ as $y \to 0^+$, after simple computations, we obtain:

$$\beta'(r) \geqslant -(1-a)\frac{(2n+a-1)}{4r^{n+2}}\int_{\partial B_r^+} y^a w^2\, dS + \frac{1}{r^n}\int_{\partial B_r^+} y^a \left|\nabla_{\mathcal{g}} w\right|^2 dS.$$

The convexity of $\{x \in B_r' : w(x,0) < 0\}$ implies that the Rayleigh quotient must be greater than $\lambda_{0,a}$ and therefore we conclude $\beta'(r) \geqslant 0$ and, in particular, $\beta(r) \leqslant \beta(1/2)$. □

3.2.5 Optimal regularity for tangentially convex global solutions

In this section we consider global solutions that represent possible asymptotic profiles, obtained by a suitable blow-up of the solution at a free boundary point.

First of all we consider functions $u : \mathbb{R}^n \times \mathbb{R} \to \mathbb{R}$, *homogeneous of degree* k, solutions of the following problem:

$$\begin{cases} u(x,0) \geqslant 0 & \text{in } \mathbb{R}^n \\ u(x,-y) = u(x,y) & \text{in } \mathbb{R}^n \times \mathbb{R} \\ L_a u = 0 & \text{in } (\mathbb{R}^n \times \mathbb{R}) \setminus \Lambda \\ L_a u \leqslant 0 & \text{in the sense of distributions in } \mathbb{R}^n \times \mathbb{R} \\ u_{\tau\tau} \geqslant 0 & \text{in } \mathbb{R}^n \times \mathbb{R}, \text{ for every tangential unit vector } \tau \end{cases} \tag{3.2.3}$$

where $\Lambda = \Lambda(u) = \{(x,0) : u(x,0) = 0\}$. The following proposition gives a lower bound for the degree k, which implies the optimal regularity of the solution.

Lemma 3.12. *If there exists a solution u of problem (3.2.3), then $k \geqslant 1 + s = (3-a)/2$.*

Proof. Apply the monotonicity formula in Lemma 3.11 to $w = u_\tau$. Then, $L_a w = 0$ in $(\mathbb{R}^n \times \mathbb{R}) \setminus \Lambda$ and, by symmetry, $w(x,0)\, w_y(x,0) = 0$. Moreover, the contact set where $w = 0$ is convex, since $u_{\tau\tau} \geqslant 0$.Therefore w satisfies all the hypotheses of that lemma. Recall that we always assume that $(0,0) \in F(u)$ so that $w(0,0) = 0$. Thus

$$\beta(r;w) = \frac{1}{r^{1-a}}\int_{B_r^+} \frac{y^a \left|\nabla w(X)\right|^2}{|X|^{n+a-1}}\, dX \leqslant \beta(1,w).$$

On the other hand, since ∇w is homogeneous of degree $k-2$, we have

$$\beta(r;w) = \frac{r^{2k-2}}{r^{1-a}}\int_{B_1^+} \frac{y^a \left|\nabla w(X)\right|^2}{|X|^{n+a-1}}\, dX = \frac{r^{2k-2}}{r^{1-a}}\beta(1,w).$$

This implies $r^{2k-2} \leqslant r^{1-a} = r^{2s}$ or $k \geqslant 1+s$. □

From Lemma 3.12 it would be possible to deduce the optimal regularity of the solution u to (3.2.1). However, to study the free boundary regularity we need to classify precisely the solutions to problem (3.2.3). For the operator L_a, we need to introduce the following subset of the coincidence set. Let

$$\Lambda_* = \left\{ (x,0) \in \mathbb{R}^n : \lim_{y \to 0+} y^a u_y(x,y) < 0 \right\}.$$

Notice that $\overline{\Lambda_*}$ is the support of $L_a u$ and since $L_a u = 0$ in $(\mathbb{R}^n \times \mathbb{R}) \setminus \Lambda$, we have $\overline{\Lambda_*} \subset \Lambda$ (Λ is closed).

The analysis depend on whether Λ_* has positive H^n measure or not. First examine the case $H^n(\Lambda_*) = 0$.

Lemma 3.13. *Let u be a solution of problem (3.2.3). If $H^n(\Lambda_*) = 0$, then u is a polynomial of degree k.*

Proof. We know from Lemma 3.7 that $|y|^a u_y(x,y)$ is locally bounded. If $\mathcal{H}^n(\Lambda_*) = 0$, then

$$\lim_{y \to 0} |y|^a u_y(x,y) = 0 \text{ a.e. } x \in \mathbb{R}^n.$$

Thus $\lim_{y \to 0} |y|^a u_y(x,y) = 0$ weak* in L^∞ and from [23] we infer that u is a global solution of $L_a u = 0$ in $\mathbb{R}^n \times \mathbb{R}$. Using Lemma A3 we conclude the proof. □

Lemma 3.14. *Let u be a solution of problem (3.2.3). If $H^n(\Lambda) \neq 0$ then, either $u \equiv 0$* or *$k = 1 + s$ and Λ is a half n–dimensional space.*

Proof (sketch). First observe that if $\mathcal{H}^n(\Lambda_*) = 0$, then $u \equiv 0$, otherwise, from Lemma 3.13, $u(x,0)$ would be a polynomial vanishing on a set of positive measure and therefore identically zero. Thus, the polynomial u must have the form

$$u(x,y) = p_1(x) y^2 + \ldots + p_j(x) y^{2j}$$

and iterating the computation of $L_a u$ one deduce $p_1 = p_2 = \ldots = p_j = 0$.

Consider now the case $\mathcal{H}^n(\Lambda_*) \neq 0$. Then Λ_* is a *thick convex* cone. Assume that e_n is a direction inside Λ_* such that a neighborhood of e_n is contained in Λ_*. Using the convexity in the e_n direction, we infer that $w = u_{x_n}$ cannot be positive at any point X. Moreover, $w = 0$ on Λ and

$$L_a w(X) = 0 \text{ in } (\mathbb{R}^n \times \mathbb{R}) \setminus \Lambda_* \supseteq (\mathbb{R}^n \times \mathbb{R}) \setminus \Lambda.$$

Thus w must coincide with the first eigenfunction of the weighted spherical Laplacian, minimizer of $\int_{S_1} |\nabla_{\mathcal{G}} v|^2 |y|^a \, dS$ over all v vanishing on Λ and such that $\int_{S_1} v^2 |y|^a \, dS = 1$.

Since Λ is convex, $\Lambda \cap B_1$ is contained in half of the sphere $B_1 \cap \{y = 0\}$. If it were exactly half of the sphere then it would be given by the first eigenfunction defined in Lemma 3.10, up to a multiplicative constant, by the explicit expression

$$w(x,y) = \left(\sqrt{x_n^2 + y^2} - x_n \right)^s \qquad s = (1-a)/2.$$

On the other hand, the above function is not a solution across $\{y = 0, x_n \geqslant 0\}$. Therefore, if $\Lambda \cap B_1$ is *strictly* contained in half of the sphere $B_1 \cap \{y = 0\}$, there must be another eigenfunction corresponding to a smaller eigenvalue and consequently to a degree of homogeneity *smaller than* s. This would imply $k < 1 + s$, contradicting Lemma 3.12. The only possibility is therefore $k = 1 + s$, with $\Lambda = \{y = 0, x_n \geqslant 0\}$. □

The next theorem gives the classification of asymptotic profiles.

Theorem 3.15. *Let u be a non trivial solution of problem (3.2.3). There are only two possibilities:*

(1) $k = 1 + s$, Λ is a half $n-$dimensional space and u depends only on two variables. Up to rotations and multiplicative constants u is unique and there is a unit vector τ such that $\Lambda = \{(x, 0) : x \cdot \tau \geqslant 0\}$ and

$$u_\tau(x, y) = \left(\sqrt{(x \cdot \tau)^2 + y^2} - (x \cdot \tau)\right)^s$$

(2) k is an integer greater than equal to 2, u is a polynomial and $\mathcal{H}^n(\Lambda) = 0$.

Proof. If $\mathcal{H}^n(\Lambda) \neq 0$, from Lemma 3.14 we deduce that, up to rotations and multiplicative constants, there is a unique solution of problem (3.2.3), homogeneous of degree $k = 1 + s$. Moreover, for this solution the free boundary $F(u)$ is flat, that is there is a unit vector (say) e_n such that

$$\Lambda = \{(x, 0) : x_n \geqslant 0\}$$

and

$$u_{x_n}(x, y) = \left(\sqrt{x_n^2 + y^2} - x_n\right)^s.$$

Integrating u_{x_n} from $F(u)$ along segments parallel to e_n we uniquely determine $u(x, 0) = u(x_n, 0)$. If we had another solution v, homogeneous of degree $1 + s$, with $v(x, 0) = u(x, 0)$, then necessarily (see the proof of the Liouville-type Lemma A3), for some constant c and $y \neq 0$, we have

$$v(x, y) - u(x, y) = c\,|y|^s\, y.$$

But the constant c must be zero, otherwise $v - u$ cannot be solution across $\{y = 0\} \setminus \Lambda$.

As a consequence, if u is a solution homogeneous of degree $1 + s$, with e_n normal to $F(u)$, then $u = u(x_n, y)$. Indeed, translating in any direction orthogonal to x_n and y we get another global solution with the same free boundary. By uniqueness, u must be invariant in those directions.

If $\mathcal{H}^n(\Lambda) = 0$, then $\mathcal{H}^n(\Lambda_*) = 0$ and from Lemma 3.13 we conclude that u is a polynomial and $k \geqslant 2$. □

3.2.6 Frequency formula

As we have already stated, a crucial tool in order to achieve optimal regularity is a frequency formula of Almgren type. If the obstacle were zero, then the frequency formula states that the quantity

$$D_a(r;u) = \frac{r\int_{B_r} |y|^a |\nabla u|^2 \, dX}{\int_{\partial B_r} |y|^a u^2 dS}$$

is bounded and monotonically increasing. The conclusion is the following.

Theorem 3.16. *Let u be a solution of the zero thin obstacle for the operator L_a in B_1. Then $D_a(r;u)$ is monotone nondecreasing in r for $r < 1$. Moreover, $D_a(r;u)$ is constant if and only if u is homogeneous.*

When the obstacle φ is non zero we cannot reduce to that case. Instead, assuming that $\varphi \in C^{2,1}$, we let

$$\tilde{u}(x,y) = u(x,y) - \varphi(x) + \frac{\Delta\varphi(0)}{2(1+a)} y^2$$

so that $L_a\tilde{u}(0) = 0$. Moreover $\Lambda = \Lambda(u) = \{\tilde{u} = 0\}$. The function $\tilde{u}$ is a solution of the following system:

$$\begin{cases} \tilde{u}(x,0) \geqslant 0 & \text{in } B_1' \\ \tilde{u}(x,-y) = \tilde{u}(x,y) & \text{in } B_1 \\ L_a\tilde{u}(x,y) = |y|^a g(x) & \text{in } B_1 \backslash \Lambda \\ L_a\tilde{u}(x,y) \leqslant |y|^a g(x) & \text{in } B_1, \text{ in the sense of distributions} \end{cases} \tag{3.2.4}$$

where

$$g(x) = \Delta\varphi(x) - \Delta\varphi(0)$$

is Lipschitz. Notice that $|y|^a g(x) \to 0$ as $x \to 0$ and in $B_1 \backslash \Lambda$

$$|L_a\tilde{u}(x,y)| \leqslant C|y|^a |x|.$$

What we expect is a small variation of Almgren's formula. Since $u - \tilde{u}$ is a $C^{2,1}$ function, it is enough to prove any regularity result for $\tilde{u}$ instead of u. *In order to simplify the notation we will still write u for $\tilde{u}$.* Define

$$F(r) = F(r;u) = \int_{\partial B_r} u^2 |y|^a \, d\sigma = r^{n+a} \int_{\partial B_1} (u(rX))^2 |y|^a \, dS.$$

We have:

$$F'(r) =$$

$$= (n+a) r^{n+a-1} \int_{\partial B_1} (u(rX))^2 |y|^a \, dS + r^{n+a} \int_{\partial B_1} 2u(rX) \nabla u(rX) \cdot X |y|^a \, dS$$

$$= \quad (n+a)\,r^{-1}\int_{\partial B_r}(u(X))^2\,|y|^a\,dS + \int_{\partial B_r} 2u(X)\,u_\nu(X)\,|y|^a\,dS$$

Thus $\log F(r)$ is differentiable for $r > 0$ and:

$$\frac{d}{dr}\log F(r) = \frac{F'(r)}{F(r)} = \frac{n+a}{r} + \frac{\int_{\partial B_r} 2uu_\nu\,|y|^a\,dS}{\int_{\partial B_r} u^2\,|y|^a\,dS}.$$

Note that the monotonicity of $D_a(r;u)$ when $\varphi = 0$ amounts to say that the function

$$r \longmapsto r\frac{d}{dr}\log F(r) = 2D(r;u) + n + a$$

is increasing, since in this case

$$\begin{aligned}\int_{\partial B_r} 2uu_\nu\,|y|^a\,dS &= \int_{B_r} L_a(u^2) = \int_{B_r}(|y|^a\,|\nabla u|^2 + 2uL_a u)dX\\ &= \int_{B_r} 2\,|y|^a\,|\nabla u|^2\,dX.\end{aligned}$$

Due to presence of a nonzero right hand side, we need to prevent the possibility that $F(r)$ become too small under rescaling when compared to the terms involving $L_a u$. It turns out that this can be realized by introducing the following modified formula:

$$\Phi(r) = \Phi(r;u) = \left(r + c_0 r^2\right)\frac{d}{dr}\log\max\left[F(r), r^{n+a+4}\right]. \qquad (3.2.5)$$

Then:

Theorem 3.17. *(Monotonicity formula). Let u be a solution of problem (3.2.4). Then, there exists a small* r_0 and a large c_0, *both depending only on a, n, $\|\varphi\|_{C^{2,1}}$, such that $\Phi(r;u)$ is monotone nondecreasing for $r < r_0$.*

For the proof, we need a Poincaré type estimate and a Rellich type identity. Recall that $u(0,0) = 0$ since the origin belongs to the free boundary.

Lemma 3.18. *Let u be a solution of problem (3.2.4), $u(0,0) = 0$. Then*

$$\int_{\partial B_r}(u(X))^2\,|y|^a\,dS \leqslant Cr\int_{B_r}|\nabla u(X)|^2\,|y|^a\,dX + c(a,n)\,r^{6+a+n}$$

and by integrating in r,

$$\int_{B_r}(u(X))^2\,|y|^a\,dX \leqslant Cr^2\int_{B_r}|\nabla u(X)|^2\,|y|^a\,dX + c(a,n)\,r^{7+a+n}$$

where c, C depend only on *a, n and $\|\varphi\|_{C^{2,1}}$.*

Proof. See [22]. □

Lemma 3.19. *The following identity holds for any $r \leqslant 1$.*

$$r\int_{\partial B_r}\left(|\nabla_g u|^2 - u_\nu^2\right)|y|^a\, dS = \int_{B_r}\left[(n+a-1)|\nabla u(X)|^2 - 2\langle X, \nabla u\rangle g(X)\right]|y|^a\, dX, \tag{3.2.6}$$

where $\nabla_g u$ denotes the tangential gradient.

Proof. Consider the vector field

$$\mathbf{F} = \frac{1}{2}y^a|\nabla u|^2 X - y^a\langle X, \nabla u\rangle\nabla u \qquad (y>0).$$

We have:

$$\operatorname{div}\mathbf{F} = \frac{1}{2}(n+a-1)y^a|\nabla u|^2 - \langle X, \nabla u\rangle L_a u.$$

Since $\langle X, \nabla u\rangle$ is a continuous function on B_r' that vanishes on $\Lambda = \{u=0\}$, we have that $\langle X, \nabla u\rangle L_a u$ has no singular part and coincides with $\langle X, \nabla u\rangle |y|^a g(x)$. An application of the divergence theorem gives (3.2.6). □

Proof of Theorem 3.17. First we observe that by taking the maximum in (3.2.5) it may happen that we get a non differentiable functions. However, $\max\left[F(r), r^{n+a+4}\right]$ is absolutely continuous (it belongs to $W^{1,1}_{loc}(0,1)$) and in any case, the jump in the derivative will be in the positive direction.

When $F(r) \leqslant r^{n+a+4}$ we have

$$\Phi(r) = \left(r + c_0 r^2\right)\frac{d}{dr}\log r^{n+a+4}$$

and $\Phi'(r) = (n+a+4)c_0 > 0$.

Thus we can concentrate on the case $F(r) > r^{n+a+4}$ where

$$\Phi(r) = \left(r + c_0 r^2\right)\frac{d}{dr}\log F(r).$$

We have:

$$\begin{aligned}\Phi(r) &= \left(r + c_0 r^2\right)\frac{\int_{\partial B_r} 2uu_\nu |y|^a\, dS}{\int_{\partial B_r} u^2 |y|^a\, dS} + (1 + c_0 r)(n+a)\\ &\equiv 2\Psi(r) + (1 + c_0 r)(n+a).\end{aligned}$$

We show that the first term is increasing, by computing its logarithmic derivative. We find:

$$\frac{d}{dr}\log\Psi(r) = \frac{1}{r} + \frac{c_0}{1+c_0 r} + \frac{\frac{d}{dr}\int_{\partial B_r} uu_\nu |y|^a\, dS}{\int_{\partial B_r} uu_\nu |y|^a\, dS} - \frac{\int_{\partial B_r} 2uu_\nu |y|^a\, dS}{\int_{\partial B_r} u^2 |y|^a\, dS} - \frac{n+a}{r}.$$

We estimate $\frac{d}{dr}\int_{\partial B_r} uu_\nu |y|^a\, dS$ from below. Since

$$\int_{\partial B_r} uu_\nu |y|^a\, dS = \int_{B_r}(|y|^a|\nabla u|^2 + uL_a u)dX$$

we can write, recalling that $|L_a u| \leqslant c\,|y|^a\,|x|$,

$$\frac{d}{dr}\int_{\partial B_r} u u_\nu\,|y|^a\,dS \geqslant \int_{\partial B_r} |y|^a\,|\nabla u|^2\,dS - cr^{(n+a+2)/2}\,[F(r)]^{1/2}.$$

We now use Lemma 3.19 to estimate $\int_{\partial B_r} |y|^a\,|\nabla u|^2\,dS$ from below. We find

$$\int_{\partial B_r} |y|^a\,|\nabla u|^2\,dS = \int_{\partial B_r} \left(|u_g|^2 + u_\nu^2\right)|y|^a\,dS =$$

$$= 2\int_{\partial B_r} u_\nu^2\,|y|^a\,dS + \frac{1}{r}\int_{B_r} [(n+a-1)\,|\nabla u(X)|^2 - 2\langle X, \nabla u\rangle\, g(X)]\,|y|^a\,dX =$$

$$\begin{aligned} &= 2\int_{\partial B_r} u_\nu^2\,|y|^a\,dS + \frac{n+a-1}{r}\int_{\partial B_r} u u_\nu\,|y|^a\,dS \\ &\quad -\frac{1}{r}\int_{B_r} [(n+a-1)u - 2\langle X, \nabla u\rangle] g(X)\,|y|^a\,dX. \end{aligned}$$

Therefore

$$\begin{aligned} \frac{d}{dr}\int_{\partial B_r} u u_\nu\,|y|^a\,dS &\geqslant 2\int_{\partial B_r} u_\nu^2\,|y|^a\,dS + \frac{n+a-1}{r}\int_{\partial B_r} u u_\nu\,|y|^a\,dS \\ &\quad -cr^{n+a+1}\left[\sqrt{G(r)} + r\sqrt{H(r)} + \sqrt{rF(r)}\right] \end{aligned}$$

where

$$G(r) = \int_{B_r} u^2\,|y|^a\,dX \quad \text{and} \quad H(r) = \int_{B_r} |\nabla u|^2\,|y|^a\,dX.$$

Collecting all the above estimates, we can write:

$$\frac{d}{dr}\log \Psi(r) = P(r) + Q(r)$$

with

$$P(r) = \frac{2\int_{\partial B_r} u_\nu^2\,|y|^a\,dS}{\int_{\partial B_r} u u_\nu\,|y|^a\,dS} - \frac{\int_{\partial B_r} 2u u_\nu\,|y|^a\,dS}{\int_{\partial B_r} u^2\,|y|^a\,dS} \geqslant 0$$

and

$$\begin{aligned} Q(r) &= \frac{c_0}{1+c_0 r} - cr^{(n+a+1)/2}\,\frac{\sqrt{G(r)} + \sqrt{rF(r)} + r\sqrt{H(r)}}{\int_{\partial B_r} u u_\nu\,|y|^a\,d\sigma} \\ &\geqslant \frac{c_0}{1+c_0 r} - cr^{(n+a+1)/2}\,\frac{\sqrt{G(r)} + \sqrt{rF(r)} + r\sqrt{H(r)}}{H(r) - r^{(n+a+2)/2}\sqrt{G(r)}}. \end{aligned}$$

First we estimate F, G, H. Since $F(r) > r^{n+4+a}$, from the Poincaré Lemma 3.17 we have:

$$r^{n+4+a} < F(r) \leqslant CrH(r) + c(a,n)\,r^{6+a+n}.$$

Integrating the above inequalities in r, we get:

$$G(r) = \int_0^r F(s)\,ds \leqslant Cr^2 H(r) + c(a,n)\,r^{7+a+n}.$$

This means that, for small enough r_0 and $r < r_0$:

$$F(r) \leqslant crH(r) \quad \text{and} \quad G(r) \leqslant Cr^2 H(r)$$

so that:

$$Q(r) \geqslant \frac{c_0}{1+cr} - cr^{(n+a+1)/2} \frac{r\sqrt{H(r)}}{H(r) - r^{(n+a+4)/2}\sqrt{H_r}}.$$

Since $r^{n+4+a} < F(r) \leqslant CrH(r)$, we also have $H(r) \geqslant cr^{n+a+3}$ and for r_0 small:

$$\begin{aligned} Q(r) &\geqslant \frac{c_0}{1+c_0 r} - cr^{(n+a+1)/2} \frac{r}{\sqrt{H(r)} - r^{(n+a+4)/2}} \\ &\geqslant \frac{c_0}{1+c_0 r} - cr^{\frac{n+a+1}{2}+1-\frac{n+a+3}{2}} = \frac{c_0}{1+cr} - c \end{aligned}$$

which is positive if c_0 is large and r_0 small. □

3.2.7 Blow-up sequences and optimal regularity

The optimal regularity of the solution can be obtained by a careful analysis of the possible values of $\Phi(0+)$. When Φ is constant and the obstacle is zero, $[\Phi(0+) - n - a]/2$ represents the degree of homogeneity at the origin. Thus, by a suitable blow-up of the solution, we will be able to classify the possible asymptotic behaviors at the origin, using the results of Section 2.5. The first result is the following.

Theorem 3.20. *Let u be a solution of problem (3.2.4), $u(0,0) = 0$. Then either $\Phi(0+;u) = n+a+2(1+s)$ or $\Phi(0+;u) \geqslant n+a+4$.*

To prove the theorem, guided by the zero obstacle case in [10], a key point is to introduce the following rescaling:

$$u_r(X) = \frac{u(rX)}{d_r} \tag{3.2.7}$$

where

$$d_r = \left(r^{-(n+a)} \int_{\partial B_r} u^2 |y|^a \, d\sigma\right)^{1/2} = \left(r^{-(n+a)} F(r)\right)^{1/2}.$$

Notice that the "natural" rescaling $u(rX)/r^{\mu}$, where $\mu = \Phi(0+;u) - n - a$, would not be appropriate, because on this kind of rescaling we have precise control of its behavior as $r \to 0$ merely from one-side. Rescaling by an average over smaller and smaller balls provides the necessary adjustments for controlling the oscillations of u around the origin.Two things can occur:

$$\liminf_{r\to 0} \frac{d_r}{r^2} \begin{cases} = +\infty & \text{first case} \\ < +\infty & \text{second case.} \end{cases} \tag{3.2.8}$$

The next lemma takes care of the first case.

Lemma 3.21. *Let u be as in Theorem 3.20. Assume that*

$$\liminf_{r\to 0} \frac{d_r}{r^2} = +\infty.$$

Then, there is a sequence $r_k \to 0$ and a function $U_0 : \mathbb{R}^{n+1} \to R$, non identically zero, such that:

1. $u_{r_k} \to U_0$ *in* $W^{1,2}\left(B_{1/2}, |y^a|\right)$
2. $u_{r_k} \to U_0$ *uniformly in* $B_{1/2}$.
3. $\nabla_x u_{r_k} \to \nabla_x U_0$ *uniformly in* $B_{1/2}$.
4. $|y|^a\, \partial_y u_{r_k} \to |y|^a\, \partial_y U_0$ *uniformly in* $B_{1/2}$.

Moreover, U_0 is a solution of system (3.2.3) and its degree of homogeneity is $[\Phi(0+;u) - n - a]/2$.

Proof. First of all, observe that $\|u_r\|_{L^2(\partial B_1,|y|^a)} = 1$ for every r. Using the frequency formula, the Poincaré type Lemma A5 below and the semiconvexity of u_r, one can show that u_r *is bounded in* $W^{1,2}\left(B_{1/2}, |y^a|\right)$.

Now, u_r^+ and u_r^- are subsolutions of the equation

$$L_a w \geqslant -cr\,|y|^a\,|x|$$

and therefore (see [30]) u_r is bounded in $L^\infty\left(B_{3/4}\right)$.

The semiconvexity of u in the x variable gives, for every tangential direction τ,

$$\partial_{\tau\tau} u_r = \frac{r^2}{d_r} u_{\tau\tau}(rX) \geqslant -C\frac{r^2}{d_r}. \tag{3.2.9}$$

From Lemma 3.9 we obtain a subsequence u_{r_k} strongly convergent in $W^{1,2}\left(B_{1/2}, |y^a|\right)$ to some function U_0 as $r_k \searrow 0$. On the other hand (Theorem 3.17) we know that $\Phi(r;u)$ is monotone and converges to $\Phi(0+;u)$ as $r \searrow 0$. We have:

$$\Phi(rs;u) \sim rs\frac{\int_{B_{rs}} |\nabla u(X)|^2\,|y|^a\,dX}{\int_{\partial B_{rs}} u^2\,|y|^a\,dS} + (n+a) = r\frac{\int_{B_r} |\nabla u_s(X)|^2\,|y|^a\,dX}{\int_{\partial B_r} u_s^2\,|y|^a\,dS} + (n+a).$$

We want to set $s = r_k \searrow 0$ and pass to the limit in the above expression to obtain:

$$\Phi(0+;u) - (n+a) = r\frac{\int_{B_r} |\nabla U_0(X)|^2\,|y|^a\,dX}{\int_{\partial B_r} U_0^2\,|y|^a\,dS}. \tag{3.2.10}$$

This is possible since one can show that $\int_{\partial B_r} u_s^2\,|y|^a\,dS \geqslant c > 0$. From (3.2.9), we have

$$\partial_{\tau\tau} u_{r_k} \geqslant -C\frac{r_k^2}{d_{r_k}} \to 0$$

and therefore U_0 is tangentially convex. On the other hand, each u_r satisfies the following conditions:

(a)
$$u_r(x,0) \geqslant 0 \quad \text{in } B_1'.$$

(b)
$$L_a u_r = \frac{r^{2-a}}{d_r} L_a u(rX) = \frac{r^2}{d_r} |y|^a g(rx) \quad \text{in } B_1 \backslash \Lambda(u_r)$$

(c)
$$|L_a u_r| \leqslant \frac{r^2}{d_r} |y|^a |g(rx)| \quad \text{in } B_1.$$

Since
$$\frac{r^2}{d_r} |y|^a |g(rx)| \leqslant c\frac{r^2}{d_r} |y|^a r|x| \to 0 \quad \text{as } r \to 0$$

it follows that U_0 is a solution of the homogeneous problem

$$\begin{cases} U_0(x,0) \geqslant 0 & \text{in } B_1' \\ U_0(x,-y) = U_0(x,y) & \text{in } B_1 \\ L_a U_0 = 0 & \text{in } B_1 \backslash \Lambda \\ L_a U_0 \leqslant 0 & \text{in } B_1\text{, in the sense of distributions} \end{cases}$$

For this problem, the frequency formula holds without any error correction. Thus we conclude that U_0 is homogeneous in $B_{1/2}$ and its degree of homogeneity is exactly $(\Phi(0+;u) - (n+a))/2$. Since it is homogeneous, then it can be extended to $\mathbb{R}^{n+1}$ as a global solution of the homogeneous problem.

Finally, from the a priori estimates in Lemma 3.7, it follows that we can choose r_k so that the sequences u_{r_k}, ∇u_{r_k} and $|y|^a \partial u_{r_k}$ converge uniformly in $B_{1/2}$. □

Proof of Theorem 3.20. In the *first case* of (3.2.8), we use Lemma 3.21 and Theorem 3.15 to find the blow-up profile U_0 and to obtain that the degree of homogeneity of U_0 is $1+s$ or at least 2. Therefore

$$\Phi(0+;u) = \Phi(0+;U_0) = n+a+2(1+s) \quad \text{or} \quad \Phi(0+;u) = \Phi(0+;U_0) \geqslant n+a+4.$$

Consider now the *second case* of (3.2.8).

If $F(r_k;u) < r_k^{n+a+4}$ for some sequence $r_k \to 0$, then, for these values of r_k,

$$\Phi(r_k;u) = (n+a+4)(1+c_0 r_k)$$

so that $\Phi(0+;U_0) = n+a+4$.

On the other hand, assume that $F(r;u) \geqslant r^{n+a+4}$ for r small. Since we are in the second case, for some sequence $r_j \searrow 0$ we have $d_{r_j}/r_j^2 \leqslant C$ so that

$$r_j^{n+a+4} \leqslant F(r_j;u) \leqslant Cr_j^{n+a+4}.$$

Taking logs in the last inequality, we get

$$(n+a+4)\log r_j \leqslant \log F(r_j;u) \leqslant C+(n+a+4)\log r_j.$$

We want to show that $\Phi(0+;u) \geqslant n+a+4$. By contradiction, assume that for small r_j we have $\Phi(r_j;u) \leqslant n+a+4-\varepsilon_0$. Take $r_m < r_j \ll 1$ and write:

$$(n+a+4)(\log r_j - \log r_m) - C \leqslant \log F(r_j;u) - \log F(r_m;u)$$

$$\begin{aligned} &= \int_{r_m}^{r_j} \frac{d}{dr}\log F(s;u)\,ds \leqslant \int_{r_m}^{r_j}\left(r+c_0 r^2\right)^{-1}\Phi(s;u)\,ds \leqslant \int_{r_m}^{r_j} r^{-1}\Phi(r_j;u)\,ds \\ &\leqslant (n+a+4-\varepsilon_0)(\log r_j - \log r_m) \end{aligned}$$

which gives a contradiction if we make $(\log r_j - \log r_m) \to +\infty$. □

From the classification of the homogeneity of a global profile, we may proceed to identify the unique limiting profile U_0, along the whole sequence u_r, modulus rotations. precisely, we have:

Proposition 3.22. *Let u be as in Theorem 3.20. Assume that $\Phi(0+;u) = n+a+2(1+s)$. There is a family of rotations A_r, with respect to x, such that $u_r \circ A_r$ converges to the unique profile U_0 of homogeneity degree $1+s$. More precisely:*

1. $u_r \circ A_r \to U_0$ *in* $W^{1,2}\left(B_{1/2}, |y^a|\right)$.
2. $u_r \circ A_r \to U_0$ *uniformly in* $B_{1/2}$.
3. $\nabla_x(u_r \circ A_r) \to \nabla_x U_0$ *uniformly in* $B_{1/2}$.
4. $|y|^a\,\partial_y(u_r \circ A_r) \to |y|^a\,\partial_y U_0$ *uniformly in* $B_{1/2}$.

We are now ready to prove the optimal $1+s$–decay of u at $(0,0)$. This is done in two steps. First, we control the decay of u in terms of the decay of $F(r;u)$. In turn, the frequency formula provides the precise control of $F(r)$ from above. This is the content of the next two lemmas.

Lemma 3.23. *If*

$$F(r;u) \leqslant cr^{n+a+2(1+\alpha)} \tag{3.2.11}$$

for every $r<1$, then $u(0,0)=0$, $|\nabla u(0,0)|=0$ and u is $C^{1,\alpha}$ at the origin in the sense that

$$|u(X)| \leqslant C_1 |X|^{1+\alpha}$$

for $|X| \leqslant 1/2$ and $C_1 = C_1(C,n,a)$.

Proof. Consider u^+ and u^-, the positive and negative parts of u. We have already noticed that

$$L_a u^+, L_a u^- \geqslant -C|y|^a|x|.$$

For some $r > 0$, let h be the L_a–harmonic replacement of u^+ in B_r. Note that

$$0 = L_a h \leqslant L_a \left(u^+ + C \frac{|X|^2 - r^2}{2(n+a+1)} \right).$$

Hence, by comparison principle

$$h \geqslant u^+ - C' r^2.$$

From (3.2.11) we have:

$$\int_{\partial B_r} h^2 |y|^a \, d\sigma = \int_{\partial B_r} (u^+)^2 |y|^a \, d\sigma \leqslant c r^{n+a+2(1+\alpha)}.$$

Since $w(X) = |y|^a$ is a A_2 weight, from the local L^∞ estimates (see [30]) we conclude

$$\sup_{B_{r/2}} |h(X)| \leqslant C_1 |r|^{1+\alpha}.$$

Then $u^+ \leqslant h + r^2 \leqslant Cr^{1+\alpha}$ in $B_{r/2}$. A similar estimate holds for u^- and the proof is complete. □

Lemma 3.24. *If $\Phi(0+; u) = \mu$ then*

$$F(r; u) \leqslant c r^\mu$$

for any $r < 1$ and $c = c(F(1; u), c_0)$.

Proof. Let $f(r) = \max\left\{F(r), r^{n+a+4}\right\} \geqslant F(r)$. Since Φ is nondecreasing:

$$\mu = \Phi(0+) \leqslant \Phi(r) = \left(r + c_0 r^2\right) \frac{d}{dr} \log f(r)$$

and then

$$\frac{d}{dr} \log f(r) \geqslant \frac{\mu}{(r + c_0 r^2)}.$$

An integration gives:

$$\begin{aligned} \log f(1) - \log f(r) &\geqslant \int_r^1 \frac{\mu}{(s + c_0 s^2)} ds \\ &\geqslant -\mu \log r - \mu \int_r^1 \left[\frac{1}{s} - \frac{\mu}{(s + c_0 s^2)} \right] ds \\ &\geqslant -\mu \log r - C_1 \mu \end{aligned}$$

so that

$$\log f(r) \leqslant \log f(1) + \mu \log r + C_1 \mu.$$

Taking the exponential of the two sides we infer

$$F(r) \leqslant f(r) \leqslant C r^\mu$$

with $C = f(1) e^{C_1 \mu}$. □

The optimal decay of a solution u of (3.2.4) follows easily. Precisely:

Theorem 3.25. *Let u be a solution of (3.2.4), with $u(0,0) = 0$. Then*

$$|u(X)| \leqslant C|X|^{1+s} \sup_{B_1} |u|,$$

where $C = C\left(n, a, \|g\|_{Lip}\right)$.

Proof. From Theorem 3.20, $\mu = \Phi(0+) \geqslant n+a+2(1+s)$ and the conclusion follows from Lemmas 3.23 and 3.24. □

The optimal regularity of the solution of the obstacle problem for the fractional laplacian is now a simple corollary.

Corrolary 3.26. *(Optimal regularity of solutions) Let $\varphi \in C^{2,1}$. Then the solution u of the obstacle problem for the operator $(-\Delta)^s$ belongs to $C^{1,s}(\mathbb{R}^n)$.*

Proof. Using the equivalence between the obstacle problem for $(-\Delta)^s$ and the thin obstacle for L_a, Theorem 3.25 shows that $u - \varphi$ has the right decay at free boundary points. This is enough to prove that $u \in C^{1,s}(\mathbb{R}^n)$. □

Remark 3.27. *Observe that it is not true that the solution of the thin obstacle problem for L_a is $C^{1,s}$ in both variables x and y. It is quite interesting however, that the optimal decay takes place in both variables at a free boundary point. In any case, we have that a solution u of one of the systems (3.2.1) or (3.2.4) belongs to $C^{1,s}(B'_{1/2})$, for every $y_0 \in (0, 1/2)$.*

3.2.8 Nondegenerate case. Lipschitz continuity of the free boundary

In analogy with what happens in the zero obstacle problem, the regularity of the free boundary can be inferred for points around which u has an asymptotic profile corresponding to the optimal homogeneity degree $\Phi(0+;u) = n+a+2(1+s)$. Accordingly, we say that $X_0 \in F(u)$ is *regular or stable* if

$$\mu(X_0) = \Phi(0+;u) = n+a+2(1+s).$$

As always, we refer to the origin ($X_0 = (0,0)$). The strategy to prove regularity of $F(u)$ follows the well known pattern first introduced by Athanasopoulos and Caffarelli in [6] and further developed by Caffarelli in [18]. The first step is to prove that in a neighborhood of $(0,0)$ there is a cone of tangential directions (cone of monotonicity) along which the derivatives of u are nonnegative and have nontrivial growth. In particular, the free boundary is the graph $x_n = f(x_1, \ldots, x_{n-1})$ of a Lipschitz function f.The second step is to prove a boundary Harnack principle assuring the Hölder continuity of the quotient of two nonnegative tangential derivatives. This implies that $F(u)$ is locally a $(n-2)$–dimensional manifold of class $C^{1,\alpha}$.

The following theorem establishes the existence of a cone of monotonicity.

Theorem 3.28. *Assume $\mu < n + a + 4$. Then, there exists a neighborhood B_ρ of the origin and a tangential cone $\Gamma'(\vartheta, e_n) \subset \mathbb{R}^n \times \{0\}$ such that, for every $\tau \in \Gamma'(\vartheta, e_n)$, we have $\partial_\tau u \geqslant 0$ in B_ρ. In particular, the free boundary is the graph $x_n = f(x_1, \ldots, x_{n-1})$ of a Lipschitz function f.*

To prove the theorem we apply the following lemma to a tangential derivative $h = \partial_\tau u_r$, where u_r is the blow-up family (3.2.7) that defines the limiting profile U_0, for r small.

Lemma 3.29. *Let Λ be a subset of $\mathbb{R}^n \times \{0\}$. Assume h is a continuous function with the following properties:*

1. $L_a h \leqslant \gamma |y|^a$ *in* $B_1 \backslash \Lambda$.
2. $h \geqslant 0$ *for* $|y| \geqslant \sigma > 0$, $h = 0$ *on* Λ.
3. $h \geqslant c_0$ *for* $|y| \geqslant \sqrt{1+a}/8n$.
4. $h \geqslant -\omega(\sigma)$ *for* $|y| < \sigma$, *where ω is the modulus of continuity of h.*

There exist $\sigma_0 = \sigma_0(n, a, c_0, \omega)$ and $\gamma_0 = \gamma_0(n, a, c_0, \omega)$ such that, if $\sigma < \sigma_0$ and $\gamma < \gamma_0$, then $h \geqslant 0$ in $B_{1/2}$.

Proof. Suppose by contradiction $X_0 = (x_0, y_0) \in B_{1/2}$ and $h(X_0) < 0$. Let

$$\Omega = \left\{ (x, y) : |x - x_0| < \frac{1}{3}, |y| \geqslant \frac{\sqrt{1+a}}{4n} \right\}$$

and $P(x, y) = |x - x_0|^2 - \frac{n}{a+1} y^2$. Observe that $L_a P = 0$. Define

$$v(X) = h(X) + \delta P(X) - \frac{\gamma}{2(a+1)} y^2.$$

Then:

- $v(X_0) = h(X_0) + \delta P(X_0) - \frac{\gamma}{2(a+1)} y_0^2 < 0$
- $v(X) \geqslant 0$ on Λ
- $L_a v = L_a h + \delta L_a P - \gamma |y|^a \leqslant 0$ outside Λ.

Thus, v must have a negative minimum on $\partial\Omega$. On the other hand, if δ, γ are small enough, then $v \geqslant 0$ on $\partial\Omega$ and we have a contradiction. Therefore $h \geqslant 0$ in $B_{1/2}$. □

Proof of Theorem 3.28. Since $\mu = \Phi(0; u) < n + a + 4$, Theorem 3.20 gives $\mu = n + a + 2(1 + s)$. Moreover, the blow-up sequence u_r converges (modulus subsequences) to the global profile U_0, whose homogeneity degree is $1 + s$ and whose free boundary is flat.

Let us assume that e_n is the normal to the free boundary of U_0. Then

$$\partial_n U_0(x, y) = c\left(\sqrt{x_n^2 + y^2} - x_n\right)^s.$$

For some $\vartheta_0 > 0$, let σ any vector orthogonal to y and x_n such that $|\sigma| < \vartheta_0$. From Theorem 3.16 we know that U_0 is constant in the direction of σ and therefore, if $\tau = e_n + \sigma$, $\partial_\tau U_0 = \partial_n U_0$.

On the other hand, $\nabla_x u_r \to \nabla_x U_0$ uniformly in every compact subset of $\mathbb{R}^{n+1}$. Thus, for every δ_0, there is an r for which

$$|\partial_\tau U_0 - \partial_\tau u_r| \leqslant \delta_0$$

where $\tau = e_n + \sigma$. If we differentiate the equation

$$L_a u_r(X) = \frac{r^2}{d_r}|y|^a g(rX)$$

we get

$$L_a[\partial_\tau u_r(X)] = \frac{r^2}{d_r}|y|^a r\partial_\tau g(rX) \leqslant Cr|y|^a \quad \text{in } B_1 \backslash \Lambda(u_r) \tag{3.2.12}$$

and the right hand side tends to zero as $r \to 0$.

Thus, for r small enough, $\partial_\tau u_r$ satisfies all the hypotheses of Lemma 3.29 and therefore is nonnegative in $B_{1/2}$. This implies that near the origin, the free boundary is a Lipschitz graph. □

3.2.9 Boundary Harnack principles and $C^{1,\alpha}$ regularity of the free boundary

3.2.9.1 Growth control for tangential derivatives

We continue to examine the regularity of $F(u)$ at stable points. As we have seen, at these point we have an exact asymptotic picture and this fact allows us to get a minimal growth for any tangential derivative when u_r is close to the blow-up limit u_0. This is needed in extending the Carleson estimate and the Boundary Harnack principle in our non-homogeneous setting.

First, we need to refine Lemma 3.29.

Lemma 3.30. *Let $\delta_0 = (12n)^{-1/2s}$. There exists $\varepsilon_0 = \varepsilon_0(n, a) > 0$ such that if v is a function satisfying the following properties:*

1. $L_a v(X) \leqslant \varepsilon_0$ *for* $X \in B_1' \times (0, \delta_0)$,
2. $v(X) \geqslant 0$ *for* $X \in B_1' \times (0, \delta_0)$,
3. $v(x, \delta_0) \geqslant \frac{1}{4n}$ *for* $x \in B_1'$,

then

$$v(x, y) \geqslant C|y|^{2s} \quad \text{in } B_{1/2}' \times [0, \delta_0].$$

Proof. Compare v with

$$w(x, y) = \left(1 + \frac{\varepsilon_0}{2}\right) y^2 - \frac{|x - x_0|^2}{n} + y^{2s}.$$

Inside $B_1' \times (0, \delta_0)$, to show that $w \leqslant v$ in $B_1' \times (0, \delta_0)$. □

Corrolary 3.31. *Let u be a solution of (3.2.4) with $|g|, |\nabla g| \leqslant \varepsilon_0$. Let u_0 the usual asymptotic nondegenerate profile and assume that $|\nabla_x u - \nabla u_0| \leqslant \varepsilon_0$.*

Then, if ε_0 is small enough, there exists $c = c(n, a)$ such that

$$u_\tau(X) \geqslant c \operatorname{dist}(X, \Lambda)^{2s}$$

for every $X \in B_{1/2}$ and every tangential direction τ such that $|\tau - e_n| < 1/2$.

Proof. From (3.2.12) and Theorem 3.28, we know that u_τ is positive in $B_{1/2}$. Applying Lemma 3.30 we get

$$u_\tau \geqslant c|y|^{2s} \quad \text{in } B_{1/4}.$$

Let now $X = (x, y) \in B_{1/8}$ and $d = \operatorname{dist}(X, \Lambda)$. Consider the ball $B_{d/2}(X)$. At the top point of this ball, say (x_T, y_T) we have $y_T \geqslant d/2$. Therefore

$$u_\tau(x_T, y_T) \geqslant cd^{2s}.$$

By Harnack's inequality,

$$u_\tau(x, y) \geqslant cu_\tau(x_T, y_T) \geqslant cd^{2s}.$$

□

3.2.9.2 Boundary Harnack

Using the growth control from below provided by Lemma 3.30 it is easy to extend the Carleson estimate to our nonhomogeneous setting.

Lemma 3.32. *(Carleson estimate). Let $D = B_1 \backslash \Lambda$ where $\Lambda \subset \{y = 0\}$. Assume that $\partial\Lambda \cap B_1'$ is given by a Lipschitz $(n-2)$–manifold with Lipschitz constant L. Let $w \geqslant 0$ in D, vanishing on Λ. Assume in addition that:*

1. $|L_a w| \leqslant c\,|y|^a$ in D;
2. *nondegeneracy:*

$$w(X) \geqslant C d_X^{\beta}$$

for some $\beta \in (0,2)$, *where* $d_X = \mathrm{dist}(X, \Lambda)$.

Then, for every $Q \in \Lambda \cap B_{1/2}$ *and r small:*

$$\sup_{B_r(Q)\cap D} w \leqslant C(n,a,L)\, w(A_r(Q)),$$

where $A_r(Q)$ is a point such that $B_{\eta r}(A_r(Q)) \subset B_r(Q) \cap D$ for some η depending only on n and L.

Proof. Let w^* be the harmonic replacement of w in $B_{2r}(Q) \cap D$, r small. Standard arguments (see e.g. [21]) give

$$w^*(X) \leqslant C w^*(A_r(Q)) \qquad \text{in } B_r(Q) \cap D.$$

On the other hand, comparing w with the function $w^*(X) + C\left(|X-Q|^2 - r^2\right)$ we get

$$\left|w^* - w\right| \leqslant c r^2 \quad \text{in } B_{2r}(Q) \cap D.$$

Thus

$$w(X) \leqslant C\left[w(A_r(Q)) + c r^2\right] \quad \text{in } B_r(Q) \cap D.$$

From the nondegeneracy condition we infer $w(A_r(Q)) \geqslant c r^{\beta}$ and since $\beta < 2$, the theorem follows. □

The following theorem expresses a boundary Harnack principle valid in our nonhomogeneous setting.

Theorem 3.33. *(Boundary Harnack principle). Let* $D = B_1 \backslash \Lambda$ *where* $\Lambda \subset \{y = 0\}$. *Let* v, w *positive functions in D satisfying the hypotheses 1 and 2 of Lemma 3.32 and symmetric in y. Then there is a constant* $c = c(n,a,L)$ *such that*

$$\frac{v(X)}{w(X)} \leqslant c \frac{v\left(0,\frac{1}{2}\right)}{w\left(0,\frac{1}{2}\right)} \quad \text{in } B_{1/2}.$$

Moreover, the ratio v/w *is Hölder continuous in* $B_{1/2}$, *uniformly up to* Λ.

Proof. Let us normalize v, w setting $v\left(0,\frac{1}{2}\right) = w\left(0,\frac{1}{2}\right) = 1$. From the Carleson estimate and Harnack inequality, for any $\delta > 0$ we get:

$$v(X) \leqslant C \quad \text{in } B_{3/4}$$

and

$$w(X) \geqslant c \quad \text{in } B_{3/4} \cap \{|y| > \delta\}.$$

This implies that, for a constant s small enough, $v - sw$ fulfills the conditions of Lemma 3.28. Therefore $v - sw \geqslant 0$ in $B_{1/2}$ or, in other words:

$$\frac{v(X)}{w(X)} \leqslant s \quad \text{in } B_{1/2}.$$

At this point, the rest of proof follows by standard iteration. □

3.2.9.3 $C^{1,\alpha}$ regularity of the free boundary

As in the case of the thin-obstacle for the Laplace operator, the $C^{1,\alpha}$ regularity of the free boundary follows by applying Theorem 3.33 to the quotient of two positive tangential derivatives. Precisely we have:

Theorem 3.34. *Let u be a solution of (3.2.4). Assume $\varphi \in C^{2,1}$ and $\Phi(0) < n + a + 4$. Then the free boundary is a $C^{1,\alpha}$ $(n-1)$ –dimensional surface around the origin.*

Remark 3.35. *As we have already noted, the boundary Harnack principle in Theorem 3.33 is somewhat weaker than the usual one. Notice the* less than quadratic *decay to zero of the solution at the boundary, necessary to control the effect of the right hand side.*

3.2.10 Non regular points on the free boundary

As we have seen, the regularity of the free boundary can be achieved around *regular or stable* points, corresponding to the optimal homogeneity $[\Phi(0+; u) - n - (a+4)]/2 = 1 + s$.

On the other hand, there are solutions of the zero-thin obstacle problem like

$$\rho^{k+1/2} \cos \frac{2k+1}{2} \vartheta \quad \text{or} \quad \rho^{2k} \cos 2k\vartheta,\, k \geqslant 2,$$

vanishing of higher order at the origin. In these cases we cannot expect any regularity of the free boundary.

The non regular points of the free boundary can be divided in two classes: the set $\Sigma(u)$ at which $\Lambda(u)$ has a vanishing density (*singular points*), that is

$$\Sigma(u) = \left\{ (x_0, 0) \in F(u) : \lim_{r \to 0+} \frac{\mathcal{H}^n\left(\Lambda(u) \cap B'_r(x_0)\right)}{r^n} = 0 \right\},$$

and the set of *non regular, non singular points*. The following example (see [37]), given by the harmonic polynomial

$$p(x_1, x_2, y) = x_1^2 x_2^2 - (x_1^2 + x_2^2) y^2 + \frac{1}{3} y^4$$

shows that the entire free boundary of the zero-thin obstacle problem (2) could be composed by singular points. In fact $F(p) = \Lambda(p)$ is given by the union of the lines $x_1 = y = 0$ and $x_2 = y = 0$.

We shall see that, as in this example, the singular set is contained in the union of C^1–manifold of suitable dimension ([37]).

3.2.10.1 Structure of the singular set (zero thin obstacle)

In this section we describe the main results and ideas from ([37]). Their techniques works also for the fractional Laplacian but, for the sake of simplicity, we present them in the (important case) of the thin obstacle problem. Consider the following problem:

$$\begin{cases} u - \varphi \geqslant 0 & \text{in } B_1' \\ \Delta u = 0 & \text{in } B_1 \setminus \{(x,0) : u(x,0) = \varphi(x)\} \\ (u-\varphi)u_{x_n} = 0,\ u_{x_n} \leqslant 0 & \text{in } B_1' \\ u(x,-y) = u(x,y) & \text{in } B_1. \end{cases}$$

where $\varphi : B' \longrightarrow \mathbb{R}$. For better clarity we outline the proofs in the special case $\varphi = 0$.

As we shall see, around the singular points a precise analysis of the behavior of u and the structure of the free boundary can be carried out. The analysis of the free boundary around the other kind of points is still an open question in general. However, in some important special cases, complete information can be given, as we shall see in the sequel.

It is convenient to classify a point on $F(u)$ according to the degree of homogeneity of u, given by the frequency formula centered at that point. In other words, set

$$\Phi^{X_0}(r;u) = r\frac{\int_{B_r(X_0)} |\nabla u|^2}{\int_{\partial B_r(X_0)} u^2 dS}$$

and define

$$\begin{aligned} F_\kappa(u) &= \left\{X_0 \in F(u) : \Phi^{X_0}(0+;u) = \kappa\right\}, \\ \Sigma_\kappa(u) &= \Sigma(u) \cap F_\kappa(u). \end{aligned}$$

According to these notations, X_0 is a regular point if it belongs to $F_{3/2}(u)$. Since $r \longmapsto \Phi^{X_0}(r;u)$ is nondecreasing, it follows that the mapping

$$X_0 \longmapsto \Phi^{X_0}(0+;u)$$

is upper-semicontinuous. Moreover, since $\Phi^{X_0}(0+;u)$ misses all the values in the interval $(3/2, 2)$, it follows that $F_{3/2}(u)$ is a relatively open subset of $F(u)$.

Before stating the structure theorems of $\Sigma(u)$, it is necessary to examine the asymptotic profiles obtained at a singular point from the rescalings $v_r(X) =$

$v(rX)/(r^{-n}\int_{\partial B_r(X_0)} u^2)^{1/2}$; indeed, saying that $X_0 = (x_0, y) \in \Sigma(u)$ is equivalent to state that

$$\lim_{r\to 0+} \mathcal{H}^n\left(\Lambda(v_r) \cap B_1'(x_0)\right) = 0. \tag{3.2.13}$$

As we see immediately, this implies that any blow-up v^* at a singular point is harmonic in $B_1(X_0)$. Moreover, it is possible to give a complete characterization of these blow-ups in terms of the value $\kappa = \Phi^{X_0}(r; u)$. In particular

$$\Sigma_\kappa(u) = F_\kappa(u) \quad \text{for } \kappa = 2m,\ m \in \mathbb{N}.$$

Theorem 3.36. *(Blow-ups at singular points). Let $(0,0) \in F_\kappa(u)$. The following statements are equivalent:*

(i) $(0,0) \in \Sigma_\kappa(u)$.

(ii) Any blow-up of u at the origin is a non zero homogeneous polynomial p_κ of degree κ satisfying

$$\Delta p_\kappa = 0, \quad p_\kappa(x,0) \geqslant 0, \quad p_\kappa(x,-y) = p_\kappa(x,y).$$

(iii) $\kappa = 2m$ for some $m \in \mathbb{N}$.

Proof. *(i)* $\Longrightarrow$ *(ii)*. Since u is harmonic in $B_1^{\pm}$, we have:

$$\Delta v_r = 2(\partial_y v_r)\mathcal{H}^n_{|\Lambda(v_r)} \quad \text{in } \mathcal{D}'(B_1). \tag{3.2.14}$$

Then v_r is equibounded in $H^1_{loc}(B_1)$, and (3.2.13) says that

$$\mathcal{H}^n\left(\Lambda(v_r) \cap B_1'\right) \to 0$$

as $r \to 0$. Thus (3.2.14) implies that $\Delta v_r \to 0$ in $\mathcal{D}'(B_1)$, and therefore any blow-up v^* must be harmonic in B_1.

From Section 2.7 we know that v^* is homogeneous and non trivial, and thus it can be extended to a harmonic function in all of $\mathbb{R}^{n+1}$. Being homogeneous, v^* has at most a polynomial growth at infinity, hence Liouville Theorem implies that v^* is a non trivial homogeneous harmonic polynomial p_κ of integer degree κ. The properties of u imply that $p_\kappa(x,0) \geqslant 0$, and $p_\kappa(x,-y) = p_\kappa(x,y)$ in $\mathbb{R}^{n+1}$.

(ii) $\Longrightarrow$ *(iii)*. We must show that κ is an even integer. If κ is odd, the nonnegativity of p_κ on $y = 0$ implies that p_κ vanishes on the hyperplane $y = 0$. On the other hand, from the even symmetry in y we infer that $\partial_y p_\kappa(x,0) \equiv 0$ in $\mathbb{R}^n$. Since p_κ is harmonic, the Cauchy-Kowaleskaya Theorem implies that $p_\kappa \equiv 0$ in $\mathbb{R}^{n+1}$. Thus $\kappa = 2m$, for some $m \in \mathbb{N}$.

(ii) $\Longrightarrow$ *(i)*. Suppose $(0,0)$ is not a singular point. Then, there exists a sequence $r_j \to 0$ such that

$$\mathcal{H}^n\left(\Lambda(v_r) \cap B_1'\right) \geqslant \delta > 0.$$

We may assume that v_{r_j} converges to a blow-up p^*. We claim that

$$\mathcal{H}^n\left(\Lambda\left(p^*\right)\cap B_1'\right)\geqslant\delta>0.$$

Indeed, otherwise, there exists an open set $U\subset\mathbb{R}^n$ with $\mathcal{H}^n(U)<\delta$ such that $\Lambda(p^*)\cap\overline{B}_1'\subset U$. Then, for j large, we must have $\Lambda(v_r)\cap\overline{B}_1'\subset U$ which is a contradiction, since $\mathcal{H}^n(\Lambda(v_{r_j})\cap\overline{B}_1')\geqslant\delta>\mathcal{H}^n(U)$.

This implies that $p^*(x,0)\equiv 0$ in $\mathbb{R}^n$ and consequently in $\mathbb{R}^{n+1}$, by the Cauchy-Kowaleskaya theorem. Contradiction to (ii).

$(iii)\implies(ii)$. From Almgren's formula, any blow-up is a κ–homogeneous solution of the zero thin obstacle problem in $\mathbb{R}^{n+1}$. Then $\Delta v=2v_y\mathcal{H}^n_{|\Lambda(v)}$ in $\mathbb{R}^{n+1}$, with $v_y\leqslant 0$ on $y=0$. Since $\kappa=2m$, the following auxiliary lemma implies that $\Delta v=0$ in $\mathbb{R}^{n+1}$ and therefore v is a polynomial. □

Lemma 3.37. *Let $v\in H^1_{loc}(\mathbb{R}^{n+1})$ satisfy $\Delta v\leqslant 0$ in $\mathbb{R}^{n+1}$ and $\Delta v=0$ in $\mathbb{R}^{n+1}\setminus\{y=0\}$. If v is homogeneous of degree $\kappa=2m$ then $\Delta v=0$ in $\mathbb{R}^{n+1}$.*

Proof. By assumption, $\mu=\Delta v$ is a nonpositive measure, supported on $\{y=0\}$. We have to show that $\mu=0$.

Let q be a $2m$–homogeneous harmonic polynomial, which is positive on $\{y=0\}\setminus(0,0)$. For instance:

$$q(X)=\sum_{j=1}^{n}\operatorname{Re}(x_j+iy)^{2m}.$$

Take $\psi\in C_0^\infty(0,+\infty)$ such that $\psi\geqslant 0$ and let $\Psi(X)=\psi(|X|)$. Then, we have:

$$\begin{aligned}
-\langle\mu,\Psi q\rangle &= -\langle\Delta v,\Psi q\rangle=\int_{\mathbb{R}^{n+1}}(\Psi\nabla v\cdot\nabla q+q\nabla v\cdot\nabla\Psi)\,dX\\
&=\int_{\mathbb{R}^{n+1}}(-\Psi v\Delta q-v\nabla q\cdot\nabla\Psi+q\nabla v\cdot\nabla\Psi)\,dX\\
&=\int_{\mathbb{R}^{n+1}}\left[-\Psi v\Delta q-v\frac{\psi'(|X|)}{|X|}(X\cdot\nabla q)+q\frac{\psi'(|X|)}{|X|}(X\cdot\nabla v)\right]dX\\
&=0
\end{aligned}$$

since $\Delta q=0$, $X\cdot\nabla q=2mq$, $X\cdot\nabla v=2mv$. This implies that μ is supported at $X=0$, that is $\mu=c\delta_{(0,0)}$. On the other hand, $\delta_{(0,0)}$ is homogeneous of degree $-(n+1)$ while μ is homogeneous of degree $2m-2$ and therefore $\mu=0$. □

Definition 3.38. *We denote by P_κ the class of homogeneous harmonic polynomials of degree $\kappa=2m$ defined in Theorem 3.36, that is:*

$$P_\kappa=\{p_\kappa:\Delta p_\kappa=0,\nabla p_\kappa\cdot X=\kappa p_\kappa,p_\kappa(x,0)\geqslant 0,p_\kappa(x,-y)=p_\kappa(x,y)\}. \tag{3.2.15}$$

Via the Cauchy-Kovaleskaya Theorem, it is easily shown that the polynomials in P_κ can be uniquely determined from their restriction to the hyperplane $y = 0$. Thus, if $p_\kappa \in P_\kappa$ is not trivial, then also its restriction to $y = 0$ must be non trivial.

The next theorem gives an exact asymptotic behavior of u near a point $X_0 \in \Sigma_\kappa(u)$.

Theorem 3.39. *($\kappa-$ differentiability at singular points) Let $X_0 \in \Sigma_\kappa(u)$, with $\kappa = 2m$, $m \in N$. Then there exists a non trivial $p_\kappa^{X_0} \in P_\kappa$ such that*

$$u(X) = p_\kappa^{X_0}(X - X_0) + o\left(|X - X_0|^\kappa\right). \tag{3.2.16}$$

Moreover, the mapping $X_0 \longmapsto p_\kappa^{X_0}$ is continuous on $\Sigma_\kappa(u)$.

The proof is given in Subsection 3.2.10.3. Note that, since P_κ is a convex subset of the finite dimensional space of the homogeneous polynomial of degree κ, all the norms on P_κ are equivalent. Thus, the continuity in Theorem 3.39 can be understood, for instance, in the $L^2(\partial B_1)$ norm.

The structure of $F(u)$ around a singular point X_0 depends on the *dimension* of the singular set at that point, as defined below in terms of the polynomial $p_\kappa^{X_0}$:

Definition 3.40. *(Dimension at a singular point) Let $X_0 \in \Sigma_\kappa(u)$. The dimension of $\Sigma_\kappa(u)$ at X_0 is defined as the integer*

$$d_\kappa^{X_0} = \dim\left\{\xi \in \mathbb{R}^n : \xi \cdot \nabla_x p_\kappa^{X_0}(x, 0) = 0 \text{ for all } x \in \mathbb{R}^n\right\}.$$

Since $p_\kappa^{X_0}(x, 0)$ is not identically zero on $\mathbb{R}^n$, we have

$$0 \leqslant d_\kappa^{X_0} \leqslant n - 1.$$

For $d = 0, 1, \dots, n-1$ we define

$$\Sigma_\kappa^d(u) = \left\{X_0 \in \Sigma_\kappa(u) : d_\kappa^{X_0} = d\right\}.$$

Here is the structure theorem:

Theorem 3.41. *(Structure of the singular set) Every set $\Sigma_\kappa^d(u)$, $\kappa = 2m$, $m \in N$, $d = 0, 1, \dots, n-1$, is contained in a countable union of $d-$ dimensional C^1 manifolds.*

For the harmonic polynomial

$$p(x_1, x_2, y) = x_1^2 x_2^2 - (x_1^2 + x_2^2)y^2 + \frac{1}{3}y^4$$

considered above, it is easy to check that $(0, 0) \in \Sigma_4^0(u)$ and the rest of the points on $F(u)$ belongs to $\Sigma_2^1(u)$ (see Figure 3.1).

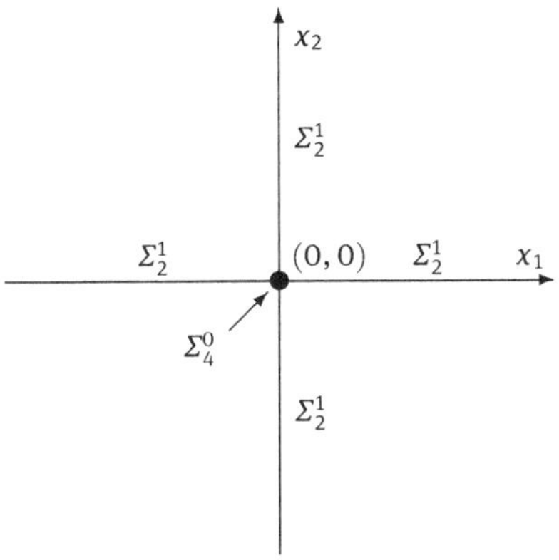

Fig. 3.1: Free boundary for $u(x_1, x_2, y) = x_1^2 x_2^2 - (x_1^2 + x_2^2)\, y^2 + \frac{1}{3} y^4$ in $\mathbb{R}^3$ with zero thin obstacle on $\mathbb{R}^2 \times \{0\}$. (Credit: Garofalo-Petrosyan, 2009)

As in the classical obstacle problem, the main difficulty in the analysis consists in establishing the uniqueness of the Taylor expansion (3.2.16), which in turn is equivalent to establish the uniqueness of the limiting profile obtained by the sequence of rescalings u_r.

A couple of monotonicity formulas, strictly related to Almgren's formula and to formulas in [66] and [54], play a crucial role in circumventing these difficulties.

3.2.10.2 Monotonicity formulas

We introduce here two main tools. We start with a one-parameter family of monotonicity formulas (see also citeW) based on the functional:

$$W_\kappa^{X_0}(r; u) = \frac{1}{r^{n-1+2\kappa}} \int_{B_r(X_0)} |\nabla u|^2 \, dX - \frac{\kappa}{r^{n+2\kappa}} \int_{\partial B_r(X_0)} u^2 \, dS$$
$$\equiv \frac{1}{r^{n-1+2\kappa}} H(r) - \frac{\kappa}{r^{n+2\kappa}} K(r) \,.$$

where $\kappa \geqslant 0$. If $X_0 = (0,0)$ we simply write $W_\kappa(r; u)$.

The functionals $W_\kappa^{X_0}(r; u)$ and $\Phi^{X_0}(r; u)$ are strictly related. Indeed, taking for brevity $X_0 = (0,0)$, we have:

$$W_\kappa(r; u) = \frac{K(r)}{r^{n+2\kappa}} \left[\Phi(r; u) - \kappa\right] . \tag{3.2.17}$$

This formula shows that $W_\kappa^{X_0}(r; u)$ is particularly suited for the analysis of asymptotic profiles at points $X_0 \in F_\kappa(u)$. Moreover, for these points, since from Almgren's frequency formula we have $\Phi^{X_0}(r; u) \geqslant \Phi^{X_0}(0+; u) = \kappa$, we deduce that

$$W_\kappa^{X_0}(r; u) \geqslant 0. \tag{3.2.18}$$

The next theorem shows the main properties of $W_\kappa(r;u)$.

Theorem 3.42. *(W–type monotonicity formula) Let u be a solution of our zero obstacle problem in B_1. Then, for $0 < r < 1$,*

$$\frac{d}{dr}W_\kappa(r;u) = \frac{1}{r^{n+2\kappa}}\int_{\partial B_r}(X\cdot\nabla u - \kappa u)^2 dS.$$

As a consequence, the function $r \longmapsto W_\kappa(r;u)$ is nondecreasing in $(0,1)$. Moreover, $W_\kappa(\cdot;u)$ is constant if and only if u is homogeneous of degree κ.

Proof. Using the identities

$$H'(r) = \int_{\partial B_r}|\nabla u|^2\, dS \text{ and } K'(r) = \frac{n}{r}K(r) + 2\int_{\partial B_r} uu_\nu dS \tag{3.2.19}$$

and

$$\int_{\partial B_r}|\nabla u|^2\, dS = \frac{(n-1)}{r}\int_{B_r}|\nabla u|^2 + 2\int_{\partial B_r}(u_\nu)^2 dS. \tag{3.2.20}$$

we get:

$$\begin{aligned}
&\frac{d}{dr}W_\kappa(r;u)\\
&= \frac{1}{r^{n-1+2\kappa}}\left\{H'(r) - \frac{n-1+2\kappa}{r}V(r) - \frac{\kappa}{r}K'(r) + \frac{\kappa(n+2\kappa)}{r^2}K(r)\right\}\\
&= \frac{1}{r^{n-1+2\kappa}}\left\{2\int_{\partial B_r}u_\nu^2 dS - \frac{4\kappa}{r}\int_{\partial B_r}uu_\nu dS + \frac{2\kappa^2}{r^2}\int_{\partial B_r}u^2 dS\right\}\\
&= \frac{2}{r^{n+1+2\kappa}}\int_{\partial B_r}(X\cdot\nabla u - \kappa u)^2 dS.
\end{aligned}$$

□

The next one is a generalization of a formula in [54], based on the functional

$$M_\kappa^{X_0}(r;u,p_\kappa) = \frac{1}{r^{n+2\kappa}}\int_{\partial B_r(X_0)}(u-p_\kappa)^2 dS.$$

We set $M_\kappa(r;u,p_\kappa) = M_\kappa^{(0,0)}(r;u,p_\kappa)$. Here $\kappa = 2m$ and p_κ is a polynomial in the class P_κ defined in (3.2.15). Since it measures the distance of u from an homogeneous polynomial of even degree, it is apparent that $M_\kappa^{X_0}(r;u,p_\kappa)$ is particularly suited for the analysis of blow-up profiles at points $X_0 \in \Sigma_\kappa(u)$. We have:

Theorem 3.43. *(M–type monotonicity formula) Let u be a solution of our zero obstacle problem in B_1. Assume $(0,0) \in \Sigma_\kappa(u)$, $\kappa = 2m$, $m \in N$. Then, for $0 < r < 1$, the function $r \longmapsto M_\kappa(r;u,p_\kappa)$ is nondecreasing in $(0,1)$.*

Proof. We show that

$$\frac{d}{dr}M_{\kappa}(r;u,p_{\kappa}) \geqslant \frac{2}{r}W_{\kappa}(r;u) \geqslant 0.$$

Let $w = u - p_{\kappa}$. We have:

$$\frac{d}{dr}\frac{1}{r^{n+2\kappa}}\int_{\partial B_r} w^2 dS = \frac{d}{dr}\frac{1}{r^{2\kappa}}\int_{\partial B_1} w(rY)^2 dS$$

$$= \frac{1}{r^{2\kappa+1}}\int_{\partial B_1} [w(rY)[\nabla w(rY)\cdot rY - \kappa w(rY)]dS$$

$$= \frac{2}{r^{n+1+2\kappa}}\int_{\partial B_r} w(X\cdot\nabla w - \kappa w)dS$$

On the other hand, since $\Phi(r;p_{\kappa}) = \kappa$, it follows that

$$W_{\kappa}(r;p_{\kappa}) = 0$$

and we can write:

$$W_{\kappa}(r;u) = W_{\kappa}(r;u) - W_{\kappa}(r;p_{\kappa})$$

$$= \frac{2}{r^{n-1+2\kappa}}\int_{B_r}(|\nabla w|^2 + 2\nabla w\cdot\nabla p_{\kappa})dX - \frac{\kappa}{r^{n+2\kappa}}\int_{\partial B_r}(w^2 + 2wp_{\kappa})dS$$

$$= \frac{2}{r^{n-1+2\kappa}}\int_{B_r}|\nabla w|^2 dX - \frac{\kappa}{r^{n+2\kappa}}\int_{\partial B_r} w^2 dS + \int_{\partial B_r} w(X\cdot\nabla p_{\kappa} - \kappa p_{\kappa})dS$$

$$= \frac{2}{r^{n-1+2\kappa}}\int_{B_r}|\nabla w|^2 dX - \frac{\kappa}{r^{n+2\kappa}}\int_{\partial B_r} w^2 dS$$

$$= -\frac{2}{r^{n-1+2\kappa}}\int_{B_r} w\Delta w dX + \frac{1}{r^{n+2\kappa}}\int_{\partial B_r} w(X\cdot\nabla w - \kappa w)dS$$

$$\leqslant \frac{1}{r^{n+2\kappa}}\int_{\partial B_r} w(X\cdot\nabla w - \kappa w)dS = \frac{r}{2}\frac{d}{dr}M_{\kappa}(r;u,p_{\kappa})$$

since $w\Delta w = (u-p_{\kappa})(\Delta u - \Delta p_{\kappa}) = -p_{\kappa}\Delta u \geqslant 0$. □

3.2.10.3 Proofs of Theorems 3.39 and 3.41

With the above monotonicity formulas at hands we are ready to prove Theorems 3.39 and 3.41. Recall that, from the frequency formula, we have the estimate

$$|u(X)| \leqslant c|X|^{\kappa} \qquad \text{in } B_{2/3} \tag{3.2.21}$$

for any solution of our free boundary problem. At a singular point, we have also a control from below.

Lemma 3.44. *Let u be a solution of our zero obstacle problem in B_1. Assume $(0,0) \in \Sigma_{\kappa}(u)$. Then,*

$$\sup_{\partial B_r}|u| \geqslant cr^{\kappa} \qquad (0 < r < 2/3). \tag{3.2.22}$$

Proof. Suppose that (3.2.22) does not hold. Then, for a sequence $r_j \to 0$ we have

$$h_j = \left(\fint_{\partial B_{r_j}} u^2 dS \right)^{1/2} = o\left(r_j^{\kappa}\right).$$

We may also assume that (see Lemma 3.36)

$$v_j(X) = \frac{u(r_j X)}{h_j} \to q_\kappa(X)$$

uniformly on ∂B_1, for some $q_\kappa \in P_\kappa$. Since $\fint_{\partial B_1} q_\kappa^2 dS = 1$, it follows that q_κ is non trivial.

Under our hypotheses, we have

$$M_\kappa(0+; u, q_\kappa) = \lim_{j\to\infty} \frac{1}{r_j^{n+2\kappa}} \int_{\partial B_{r_j}} (u - q_\kappa)^2 dS = \int_{\partial B_1} q_\kappa^2 dS = \frac{1}{r_j^{n+2\kappa}} \int_{\partial B_{r_j}} q_\kappa^2 dS.$$

Hence,

$$\int_{\partial B_{r_j}} (u - q_\kappa)^2 dS \geqslant \int_{\partial B_{r_j}} q_\kappa^2 dS$$

or

$$\int_{\partial B_{r_j}} (u^2 - 2uq_\kappa) dS \geqslant 0.$$

Rescaling, we obtain

$$h_j r_j^{\kappa} \int_{\partial B_{r_j}} \left(\frac{h_j}{r_j^{\kappa}} v_j^2 - 2v_j q_\kappa\right) dS \geqslant 0.$$

Dividing by $h_j r_j^{\kappa}$ and letting $j \to \infty$ we get

$$-\int_{\partial B_{r_j}} q_\kappa^2 dS \geqslant 0$$

which gives a contradiction, since q_κ is non trivial. □

Given the estimates (3.2.21) and (3.2.22) around a point $X_0 \in \Sigma_\kappa(u)$, it is natural to introduce the family of *homogeneous rescalings* given by

$$u_r^{(\kappa)}(X) = \frac{u(rX + X_0)}{r^\kappa}.$$

From the estimate (3.2.21) we have that, along a sequence $r = r_j$, $u_r^\kappa \to u_0$ in $C^{1,\alpha}_{loc}(\mathbb{R}^n)$. We call u_0 *homogeneous blow-up*. Lemma 3.44 assures that u_0 is non trivial. The next results establishes the uniqueness of these asymptotic profiles and proves the first part of Theorem 3.39.

Theorem 3.45. *(Uniqueness of homogeneous blow-up at singular points) Assume* $(0,0) \in \Sigma_\kappa(u)$. *Then there exists a unique non trivial* $p_\kappa \in P_\kappa$ *such that*

$$u_r^{(\kappa)}(X) = \frac{u(rX)}{r^\kappa} \to p_\kappa(X).$$

As a consequence, (3.2.16) holds.

Proof. Consider a homogeneous blow-up u_0. For any $r > 0$ we have:

$$W_\kappa(r; u_0) = \lim_{r_j \to 0} W_\kappa(r; u_j^{(\kappa)}) = \lim_{r_j \to 0} W_\kappa(rr_j; u) = \lim_{r_j \to 0} W_\kappa(0+; u).$$

From Theorem 3.39 we infer that u_0 is homogeneous of degree κ. The same arguments in the proof of Lemma 3.37 give that u_0 must be a polynomial $p_\kappa \in P_\kappa$.

To prove the uniqueness of u_0, apply the M–monotonicity formula to u and u_0. We have:

$$M_\kappa(0+; u, u_0) = c_n \lim_{j \to \infty} \int_{\partial B_1} (u_j^{(\kappa)} - u_0)^2 dS = 0.$$

In particular, by monotonicity, we obtain also that

$$c_n \int_{\partial B_1} (u_r^{(\kappa)} - u_0)^2 dS = M_\kappa(r; u, u_0) \to 0$$

as $r \to 0$, and not just over a subsequence r_j. Thus, if u_0' is a homogeneous blow-up, obtained over another sequence $r_j' \to 0$, we deduce that

$$\int_{\partial B_1} (u_0' - u_0)^2 dS = 0.$$

Since u_0 and u_0' are both homogeneous of degree κ, they must coincide in $\mathbb{R}^{n+1}$. □

The next lemma gives the second part of Theorem 3.39.

Lemma 3.46. *(Continuous dependence of the blow-ups) For* $X_0 \in \Sigma_\kappa(u)$ *denote by* $p_\kappa^{X_0}$ *the blow-up of* u *obtained in Theorem 3.45 so that:*

$$u(X) = p_\kappa^{X_0}(X - X_0) + o(|X - X_0|^\kappa).$$

Then, the mapping $X_0 \longmapsto p_\kappa^{X_0}$ *from* $\Sigma_\kappa(u)$ *to* P_κ *is continuous. Moreover, for any compact* $K \subset \Sigma_\kappa(u) \cap B_1$, *there exists a modulus of continuity* σ_K, $\sigma_K(0+) = 0$, *such that*

$$\left| u(X) - p_\kappa^{X_0}(X - X_0) \right| \leqslant \sigma_K(|X - X_0|)\, |X - X_0|^\kappa$$

for every $X_0 \in K$.

Proof. As we have already observed, we endow P_κ with the $L^2(\partial B_1)$ norm. The first part of the lemma follows as in Theorem 3.45. Indeed, fix $\varepsilon > 0$ and r_ε such that

$$M_\kappa^{X_0}(r_\varepsilon; u, p_\kappa^{X_0}) = \frac{1}{r_\varepsilon^{n+2\kappa}} \int_{\partial B_{r_\varepsilon}} (u(X + X_0) - p_\kappa^{X_0})^2 dS < \varepsilon.$$

There exists δ_ε such that if $X_0' \in \Sigma_\kappa(u)$ and $|X_0 - X_0'| < \delta_\varepsilon$, then

$$M_\kappa^{X_0'}(r_\varepsilon; u, p_\kappa^{X_0}) = \frac{1}{r_\varepsilon^{n+2\kappa}} \int_{\partial B_{r_\varepsilon}} (u\left(X + X_0'\right) - p_\kappa^{X_0})^2 dS < 2\varepsilon.$$

By monotonicity, we deduce that, for $0 < r < r_\varepsilon$,

$$M_\kappa^{X_0'}(r; u, p_\kappa^{X_0}) = \frac{1}{r^{n+2\kappa}} \int_{\partial B_r} (u\left(X + X_0'\right) - p_\kappa^{X_0})^2 dS < 2\varepsilon.$$

Letting $r \to 0$ we obtain

$$M_\kappa^{X_0'}(0+; u, p_\kappa^{X_0}) = c_n \int_{\partial B_1} (p_\kappa^{X_0'} - p_\kappa^{X_0})^2 dS < 2\varepsilon$$

and therefore the first part of the lemma is proved.

To show the second part, note that if $|X_0 - X_0'| < \delta_\varepsilon$ and $0 < r < r_\varepsilon$, we have:

$$\begin{aligned} \left\| u\left(\cdot + X_0'\right) - p_\kappa^{X_0'} \right\|_{L^2(\partial B_r)} &\leqslant \left\| u\left(\cdot + X_0'\right) - p_\kappa^{X_0} \right\|_{L^2(\partial B_r)} + \left\| p_\kappa^{X_0} - p_\kappa^{X_0'} \right\|_{L^2(\partial B_r)} \\ &\leqslant 2(2\varepsilon)^{1/2} r^{\kappa+(n-1)/2}. \end{aligned}$$

This is equivalent to

$$\left\| w_r^{X_0'} - p_\kappa^{X_0'} \right\|_{L^2(\partial B_1)} \leqslant 2(2\varepsilon)^{1/2} \tag{3.2.23}$$

where

$$w_r^{X_0'}(X) = \frac{u(rX + X_0')}{r^\kappa}.$$

Covering K with a finite number of balls $B_{\delta_\varepsilon(X_0')}(X_0')$ for some $X_0' \in K$, we obtain that (3.2.23) holds for all $X_0' \in K$ with $r \leqslant r_\varepsilon^K$.

We claim that, if $X_0' \in K$ and $0 < r < r_\varepsilon^K$ then

$$\left\| w_r^{X_0'} - p_\kappa^{X_0'} \right\|_{L^\infty(B_{1/2})} \leqslant C_\varepsilon \quad \text{with } C_\varepsilon \to 0 \text{ as } \varepsilon \to 0. \tag{3.2.24}$$

To prove the claim, observe that the two functions $w_r^{X_0'}$ and $p_\kappa^{X_0'}$ are both solution of our zero thin obstacle problem, uniformly bounded in $C^{1,\alpha}(\overline{B}_1^\pm)$. If (3.2.24) were not true, by compactness, we can construct a sequence of solutions converging to a non trivial zero trace solutions (from (3.2.23)). The uniqueness of the solution of the thin obstacle problem with Dirichlet data implies a contradiction.

It is easy to check that the claim implies the second part of the lemma. □

We are now in position to prove Theorem 3.41. The proof uses Whitney's Extension Theorem (see [67] or [68]) and the implicit function theorem. We recall that the extension theorem prescribes the compatibility conditions under which there exists a C^k function f in $\mathbb{R}^N$ having prescribed derivatives up to the order k on a given closed set.

Since our reference set is $\Sigma_\kappa(u)$, we first need to show that $\Sigma_\kappa(u)$ is a *countable union of closed sets* (an F_σ set). This is done in the next Lemma.

Lemma 3.47. *(Topological structure of $\Sigma_\kappa(u)$)* $\Sigma_\kappa(u)$ is a F_σ set.

Proof. Let E_j be the set of points $X_0 \in \Sigma_\kappa(u) \cap \overline{B_{1-1/j}}$ such that

$$\frac{1}{j}\rho^\kappa \leqslant \sup_{|X-X_0|=\rho} |u(X)| < j\rho^\kappa \tag{3.2.25}$$

for $0 < \rho < 1 - |X_0|$. By non degeneracy and (3.2.21) we know that

$$\Sigma_\kappa(u) \subset \cup_{j\geqslant 1} E_j.$$

We want to show that E_j is a closed set. Indeed, if $X_0 \in \overline{E}_j$ then X_0 satisfies (3.2.25), and we only need to show that $X_0 \in \Sigma_\kappa(u)$, i.e., from Theorem 2.9.1, that $\Phi^{X_0}(0+;u) = \kappa$.

Since the function $X \mapsto \Phi^X(0+;u)$ is upper-semicontinuous we deduce that $\Phi^{X_0}(0+;u) = \kappa' \geqslant \kappa$. If we had $\kappa' > \kappa$, we would have

$$|u(X)| \leqslant |X - X_0|^{\kappa'} \quad \text{in } B_{1-|X_0|}(X_0),$$

which contradicts the estimate from below in (3.2.25). Thus $\kappa' = \kappa$ and $X_0 \in \Sigma_\kappa(u)$. □

We are now in position for the proof of Theorem 3.41.

Proof of Theorem 3.41. We divide the proof into two steps. Recall that $\Sigma_\kappa(u) = F_\kappa(u)$ if $\kappa = 2m$.

Step 1. Whitney's extension. For simplicity it is better to make a slight change of notations, letting $y = x_{n+1}$ and $X = (x_1, \ldots, x_n, x_{n+1})$. Let $K = E_j$ be one of the compact subsets of $\Sigma_\kappa(u)$ constructed in Lemma 3.47. We can write

$$p_\kappa^{X_0}(X) = \sum_{|\alpha|=\kappa} \frac{a_\alpha(X_0)}{\alpha!} X^\alpha.$$

The coefficients $a_\alpha(X)$ are continuous on $\Sigma_\kappa(u)$ by Theorem 3.39. Since $u = 0$ on $\Sigma_\kappa(u)$ we have

$$\left|p_\kappa^{X_0}(X - X_0)\right| \leqslant \sigma(|X - X_0|)\,|X - X_0|^\kappa \qquad X \in K.$$

For every multi-index α, $0 \leqslant |\alpha| \leqslant \kappa$, define:

$$f_\alpha(X) = \begin{cases} 0 & \text{if } 0 < |\alpha| < \kappa \\ a_\alpha(X) & \text{if } |\alpha| = \kappa \end{cases} \qquad X \in \Sigma_\kappa(u).$$

We want to construct a function $f \in C^{\kappa}\left(\mathbb{R}^{n+1}\right)$, whose derivatives $\partial^{\alpha} f$ up to the order κ are prescribed and equal to f_{α} on K. The Whitney extension theorem states that this is possible if, for all $X, X_0 \in K$, the following coherence conditions hold for every multi-index α, $0 \leqslant |\alpha| \leqslant \kappa$:

$$f_{\alpha}(X) = \sum_{|\beta| \leqslant \kappa - |\alpha|} \frac{f_{\alpha+\beta}(X_0)}{\beta!}(X - X_0)^{\beta} + R_{\alpha}(X, X_0) \tag{3.2.26}$$

with

$$|R_{\alpha}(X, X_0)| \leqslant \sigma_{\alpha}^{K}(|X - X_0|)\,|X - X_0|^{\kappa - |\alpha|}, \tag{3.2.27}$$

where σ_{α}^{K} is a modulus of continuity.

Claim: (3.2.26) and (3.2.27) *hold in our case.*

Proof. Case $|\alpha| = \kappa$. Then we have

$$R_{\alpha}(X, X_0) = a_{\alpha}(X) - a_{\alpha}(X_0)$$

and therefore $|R_{\alpha}(X, X_0)| \leqslant \sigma_{\alpha}(|X - X_0|)$ by the continuity on K of the map $X \mapsto p_{\kappa}^{X}$.

Case $0 \leqslant |\alpha| < \kappa$. We have

$$R_0(X, X_0) = -\sum_{\gamma > \alpha, |\gamma| = \kappa} \frac{a_{\gamma}(X_0)}{(\gamma - \alpha)!}(X - X_0)^{\gamma - \alpha} = -\partial^{\alpha} p_{\kappa}^{X_0}(X - X_0).$$

Now, suppose that there exists no modulus of continuity σ_{α} such that (3.2.27) holds for all $X, X_0 \in K$. Then, there is $\delta > 0$ and sequences $X^i, X_0^i \in K$ with

$$\left|X^i - X_0^i\right| = \rho_i \to 0$$

and such that

$$\left|\partial^{\alpha} p_{\kappa}^{X_0}(X - X_0)\right| \geqslant \delta \left|X^i - X_0^i\right|^{\kappa - |\alpha|}. \tag{3.2.28}$$

Consider the rescalings

$$w^i(X) = \frac{u\left(X_0^i + \rho_i X\right)}{\rho_i^{\kappa}}, \qquad \xi^i = \frac{X^i - X_0^i}{\rho_i}.$$

We may assume that $X_0^i \to X_0 \in K$ and $\xi^i \to \xi_0 \in \partial B_1$. From Lemma 3.46 we have that

$$\left|w^i(X) - p_{\kappa}^{X_0^i}(X)\right| \leqslant \sigma(\rho_i |X|)\,|X|^{\kappa}$$

and therefore $w^i(X)$ converges to $p_{\kappa}^{X_0^i}(X)$, uniformly in every compact subset of $\mathbb{R}^{n+1}$.

Note that, since $X^i, X_0^i \in K$, the inequalities (3.2.25) are satisfied there. Moreover, we also have that similar inequalities are satisfied for the rescaled function w^i at 0 and ξ^i.

Thus, passing to the limit, we deduce that

$$\frac{1}{j}\rho^{\kappa} \leqslant \sup_{|X-X_0|=\rho} \left|p_{\kappa}^{X_0}(X)\right| < j\rho^{\kappa}, \qquad 0 < \rho < +\infty.$$

This implies that ξ_0 is a point of frequency $\kappa = 2m$ for the polynomial $p_{\kappa}^{X_0}$ so that, from Theorem 3.39, we infer that $\xi_0 \in \Sigma_{\kappa}(p_{\kappa}^{\xi_0})$. In particular,

$$\partial^{\alpha} p_{\kappa}^{\xi_0} = 0 \qquad \text{for } |\alpha| < \kappa.$$

However, dividing both sides of (3.2.28) by $\rho_i^{\kappa-|\alpha|}$ and passing to the limit, we obtain

$$\left|\partial^{\alpha} p_{\kappa}^{\xi_0}\right| \geqslant \delta,$$

a contradiction.

This ends the proof of the claim.

Step 2. Implicit function theorem. Applying Whitney's Theorem we deduce the existence of a function $f \in C^{\kappa}\left(\mathbb{R}^{n+1}\right)$ such that

$$\partial^{\alpha} f = f_{\alpha} \quad \text{on } E_j$$

for every $|\alpha| \leqslant \kappa$. Suppose now $X_0 \in \Sigma_{\kappa}^{d}(u)$. This means that

$$d = \dim\left\{\xi \in \mathbb{R}^n : \xi \cdot \nabla_X p_{\kappa}^{X_0}(x,0) \equiv 0\right\}.$$

Then there are $n-d$ linearly independent unit vectors $\nu_i \in \mathbb{R}^n$, such that

$$\nu_i \cdot \nabla_X p_{\kappa}^{X_0}(x,0) \text{ is not identically zero.}$$

This implies that there exist multi-indices β^i of order $\left|\beta^i\right| = \kappa - 1$ such that

$$\partial_{\nu_i}(\partial^{\beta^i} p_{\kappa}^{X_0}(X_0)) \neq 0.$$

This can be written as

$$\partial_{\nu_i}(\partial^{\beta^i} f(X_0)) \neq 0, \qquad i = 1, \ldots, n-d. \tag{3.2.29}$$

On the other hand, we have

$$\Sigma_{\kappa}^{d}(u) \cap E_j \subset \cup_{i=1,\ldots,n-d}\left\{\partial^{\beta^i} f = 0\right\}.$$

From (3.2.29) and the implicit function theorem, we deduce that $\Sigma_{\kappa}^{d}(u) \cap E_j$ is contained in a $d-$dimensional C^1 manifold in a neighborhood of X_0. Since $\Sigma_{\kappa}(u) = \cup E_j$ we conclude the proof. ▫

3.2.11 Non zero obstacle

The above differentiability and the structure theorems can be extended to the non zero obstacle case, if $\varphi \in C^{k,1}(B_1')$ for some integer $k \geqslant 2$ ([37]). A crucial tool is once more a frequency formula which generalizes the one in Theorem 3.17. First, we introduce a useful change of variable to reduce to the case in which the Laplacian is very small outside the coincidence set. Let u be a solution of the obstacle problem and set:

$$v(x,y) = u(x,y) - \tilde{Q}_k(x,y) - (\varphi(x) - Q_k(x)),$$

where Q_k is the k-th Taylor polynomial of φ at the origin and $\tilde{Q}_k$ is its even harmonic extension to all of $\mathbb{R}^{n+1}$. Now v can be evenly extended to $y < 0$ and then

$$|\Delta v| \leqslant M|x|^{k-1} + 2|u_y|\mathcal{H}^n_{|B_1'} \quad \text{in } \mathcal{D}'(B_1).$$

The generalized frequency formula takes the following form: *If* $\|v\|_{C^1(\overline{B_1^+})} \leqslant M/2$, *then there exists* r_M *and* C_M *such that the function*

$$r \longmapsto \Phi_k(r;u) = \left(r + C_M r^2\right)\frac{d}{dr}\log\max\left\{K(r), r^{n+2k}\right\}$$

is nondecreasing for $0 < r < r_M$.

Also the Weiss and Monneau monotonicity formulas have to be modified accordingly to take into account the perturbation introduced by the non-zero obstacle. Indeed we have: *For* $\kappa \leqslant k$, $0 < r < r_M'$, $C_M' > 0$,

$$\frac{d}{dr}W_\kappa(r;u) \geqslant -C_M'$$

and for $\kappa = 2m < k$, $0 < r < r_m''$, $C_M'' > 0$,

$$\frac{d}{dr}M_\kappa(r;u,p_\kappa) \geqslant -C_M''\left(1 + \|p_\kappa\|_{L^2(B_1)}\right).$$

Coherently, the Differentiability Theorem 3.39 and the Structure Theorem 3.41 hold for $\kappa = 2m < k$. This limitation is due to the fact that for points in $F_\kappa(u)$ even the blow-ups are not properly defined.

Thus, the analysis of the free boundary around a singular point is rather satisfactory. It remains open the study of the nonregular, nonsingular points, i.e. the set

$$\Xi(u) = \cup\left\{F_\kappa(u) : \kappa > \frac{3}{2}, \kappa \neq 2m, m \geqslant 1, \text{ integer}\right\}.$$

It should be noted that $\Xi(u)$ could in principle be a large part of $F(u)$. Another important issue that remains open is whether $F(u)$ has Hausdorff dimension $n-1$.

Further analysis of $F(u)$ clearly depends on the possible values that the frequency κ may attain. A partial classification of convex global solutions excludes the interval $(3/2, 2)$ from the range of possible values of κ.

As observed in ([37]) it is plausible that the only possible values are

$$\kappa = 2m - \frac{1}{2} \text{ or } \kappa = 2m, \ m \geqslant 1, \text{ integer}.$$

This is indeed true in dimension 2 ($n = 1$), see remark 1.2.8. in [37].

3.2.12 A global regularity result (fractional Laplacian)

In ([12]), Barrios, Figalli and Ros-Oton consider both the global and the local version of the Signorini problem for the operator L_a, thus covering also the case of the obstacle problem for the fractional Laplacian. We limit ourselves to briefly describe their main results and ideas in the case of the local Signorini problem.

Under basically two main assumptions the authors can give a complete picture of the free boundary, recovering a result completely analogous to the classical case $(s = 1)$. The first assumption is a strict concavity of the obstacle, the same assumption needed in the case of the classical obstacle problem. The second one prescribes zero boundary values of the solution and it turns out to be a crucial assumption. Precisely their main result in the case of the Signorini problem is the following. As in Section 2.7 we define the blow-ups of v at X_0 by

$$v_r^{X_0}(X) = \frac{v(r(X - X_0))}{d_r} \qquad d_r = \left(r^{-n-a} \int_{\partial B_r(X_0)} u^2 |y|^a \, d\sigma\right)^{1/2}. \tag{3.2.30}$$

Theorem 3.48. *Let $\varphi : B_1' \to \mathbb{R}$ with $\varphi_{|\partial B_1'} < 0$, and $u : B_1 \subset \mathbb{R}^{n+1} \to \mathbb{R}$ be a solution of the following problem:*

$$\begin{cases} u(x,0) \geqslant \varphi(x) & \text{on } B_1' \\ L_a u = 0 & \text{in } B_1 \setminus (\{y = 0\} \cap \{u = \varphi\}) \\ L_a u \geqslant 0 & \text{in } B_1 \\ u = 0 & \text{on } \partial B \end{cases}$$

with $u(x,-y) = u(x,y)$. Assume that

$$\varphi \in C^{3,\gamma}\left(B_1'\right), \ \Delta\varphi \leqslant -c_0 < 0 \ \text{in } \{\varphi > 0\}, \ \varnothing \neq \{\varphi > 0\} \subset\subset B_1' \tag{3.2.31}$$

for some $c_0 > 0$ and $\gamma > 0$. Then, at every singular point the blow-up of u is a homogeneous polynomial of degree 2, and the free boundary can be decomposed as

$$F(u) = F_{1+s}(u) \cup F_2(u)$$

where $F_{1+s}(u)$ (resp. $F_2(u)$) is an open (resp. closed) subset of $F(u)$. Moreover, $F_{1+s}(u)$ is a $(n-1)$-dimensional manifold of class $C^{1,\alpha}$, while $F_2(u)$ can be stratified as the union of sets $\left\{F_2^k(u)\right\}_{k=0,1,\dots,n-1}$, where $F_2^k(u)$ is contained in a k-dimensional manifold of class C^1.

The regularity of $F_{1+s}(u)$ comes from [22]. Here the main points are that the nonregular points are only singular points, that the blow-up at these points has homogeneity 2, that $F_2(u)$ can be stratified into C^1 manifolds and that each $F_2^k(u)$ is completely contained in a k-dimensional C^1–manifold.

The key lemma is the following nondegeneracy result. It says that u detaches quadratically from the obstacle putting out of game all possible frequencies greater than 2.

Lemma 3.49. *Let u be as in Theorem 3.48. Then there exists constants $c_1, r_1 > 0$ such that the following holds: for any $X_0 \in F(u)$ we have*

$$\sup_{x \in B_r'(x_0)} \left(u(x,0) - \varphi(x)\right) \geqslant c_1 r^2, \qquad 0 < r < r_1.$$

Proof. Since $u > 0$ on the contact set, compactly contained in B_1, we deduce that

$$\varphi \geqslant h_0 > 0 \quad \text{on } \{u = \varphi\}.$$

For $r_1 > 0$ small, from the properties of φ, in the set

$$U_1 = \{x : u(x,0) > \varphi(x), \operatorname{dist}(x, F(u)) \leqslant 2r_1\} \subset\subset B_1',$$

we have $\varphi > 0$. Take $x_1 \in U_1$, with $\operatorname{dist}(x, F(u)) \leqslant r_1$ and consider the barrier function given by

$$w(x,y) = u(x,y) - \varphi(x) - \frac{c_0}{2n + 2(1+a)}\left(|x - x_1| + y^2\right)$$

where c_0 is as in (3.2.31). Note that $w(x_1, 0) > 0$ and $w < 0$ on $\{u = \varphi\} \cap \{y = 0\}$. We want to apply the maximum principle in the set

$$U = B_r(x_1, 0) \setminus (\{u = \varphi\} \cap \{y = 0\}) \subset\subset B_1.$$

Since $L_a u = 0$ and $\Delta\varphi(x) \leqslant -c_0$, we have

$$L_a w(x,y) = L_a u(x,y) - |y|^a (\Delta\varphi(x) + c_0) \geqslant 0.$$

Noticing that $\partial U = \partial B_r(x_1, 0) \cup \{u = \varphi\} \cap \{y = 0\}$, by the maximum principle we infer that

$$\begin{aligned} 0 &< w(x_1, 0) \leqslant \sup_U w = \sup_{\partial U} w = \sup_{\partial B_r(x_1,0)} w \\ &= \sup_{\partial B_r(x_1,0)} (u - \varphi) - \frac{c_0}{2n + 2(1+a)} r^2. \end{aligned}$$

Letting $x_1 \to x_0$ we get

$$\sup_{B_r(x_0,0)} (u - \varphi) \geqslant \sup_{\partial B_r(x_0,0)} (u - \varphi) \geqslant \frac{c_0}{2n + 2(1+a)} r^2. \tag{3.2.32}$$

To conclude the proof we must show that the above supremum is attained on $y = 0$. To prove it we show that $u_y \leqslant 0$ in B_1^+. Here comes into play the zero boundary condition.

Since $L_a u \geqslant 0$ and $L_a u = 0$ outside the contact set $\{u = \varphi\}$, by the maximum principle it follows that $u \geqslant 0$ in B_1 and u attains its maximum on $\{u = \varphi\}$. Using again that $u = 0$ on ∂B_1 and that $u(x, -y) = u(x, y)$, we deduce that

$$|y|^a u_y \leqslant 0 \quad \text{and} \quad \lim_{y \to 0+} |y|^a u_y(x, y) \leqslant 0 \quad \text{on } \{u = \varphi\}.$$

Moreover

$$\lim_{y \to 0+} |y|^a u_y(x, y) = 0 \quad \text{on } \{u > \varphi\}.$$

Now, by direct computation, one can check that $L_{-a}(y^a u_y) = 0$ in B_1^+. By the maximum principle we infer that $y^a u_y \leqslant 0$ in B_1^+. Thus, $u(x, y)$ is decreasing with respect to y in B_1^+. Since u is even in y, this yields

$$u(x, y) \leqslant u(x, 0) \qquad \text{in } B_1$$

and (3.2.32) finally gives

$$\sup_{x \in B_r'(x_0)} (u(x, 0) - \varphi(x)) = \sup_{B_r(x_0, 0)} (u - \varphi) \geqslant \frac{c_0}{2n + 2(1 + a)} r^2.$$

□

Having the above nondegeneracy at hands, to proceed further once again the main tools are frequency and monotonicity formulas, adapted to take into account the homogeneity of the solution. For a free boundary point $x_0 \in F(u)$, the change of variable

$$v^{x_0}(x, y) = u(x, y) - \varphi(x) + \frac{1}{2(1 + a)} \left\{ \Delta\varphi(x_0) y^2 + \nabla\Delta\varphi(x_0) \cdot (x - x_0) y^2 \right\}$$

reduces to the case

$$\left| L_a v^{x_0} \right| \leqslant C |y|^a |x - x_0|^{1+\gamma} \tag{3.2.33}$$

outside the coincidence set and takes care of the errors in the frequency/monotonicity formulas, due to the presence of a non-zero obstacle. Note that

$$v^{x_0}(x_0, 0) = u(x, 0) - \varphi(x)$$

and that v^{x_0} depends continuously on x_0 since $\varphi \in C^{3,\gamma}$. The non-degeneracy of u translates into

$$\sup_{B_r'(x_0)} v^{x_0}(x, 0) \geqslant cr^2.$$

Moreover, exploiting (3.2.33) and a weak Harnack inequality (see [30]) one also gets, for small r,

$$\int_{B_r(x_0, 0)} |y^a| \left| v^{x_0}(x, y) \right|^2 dxdy \geqslant c_1 r^{n+a+5}. \tag{3.2.34}$$

Setting $x_0 = 0$, $v = v^0$ and

$$K(r) = \int_{\partial B_r(0,0)} \left| y^2 \right| v^2,$$

the function

$$r \longmapsto \Phi(r; v) = \left(r + C_0 r^2\right) \frac{d}{dr} \log \max\left\{K(r), r^{n+a+4+2\gamma}\right\}$$

is monotone non-decreasing for small r, and a suitable constant $C_0 > 0$. Let

$$\Phi(0+; v) = n + a + 2m.$$

Now, the above non-degeneracy estimates implies that either $m = 1 + s$ or $m = 2$. It is clearly enough to show that $m \leqslant 2$. Indeed, the monotonicity of Φ gives $n + a + 2m \leqslant \Phi(r; v)$ and integrating we get

$$K(r) \leqslant Cr^{n+a+2m}.$$

A further integration gives

$$\int_{B_r(0,0)} |y^a| \left|v^{x_0}(x, y)\right|^2 dxdy \leqslant c_1 r^{n+a+2m+1}$$

that together with (3.2.34) implies $m \leqslant 2$. At this point, following the strategy in [22], it is possible to show that, up to a subsequence, the blow-ups v_r in (3.2.30) converge as $r \to 0^+$ to a function, which is nonnegative on $y = 0$ and homogeneous of degree m. Moreover:

a) either there exist $c > 0$ and a unit vector ν such that

$$m = 1 + s \quad \text{and} \quad v_0(x, 0) = c(x \cdot \nu)_+^{1+s}$$

b) or

$$m = 2 \quad \text{and} \quad v_0(x, 0) \text{ is a polynomial of degree 2}$$

and the origin is a singular point.

Uniqueness of the blow-ups can be proved by a suitable modification of the Weiss and Monneau monotonicity formulas. As consequence, *there exists a modulus of continuity $\omega : \mathbb{R}^+ \to \mathbb{R}^+$ such that, for all x_0 belonging to the singular set $F_2(u)$ of the free boundary, we have*

$$u(x) - \varphi(x) = p_2^{x_0}(x - x_0) + \omega(|x - x_0|)\left|x - x_0^2\right| \tag{3.2.35}$$

for some polynomial $p_2^{x_0}(x) = (A^{x_0}x, x)$, $A \in \mathbb{R}^{n\times n}$, symmetric and nonnegative, $A \neq 0$. In addition, the mapping $F_2(u) \ni x_0 \mapsto p_2^{x_0}$ is continuous with

$$\int_{\partial B_1} |y|^a \left(p_2^{x_1} - p_2^{x_0}\right) \leqslant \omega(|x_1 - x_0|) \qquad \forall x_1, x_0 \in F_2(u).$$

Concerning the regularity of $F_2(u)$, given any point x_0 in this set, from (3.2.35), the blow-up of $u - \varphi$ coincides with the polynomial $p_2^{x_0}(x) = (A^{x_0}x, x)$. We stratify $F_2(u)$ according to the dimension of $\ker A$:

$$F_2^k(u) = \left\{x_0 \in F_2(u) : \dim \ker A^{x_0} = k\right\} \quad k = 0, 1, \ldots, n-1.$$

Then, reasoning as in the classical obstacle case (see [18]), one can show that for any $x_0 \in F_2^k(u)$ there exists $r = r_{x_0} > 0$ such that $F_2^k(u) \cap B_r(x_0)$ is contained in a connected k–dimensional C^1 manifold. This concludes the proof of Theorem 3.48.

3.3 Comments and Further Reading

In recent times and while we were writing this survey several important results have been established. We briefly mention some of the more strictly related to our presentation.

• *Different approaches to the regularity of the free boundary.* When the thin manifold is non-flat, by a standard procedure one is lead to a Signorini problem, but for a divergence form operator with variable coefficient matrix $A(x) = [a_{ij}(x)]$. For instance, when $\mathcal{M}$ satisfies the minimal smoothness assumption $C^{1,1} = W^{2,\infty}$, then $A(x)$ is $C^{0,1} = W^{1,\infty}$, i.e., Lipschitz continuous.

The approach described above in the case $A(x) \equiv I_n$ to establish the $C^{1,\alpha}$ regularity of the regular free boundary, based on differentiating the equation for the solution v in tangential directions $e \in \mathbb{R}^{n-1}$ and establishing directional monotonicity, does not work well for variable coefficients, particularly when the obstacle $\varphi \neq 0$. For coefficients $A(x) \in W^{1,\infty}$, and the obstacle $\varphi \in W^{2,\infty}$, the optimal $C^{1,1/2}_{\text{loc}}(\Omega_\pm \cup \mathcal{M})$ regularity of the solutions was established by Garofalo and Smit Vega Garcia in [40] by means of monotonicity formulas of Almgren's type. The $C^{1,\alpha}$ smoothness of the regular free boundary has been obtained by Garofalo, Petrosyan and Smit Vega Garcia in [38]. The two central tools are a Weiss type monotonicity formula and an epiperimetric inequality, which allow to control the homogeneous blow-ups. The latter results, in the special case $A(x) \equiv I_n$, have also been established by Focardi and Spadaro ([32]). One should also see the recent paper by Colombo, Spolaor and Velichkov [25], where they introduce a logarithmic epiperimetric inequality for the 2m-Weiss energy. A different approach, based on Carleman estimates, to the optimal regularity of the solutions and $C^{1,\alpha}$ regularity of the free boundary for $A(x) \in W^{1,p}$, $p > 2n$, and vanishing obstacle is used by Koch, Rüland and Shi in [45] and [46] (more recently the authors were able to extend these results to non-zero obstacles in $W^{2,p}$, $p > 2n$).

• *Higher regularity.* Real analyticity of the regular part of the free boundary for the thin obstacle (Signorini) problem has been proved by Koch, Petrosyan and Shi in [44] via a partial hodograph-Legendre transformation, subsequently extended by Koch, Rüland and Shi to the fractional Laplacian operator in ([47]). The lack of regularity of this map can be overcome by providing a precise asymptotic behavior at a regular free boundary point. The Legendre transforms (on which one reads the regularity of the free boundary) satisfies a subelliptic equation of Baouendi-Grushin type and its analyticity is achieved by using the L^p theory available for this kind of operator. A different approach to higher regularity for the thin obstacle and also to one-phase free boundary problems is due to De Silva and Savin in [28] and is based on a higher order Boundary Harnack principle. Jhaveri and Neumayer in [42] extend this approach to the fractional Laplacian obstacle problem. The higher regularity of the free boundary in the variable coefficient case has been established in [48].

• *More general problems and operators.* Allen, Lindgren and Petrosyan in [1] consider the two phase problem for the fractional Laplacian and prove optimal regularity of the solution and separation of the positive and negative phase.

Operators with drift in the subcritical regime $s \in (1/2, 1)$ are considered by Petrosyan and Pop in [55] where optimal regularity of the solution is proved. The regularity of the free boundary is addressed by Garofalo et al. in [39]. We emphasize that the presence of the drift the extension operator exhibits some singularities and makes the problem quite delicate.

The results in [22] have been extended to obstacle problem for integro-differential operators by Caffarelli, Ros-Oton and Serra in [20]. Here the authors develop an entirely non local powerful approach, independent of any monotonicity formulas.

Korvenpää, Kuusi and Palatucci in [49] consider the obstacle problem for a class of nonlinear integro-differential operators including the fractional p–Laplacian. The solution exists, it is unique and inherits regularity properties (such as Hölder continuity) from the obstacle.

3.4 Parabolic Obstacle Problems

In this section we will focus on time-dependent models, which can be thought of as parabolic counterparts of the systems (3.1.1) and (3.1.2). We emphasize that, although their time-independent versions are locally equivalent (for $s = 1/2$), the problems we are about to describe are very different from each other.

3.4.1 The parabolic fractional obstacle problem

As mentioned above, one of the motivations behind the recent increased interest in studying constrained variational problems with a fractional diffusion comes from mathematical finance. Jump-diffusion processes allow, in fact, to take into account large price changes, and appear to be better suited to model market fluctuations. An American option gives its holder the right to buy a stock or basket of stocks at a given price prior to, but not later than, a given time $T > 0$ from the time of inception of the contract. If $v(x, t)$ denotes the price of an American option with a payoff ψ at time T, then v will be a viscosity solution to the obstacle problem

$$\begin{cases} \min\{\mathcal{L}v, v - \psi\} = 0 \\ v(T) = \psi. \end{cases} \tag{3.4.1}$$

Here $\mathcal{L}$ is a backward parabolic integro-differential operator of the form

$$\mathcal{L}v = -v_t - rv - b \cdot \nabla v + (-\Delta)^s v + \mathcal{H}v, \quad s \in (0, 1),$$

where $r > 0$, $b \in \mathbb{R}^n$, and $\mathcal{H}$ is a non-local operator of lower order with respect to $(-\Delta)^s$.

This problem was studied by Caffarelli and Figalli in the paper [19], where regularity properties are established for the model equation

$$\begin{cases} \min\{-v_t + (-\Delta)^s v, v - \psi\} = 0 & \text{on } [0,T] \times \mathbb{R}^n \\ v(T) = \psi & \text{on } \mathbb{R}^n. \end{cases} \tag{3.4.2}$$

The advantage of considering (3.4.2) is twofold. On the one hand, the absence of the transport term allows to prove that solutions have the same regularity as in the stationary case for all values of $s \in (0,1)$, even if, when $s \in (0,1/2)$, the time derivative is of higher order with respect to the elliptic term $(-\Delta)^s v$. In addition, it is feasible that the regularity theory established for the model equation (3.4.2) could be adapted to the solutions to (3.4.1) when $s > 1/2$. It should be noted that when $s < 1/2$, the leading term becomes $b \cdot \nabla v$, and there is no expectation of a regularity theory.

In order to proceed, we need to introduce the relevant function spaces.

Definition 3.50. *Given $\alpha, \beta \in (0,1)$ and $[a,b] \subset \mathbb{R}$, we say that*

- $w \in C^{\alpha,\beta}_{t,x}([a,b] \times \mathbb{R}^n)$ *if*

$$\begin{aligned} \|w\|_{C^{\alpha,\beta}_{t,x}([a,b]\times\mathbb{R}^n)} &:= \|w\|_{L^\infty([a,b]\times\mathbb{R}^n)} + [w]_{C^{\alpha,\beta}_{t,x}([a,b]\times\mathbb{R}^n)} \\ &= \|w\|_{L^\infty([a,b]\times\mathbb{R}^n)} + \sup_{[a,b]\times\mathbb{R}^n} \frac{|w(t,x) - w(t',x')|}{|t-t'|^\alpha + |x-x'|^\beta} < \infty. \end{aligned}$$

- $w \in \log\mathrm{Lip}_t C^\beta_x([a,b] \times \mathbb{R}^n)$ *if*

$$\begin{aligned} \|w\|_{\log\mathrm{Lip}_t C^\beta_x([a,b]\times\mathbb{R}^n)} :=& \|w\|_{L^\infty([a,b]\times\mathbb{R}^n)} \\ &+ \sup_{[a,b]\times\mathbb{R}^n} \frac{|w(t,x) - w(t',x')|}{|t-t'|(1 + |\log|t-t'||) + |x-x'|^\beta} < \infty. \end{aligned}$$

We will also say that

- $w \in C^{\alpha-0^+,\beta}_{t,x}([a,b] \times \mathbb{R}^n)$ *if* $w \in C^{\alpha-\varepsilon,\beta}_{t,x}([a,b] \times \mathbb{R}^n)$ *for all* $\varepsilon > 0$;
- $w \in C^{\alpha,\beta}_{t,x}((a,b] \times \mathbb{R}^n)$ *if* $w \in C^{\alpha,\beta}_{t,x}([a+\varepsilon,b] \times \mathbb{R}^n)$ *for all* $\varepsilon > 0$.

In what follows, $\psi : \mathbb{R}^n \to \mathbb{R}^+$ will be a globally Lipschitz function of class C^2 satisfying $\int_{\mathbb{R}^n} \frac{\psi}{(1+|x|)^{n+2s}} < \infty$ and $(-\Delta)^s \psi \in L^\infty(\mathbb{R}^n)$. For $s \in (0,1)$ we let $u : [0,T] \times \mathbb{R}^n \to \mathbb{R}$ be a continuous viscosity solution to the obstacle problem

$$\begin{cases} \min\{u_t + (-\Delta)^s u, u - \psi\} = 0 & \text{on } [0,T] \times \mathbb{R}^n \\ u(0) = \psi & \text{on } \mathbb{R}^n. \end{cases} \tag{3.4.3}$$

Existence and uniqueness of solutions can be shown either by probabilistic techniques, or by approximating the equation via a penalization method.

We are now ready to state the main result from [19].

Theorem 3.51. *Assume that* $\psi \in C^2(\mathbb{R}^n)$ *satisfies*

$$\|\nabla\psi\|_{L^\infty(\mathbb{R}^n)} + \|D^2\psi\|_{L^\infty(\mathbb{R}^n)} + \|(-\Delta)^s\psi\|_{C_x^{1-s}(\mathbb{R}^n)} < \infty,$$

and that u *solves* (3.4.3). *Then* u *is globally Lipschitz in space-time on* $[0,T]\times\mathbb{R}^n$, *and satisfies*

$$\begin{cases} u_t \in \log\mathrm{Lip}_t C_x^{1-s}((0,T]\times\mathbb{R}^n), & \\ \qquad (-\Delta)^s u \in \log\mathrm{Lip}_t C_x^{1-s}((0,T]\times\mathbb{R}^n) & \text{if } s \leqslant 1/3, \\ u_t \in C_{t,x}^{\frac{1-s}{2s}-0^+}((0,T]\times\mathbb{R}^n), & \\ \qquad (-\Delta)^s u \in C_{t,x}^{\frac{1-s}{2s}}((0,T]\times\mathbb{R}^n) & \text{if } s > 1/3. \end{cases}$$

Some remarks are in order. First of all, comparison with Corollary 3.26 shows that, at least in terms of spacial regularity, this result is optimal. Secondly, the criticality of $s = 1/3$ is a consequence of the invariance of the operator $\partial_t + (-\Delta)^s$ under the scaling $(t,x) \to (\lambda^{2s}t, \lambda x)$. Hence, the spacial regularity C_x^{1-s} naturally corresponds to time regularity $C_t^{\frac{1-s}{2s}}$ when $\frac{1-s}{2s} < 1$, or $s > 1/3$. In addition, the time regularity is almost optimal in the case $s = 1/2$ (as it is possible to construct traveling waves which are $C^{1+1/2}$ both in space and time), as well as in the limit $s \to 1$ (since, when $s = 1$, it is well known that solutions are $C^{1,1}$ in space and C^1 in time).

Several basic properties of the solution u, such as global regularity in space-time, semi-convexity in space, and the boundedness of $(-\Delta)^s u$, follow from a comparison principle, which in turn is established using a penalization method. Combining the semi-convexity of $u(t)$ with the L^∞ bound on $(-\Delta)^s u(t)$, it is possible then to deduce the C^1 regularity in time of solutions when $s \geqslant 1/2$. The next step toward the proof of Theorem 3.51 is to prove a $C^{\alpha+2s}$-regularity result in space, which, roughly speaking, says the following. Let $v : \mathbb{R}^n \to \mathbb{R}$ be a semiconvex function touching from above an obstacle $\psi : \mathbb{R}^n \to \mathbb{R}$ of class C^2. If $(-\Delta)^s v$ is non-positive outside the contact set and non-negative on it, then v detaches from ψ in a $C^{\alpha+2s}$ fashion, with $\alpha > 0$ depending exclusively on s. More precisely, assume that $v, \psi : \mathbb{R}^n \to \mathbb{R}$ are two globally Lipschitz functions with $v \geqslant \psi$ satisfying the following conditions:

A1. $\|D^2\psi\|_{L^\infty(\mathbb{R}^n)} := C_0 < \infty$;
A2. $\|(-\Delta)^s\psi\|_{C_x^{1-s}(\mathbb{R}^n)} < \infty$;
A3. $v - \psi,\ (-\Delta)^s v \in L^\infty(\mathbb{R}^n)$;
A4. The function $v + C_0|x^2|/2$ is convex (we say that v is C_0*-semiconvex*);
A5. v is smooth and $(-\Delta)^s v \leqslant 0$ inside the open set $\{v > \psi\}$;
A6. $\|(-\Delta)^s\psi\|_{L^\infty(\mathbb{R}^n)} \geqslant (-\Delta)^s v \geqslant 0$ on $\{v = \psi\}$.

One has then the following:

Theorem 3.52. *There exist $C > 0$ and $\alpha \in (0,1)$, depending only on C_0, $\|(-\Delta)^s\psi\|_{C_x^{1-s}(\mathbb{R}^n)}$, $\|v - \psi\|_{L^\infty(\mathbb{R}^n)}$, and $\|(-\Delta)^s v\|_{L^\infty(\mathbb{R}^n)}$, such that*

$$\sup_{B_r(x)} |v-\psi| \leqslant Cr^{\alpha+2s}, \qquad \sup_{B_r(x)} |(-\Delta)^s v\chi_{\{v=\psi\}}| \leqslant Cr^{\alpha} \tag{3.4.4}$$

for all $r \leqslant 1$ and for every $x \in \partial\{u=\psi\}$.

The strategy of the proof is analogous to the one used in [7]. It is based on growth estimates for the L_a-harmonic extension of v, i.e. the solution to (3.1.4), satisfying (with slight abuse of notation) $v(x,0) = v(x)$, with $v(x)$ as above.

Having shown that $(-\Delta)^s v\chi_{\{v=\psi\}}$ grows at most as r^α near any free boundary point, it is not difficult to show that $(-\Delta)^s v\chi_{\{v=\psi\}} \in C_x^\alpha(\mathbb{R}^n)$:

Corrolary 3.53. *There exist $C' > 0$ and $\alpha \in (0, 1-s]$, depending only on C_0, $\|(-\Delta)^s\psi\|_{C_x^{1-s}(\mathbb{R}^n)}$, $\|v-\psi\|_{L^\infty(\mathbb{R}^n)}$, and $\|(-\Delta)^s v\|_{L^\infty(\mathbb{R}^n)}$, such that*

$$\|(-\Delta)^s v\chi_{\{v=\psi\}}\|_{C_x^\alpha(\mathbb{R}^n)} \leqslant C'.$$

At this point we consider the function $w : \mathbb{R}^n \times \mathbb{R}^+ \to \mathbb{R}$, which solves the Dirichlet problem

$$\begin{cases} L_{-a}w = 0 & \text{on } \mathbb{R}^n \times \mathbb{R}^+, \\ w(x,0) = (-\Delta)^s v(x)\chi_{\{v=\psi\}}(x) & \text{on } \mathbb{R}^n. \end{cases}$$

Since $w(x,0) \geqslant 0$, the maximum principle implies $w \geqslant 0$ everywhere. Assume now that 0 is a free boundary point. Given that $(-\Delta)^s v(x)$ is globally bounded by (A3) above, it follows from the Poisson representation formula for w (see [23]) and Corollary 3.53 that

$$\sup_{|x|^2+y^2\leqslant r^2} w(x,y) \leqslant Cr^\alpha,$$

for some uniform constant C and some $\alpha \in (0, 1-s]$. Our next goal is to establish the following sharp growth estimate:

Proposition 3.54. *There exists $\tilde{C} > 0$ depending only on C_0, $\|(-\Delta)^s\psi\|_{C_x^{1-s}(\mathbb{R}^n)}$, $\|v - \psi\|_{L^\infty(\mathbb{R}^n)}$, and $\|(-\Delta)^s v\|_{L^\infty(\mathbb{R}^n)}$, such that*

$$\sup_{|x|^2+y^2\leqslant r^2} w(x,y) \leqslant \tilde{C}r^{1-s}.$$

A crucial ingredient in the proof of Proposition 3.54 is the following monotonicity formula (compare with Lemma 3.11).

Lemma 3.55. *Let w be as above and define*

$$\varphi(r) = \frac{1}{r^{2(1-s)}} \int_{B_r^+} \frac{y^{-a}\,|\nabla w(X)|^2}{|X|^{n-a-1}}\,dX.$$

Then there exists a constant C'', depending only on C_0, $\|(-\Delta)^s\psi\|_{C_x^{1-s}(\mathbb{R}^n)}$, $\|v-\psi\|_{L^\infty(\mathbb{R}^n)}$, and $\|(-\Delta)^s v\|_{L^\infty(\mathbb{R}^n)}$, such that

$$\varphi(r) \leqslant C''[1 + r^{2\alpha+\delta_\alpha-a-1}]$$

for all $r \leqslant 1$. Here, $\delta_\alpha = \frac{1}{4}\left(\frac{\alpha}{\alpha+2s} - \frac{\alpha}{2}\right)$.

Proof of Proposition 3.54. Using an approximation argument, it is possible to show that the function

$$\tilde{w}(x,y) = \left(w(x,|y|) - r^{\alpha+\delta_\alpha}\right)^+ + \left(1 + \frac{nC_0}{1+a}\right)|y|^{1+a}$$

is globally L_{-a}-subharmonic. Moreover, because of Lemma 3.55, it vanishes on more than half of the n-dimensional disc $B_r \times \{0\}$. One can then apply the weighted Poincaré inequality established in [30] to show

$$\int_{B_r^+} (\tilde{w})^2 y^{-a}\, dX \leqslant Cr^{n+2}[\varphi(r) + 1]$$

for all $r \leqslant 1$. Combining this estimate with the L_{-a}-subharmonicity of $\tilde{w}$, and Lemma 3.55 again, we infer

$$\sup_{B_{r/2}^+}(\tilde{w})^2 \leqslant \frac{C}{r^{n+1-a}} \int_{B_r^+} (\tilde{w})^2 y^{-a}\, dX \leqslant C[r^{1+a} + r^{2\alpha+\delta_\alpha}].$$

But $1 + a = 2(1-s)$, and therefore

$$\sup_{B_{r/2}^+} w \leqslant C\left[\sup_{B_{r/2}^+} \tilde{w} + r^{\alpha+\delta_\alpha} + r^{1+a}\right] \leqslant C[r^{1-s} + r^{\alpha+\delta_\alpha/2}].$$

Arguments similar to the ones in the proof of Corollary 3.53 yield

$$\|w\|_{C_x^{\beta_\alpha}(\mathbb{R}^n)} \leqslant C,$$

with $\beta_\alpha = \min\{\alpha + \delta_\alpha/2, 1-s\}$. An iteration gives the desired conclusion. □

Arguing as in the proof of Corollary 3.53, we obtain

$$\|(-\Delta)^s v \chi_{\{v=\psi\}}\|_{C_x^{1-s}(\mathbb{R}^n)} \leqslant C''. \tag{3.4.5}$$

Next, we apply (3.4.5) to $v = u(t)$, obtaining the following result.

Proposition 3.56. *Let u be a solution to* (3.4.3), *with $\psi \in C^2(\mathbb{R}^n)$ satisfying* (A1) *and* (A2) *above. Then there exists a constant $C_T > 0$, depending on T, $\|D^2\psi\|_{L^\infty(\mathbb{R}^n)}$, and $\|(-\Delta)^s\psi\|_{C_x^{1-s}(\mathbb{R}^n)}$, such that*

$$\sup_{t\in[0,T]} \|(-\Delta)^s u(t)\chi_{\{u(t)=\psi\}}\|_{C_x^{1-s}(\mathbb{R}^n)} \leqslant C_T.$$

Finally, to conclude the proof of Theorem 3.51 one exploits the fact that u is a solution of the parabolic equation

$$u_t + (-\Delta)^s u = ((-\Delta)^s u)\chi_{\{u=\psi\}} \qquad \text{on } (0, T] \times \mathbb{R}^n. \tag{3.4.6}$$

Proposition 3.56 ensures that the right-hand side of (3.4.6) is in $L^\infty((0, T]; C_x^{1-s}(\mathbb{R}^n))$. By parabolic regularity theory we infer

$$u_t, (-\Delta)^s u \in L^\infty((0, T]; C_x^{1-s-0^+}(\mathbb{R}^n)).$$

The desired Hölder regularity in time follows from a bootstrap argument which uses equation (3.4.6) again.

3.4.2 The parabolic Signorini problem

In this section we will give an overview of the time-dependent analogue of (3.1.2), i.e., the parabolic Signorini problem, following the ideas in [27].

3.4.2.1 Statement of the problem

Given a domain Ω in $\mathbb{R}^n$, $n \geqslant 2$, with a sufficiently regular boundary $\partial\Omega$, let $\mathcal{M}$ be a relatively open subset of $\partial\Omega$ (in its relative topology), and set $\mathcal{S} = \partial\Omega \backslash \mathcal{M}$. We consider the solution of the problem

$$\Delta v - \partial_t v = 0 \quad \text{in } \Omega_T := \Omega \times (0, T], \tag{3.4.7}$$

$$v \geqslant \varphi, \quad \partial_\nu v \geqslant 0, \quad (v - \varphi)\partial_\nu v = 0 \quad \text{on } \mathcal{M}_T := \mathcal{M} \times (0, T], \tag{3.4.8}$$

$$v = g \quad \text{on } \mathcal{S}_T := \mathcal{S} \times (0, T], \tag{3.4.9}$$

$$v(\cdot, 0) = \varphi_0 \quad \text{on } \Omega_0 := \Omega \times \{0\}, \tag{3.4.10}$$

where ∂_ν is the outer normal derivative on $\partial\Omega$, and $\varphi : \mathcal{M}_T \to \mathbb{R}$, $\varphi_0 : \Omega_0 \to \mathbb{R}$, and $g : \mathcal{S}_T \to \mathbb{R}$ are prescribed functions satisfying the compatibility conditions: $\varphi_0 \geqslant \varphi$ on $\mathcal{M} \times \{0\}$, $g \geqslant \varphi$ on $\partial\mathcal{S} \times (0, T]$, and $g = \varphi$ on $\mathcal{S} \times \{0\}$, see Fig. 3.2. Classical examples where Signorini-type boundary conditions appear are the problems with unilateral constraints in elastostatics (including the original Signorini problem [62, 31]), problems with semipermeable membranes in fluid mechanics (including the phenomenon of osmosis and osmotic pressure in biochemistry), and the problems on the temperature control on the boundary in thermics. We refer to the book of Duvaut and Lions [29] for further details.

Another historical importance of the parabolic Signorini problem is that it serves as one of the prototypical examples of evolutionary *variational inequalities*. We thus

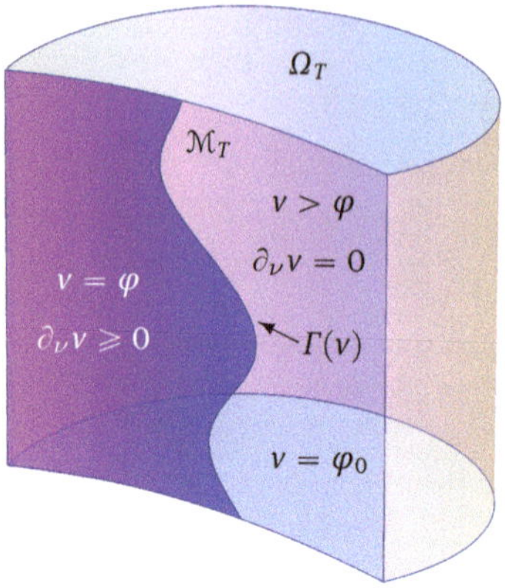

Fig. 3.2: The parabolic Signorini problem

say that a function $v \in W_2^{1,0}(\Omega_T)$ solves (3.4.7)–(3.4.10) if

$$\int_{\Omega_T} \nabla v \nabla(w - v) + \partial_t v(w - v) \geqslant 0 \quad \text{for every } w \in \mathfrak{K},$$

$$v \in \mathfrak{K}, \quad \partial_t v \in L_2(\Omega_T), \quad v(\cdot, 0) = \varphi_0,$$

where $\mathfrak{K} = \{w \in W_2^{1,0}(\Omega_T) \mid w \geqslant \varphi \text{ on } \mathcal{M}_T,\ w = g \text{ on } \mathcal{S}_T\}$ (please see below for the definitions of the relevant parabolic functional classes). The existence and uniqueness of such v, under some natural assumptions on φ, φ_0, and g can be found in [14, 29, 3, 4].

Similarly to the elliptic case, we are interested in:

- the regularity properties of v;
- the structure and regularity of the *free boundary*

$$\Gamma(v) = \partial_{\mathcal{M}_T}\{(x, t) \in \mathcal{M}_T \mid v(x, t) > \varphi(x, t)\},$$

where $\partial_{\mathcal{M}_T}$ indicates the boundary in the relative topology of $\mathcal{M}_T$.

Concerning the regularity of v, it has long been known that the spatial derivatives $\partial_{x_i} v$, $i = 1, \dots, n$, are α-Hölder continuous on compact subsets of $\Omega_T \cup \mathcal{M}_T$, for some unspecified $\alpha \in (0, 1)$. In the parabolic case, this was first proved by Athanasopoulos [5], and subsequently by Uraltseva [65] (see also [3]), under certain regularity assumptions on the boundary data, which were further relaxed by Arkhipova and Uraltseva [4].

One of the main objectives of this section is to establish, in the parabolic Signorini problem, and for a flat thin manifold $\mathcal{M}$, that $v \in H^{3/2,3/4}_{\text{loc}}(\Omega_T \cup \mathcal{M}_T)$, see Theorem 3.69 below. The proof, which we only sketch here, is inspired by the works [10] and [22] described in Section 3.2. For further details, we refer to [27].

Before proceeding, we introduce the relevant parabolic function spaces and notations. We use notations similar to those in the classical book of Ladyzhenskaya, Solonnikov, and Uraltseva [50]. The class $C(\Omega_T) = C^{0,0}(\Omega_T)$ is the class of functions continuous in Ω_T with respect to parabolic (or Euclidean) distance. Further, given for $m \in \mathbb{Z}_+$ we say $u \in C^{2m,m}(\Omega_T)$ if for $|\alpha| + 2j \leqslant 2m$ $\partial_x^\alpha \partial_t^j u \in C^{0,0}(\Omega_T)$, and define the norm

$$\|u\|_{C^{2m,m}(\Omega_T)} = \sum_{|\alpha|+2j\leqslant 2m} \sup_{(x,t)\in\Omega_T} |\partial_x^\alpha \partial_t^j u(x,y)|.$$

The parabolic Hölder classes $H^{\ell,\ell/2}(\Omega_T)$, for $\ell = m+\gamma$, $m \in \mathbb{Z}_+$, $0 < \gamma \leqslant 1$ are defined as follows. First, we let

$$\begin{aligned}
\langle u\rangle_{\Omega_T}^{(0)} &= |u|_{\Omega_T}^{(0)} = \sup_{(x,t)\in\Omega_T} |u(x,t)|,\\
\langle u\rangle_{\Omega_T}^{(m)} &= \sum_{|\alpha|+2j=m} |\partial_x^\alpha \partial_t^j u|_{\Omega_T}^{(0)},\\
\langle u\rangle_{x,\Omega_T}^{(\beta)} &= \sup_{\substack{(x,t),(y,t)\in\Omega_T\\ 0<|x-y|\leqslant\delta_0}} \frac{|u(x,t)-u(y,t)|}{|x-y|^\beta}, \quad 0<\beta\leqslant 1,\\
\langle u\rangle_{t,\Omega_T}^{(\beta)} &= \sup_{\substack{(x,t),(x,s)\in\Omega_T\\ 0<|t-s|<\delta_0^2}} \frac{|u(x,t)-u(x,s)|}{|t-s|^\beta}, \quad 0<\beta\leqslant 1,\\
\langle u\rangle_{x,\Omega_T}^{(\ell)} &= \sum_{|\alpha|+2j=m} \langle \partial_x^\alpha \partial_t^j u\rangle_{x,\Omega_T}^{(\gamma)},\\
\langle u\rangle_{t,\Omega_T}^{(\ell/2)} &= \sum_{m-1\leqslant|\alpha|+2j\leqslant m} \langle \partial_x^\alpha \partial_t^j u\rangle_{t,\Omega_T}^{((\ell-|\alpha|-2j)/2)},\\
\langle u\rangle_{\Omega_T}^{(\ell)} &= \langle u\rangle_{x,\Omega_T}^{(\ell)} + \langle u\rangle_{t,\Omega_T}^{(\ell/2)}.
\end{aligned}$$

Then, we define $H^{\ell,\ell/2}(\Omega_T)$ as the space of functions u for which the following norm is finite:

$$\|u\|_{H^{\ell,\ell/2}(\Omega_T)} = \sum_{k=0}^{m} \langle u\rangle_{\Omega_T}^{(k)} + \langle u\rangle_{\Omega_T}^{(\ell)}.$$

The parabolic Lebesgue space $L_q(\Omega_T)$ indicates the Banach space of those measurable functions on Ω_T for which the norm

$$\|u\|_{L_q(\Omega_T)} = \Big(\int_{\Omega_T} |u(x,t)|^q dxdt\Big)^{1/q}$$

is finite. The parabolic Sobolev spaces $W_q^{2m,m}(\Omega_T)$, $m \in \mathbb{Z}_+$, denote the spaces of those functions in $L_q(\Omega_T)$, whose distributional derivative $\partial_x^\alpha \partial_t^j u$ belongs to $L_q(\Omega_T)$, for $|\alpha| + 2j \leqslant 2m$.

For $x, x_0 \in \mathbb{R}^n$, $t_0 \in \mathbb{R}$ we let

$$x' = (x_1, x_2, \ldots, x_{n-1}),\ x'' = (x_1, x_2, \ldots, x_{n-2}),\ x = (x', x_n),\ x' = (x'', x_{n-1}),$$

and

$$
\begin{aligned}
&B_r(x_0) = \{x \in \mathbb{R}^n \mid |x| < r\} && \text{(Euclidean ball)}\\
&B_r^{\pm}(x_0) = B_r(x_0) \cap \mathbb{R}^n_{\pm} && \text{(Euclidean halfball)}\\
&B_r'(x_0) = B_r(x) \cap \mathbb{R}^{n-1} && \text{(``thin'' ball)}\\
&Q_r(x_0, t_0) = B_r(x_0) \times (t_0 - r^2, t_0] && \text{(parabolic cylinder)}\\
&Q_r'(x_0, t_0) = B_r'(x_0) \times (t_0 - r^2, t_0] && \text{(``thin'' parabolic cylinder)}\\
&Q_r^{\pm}(x_0, t_0) = B_r^{\pm}(x_0) \times (t_0 - r^2, t_0] && \text{(parabolic halfcylinders)}\\
&Q_r''(x_0, t_0) = B_r''(x_0) \times (t_0 - r^2, t_0] && \\
&S_r = \mathbb{R}^n \times (-r^2, 0] && \text{(parabolic strip)}\\
&S_r^{\pm} = \mathbb{R}^n_{\pm} \times (-r^2, 0] && \text{(parabolic halfstrip)}\\
&S_r' = \mathbb{R}^{n-1} \times (-r^2, 0] && \text{(``thin'' parabolic strip)}
\end{aligned}
$$

When $x_0 = 0$ and $t_0 = 0$, we omit the centers x_0 and (x_0, t_0) in the above notations.

Since we are mostly interested in local properties of the solution v of the parabolic Signorini problem and of its free boundary, we focus our attention on solutions in parabolic (half-)cylinders.

Definition 3.57. *Given $\varphi \in H^{2,1}(Q_1')$, we say that $v \in \mathfrak{S}_\varphi(Q_1^+)$ if $v \in W_2^{2,1}(Q_1^+) \cap L_\infty(Q_1^+)$, $\nabla v \in H^{\alpha,\alpha/2}(Q_1^+ \cup Q_1')$ for some $0 < \alpha < 1$, and v satisfies*

$$\Delta v - \partial_t v = 0 \quad \text{in } Q_1^+, \tag{3.4.11}$$

$$v - \varphi \geqslant 0, \quad -\partial_{x_n} v \geqslant 0, \quad (v - \varphi)\partial_{x_n} v = 0 \quad \text{on } Q_1', \tag{3.4.12}$$

and

$$(0,0) \in \Gamma_*(v) := \partial_{Q_1'}\{(x', t) \in Q_1' \mid v(x', 0, t) = \varphi(x', t),\ \partial_{x_n} v(x', 0, t) = 0\}, \tag{3.4.13}$$

where $\partial_{Q_1'}$ is the boundary in the relative topology of Q_1'.

Our very first step is the reduction to vanishing obstacle. The difference $v(x,t) - \varphi(x',t)$ satisfies the Signorini conditions on Q_1' with zero obstacle, but at an expense of solving a nonhomogeneous heat equation instead of the homogeneous one. This difference may then be extended to the strip S_1^+ by multiplying it by a suitable cutoff function ψ. The resulting function will satisfy

$$\Delta u - \partial_t u = f(x,t) \quad \text{in } S_1^+,$$

with

$$f(x,t) = -\psi(x)[\Delta' \varphi - \partial_t \varphi] + [v(x,t) - \varphi(x',t)]\Delta\psi + 2\nabla v \nabla \psi. \tag{3.4.14}$$

Remark 3.58. *It is important to observe that, for smooth enough φ, the function f is bounded in S_1^+.*

With this process, we arrive at the following notion of solution.

Definition 3.59. *A function u is in the class $\mathfrak{S}^f(S_1^+)$, for $f \in L_\infty(S_1^+)$, if $u \in W_2^{2,1}(S_1^+)$, $\nabla u \in H^{\alpha,\alpha/2}(S_1^+ \cup S_1')$, u has compact support and solves*

$$\Delta u - \partial_t u = f \quad \text{in } S_1^+, \tag{3.4.15}$$

$$u \geqslant 0, \quad -\partial_{x_n} u \geqslant 0, \quad u\partial_{x_n} u = 0 \quad \text{on } S_1', \tag{3.4.16}$$

$$(0,0) \in \Gamma(u) = \partial\{(x',t) \in S_1' : u(x',0,t) > 0\}. \tag{3.4.17}$$

3.4.2.2 Monotonicity of the generalized frequency function

Similarly to the elliptic case, one of the central results towards the study of regularity properties (both of the solution and of the free boundary) is a generalization of Almgren's frequency formula, see Theorem 3.61 below. As it is well known, the parabolic counterpart of Almgren's formula was established by Poon [59], for functions which are caloric in an infinite strip $S_\rho = \mathbb{R}^n \times (-\rho^2, 0]$. Poon's parabolic frequency function is given by

$$N_u(r) = \frac{i_u(-r^2)}{h_u(-r^2)}, \tag{3.4.18}$$

where

$$h_u(t) = \int_{\mathbb{R}^n_+} u(x,t)^2 G(x,t)\,dx,$$

$$i_u(t) = -t \int_{\mathbb{R}^n_+} |\nabla u(x,t)|^2 G(x,t)\,dx,$$

for any function u in the parabolic half-strip S_1^+ for which the integrals involved are finite. Here G denotes the *backward heat kernel* on $\mathbb{R}^n \times \mathbb{R}$

$$G(x,t) = \begin{cases} (-4\pi t)^{-\frac{n}{2}} e^{\frac{x^2}{4t}}, & t < 0, \\ 0, & t \geqslant 0. \end{cases}$$

We explicitly remark that Poon's monotonicity formula cannot be directly applied in the present context, since functions in the class $\mathfrak{S}^f(S_1^+)$ (see Definition 3.59) are *not* caloric functions. It should also be noted that the time dependent case of the Signorini problem presents substantial novel challenges with respect to the stationary setting. These are mainly due to the lack of regularity of the solution in the t-variable, a fact which makes the justification of differentiation formulas and the control of error terms quite difficult. To overcome these obstructions, (Steklov-type) averaged versions of the quantities involved are introduced in the main monotonicity formulas. This basic idea allows to successfully control the error terms. More precisely, we introduce the quantities

$$H_u(r)= \frac{1}{r^2}\int_{-r^2}^{0} h_u(t)dt = \frac{1}{r^2}\int_{S_r^+} u(x,t)^2 G(x,t)dxdt,$$

$$I_u(r)= \frac{1}{r^2}\int_{-r^2}^{0} i_u(t)dt = \frac{1}{r^2}\int_{S_r^+} |t||\nabla u(x,t)|^2 G(x,t)dxdt,$$

One further obstruction is represented by the fact that the above integrals may become unbounded near the endpoint $t = 0$, where G becomes singular. To remedy this problem we introduce truncated versions of H_u and I_u:

$$H_u^\delta(r) = \frac{1}{r^2}\int_{-r^2}^{-\delta^2 r^2} h_u(t)dt = \frac{1}{r^2}\int_{S_r^+\setminus S_{\delta r}^+} u(x,t)^2 G(x,t)dxdt,$$

$$I_u^\delta(r) = \frac{1}{r^2}\int_{-r^2}^{-\delta^2 r^2} i_u(t)dt = \frac{1}{r^2}\int_{S_r^+\setminus S_{\delta r}^+} |t||\nabla u(x,t)|^2 G(x,t)dxdt$$

for $0 < \delta < 1$.

The idea at this point is to obtain differentiation formulas for $H_u^\delta(r)$ and $I_u^\delta(r)$, which - by means of a delicate limiting process - will yield corresponding ones for $H_u(r)$ and $I_u(r)$. In turn, such formulas will allow to establish almost-monotonicity of a suitably defined frequency function. To state this result we need the following notion.

Definition 3.60. *We say that a positive function $\mu(r)$ is a* log-convex function *of* $\log r$ *on $\mathbb{R}^+$ if $\log\mu(e^t)$ is a convex function of t. In other words*

$$\mu(e^{(1-\lambda)s+\lambda t}) \leqslant \mu(e^s)^{1-\lambda}\mu(e^t)^\lambda, \quad 0 \leqslant \lambda \leqslant 1.$$

This is equivalent to saying that μ is locally absolutely continuous on $\mathbb{R}_+$ and $r\mu'(r)/\mu(r)$ is nondecreasing. For instance, $\mu(r) = r^\kappa$ is a log-convex function of $\log r$ for any κ. The importance of this notion in our context is that Almgren's and Poon's frequency formulas can be regarded as log-convexity statements in $\log r$ for the appropriately defined quantities $H_u(r)$.

Theorem 3.61. *(Monotonicity of the truncated frequency) Let $u \in \mathfrak{S}^f(S_1^+)$ with f satisfying the following condition: there is a positive monotone nondecreasing* log-convex *function $\mu(r)$ of $\log r$, and constants $\sigma > 0$ and $C_\mu > 0$, such that*

$$\mu(r) \geqslant C_\mu r^{4-2\sigma}\int_{\mathbb{R}^n} f^2(\cdot,-r^2)G(\cdot,-r^2)\,dx.$$

Then, there exists $C > 0$, depending only on σ, C_μ and n, such that the function

$$\Phi_u(r) = \frac{1}{2}re^{Cr^\sigma}\frac{d}{dr}\log\max\{H_u(r),\mu(r)\} + 2(e^{Cr^\sigma}-1)$$

is nondecreasing for $r \in (0,1)$.

Remark 3.62. *On the open set where $H_u(r) > \mu(r)$ we have $\Phi_u(r) \sim \frac{1}{2} r H_u'(r)/H_u(r)$, which coincides, when $f = 0$, with $2N_u$ (N_u as in (3.4.18)). The purpose of the "truncation" of $H_u(r)$ with $\mu(r)$ is to control the error terms in computations that appear from the right-hand-side f.*

Proof of Theorem 3.61. First, we observe that the functions $H_u(r)$ and $\mu(r)$ are absolutely continuous and therefore so is $\max\{H_u(r), \mu(r)\}$. It follows that Φ_u is uniquely identified only up to a set of measure zero. The monotonicity of Φ_u should be understood in the sense that there exists a monotone increasing function which equals Φ_u almost everywhere. Therefore, without loss of generality we may assume that

$$\Phi_u(r) = \frac{1}{2} r e^{Cr^\sigma} \frac{\mu'(r)}{\mu(r)} + 2(e^{Cr^\sigma} - 1)$$

on $\mathcal{F} = \{H_u(r) \leqslant \mu(r)\}$ and

$$\Phi_u(r) = \frac{1}{2} r e^{Cr^\sigma} \frac{H_u'(r)}{H_u(r)} + 2(e^{Cr^\sigma} - 1)$$

in $\mathcal{O} = \{H_u(r) > \mu(r)\}$. Following an idea introduced in [34, 35] we now note that it will be enough to check that $\Phi_u'(r) > 0$ in $\mathcal{O}$. Indeed, from the assumption on μ, it is clear that Φ_u is monotone on $\mathcal{F}$. Next, if (r_0, r_1) is a maximal open interval in $\mathcal{O}$, then $H_u(r_0) = \mu(r_0)$ and $H_u(r_1) = \mu(r_1)$ unless $r_1 = 1$. Besides, if Φ_u is monotone in (r_0, r_1), it is easy to see that the limits $H_u'(r_0+)$ and $H_u'(r_1-)$ will exist and satisfy

$$\mu'(r_0+) \leqslant H_u'(r_0+), \quad H_u'(r_1-) \leqslant \mu'(r_1-) \quad (\text{unless } r_1 = 1)$$

and therefore we will have

$$\Phi_u(r_0) \leqslant \Phi_u(r_0+) \leqslant \Phi_u(r_1-) \leqslant \Phi_u(r_1),$$

with the latter inequality holding when $r_1 < 1$. This will imply the monotonicity of Φ_u in $(0, 1)$.

Therefore, we will concentrate only on the set $\mathcal{O} = \{H_u(r) > \mu(r)\}$, where the monotonicity of $\Phi_u(r)$ is equivalent to that of

$$\left(r \frac{H_u'(r)}{H_u(r)} + 4\right) e^{Cr^\sigma} = 2\Phi_u(r) + 4.$$

The latter will follow, once we show that

$$\frac{d}{dr}\left(r \frac{H_u'(r)}{H_u(r)}\right) \geqslant -C\left(r \frac{H_u'(r)}{H_u(r)} + 4\right) r^{-1+\sigma}$$

in $\mathcal{O}$. Now, one can show that

$$r \frac{H_u'(r)}{H_u(r)} = 4 \frac{I_u(r)}{H_u(r)} - \frac{4}{r^2} \frac{\int_{S_r^+} tufG}{H_u(r)}.$$

The desired result will be obtained by direct differentiation of this expression, and using the aforementioned formulas for the derivatives of $H_u(r)$ and $I_u(r)$. We omit the details. □

3.4.2.3 Blow-ups and regularity of solutions

Similarly to the elliptic case, the generalized frequency formula in Theorem 3.61 can be used to study the behavior of the solution u near the origin. The central idea is again to consider some appropriately normalized rescalings of u, indicated with u_r (see Definition 3.64), and then pass to the limit as $r \to 0+$ (see Theorem 3.65).

Henceforth, we assume that $u \in \mathfrak{S}^f(S_1^+)$, and that $\mu(r)$ be such that the conditions of Theorem 3.61 are satisfied. In particular, we assume that

$$r^4 \int_{\mathbb{R}^n} f^2(\cdot, -r^2) G(\cdot, -r^2)\, dx \leqslant \frac{r^{2\sigma}\mu(r)}{C_\mu}.$$

Consequently, Theorem 3.61 implies that the function

$$\Phi_u(r) = \frac{1}{2} r e^{Cr^\sigma} \frac{d}{dr} \log \max\{H_u(r), \mu(r)\} + 2(e^{Cr^\sigma} - 1)$$

is nondecreasing for $r \in (0, 1)$. Hence, there exists the limit

$$\kappa := \Phi_u(0+) = \lim_{r \to 0+} \Phi_u(r). \tag{3.4.19}$$

It is possible to show that κ is independent of the choice of the cut-off ψ introduced in the extension procedure. Since we assume that $r\mu'(r)/\mu(r)$ is nondecreasing, the limit

$$\kappa_\mu := \frac{1}{2} \lim_{r \to 0+} \frac{r\mu'(r)}{\mu(r)} \tag{3.4.20}$$

also exists. We then have the following basic proposition concerning the values of κ and κ_μ.

Lemma 3.63. *Let $u \in \mathfrak{S}^f(S_1^+)$ and μ satisfy the conditions of Theorem 3.61. With κ, κ_μ as above, we have*

$$\kappa \leqslant \kappa_\mu.$$

Moreover, if $\kappa < \kappa_\mu$, then there exists $r_u > 0$ such that $H_u(r) \geqslant \mu(r)$ for $0 < r \leqslant r_u$. In particular,

$$\kappa = \frac{1}{2} \lim_{r \to 0+} \frac{rH_u'(r)}{H_u(r)} = 2 \lim_{r \to 0+} \frac{I_u(r)}{H_u(r)}.$$

We now define the appropriate notion of rescalings that works well with the generalized frequency formula.

Definition 3.64. *For $u \in \mathfrak{S}^f(S_1^+)$ and $r > 0$ define the* rescalings

$$u_r(x, t) := \frac{u(rx, r^2 t)}{H_u(r)^{1/2}}, \quad (x, t) \in S_{1/r}^+ = \mathbb{R}_+^n \times (-1/r^2, 0].$$

It is easy to see that the function u_r solves the nonhomogeneous Signorini problem

$$\begin{aligned} \Delta u_r - \partial_t u_r = f_r(x,t) \quad &\text{in } S^+_{1/r}, \\ u_r \geqslant 0, \quad -\partial_{x_n} u_r \geqslant 0, \quad u_r \partial_{x_n} u_r = 0 \quad &\text{on } S'_{1/r}, \end{aligned}$$

with

$$f_r(x,t) = \frac{r^2 f(rx, r^2 t)}{H_u(r)^{1/2}}.$$

In other words, $u_r \in \mathfrak{S}^{f_r}(S^+_{1/r})$. We next show that, unless we are in the borderline case $\kappa = \kappa_\mu$, we will be able to study the blowups of u at the origin. The condition $\kappa < \kappa_\mu$ below can be understood, in a sense, that we can "detect" the growth of u near the origin.

Theorem 3.65. *(Existence and homogeneity of blowups) Let $u \in \mathfrak{S}^f(S^+_1)$, μ satisfy the conditions of Theorem 3.61, and*

$$\kappa := \Phi_u(0+) < \kappa_\mu = \frac{1}{2} \lim_{r \to 0+} r \frac{\mu'(r)}{\mu(r)}.$$

Then, we have:

i) *For any $R > 0$, there is $r_{R,u} > 0$ such that*

$$\int_{S^+_R} (u_r^2 + |t||\nabla u_r|^2 + |t|^2 |D^2 u_r|^2 + |t|^2 (\partial_t u_r)^2) G \leqslant C(R), \quad 0 < r < r_{R,u}.$$

ii) *There is a sequence $r_j \to 0+$, and a function u_0 in $S^+_\infty = \mathbb{R}^n_+ \times (-\infty, 0]$, such that*

$$\int_{S^+_R} (|u_{r_j} - u_0|^2 + |t||\nabla(u_{r_j} - u_0)|^2) G \to 0.$$

We call any such u_0 a blowup *of u at the origin.*

iii) *u_0 is a nonzero* global solution *of Signorini problem:*

$$\begin{aligned} \Delta u_0 - \partial_t u_0 = 0 \quad &\text{in } S^+_\infty \\ u_0 \geqslant 0, \quad -\partial_{x_n} u_0 \geqslant 0, \quad u_0 \partial_{x_n} u_0 = 0 \quad &\text{on } S'_\infty, \end{aligned}$$

in the sense that it solves the Signorini problem in every Q^+_R.

iv) *u_0 is parabolically homogeneous of degree κ:*

$$u_0(\lambda x, \lambda^2 t) = \lambda^\kappa u_0(x,t), \quad (x,t) \in S^+_\infty, \ \lambda > 0$$

In addition to Theorem 3.61 and Lemma 3.63, the main ingredients in the proof of Theorem 3.65 are growth estimates for $H_{u_r}(\rho)$ and log-Sobolev inequalities.

Remark 3.66. *Using growth estimates for $H_u(r)$ (where $u \in \mathfrak{S}^f(S^+_1)$), it is possible to show that necessarily $\kappa > 1$.*

In addition, we have the following:

Proposition 3.67. *Let u_0 be a nonzero κ-parabolically homogeneous solution of the Signorini problem in $S_\infty^+ = \mathbb{R}^n_+ \times (-\infty, 0]$ with $1 < \kappa < 2$. Then, $\kappa = 3/2$ and*

$$u_0(x,t) = C\,\mathrm{Re}(x' \cdot e + ix_n)_+^{3/2} \quad \textit{in } S_\infty^+$$

for some tangential direction $e \in \partial B_1'$.

Proof. Extend u_0 by even symmetry in x_n to the strip S_∞, i.e., by putting

$$u_0(x', x_n, t) = u_0(x', -x_n, t).$$

Take any $e \in \partial B_1'$, and consider the positive and negative parts of the directional derivative $\partial_e u_0$

$$v_e^\pm = \max\{\pm \partial_e u_0, 0\}.$$

It can be shown that they satisfy the following conditions

$$(\Delta - \partial_t) v_e^\pm \geqslant 0, \quad v_e^\pm \geqslant 0, \quad v_e^+ \cdot v_e^- = 0 \quad \text{in } S_\infty.$$

Hence, we can apply Caffarelli's monotonicity formula [17] to the pair $v_e^\pm$, obtaining that the functional

$$\varphi(r) = \frac{1}{r^4} \int_{S_r} |\nabla v_e^+|^2 G \int_{S_r} |\nabla v_e^-|^2 G,$$

is monotone nondecreasing in r. On the other hand, from the homogeneity of u, it is easy to see that

$$\varphi(r) = r^{4(\kappa-2)} \varphi(1), \quad r > 0.$$

Since $\kappa < 2$, $\varphi(r)$ can be monotone increasing if and only if $\varphi(1) = 0$ and consequently $\varphi(r) = 0$ for all $r > 0$. It follows that one of the functions $v_e^\pm$ is identically zero, which is equivalent to $\partial_e u_0$ being either nonnegative or nonpositive on the entire $\mathbb{R}^n \times (-\infty, 0]$. Since this is true for any tangential direction $e \in \partial B_1$, it thus follows that u_0 depends only on one tangential direction, and is monotone in that direction. Without loss of generality, we may thus assume that $n = 2$ and that the coincidence set at $t = -1$ is an infinite interval $\Lambda_{-1} = \{(x', 0) \in \mathbb{R}^2 \mid u_0(x', 0, -1) = 0\} = (-\infty, a] \times \{0\} =: \Sigma_a^-$. Repeating the monotonicity formula argument above for the pair of functions $\max\{\pm w, 0\}$, where

$$w(x,t) = \begin{cases} -\partial_{x_2} u_0(x_1, x_2, t), & x_2 \geqslant 0 \\ \partial_{x_2} u_0(x_1, x_2, t), & x_2 < 0 \end{cases}$$

we obtain that also w does not change sign. Hence, we get

$$\partial_{x_1} u_0 \geqslant 0, \quad -\partial_{x_2} u_0(x_1, x_2, t) \geqslant 0 \quad \text{in } \mathbb{R}^2_+ \times (-\infty, 0].$$

Let $g_1(x) = \partial_{x_1} u_0(x, -1)$, and $g_2(x) = -\partial_{x_2} u_0(x_1, x_2, -1)$ in $x_2 \geqslant 0$ and $g_2(x) = \partial_{x_2} u_0(x_1, x_2, -1)$ for $x_2 < 0$. Exploiting the fact that g_1 and g_2 are the ground states for the Ornstein-Uhlenbeck operator in $\mathbb{R}^2 \backslash \Sigma_a^-$ and $\mathbb{R}^2 \backslash \Sigma_a^+$, (with $\Sigma_a^+ := [a, \infty) \times \{0\}$) respectively, one reaches the conclusion that $\kappa = 3/2$ and that $g_1(x)$ must be a multiple of $\operatorname{Re}(x_1 + i|x_2|)^{1/2}$. From this, the desired conclusion easily follows. □

From Remark 3.66 and Proposition 3.67, we immediately obtain the following

Theorem 3.68. *Let $u \in \mathfrak{S}^f(S_1^+)$ and μ satisfies the conditions of Theorem 3.61. Assume also $\kappa_\mu = \frac{1}{2} \lim_{r \to 0+} r\mu'(r)/\mu(r) \geqslant 3/2$. Then*

$$\kappa := \Phi_u(0+) \geqslant 3/2.$$

More precisely, we must have

$$\textit{either} \quad \kappa = 3/2 \quad \textit{or} \quad \kappa \geqslant 2.$$

We are now ready to state the optimal regularity of solutions of the parabolic Signorini problem with sufficiently smooth obstacles.

Theorem 3.69. *Let $\varphi \in H^{2,1}(Q_1^+)$, $f \in L_\infty(Q_1^+)$. Assume that $v \in W_2^{2,1}(Q_1^+)$ be such that $\nabla v \in H^{\alpha, \alpha/2}(Q_1^+ \cup Q_1')$ for some $0 < \alpha < 1$, and satisfy*

$$\Delta v - \partial_t v = f \quad \textit{in } Q_1^+, \tag{3.4.21}$$

$$v - \varphi \geqslant 0, \quad -\partial_{x_n} v \geqslant 0, \quad (v - \varphi)\partial_{x_n} v = 0 \quad \textit{on } Q_1'. \tag{3.4.22}$$

Then, $v \in H^{3/2,3/4}(Q_{1/2}^+ \cup Q_{1/2}')$ with

$$\|v\|_{H^{3/2,3/4}(Q_{1/2}^+ \cup Q_{1/2}')} \leqslant C_n \left(\|v\|_{W_\infty^{1,0}(Q_1^+)} + \|f\|_{L_\infty(Q_1^+)} + \|\varphi\|_{H^{2,1}(Q_1')} \right).$$

The proof of Theorem 3.69 will follow from the interior parabolic estimates and the following growth bound of u away from the free boundary $\Gamma(v)$ (which, in turn, can be deduced from Theorem 3.61 by fixing $\sigma = 1/4$ and choosing $\mu(r) = M^2 r^{4-2\sigma}$).

Lemma 3.70. *Let $u \in \mathfrak{S}^f(S_1^+)$ with $\|u\|_{L_\infty(S_1^+)}, \|f\|_{L_\infty(S_1^+)} \leqslant M$. Then,*

$$H_r(u) \leqslant C_n M^2 r^3.$$

3.4.2.4 Regular free boundary points

At this point we turn our attention to the study of points on the free boundary having minimal frequency $\kappa = 3/2$.

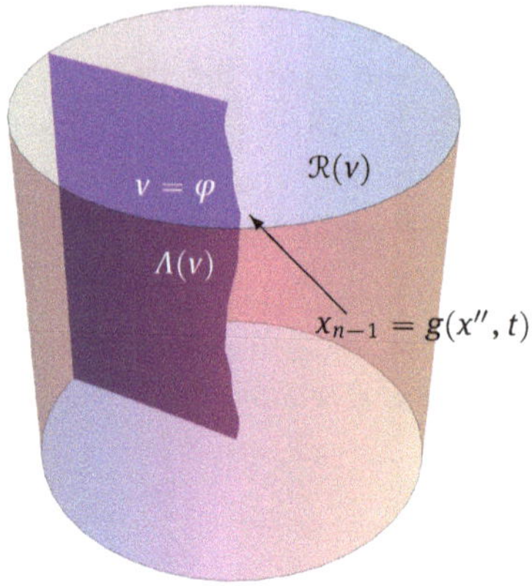

Fig. 3.3: The regular set $\mathcal{R}(v)$ in Q'_δ, given by the graph $x_{n-1} = g(x'')$ with $g \in H^{1,1/2}(Q''_\delta)$ and $\nabla''^{\alpha,\alpha/2}(Q''_\delta)$ by Theorems 3.73 and 3.75

Definition 3.71. *Let $v \in \mathfrak{S}_\varphi(Q_1^+)$ with $\varphi \in H^{\ell,\ell/2}(Q'_1)$, $\ell \geqslant 2$. We say that $(x_0, t_0) \in \Gamma_*(v)$ is a* regular *free boundary point if it has a minimal homogeneity $\kappa = 3/2$. The collection $\mathcal{R}(v)$ of regular free boundary points will be called the* regular set *of v.*

The following basic property about $\mathcal{R}(v)$ is a consequence of the fact that κ does not take any values between $3/2$ and 2 .

Proposition 3.72. *The regular set $\mathcal{R}(v)$ is a relatively open subset of $\Gamma(v)$. In particular, for any $(x_0, t_0) \in \mathcal{R}(v)$ there exists $\delta_0 > 0$ such that*

$$\Gamma(v) \cap Q'_{\delta_0}(x_0, t_0) = \mathcal{R}(v) \cap Q'_{\delta_0}(x_0, t_0).$$

Our goal is to show that, if the thin obstacle φ is sufficiently smooth, then the regular set can be represented locally as a $(n-2)$-dimensional graph of a parabolically Lipschitz function. Further, such function can be shown to have Hölder continuous spatial derivatives. We begin with the following basic result.

Theorem 3.73. *(Lipschitz regularity of $\mathcal{R}(v)$) Let $v \in \mathfrak{S}_\varphi(Q_1^+)$ with $\varphi \in H^{\ell,\ell/2}(Q'_1)$, $\ell \geqslant 3$ and that $(0,0) \in \mathcal{R}(v)$. Then, there exist $\delta = \delta_v > 0$, and $g \in H^{1,1/2}(Q''_\delta)$ (i.e., g is a parabolically Lipschitz function), such that possibly after a rotation in $\mathbb{R}^{n-1}$, one has*

$$\Gamma(v) \cap Q'_\delta = \mathcal{R}(v) \cap Q'_\delta = \{(x', t) \in Q'_\delta \mid x_{n-1} = g(x'', t)\},$$
$$\Lambda(v) \cap Q'_\delta = \{(x', t) \in Q'_\delta \mid x_{n-1} \leqslant g(x'', t)\},$$

The proof of the space regularity follows the same circle of ideas illustrated, for the elliptic case, in Section 3.2.8. To show the $1/2$-Hölder regularity in t (actually better than that), we will use the fact that the $3/2$-homogeneous solutions of the parabolic Signorini problem are t-independent (see Proposition 3.67).

However, in order to carry out the program outlined above, in addition to (i) and (ii) in Theorem 3.65 above, we will need a stronger convergence of the rescalings u_r to the blowups u_0. This will be achieved by assuming a slight increase in the regularity assumptions on the thin obstacle φ, and, consequently, on the regularity of the right-hand side f in (3.4.14).

Lemma 3.74. *Let $u \in \mathfrak{S}^f(S_1^+)$, and suppose that for some $\ell_0 \geqslant 2$*

$$|f(x,t)| \leqslant M\|(x,t)\|^{\ell_0-2} \quad \text{in } S_1^+,$$
$$|\nabla f(x,t)| \leqslant L\|(x,t)\|^{(\ell_0-3)^+} \quad \text{in } Q_{1/2}^+,$$

and

$$H_u(r) \geqslant r^{2\ell_0}, \quad \text{for } 0 < r < r_0.$$

Then, for the family of rescalings $\{u_r\}_{0<r<r_0}$ we have the uniform bounds

$$\|u_r\|_{H^{3/2,3/4}(Q_R^+ \cup Q_R')} \leqslant C_u, \quad 0 < r < r_{R,u}.$$

In particular, if the sequence of rescalings u_{r_j} converges to u_0 as in Theorem 3.65, then over a subsequence

$$u_{r_j} \to u_0, \quad \nabla u_{r_j} \to \nabla u_0 \quad \text{in } H^{\alpha,\alpha/2}(Q_R^+ \cup Q_R'),$$

for any $0 < \alpha < 1/2$ and $R > 0$.

Proof of Theorem 3.73. We only sketch the main ideas of the proof, beginning with the space regularity. For $u \in \mathfrak{S}^f(S_1^+)$ and $r > 0$ define the rescalings

$$u_r(x,t) := \frac{u(rx, r^2 t)}{H_u(r)^{1/2}}, \quad (x,t) \in S_{1/r}^+ = \mathbb{R}_+^n \times (-1/r^2, 0].$$

Theorem 3.65 guarantees the existence of u_0 such that

$$\int_{S_R^+} (|u_{r_j} - u_0|^2 + |t||\nabla(u_{r_j} - u_0)|^2)\rho \to 0.$$

Since $\kappa = 3/2$, Proposition 3.67 ensures that $u_0(x,t) = C\,\mathrm{Re}(x' \cdot e + ix_n)_+^{3/2}$ in S_∞^+. For given $\eta > 0$, define now the thin cone

$$C'(\eta) = \{x' = (x'', x_{n-1}) \in \mathbb{R}^{n-1} \mid x_{n-1} \geqslant \eta |x''|\}.$$

A direct computation shows that for any unit vector $e \in C'(\eta)$

$$\partial_e u_0 \geqslant 0 \text{ in } Q_1^+, \quad \partial_e u_0 \geqslant \delta_{n,\eta} > 0 \text{ in } Q_1^+ \cap \{x_n \geqslant c_n\}.$$

Thanks to Lemma 3.74, we know that $\partial_e u_{r_j} \geqslant 0$ in $Q_{1/2}^+$, and thus, undoing the scaling, $\partial_e u \geqslant 0$ in $Q_{r_\eta}^+$ for any unit vector $e \in C'(\eta)$. The Lipschitz continuity in space follows in a standard fashion.

As for the regularity in time, our goal is to show

$$|g(x,t)-g(x,s)| = o(|t-s|^{1/2}).$$

uniformly in $Q''_{r_\eta/2}$. Arguing by contradiction, we assume

$$|g(x''_j,t_j)-g(x''_j,s_j)| \geqslant C|t_j-s_j|^{1/2}.$$

Let $x'_j = (x''_j, g(x''_j,t_j))$, $y'_j = (x''_j, g(x''_j,s_j))$ and

$$\delta_j = \max\{|g(x''_j,t_j)-g(x''_j,s_j)|, |t_j-s_j|^{1/2}\}.$$

We consider the rescalings of u at (x_j,t_j) by the factor of δ_j:

$$w_j(x,t) = \frac{u(x_j+\delta_j x, t_j+\delta_j^2 t)}{H_{u^{(x_j,t_j)}}(\delta_j)^{1/2}}. \tag{3.4.23}$$

The sequence w_j converges to a homogeneous global solution in S_∞, time-independent, of homogeneity $3/2$. Combining this information with the fact that $\partial_e u \geqslant 0$ in a thin cone, we obtain a contradiction. □

We next observe that the regularity of the function g can be improved with an application of a boundary Harnack principle.

Theorem 3.75. *(Hölder regularity of $\nabla'' g$)*
In the conclusion of Theorem 3.73, one can take $\delta > 0$ so that $\nabla'' g \in H^{\alpha,\frac{\alpha}{2}}(Q''_\delta)$ for some $\alpha > 0$.

The proof of this result relies on two crucial ingredients. The first one is the following non-degeneracy of $\partial_e u$:

$$\partial_e u \geqslant c d(x,t) \text{ in } Q^+_\delta,$$

where $d(x,t)$ denotes the parabolic distance from coincidence set $\Lambda(v) \cap Q'_\delta$. This property, in turn, relies on a suitable parabolic version of Lemma 3.48. The second ingredient in the proof of Theorem 3.75 is the following version of the parabolic boundary Harnack principle for domains with thin Lipschitz complements established in [57]*Section 7. To state the result, we will need the following notations. For a given $L \geqslant 1$ and $r > 0$ denote

$$\begin{aligned}
\Theta''_r &= \{(x'',t) \in \mathbb{R}^{n-2} \times \mathbb{R} \mid |x_i| < r, i = 1,\dots,n-2, -r^2 < t \leqslant 0\},\\
\Theta'_r &= \{(x',t) \in \mathbb{R}^{n-1} \times \mathbb{R} \mid (x'',t) \in \Theta''_r, |x_{n-1}| < 4nLr\},\\
\Theta_r &= \{(x,t) \in \mathbb{R}^{n} \times \mathbb{R} \mid (x'',t) \in \Theta'_r, |x_n| < r\}.
\end{aligned}$$

Lemma 3.76. *(Boundary Harnack principle) Let*

$$\Lambda = \{(x',t) \in \Theta'_1 \mid x_{n-1} \leqslant g(x'',t)\}$$

for a parabolically Lipschitz function g in Θ_1'' with Lipschitz constant $L \geqslant 1$ such that $g(0,0) = 0$. Let u_1, u_2 be two continuous nonnegative functions in Θ_1 such that for some positive constants c_0, C_0, M, and $i = 1, 2$,

i) $0 \leqslant u_i \leqslant M$ *in* Θ_1 *and* $u_i = 0$ *on* Λ,

ii) $|(\Delta - \partial_t)u_i| \leqslant C_0$ *in* $\Theta_1 \backslash \Lambda$,

iii) $u_i(x,t) \geqslant c_0\, d(x,t)$ *in* $\Theta_1 \backslash \Lambda$, *where* $d(x,t) = \sup\{r \mid \tilde{\Theta}_r(x,t) \cap \Lambda = \varnothing\}$.

Assume additionally that u_1 and u_2 are symmetric in x_n. Then, there exists $\alpha \in (0,1)$ such that

$$\frac{u_1}{u_2} \in H^{\alpha,\alpha/2}(\Theta_{1/2}).$$

Furthermore, α and the bound on the corresponding norm $\|u_1/u_2\|_{H^{\alpha,\alpha/2}(\Theta_{1/2})}$ depend only on n, L, c_0, C_0, and M. □

Remark 3.77. *Lemma 3.76 is the parabolic version of Theorem 3.33. Unlike the elliptic case, it cannot be reduced to the other known results in the parabolic setting. We also note that this version of the boundary Harnack is for functions with nonzero right-hand side and therefore the nondegeneracy condition as in iii) is necessary.*

3.4.2.5 Singular free boundary points

The main goal of this section is to establish a structural theorem for the set of the so-called *singular points*, i.e. the points where the coincidence set $\{v = \varphi\}$ has zero $\mathcal{H}^n$-density in the thin manifold with respect to the thin parabolic cylinders. This corresponds to free boundary points with frequency $\kappa = 2m$, $m \in \mathbb{N}$. We will show that the blowups at those points are parabolically κ-homogeneous polynomials.

As in the approach in [37], described in Sections 3.2.10.1-3.2.10.3, the main tools are parabolic versions of monotonicity formulas of Weiss and Monneau type. These are instrumental in proving the uniqueness of the blowups at singular free boundary points (x_0, t_0), and consequently obtain a Taylor expansion of the type

$$v(x,t) - \varphi(x',t) = q_\kappa(x - x_0, t - t_0) + o(\|(x - x_0, t - t_0)\|^\kappa), \quad t \leqslant t_0,$$

where q_κ is a polynomial of parabolic degree κ that depends continuously on the singular point (x_0, t_0) with frequency κ. We note explicitely that such expansion holds only for $t \leqslant t_0$ and may fail for $t > t_0$ (see Remark 3.81 below). Nevertheless, this expansion essentially holds when restricted to singular points (x,t), even for $t \geqslant t_0$. This is necessary in order to verify the compatibility condition in a parabolic version of the Whitney's extension theorem. Using the latter we are then ready to prove a structural theorem for the singular set. It should be mentioned at this moment that one difference between the parabolic case treated in this section and its elliptic counterpart is the presence of new types of singular points, which we call *time-like*. At such points the blowup may become independent of the space variables x'. We show that such singular points are contained in a countable union of graphs of the type

$$t = g(x_1, \dots, x_{n-1}),$$

where g is a C^1 function. The other singular points, which we call *space-like*, are contained in countable union of d-dimensional $C^{1,0}$ manifolds ($d < n-1$). After a possible rotation of coordinates in $\mathbb{R}^{n-1}$, such manifolds are locally representable as graphs of the type

$$(x_{d+1}, \dots, x_{n-1}) = g(x_1, \dots, x_d, t),$$

with g and $\partial_{x_i} g$, $i = 1, \dots, d$, continuous.

We now proceed to make these statements more precise. Since the overall strategy is similar to the one described in Sections 3.2.10.1-3.2.10.3, we will omit the proofs and focus only on the main differences between the elliptic and parabolic cases. For futher details, we refer the interested reader to [27].

Definition 3.78. *Let $v \in \mathfrak{S}_\varphi(Q_1^+)$ with $\varphi \in H^{\ell,\ell/2}(Q_1')$, $\ell \geqslant 2$. We say that $(x_0, t_0) \in \Gamma_*(v)$ is* singular *if*

$$\lim_{r\to 0+} \frac{\mathcal{H}^n(\Lambda(v) \cap Q_r'(x_0, t_0))}{\mathcal{H}^n(Q_r')} = 0.$$

We will denote the set of singular points by $\Sigma(u)$ and call it the singular set. *We can further classify singular points according to the homogeneity of their blowup, by defining*

$$\Sigma_\kappa(v) := \Sigma(v) \cap \Gamma_\kappa^{(\ell)}(v), \quad \kappa \leqslant \ell.$$

The following proposition gives a complete characterization of the singular points in terms of the blowups and the generalized frequency. In particular, it establishes that

$$\Sigma_\kappa(u) = \Gamma_\kappa(u) \quad \text{for } \kappa = 2m < \ell,\ m \in \mathbb{N}.$$

Proposition 3.79. *Let $u \in \mathfrak{S}^f(S_1^+)$ with $|f(x,t)| \leqslant M\|(x,t)\|^{\ell-2}$ in S_1^+, $|\nabla f(x,t)| \leqslant L\|(x,t)\|^{\ell-3}$ in $Q_{1/2}^+$, $\ell \geqslant 3$ and $0 \in \Gamma_\kappa^{(\ell)}(u)$ with $\kappa < \ell$. Then, the following statements are equivalent:*

(i) $0 \in \Sigma_\kappa(u)$.

(ii) *any blowup of u at the origin is a nonzero parabolically κ-homogeneous polynomial p_κ in S_∞ satisfying*

$$\Delta p_\kappa - \partial_t p_\kappa = 0, \quad p_\kappa(x', 0, t) \geqslant 0, \quad p_\kappa(x', -x_n, t) = p_\kappa(x', x_n, t).$$

We denote this class by $\mathbb{P}_\kappa$.

(iii) $\kappa = 2m$, $m \in \boldsymbol{N}$.

We now state the two main results of this section.

Theorem 3.80. *Let $u \in \mathfrak{S}^f(S_1^+)$ with $|f(x,t)| \leqslant M\|(x,t)\|^{\ell-2}$ in S_1^+, $|\nabla f(x,t)| \leqslant L\|(x,t)\|^{\ell-3}$ in $Q_{1/2}^+$, $\ell \geqslant 3$, and $0 \in \Sigma_\kappa(u)$ for $\kappa = 2m < \ell$, $m \in \mathbf{N}$. Then, there exists a nonzero $p_\kappa \in \mathbb{P}_\kappa$ such that*

$$u(x,t) = p_\kappa(x,t) + o(\|(x,t)\|^\kappa), \quad t \leqslant 0.$$

Moreover, if $v \in \mathfrak{S}_\varphi(Q_1^+)$ with $\varphi \in H^{\ell,\ell/2}(Q_1')$, $(x_0,t_0) \in \Sigma_\kappa(v)$ and we let $v^{(x_0,t_0)} = v(x_0+\cdot, t_0+\cdot)$, then the mapping $(x_0,t_0) \mapsto p_\kappa^{(x_0,t_0)}$ from $\Sigma_\kappa(v)$ to $\mathbb{P}_\kappa$ is continuous.

Remark 3.81. *We want to emphasize here that the asymptotic development, as stated in Theorem 3.80, does not generally hold for $t > 0$. Indeed, consider the following example. Let $u : \mathbb{R}^n \times \mathbb{R} \to \mathbb{R}$ be a continuous function such that*
- *$u(x,t) = -t - x_n^2/2$ for $x \in \mathbb{R}^n$ and $t \leqslant 0$.*
- *In $\{x_n \geqslant 0, t \geqslant 0\}$, u solves the Dirichlet problem*

$$\begin{aligned} \Delta u - \partial_t u &= 0, && x_n > 0, t > 0,\\ u(x,0) &= -x_n^2, && x_n \geqslant 0,\\ u(x',0,t) &= 0 && t \geqslant 0. \end{aligned}$$

- *In $\{x_n \leqslant 0, t \geqslant 0\}$, we extend the function by even symmetry in x_n:*

$$u(x',x_n,t) = u(x',-x_n,t).$$

It is easy to see that u solves the parabolic Signorini problem with zero obstacle and zero right-hand side in all of $\mathbb{R}^n \times \mathbb{R}$. Moreover, u is homogeneous of degree two and clearly $0 \in \Sigma_2(u)$. Now, if $p(x,t) = -t - x_n^2/2$, then $p \in \mathbb{P}_2$ and we have the following equalities:

$$\begin{aligned} u(x,t) &= p(x,t), && &&\text{for } t \leqslant 0,\\ u(x',0,t) &= 0, && p(x',0,t) = -t && \text{for } t \geqslant 0. \end{aligned}$$

So for $t \geqslant 0$ the difference $u(x,t) - p(x,t)$ is not $o(\|(x,t)\|^2)$, despite being zero for $t \leqslant 0$.

In order to state the aforementioned structural theorem, we need the following definitions.

Definition 3.82. *For a singular point $(x_0,t_0) \in \Sigma_\kappa(v)$ we define*

$$\begin{aligned} d_\kappa^{(x_0,t_0)} = \dim \{\xi \in \mathbb{R}^{n-1} \mid \xi \cdot \nabla_{x'} \partial_{x'}^{\alpha'} \partial_t^j p_\kappa^{(x_0,t_0)} = 0 \\ \text{for any } \alpha' = (\alpha_1, \dots, \alpha_{n-1}) \text{ and } j \geqslant 0 \text{ such that } |\alpha'| + 2j = \kappa - 1\}, \end{aligned}$$

which we call the spatial dimension *of $\Sigma_\kappa(v)$ at (x_0,t_0). Clearly, $d_\kappa^{(x_0,t_0)}$ is an integer between 0 and $n-1$. Then, for any $d = 0, 1, \dots, n-1$ define*

$$\Sigma_\kappa^d(v) := \{(x_0,t_0) \in \Sigma_\kappa(v) \mid d_\kappa^{(x_0,t_0)} = d\}.$$

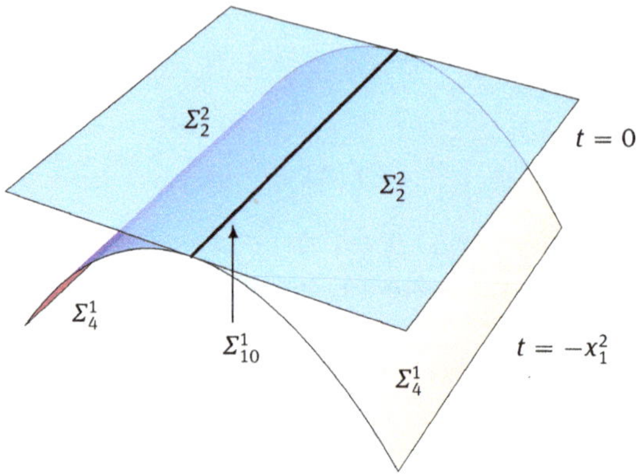

Fig. 3.4: Structure of the singular set $\Sigma(v) \subset \mathbb{R}^2 \times (-\infty, 0]$ for the solution v with $v(x_1, x_2, 0, t) = -t(t + x_1^2)^4$, $t \leqslant 0$ with zero thin obstacle. Note that the points on Σ_4^1 and Σ_{10}^1 are space-like, and the points on Σ_2^2 are time-like.

Definition 3.83. *We say that a $(d + 1)$-dimensional manifold $\mathcal{S} \subset \mathbb{R}^{n-1} \times \mathbb{R}$, $d = 0, \dots, n - 2$, is* space-like *of class $C^{1,0}$, if locally, after a rotation of coordinate axes in $\mathbb{R}^{n-1}$ one can represent it as a graph*

$$(x_{d+1}, \dots, x_{n-1}) = g(x_1, \dots, x_d, t),$$

where g is of class $C^{1,0}$, i.e., g and $\partial_{x_i} g$, $i = 1, \dots, d$ are continuous.

We say that $(n - 1)$-dimensional manifold $\mathcal{S} \subset \mathbb{R}^{n-1} \times \mathbb{R}$ is time-like *of class C^1 if it can be represented locally as*

$$t = g(x_1, \dots, x_{n-1}),$$

where g is of class C^1.

Theorem 3.84. *(Structure of the singular set) Let $v \in \mathfrak{S}_\varphi(Q_1^+)$ with $\varphi \in H^{\ell,\ell/2}(Q_1')$, $\ell \geqslant 3$. Then, for any $\kappa = 2m < \ell$, $m \in \mathbb{N}$, we have $\Gamma_\kappa(v) = \Sigma_\kappa(v)$. Moreover, for every $d = 0, 1, \dots, n - 2$, the set $\Sigma_\kappa^d(v)$ is contained in a countable union of $(d + 1)$-dimensional space-like $C^{1,0}$ manifolds and $\Sigma_\kappa^{n-1}(v)$ is contained in a countable union of $(n - 1)$-dimensional time-like C^1 manifolds.*

For a small illustration, see Fig. 3.4.

The following two monotonicity formulas of Weiss and Monneau type play a crucial role in the proofs of Theorems 3.80 and 3.84.

Theorem 3.85. *(Weiss-type monotonicity formula) Let $u \in \mathfrak{S}^f(S_1^+)$ with $|f(x,t)| \leqslant M\|(x,t)\|^{\ell-2}$ in S_1^+, $\ell \geqslant 2$. For any $\kappa \in (0,\ell)$, define the Weiss energy functional*

$$W_u^\kappa(r) := \frac{1}{r^{2\kappa+2}} \int_{S_r^+} \left(|t||\nabla u|^2 - \frac{\kappa}{2}u^2\right) G$$
$$= \frac{1}{r^{2\kappa}} \left(I_u(r) - \frac{\kappa}{2} H_u(r)\right), \quad 0 < r < 1.$$

Then, for any $\sigma < \ell - \kappa$ there exists $C > 0$ depending only on σ, ℓ, M, and n, such that

$$\frac{d}{dr} W_u^\kappa \geqslant -C r^{2\sigma-1}, \quad \text{for a.e. } r \in (0,1).$$

In particular, the function

$$r \mapsto W_u^\kappa(r) + C r^{2\sigma}$$

is monotonically nondecreasing for $r \in (0,1)$.

The proof is by direct computation. Note that in Theorem 3.85 we do not require $0 \in \Gamma_\kappa^{(\ell)}(u)$. However, if we do so, then we will have the following fact.

Lemma 3.86. *Let u be as in Theorem 3.85, and assume additionally that $0 \in \Gamma_\kappa^{(\ell)}(u)$, $\kappa < \ell$. Then,*

$$W_u^\kappa(0+) = 0.$$

The proof of this result uses the following growth estimate.

Lemma 3.87. *Let $u \in \mathfrak{S}^f(S_1^+)$ with $|f(x,t)| \leqslant M\|(x,t)\|^{\ell-2}$, $\ell \geqslant 2$, and $0 \in \Gamma_\kappa^{(\ell)}(u)$ with $\kappa < \ell$ and let $\sigma < \ell - \kappa$. Then*

$$H_r(u) \leqslant C(\|u\|^2_{L_2(S_1^+,G)} + M^2) r^{2\kappa}, \quad 0 < r < 1,$$

with C depending only on σ, ℓ, n.

Theorem 3.88. *(Monneau-type monotonicity formula) Let $u \in \mathfrak{S}^f(S_1^+)$ with $|f(x,t)| \leqslant M\|(x,t)\|^{\ell-2}$ in S_1^+, $|\nabla f(x,t)| \leqslant L\|(x,t)\|^{\ell-3}$ in $Q_{1/2}^+$, $\ell \geqslant 3$. Suppose that $0 \in \Sigma_\kappa(u)$ with $\kappa = 2m < \ell$, $m \in \mathbb{N}$. Further, let p_κ be any parabolically κ-homogeneous caloric polynomial from class $\mathbb{P}_\kappa$ as in Proposition 3.79. For any such p_κ, define Monneau's functional as*

$$M_{u,p_\kappa}^\kappa(r) := \frac{1}{r^{2\kappa+2}} \int_{S_r^+} (u - p_\kappa)^2 G, \quad 0 < r < 1,$$
$$= \frac{H_w(r)}{r^{2\kappa}}, \quad \text{where } w = u - p_\kappa.$$

Then, for any $\sigma < \ell - \kappa$ there exists a constant C, depending only on σ, ℓ, M, and n, such that

$$\frac{d}{dr} M_{u,p_\kappa}^\kappa(r) \geqslant -C\left(1 + \|u\|_{L_2(S_1^+,G)} + \|p_\kappa\|_{L_2(S_1^+,G)}\right) r^{\sigma-1}.$$

In particular, the function

$$r \mapsto M^{\kappa}_{u,p_\kappa}(r) + Cr^{\sigma}$$

is monotonically nondecreasing for $r \in (0,1)$ *for a constant* C *depending* σ, ℓ, M, n, $\|u\|_{L_2(S_1^+,G)}$, *and* $\|p_\kappa\|_{L_2(S_1^+,G)}$.

The proof of Theorem 3.88 relies on Theorem 3.85, and Lemmas 3.86 and 3.87. To proceed with the proofs of Theorems 3.80 and 3.84, we observe that, similarly to the elliptic case, it will be more convenient to work with a slightly different type of rescalings and blowups than the ones used up to now. Namely, we will work with the following *κ-homogeneous rescalings*

$$u_r^{(\kappa)}(x,t) := \frac{u(rx, r^2 t)}{r^{\kappa}}$$

and their limits as $r \to 0+$. The next lemma, which is the parabolic analogue of Lemma 3.44, shows the viability of this approach.

Lemma 3.89. *Let* $u \in \mathfrak{S}^f(S_1^+)$ *with* $|f(x,t)| \leqslant M\|(x,t)\|^{\ell-2}$ *in* S_1^+, $|\nabla f(x,t)| \leqslant L\|(x,t)\|^{\ell-3}$ *in* $Q^+_{1/2}$, $\ell \geqslant 3$, *and* $0 \in \Sigma_\kappa(u)$ *for* $\kappa < \ell$. *Then, there exists* $c = c_u > 0$ *such that*

$$H_u(r) \geqslant c\, r^{2\kappa}, \quad \textit{for any } 0 < r < 1.$$

We explicitely observe that, by combining Lemmas 3.87 and 3.89, we obtain that the κ-homogeneous rescalings are essentially equivalent to the ones introduced in Definition 3.64. Using this fact and Theorem 3.88, it is possible to show the uniqueness of the homogeneous blow-ups. The proofs of Theorems 3.80 and 3.84, at this point, proceed along the lines of their elliptic counterparts (see Section 3.2.10.3), and therefore we omit them. We only point out that, in the first step of the proof of Theorem 3.84, one needs the following parabolic version of Whitney's extension theorem.

Theorem 3.90. *Let* $\{f_{\alpha,j}\}_{|\alpha|+2j\leqslant m}$ *be a family of functions on* E, *with* $f_{0,0} \equiv f$, *satisfying the following compatibility conditions: there exists a family of moduli of continuity* $\{\omega_{\alpha,j}\}_{|\alpha|+2j\leqslant 2m}$, *such that*

$$f_{\alpha,j}(x,t) = \sum_{|\beta|+2k\leqslant 2m-|\alpha|-2j} \frac{f_{\alpha+\beta,j+k}(x_0,t_0)}{\beta!k!}(x-x_0)^{\beta}(t-t_0)^k + R_{\alpha,j}(x,t;x_0,t_0)$$

and

$$|R_{\alpha,j}(x,t;x_0,t_0)| \leqslant \omega_{\alpha,j}(\|(x-x_0,t-t_0)\|)\|(x-x_0,t-t_0)\|^{2m-|\alpha|-2j}.$$

Then, there exists a function $F \in C^{2m,m}(\mathbb{R}^n \times \mathbb{R})$ *such that* $F = f$ *on* E *and moreover* $\partial_x^{\alpha}\partial_t^{j} F = f_{\alpha,j}$ *on* E, *for* $|\alpha| + 2j \leqslant 2m$.

Finally, in the second step of the proof of Theorem 3.84, some care needs to be applied in the use of the implicit function theorem, as two cases need to be considered. When

$d \in \{0, 1, \ldots, n-2\}$, $\Sigma_\kappa^d(v)$ is contained in the countable union of d-dimensional space-like $C^{1,0}$ manifolds. When instead $d = n-1$, $\Sigma_\kappa^d(v)$ is contained in a time-like $(n-1)$-dimensional C^1 manifold, as required.

Comments and further readings

– The continuity of the temperature, in space and time variables, in boundary heat control problems (both of Stefan and porous media type) has been established by Athanasopoulos and Caffarelli in [8]. In addition, the authors extend this result to the fractional order case. The proof is based on a combination of penalization and De Giorgi's method.

– Higher regularity of the time derivative in the parabolic thin obstacle problem has recently been established by Petrosyan and Zeller in [58] and Athanasopoulos, Caffarelli and Milakis in [9]. It was already known from Arkhipova A, Uraltseva ([4]) that if the initial data $\varphi_0 \in W^2_\infty(B_1^+)$, then the time derivative $\partial_t v$ of the solution to (3.4.7)-(3.4.10) is locally bounded in $Q_1^+ \cap Q_1'$. This assumption on the initial data, however, is rather restrictive and excludes time-independent solutions as in Proposition 3.67. The first main result in [58] shows that $\partial_t v$ is in fact bounded, without any extra assumptions on the initial data, even though some more regularity on the thin obstacle φ is required. The key observation in the proof is that incremental quotients of the solution in the time variable satisfy a differential inequality. In addition, the authors prove the Hölder continuity of $\partial_t v$ at regular free boundary points, which in turn allows them to show that the free boundary, at regular points, is a $C^{1,\alpha}$ surface both in space and time. The Hölder continuity in time for solutions (and the ensuing $C^{1,\alpha}$ regularity of the free boundary) has also been established in [9], with a different approach based on quasi-convexity properties of the solution. The regularity of the time derivative plays also a crucial role in [11], where Banerjee, Smit Vega Garcia and Zeller show that the free boundary near a regular point is C^∞ in space and time when the obstacle is zero.

Some auxiliary tools

In this section we collect some results on the fractional Laplacian and the extension operator L_a from [23] and [22]. Standard reference for the fractional Laplacian is Landkof's book [51]. See also the recent book by Molica Bisci, Radulescu and Servadei [53].

Supersolutions and comparison for the fractional Laplacian
The definition of supersolution is given for $u \in \mathcal{P}_s'$ the dual of

$$\mathcal{P}_s = \left\{ f \in C^\infty(\mathbb{R}^n) : \left(1 + |x|^{n+2s}\right) D^k f \text{ is bounded, } \forall k \geqslant 0 \right\},$$

endowed with the topology induced by the seminorms

$$|f|_{k,s} = \sup \left| \left(1 + |x|^{n+2s}\right) D^k f \right| .$$

and the meaning is that it is a nonnegative measure:

Let $u \in \mathcal{P}_s'$. We say that $(-\Delta)^s u \geqslant 0$ in an open set Ω if $\langle (-\Delta)^s u, \varphi \rangle \geqslant 0$, for every nonnegative test function $\varphi \in C^\infty(\mathbb{R}^n)$, rapidly decreasing at infinity.

Every supersolution shares some properties of superharmonic functions. For instance, u is lower-semicontinuous.

Proposition A1. Let $(-\Delta)^s u \geqslant 0$ in an open set Ω, then u is lower-semicontinuous in Ω.

Moreover, if the restriction of u on $\operatorname{supp}((-\Delta)^s u)$ is continuous then, u is continuous everywhere. Precisely, we have:

Proposition A2. Let v be a bounded function in $\mathbb{R}^n$ such that $(-\Delta)^s v \geqslant 0$. If $E = \operatorname{supp}((-\Delta)^s v)$ and $v_{|E}$ is continuous, then v is continuous in $\mathbb{R}^n$.

Due to the nonlocal nature of $(-\Delta)^s$, a comparison theorem in a domain Ω must take into account what happens outside Ω. Indeed we have:

Proposition A3. (Comparison) Let $\Omega \Subset \mathbb{R}^n$ be an open set. Let $(-\Delta)^s u \geqslant 0$ and $(-\Delta)^s v \leqslant 0$ in Ω, such that $u \geqslant v$ in $\mathbb{R}^n \backslash \Omega$ and $u - v$ is lower-semicontinuous in $\overline{\Omega}$. Then $u \geqslant v$ in $\mathbb{R}^n$. Moreover, if $x \in \Omega$ and $u(x) = v(x)$ then $u = v$ in $\mathbb{R}^n$.

Also, the set of supersolutions is a directed set, as indicated by the following proposition.

Proposition A4. Let $\Omega \Subset \mathbb{R}^n$ be an open set. Let $(-\Delta)^s u_1 \geqslant 0$ and $(-\Delta)^s u_2 \geqslant 0$ in Ω, such that $u \geqslant v$ in $\mathbb{R}^n \backslash \Omega$. Then $u = \min\{u_1, u_2\}$ is a supersolution in Ω.

Estimates for the operator L_a
As we have already noted, the operator L_a is a particular case of the class of degenerate elliptic operators considered in [30]. For the following result see Theorems 2.3.8 and 2.3.12 in that paper.

(2) $v \in C^{0,\alpha}\left(B_{r/2}\right)$ *for some* $\alpha \leqslant 1$ *and if* $f = 0$

$$\|v\|_{C^{0,\alpha}(B_{r/2})} \leqslant \frac{C}{r^\alpha} \operatorname*{osc}_{B_r} v.$$

Using the translational invariance of the equation in the x variable, we obtain the following result.

Lemma A1. (Schauder type estimates) *Assume $L_a v = 0$ in B_r. Then, for every $k \geqslant 1$, integer:*

$$\left\| D_x^k v \right\|_{L^\infty(B_{r/2})} \leqslant \frac{C}{r^k} \operatorname*{osc}_{B_r} v$$

and[3.3]

$$\left| D_x^k v \right|_{C^{0,\alpha}(B_{r/2})} \leqslant \frac{C}{r^{k+\alpha}} \operatorname*{osc}_{B_r} v.$$

Using the above Theorem and the equation in the form

$$\Delta_x v = -v_{yy} - \frac{a}{y} v_y$$

we get:

Lemma A2. *Assume $L_a v = 0$ in B_r. Then, for every $r \leqslant 1$:*

$$\left\| v_{yy} + \frac{a}{y} v_y \right\|_{L^\infty(B_{r/2})} \leqslant \frac{C}{r^2} \operatorname*{osc}_{B_r} v.$$

The next is a Liouville type result.

Lemma A3. Let *v a solution of $L_a v(X) = 0$ in $\mathbb{R}^{n+1}$. Assume that*

$$v(x, y) = v(x, -y) \quad \text{and} \quad |v(X)| \leqslant C\left(1 + |X|^\gamma\right), \gamma \geqslant 0.$$

Then v is a polynomial.

We now state a mean value property for supersolution of a nonhomogeneous solution:

Lemma A4. (Mean value property) *Let v be a solution of*

$$L_a v(X) \leqslant C\left|y^a\right| |X|^k \quad \text{in } B_1.$$

Then, for every $r \leqslant 1$,

$$v(0) \geqslant \frac{1}{\omega_{n+a} r^{n+a}} \int_{\partial B_r} v(X) |y|^a \, dS - Cr^{k+2}$$

where

$$\omega_{n+a} = \int_{\partial B_1} |y|^a \, dS.$$

3.3 $|w|_{C^{0,\alpha}(D)}$ denotes the seminorm

$$\sup_{x, y \in D} \frac{|w(x) - w(y)|}{|x - y|^\alpha}.$$

Poincaré type inequalities in the context of weighted Sobolev spaces can be found in [30]. In our case, letting $\overline{v} = \frac{1}{\omega_{n+a} r^{n+a}} \int_{\partial B_r} v(X) |y|^a \, d\sigma$,we have[3.4]:

Lemma A5. (Poincaré inequalities) *Let* $v \in W^{1,2}(B_1, |y|^a)$ *. Then, for* $r \leqslant 1$ *:*

$$\int_{\partial B_r} (v(X) - \overline{v})^2 |y|^a \, d\sigma \leqslant C(a, n) \, r \int_{B_r} |\nabla v(X)| \, |y|^a \, dX.$$

and

$$\int_{\partial B_1} (v(X) - v(rX))^2 |y|^a \, d\sigma \leqslant C(a, n, r) \int_{B_1} |\nabla v(X)| \, |y|^a \, dX. \qquad (.0.24)$$

The first inequality is standard, The second one can be proved by integrating ∇v along the lines sX with $s \in (r, 1)$.

Acknowledgment: D.D. was supported in part by NSF grant DMS-1101246.

Bibliography

[1] Allen M, Lindgren E, Petrosyan A: *The two-phase fractional obstacle problem*, SIAM J. Math. Anal. 47 (3), 1879–1905, (2015).

[2] Alt W, Caffarelli L. A, Friedman, A: *Variational problems with two phases and their free boundaries*, Trans. A.M.S. 282 (2), 431-461, (1984).

[3] Arkhipova A. A, Uraltseva N. N: *Regularity of the solution of a problem with a two-sided limit on a boundary for elliptic and parabolic equations*, Trudy Mat. Inst. Steklov. 179, 5–22, 241, (1988) (Russian); Translated in Proc. Steklov Inst. Math. 2, 1–19, (1989) Boundary value problems of mathematical physics, 13 (Russian).

[4] Arkhipova A, Uraltseva, N. N: *Sharp estimates for solutions of a parabolic Signorini problem*, Math. Nachr., 177, 11–29, (1996).

[5] Athanasopoulous I: *Regularity of the solution of an evolution problem with inequalities on the boundary*, Comm. Partial Differential Equations 7 (12), 1453–1465, (1982).

[6] Athanasopoulos I, Caffarelli L.A: *A Theorem of Real Analysis and its Application to Free Boundary problems*, C.P.A.M. 38, 499-502, (1985).

[7] Athanasopoulos I, Caffarelli L.A: *Optimal regularity of lower dimensional obstacle problems*, Zap. Nauchn. Semin. POMI 310, 49-66,226 (2004).

[8] Athanasopoulos I, Caffarelli L.A: *Continuity of the temperature in boundary heat control problems*, Adv. Math. 224 (1), 293–315, (2010).

[9] Athanasopoulos I, Caffarelli L. A, Milakis E: *Parabolic Obstacle Problems. Quasi-convexity and Regularity*, preprint arXiv: 1601.01516, (2016).

3.4 A function $v \in W^{1,2}(B_1, |y|^a)$ has a trace in $L^2(\partial B_1, |y|^a)$ and the trace operator is compact.

[10] Athanasopoulos I, Caffarelli L. A, Salsa S: *The structure of the free boundary for lower dimensional obstacle problems*, American J. Math. 130 (2), 485-498, (2008).
[11] Banerjee A, Smit Vega Garcia M, Zeller A: *Higher regularity of the free boundary in the parabolic Signorini problem*, preprint arXiv:1601.02976, (2016).
[12] Barrios I, Figalli A, Ros-Oton X: *Global regularity for the free boundary in the obstacle problem for the fractional Laplacian*, Amer. J. Math., in press (2017).
[13] Bouchaud J.P, Georges, A: *Anomalous diffusion in disordered media: Statistical mechanics models and physical interpretations.* Physic reports, 195 (4&5), (1990).
[14] Brézis H: *Problèmes unilatéraux*, J. Math. Pures Appl. 51 (9), 1–168, (1972).
[15] Brézis H, Kinderlehrer D: *The Smoothness of Solutions to Nonlinear Variational Inequalities*, Indiana Univ. Math. Journal 23 (9), 831-844, (1974).
[16] Caffarelli L.A: *Further regularity for the Signorini problem*, Comm. P.D.E. 4 (9), 1067-1075, (1979).
[17] Caffarelli L. A:, *A monotonicity formula for heat functions in disjoint domains*, Boundary value problems for partial differential equations and applications, RMA Res. Notes Appl. Math., 29, Masson, Paris, 53–60, (1993).
[18] Caffarelli L.A: *The Obstacle Problem Revisited*, J. Fourier Analysis Appl. 4, 383-402, (1998).
[19] Caffarelli L. A, Figalli A.: *Regularity of solutions to the parabolic fractional obstacle problem*, J. Reine angew. Math. 680, 191-233, (2013).
[20] Caffarelli L, Ros-Oton X, Serra J: *Obstacle problems for integro-differential operators: regularity of solutions and free boundaries*, Inventiones Mathematicae 208 (2017), 1155-1211.
[21] Caffarelli L. A, Salsa S: *A geometric approach to free boundary problems*, A.M.S. Providence, G.S.M., vol 68.
[22] Caffarelli L.A, Salsa S, Silvestre L: *Regularity estimates for the solution and the free boundary of the obstacle problem for the fractional Laplacian*, Inventiones Mathematicae 171, 425-461, (2008).
[23] Caffarelli L. A, Silvestre L: *An extension problem related to the fractional Laplacian*, Comm. P.D.E. 32(8), 1245-1260 (2007).
[24] Carrillo J. A, Delgadino M. G., Mellet A.: *Regularity of local minimizers of the interaction energy via obstacle problems, Comm. Math. Phys.* 343 (3), 747-781, (2016).
[25] Colombo M, Spolaor L, Velichkov B: *Direct epiperimetric inequalities for the thin obstacle problem and applications*, preprint arXiv:1709.03120, (2017).
[26] Cont R, Tankov P: *Finantial modelling with jump processes*, Chapman hall/CRC Financ. Math. Ser., Boca Raton , FL (2004).
[27] Danielli D, Garofalo N, Petrosyan A, To T: *Optimal Regularity and the Free Boundary*
in the Parabolic Signorini Problem, to appear in Memoirs A.M.S.
[28] De Silva D, Savin O: *Boundary Harnack estimates in slit domains and applications to thin free boundary problems*, Rev. Mat. Iberoam., to appear.

[29] Duvaut G, Lions J.L: *Les inequations en mechanique et en physique*, Paris, Dunod (1972).

[30] Fabes E, Kenig C, Serapioni R: *The local regularity of solutions of degenerate elliptic equations*, Comm. P.D.E. 7(1), 77-116, (1982).

[31] Fichera G: *Problemi elastostatici con vincoli unilaterali: Il problema di Signorini con ambigue condizioni al contorno*, Atti Accad. Naz. Lincei Mem. Cl. Sci. Fis. Mat. Natur. Sez. I, 7 (8), 91–140, (1963/1964).

[32] Focardi M, Spadaro E: *An epiperimetric inequality for the thin obstacle problem*, Adv. Differential Equations 21 (1-2), 91–140, (2016).

[33] Frehse J: *On Signorini's problem and variational problems with thin ostacles*, Ann. Sc. Norm. Sup. Pisa 4, 343-362, (1977).

[34] Garofalo N, Lin F:, *Monotonicity properties of variational integrals, A_p weights and unique continuation*, Indiana Univ. Math. J. 35 (2), 245–268, (1986).

[35] Garofalo N, Lin F:, *Unique continuation for elliptic operators: a geometric-variational approach*, Comm. Pure Appl. Math. 40 (3), 347–366, (1987).

[36] Garofalo N: *Unique continuation for a class of elliptic operators which degenerate on a manifold of arbitrary codimension.* J. Diff. Eq. 104 (1), 117-146, (1993).

[37] Garofalo N, Petrosyan A: *Some new monotonicity formulas and the singular set in the lower dimensional obstacle problem*, Inventiones Mathematicae 177, 415-464 (2009).

[38] Garofalo N, Petrosyan A, and Smit Vega Garcia M: *An epiperimetric inequality approach to the regularity of the free boundary in the Signorini problem with variable coefficients*, J. Math. Pures Appl. (9) 105 (6), 745–787, (2016).

[39] Garofalo N, Petrosyan A, Pop C., and Smit Vega Garcia M: *Regularity of the free boundary for the obstacle problem for the fractional Laplacian with drift,*, to appear in Ann. Inst. H. Poincaré Anal. Non Linéaire.

[40] Garofalo N, Smit Vega Garcia M: *New monotonicity formulas and the optimal regularity in the Signorini problem with variable coefficients*, Adv. Math. 262, 682–750, (2014).

[41] Guillen N: *Optimal regularity for the Signorini problem*, Calculus of Variations 36 (4), 533–546, (2009).

[42] Jhaveri Y, Neumayer R: *Higher Regularity of the Free Boundary in the Obstacle Problem for the Fractional Laplacian*, preprint arXiv:1605.01222, (2016).

[43] Kilpeläinen T: *Smooth approximation in weighted Sobolev spaces*, Commentat. Math. Univ. Carol., 38(1), 29-35, (1997).

[44] Koch H, Petrosyan A, Shi W: *Higher regularity of the free boundary in the elliptic Signorini problem*, Nonlinear Anal., 126,3–44, (2015).

[45] Koch H., Rüland A., Shi W.: *The Variable Coefficient Thin Obstacle Problem: Carleman inequalities*, preprint arXiv:1501.04496, (2015).

[46] Koch H., Rüland A., Shi W.: *The Variable Coefficient Thin Obstacle Problem: Optimal Regularity and Regularity of the Regular Free Boundary*, preprint arXiv:1504.03525, (2015).

[47] Koch H, Rüland A , Shi W: *Higher Regularity for the fractional thin obstacle problem*, preprint arXiv:1605.06662, (2016).
[48] Koch H, Rüland A, Shi W: *The Variable Coefficient Thin Obstacle Problem: Higher Regularity*, preprint arXiv:1606.02002, (2016).
[49] Korvenpää J., Kuusi T., Palatucci G.: *The obstacle problem for nonlinear integro-differential operators,* Calc. Var. Partial Differential Equations, 55 (3), Art. 63, (2016).
[50] Ladyženskaja O. A, Solonnikov V. A, Uraltseva N. N: *Linear and quasilinear equations of parabolic type*, Translated from the Russian by S. Smith. Translations of Mathematical Monographs, Vol. 23, American Mathematical Society, Providence, R.I., (1967).
[51] Landkof N.S: *Fundation of modern potential theory,,* Die Grundlagen der Mathematischen Wisswnshaften, band 180, Springer, New York, (1972).
[52] Milakis E, Silvestre L: *Regularity for the non linear Signorini problem,* Advances in Mathematics 217, 1301-1312, (2008).
[53] Molica Bisci G, Radulescu V, Servadei R: *Variational Methods for Nonlocal Fractional Problems,* Encyclopedia of Mathematics and its Applications, 162. Cambridge University Press, Cambridge, (2016).
[54] Monneau R: *On the number of singularities for the obstacle problem in two dimensions,* J. Geom. Anal. 13 (2), 359-389, (2003).
[55] Petrosyan A, Pop C. A: *Optimalregularity of solutions to the obstacle problem for the fractional Laplacian with drift,* J. Funct. Anal. 268 (2), 417-472, (2015).
[56] Petrosyan A, Shahgholian H, Uraltseva N: Regularity of free boundaries in obstacle type problems, volume 136 of Graduate Studies in Mathematics, AMS, Providence, RI, (2012).
[57] Petrosyan A, Shi W: *Parabolic Boundary Harnack Principle in Domains with Thin Lipschitz Complement*, Anal. PDE 7 (6), 1421–1463, (2014).
[58] Petrosyan A, Zeller A: *Boundedness and continuity of the time derivative in the parabolic Signorini problem*, preprint arXiv:1512.09173, (2015).
[59] Poon C:, *Unique continuation for parabolic equations*, Comm. Partial Differential Equations, 21 (3-4), 521–539, (1996).
[60] Richardson D: *Variational problems with thin obstacles* (Thesis), University of British Columbia, (1978).
[61] Salsa S: *The Problems of the Obstacle in Lower Dimension and for the Fractional Laplacian,* Springer Lecture Notes in Mathematics, 2045, Cetraro, Italy, (2009).
[62] Signorini A: *Questioni di elasticità non linearizzata e semilinearizzata*, Rend. Mat. Pura e Appl. 18 (5), 95–139, (1959).
[63] Silvestre L: *Regularity of the obstacle problem for a fractional power of the Laplace operator,* C.P.A.M. 60 (1), 67-112, (2007).
[64] Stein E: *Singular integrals and differentiability properties of functions,* Princeton U. Press, (1970).
[65] Uraltseva N.N: *Hoelder continuity of gradients of solutions of parabolic equations*

with boundary conditions of Signorini type, Dokl. Akad. Nauk. SSSR 280 (3), 563-565, (1985).
[66] Weiss G: *A homogeneity improvement approach to the obstacle problem,* Inventiones Mathematicae 138 (1), 23-50, (1999).
[67] Whitney H: *Analytic extensions of differentiable functions defined in closed sets,* Trans. A.M.S. 36 (1), 63-89, (1934).
[68] Ziemer W. P: *Weakly Differentiable Functions,* Graduate Text in Math, Springer-Verlag New York, (1989).

Serena Dipierro and Enrico Valdinoci

Nonlocal Minimal Surfaces: Interior Regularity, Quantitative Estimates and Boundary Stickiness

Abstract: We consider surfaces which minimize a nonlocal perimeter functional and we discuss their interior regularity and rigidity properties, in a quantitative and qualitative way, and their (perhaps rather surprising) boundary behavior. We present at least a sketch of the proofs of these results, in a way that aims to be as elementary and self contained as possible, referring to the papers [8, 27, 10, 3, 20, 19, 12] for full details.

...taurino quantum possent circumdare tergo...

4.1 Introduction

The study of surfaces which minimize the perimeter is a classical topic in analysis and geometry and probably one of the oldest problems in the mathematical literature: according to the first book of Virgil's Aeneid, Dido, the legendary queen of Carthage, needed to study problems of geometric minimization in order to found her reign in 814 B.C. (in spite of the great mathematical talent of Dido and of her vivid geometric intuition, Aeneas broke his betrothal with her after a short time to sail the Mediterranean towards the coasts of Italy, but this is another story).

Serena Dipierro, Dipartimento di Matematica, Università degli studi di Milano, Via Saldini 50, 20133 Milan, Italy,
and School of Mathematics and Statistics, University of Melbourne, 813 Swanston St, Parkville VIC 3010, Australia,
and School of Mathematics and Statistics, University of Western Australia, 35 Stirling Highway, Crawley, Perth WA 6009, Australia, E-mail: serena.dipierro@unimi.it
Enrico Valdinoci, School of Mathematics and Statistics, University of Melbourne, 813 Swanston St, Parkville VIC 3010, Australia,
School of Mathematics and Statistics, University of Western Australia, 35 Stirling Highway, Crawley, Perth WA 6009, Australia,
Weierstraß-Institut für Angewandte Analysis und Stochastik, Hausvogteiplatz 5/7, 10117 Berlin, Germany,
and Dipartimento di Matematica, Università degli studi di Milano, Via Saldini 50, 20133 Milan, Italy

https://doi.org/10.1515/9783110571561-006

The first problem in the study of these surfaces of minimal perimeter (minimal surfaces, for short) lies in proving that minimizers do exist. Indeed "nice" sets, for which one can compute the perimeter using an intuitive notion known from elementary school, turn out to be a "non compact" family (roughly speaking, for instance, an "ugly" set can be approximated by a sequence of "nice" sets, thus the limit point of the sequence may end up outside the family). To overcome this difficulty, a classical tool of the Calculus of Variations is to look for minimizers in a wider family of candidates: this larger family has to be chosen to satisfy the desired compactness property to ensure the existence of a minimum, and then the regularity of the minimal candidate can be (hopefully) proved a posteriori.

To this end, one needs to set up an appropriate notion of perimeter for the sets in the enlarged family of candidates, since no intuitive notion of perimeter is available, in principle, in this generality. The classical approach of Caccioppoli (see e.g. [7]) to this question lies in the observation that if Ω and E are[4.1] smooth sets and ν is the external normal of E, then, for any vector field $T \in C^1_0(\Omega, \mathbb{R}^n)$ with $|T(x)| \leqslant 1$ for any $x \in \Omega$, we have that

$$T \cdot \nu \leqslant |T|\,|\nu| \leqslant 1.$$

Consequently, the perimeter of E in Ω, i.e. the measure of the boundary of E inside Ω (that is, the $(n-1)$-dimensional Hausdorff measure of ∂E in Ω), satisfies the inequality

$$\operatorname{Per}(E, \Omega) = \mathcal{H}^{n-1}((\partial E) \cap \Omega) \geqslant \int_{\partial E} T \cdot \nu \, d\mathcal{H}^{n-1} = \int_E \operatorname{div} T(x)\, dx, \tag{4.1.1}$$

for every vector field $T \in C^1_0(\Omega, \mathbb{R}^n)$ with $\|T\|_{L^\infty(\mathbb{R}^n,\mathbb{R}^n)} \leqslant 1$, where the Divergence Theorem has been used in the last identity.

Viceversa, if E is a smooth set, its normal vector can be extended near ∂E, and then to the whole of $\mathbb{R}^n$, to a vector field $\nu_* \in C^1(\mathbb{R}^n, \mathbb{R}^n)$, with $|\nu_*(x)| \leqslant 1$ for any $x \in \mathbb{R}^n$. Then, if $\eta \in C^\infty_0(\Omega, [0, 1])$, with $\eta = 1$ in an interior ε-neighborhood of Ω, one can take $T := \eta\nu_*$ and find that $T \in C^1_0(\Omega, \mathbb{R}^n)$, $|T(x)| \leqslant 1$ for any $x \in \mathbb{R}^n$ and

$$\begin{aligned} \int_E \operatorname{div} T(x)\, dx &= \int_{\partial E} T \cdot \nu \, d\mathcal{H}^{n-1} \\ &= \int_{\partial E} \eta\nu_* \cdot \nu \, d\mathcal{H}^{n-1} = \int_{\partial E} \eta \, d\mathcal{H}^{n-1} \\ &\geqslant \mathcal{H}^{n-1}((\partial E) \cap \Omega) - O(\varepsilon) = \operatorname{Per}(E, \Omega) - O(\varepsilon). \end{aligned}$$

By taking ε as small as we wish and recalling (4.1.1), we obtain that

$$\operatorname{Per}(E, \Omega) = \sup_{\substack{T \in C^1_0(\Omega, \mathbb{R}^n) \\ \|T\|_{L^\infty(\mathbb{R}^n,\mathbb{R}^n)} \leqslant 1}} \int_E \operatorname{div} T(x)\, dx. \tag{4.1.2}$$

4.1 From now on, we reserve the name of Ω to an open set, possibly with smooth boundary, which can be seen as the "ambient space" for our problem.

While (4.1.2) was obtained for smooth sets E, the classical approach for minimal surfaces is in fact to take (4.1.2) as *definition* of perimeter of a (not necessarily smooth) set E in Ω. The class of sets obtained in this way indeed has the necessary compactness property (and the associated functional has the desired lower semicontinuity properties) to give the existence of minimizers: that is, one finds (at least) one set $E \subseteq \mathbb{R}^n$ satisfying

$$\operatorname{Per}(E, \Omega) \leqslant \operatorname{Per}(F, \Omega) \tag{4.1.3}$$

for any $F \subseteq \mathbb{R}^n$ such that F coincides with E in a neighborhood of Ω^c.

The boundary of this minimal set E satisfies, a posteriori, a bunch of additional regularity properties – just to recall the principal ones:

$$\text{If } n \leqslant 7 \text{ then } (\partial E) \cap \Omega \text{ is smooth;} \tag{4.1.4}$$

$$\text{If } n \geqslant 8 \text{ then } ((\partial E) \cap \Omega) \setminus \Sigma \text{ is smooth,} \tag{4.1.5}$$

being Σ a closed set of Hausdorff dimension at most $n - 8$;

The statement in (4.1.5) is sharp, since there exist (4.1.6)
examples in which the singular set Σ
has Hausdorff dimension $n - 8$.

We refer to [21] for complete statements and proofs (in particular, the claim in (4.1.4) here corresponds to Theorem 10.11 in [21], the claim in (4.1.5) here to Theorem 11.8 there, and the claim in (4.1.6) here to Theorem 16.4 there).

A natural problem that is closely related to these regularity results is the complete description of classical minimal surfaces in the whole of the space which are also graphs in some direction (the so-called minimal graphs). These questions, that go under the name of Bernstein's problem, have, in the classical case, the following positive answer:

If $n \leqslant 8$ and E is a minimal graph, then E is a halfspace; (4.1.7)
The statement in (4.1.7) is sharp, since there exist (4.1.8)
examples of minimal graphs in dimension 9 and higher
that are not halfspaces.

We refer to Theorems 17.8 and 17.10 in [21] for further details on the claims in (4.1.7) and (4.1.8), respectively.

It is also worth recalling that

surfaces minimizing perimeters have zero mean curvature, (4.1.9)

see e.g. Chapter 10 in [21].

Recently, and especially in light of the seminal paper [8], some attention has been devoted to a variation of the classical notion of perimeters which takes into account also long-range interactions between sets, as well as the corresponding minimization problem. This type of *nonlocal minimal surfaces* arises naturally, for instance, in the study of fractals [28], cellular automata [22, 11] and phase transitions [26] (see also [6] for a detailed introduction to the topic).

A simple idea for defining a notion of nonlocal perimeter may be described as follows. First of all, such nonlocal perimeter should compute the interaction I of all the points of E against all the points of the complement of E, which we denote by E^c.

On the other hand, if we want to localize these contributions inside the domain Ω, it is convenient to split E into $E\cap\Omega$ and $E\setminus\Omega$, as well as the set E^c into $E^c\cap\Omega$ and $E^c\setminus\Omega$, and so consider the four possibilities of interaction between E and E^c given by

$$\begin{aligned} &I(E\cap\Omega, E^c\cap\Omega), \qquad I(E\cap\Omega, E^c\setminus\Omega), \\ &I(E\setminus\Omega, E^c\cap\Omega), \qquad \text{and} \qquad I(E\setminus\Omega, E^c\setminus\Omega). \end{aligned} \tag{4.1.10}$$

Among these interactions, we observe that the latter one only depends on the configuration of the set outside Ω, and so

$$I(E\setminus\Omega, E^c\setminus\Omega) = I(F\setminus\Omega, F^c\setminus\Omega)$$

for any $F\subseteq\mathbb{R}^n$ such that $F\setminus\Omega = E\setminus\Omega$. Therefore, in a minimization process with fixed data outside Ω, the term $I(E\setminus\Omega, E^c\setminus\Omega)$ does not change the minimizers. It is therefore natural to omit this term in the energy functional (and, as a matter of fact, omitting this term may turn out to be important from the mathematical point of view, since this term may provide an infinite contribution to the energy). For this reason, the nonlocal perimeter considered in [8] is given by the sum of the first three terms in (4.1.10), namely one defines

$$\operatorname{Per}_s(E,\Omega) := I(E\cap\Omega, E^c\cap\Omega) + I(E\cap\Omega, E^c\setminus\Omega) + I(E\setminus\Omega, E^c\cap\Omega).$$

As for the interaction $I(\cdot,\cdot)$, of course some freedom is possible, and basically any interaction for which $\operatorname{Per}_s(E,\Omega)$ is finite, say, for smooth sets E makes perfect sense. A natural choice performed in [8] is to take the interaction as a weighted Lebesgue measure, where the weight is translation invariant, isotropic and homogeneous: more precisely, for any disjoint sets S_1 and S_2, one defines[4.2]

$$I(S_1,S_2) := \iint_{S_1\times S_2} \frac{dx\,dy}{|x-y|^{n+2s}}, \tag{4.1.11}$$

with $s\in\left(0,\frac{1}{2}\right)$. With this choice of the fractional parameter s, one sees that

$$[\chi_E]_{W^{\sigma,p}(\mathbb{R}^n)} := \iint_{\mathbb{R}^n\times\mathbb{R}^n} \frac{|\chi_E(x)-\chi_E(y)|^p}{|x-y|^{n+p\sigma}}\,dx\,dy$$

4.2 We remark that (4.1.11) gives that the "natural scaling" of the interaction I is "meters to the power $n-2s$" (where $2n$ comes from $dx\,dy$ and $-n-2s$ comes from $|x-y|^{-n-2s}$). When $s=1/2$, this scaling boils down to the one of the classical perimeter.

$$= 2 \iint\limits_{E\times E^c} \frac{dx\,dy}{|x-y|^{n+p\sigma}} = 2I(E,E^c) = 2\,\mathrm{Per}_s\,(E,\mathbb{R}^n)$$

as long as $p\sigma = 2s$, that is the fractional perimeter of a set coincides (up to normalization constants) to a fractional Sobolev norm of the corresponding characteristic function (see e.g. [16] for a simple introduction to fractional Sobolev spaces).

Moreover, for any fixed $y \in \mathbb{R}^n$,

$$\mathrm{div}_x \frac{x-y}{|x-y|^{n+2s}} = -\frac{2s}{|x-y|^{n+2s}}. \tag{4.1.12}$$

Also, for any fixed $x \in \mathbb{R}^n$,

$$\mathrm{div}_y \frac{\nu(x)}{|x-y|^{n+2s-2}} = (n+2s-2)\frac{\nu(x)\cdot(x-y)}{|x-y|^{n+2s}}.$$

Accordingly, by the Divergence Theorem[4.3]

$$\begin{aligned}
&\mathrm{Per}_s\,(E,\mathbb{R}^n)\\
&= -\frac{1}{2s}\int_{E^c} dy\left[\int_E \mathrm{div}_x \frac{x-y}{|x-y|^{n+2s}}\,dx\right]\\
&= -\frac{1}{2s}\int_{E^c} dy\left[\int_{\partial E} \frac{\nu(x)\cdot(x-y)}{|x-y|^{n+2s}}\,d\mathcal{H}^{n-1}(x)\right]\\
&= -\frac{1}{2s\,(n+2s-2)}\int_{\partial E} d\mathcal{H}^{n-1}(x)\left[\int_{E^c} \mathrm{div}_y \frac{\nu(x)}{|x-y|^{n+2s-2}}\,dy\right]\\
&= \frac{1}{2s\,(n+2s-2)}\iint\limits_{(\partial E)\times(\partial E)} \frac{\nu(x)\cdot\nu(y)}{|x-y|^{n+2s-2}}\,d\mathcal{H}^{n-1}(x)\,d\mathcal{H}^{n-1}(y).
\end{aligned} \tag{4.1.13}$$

That is,

$$\mathrm{Per}_s\,(E,\mathbb{R}^n) = \frac{1}{4s\,(n+2s-2)}\iint\limits_{(\partial E)\times(\partial E)} \frac{2-|\nu(x)-\nu(y)|^2}{|x-y|^{n+2s-2}}\,d\mathcal{H}^{n-1}(x)\,d\mathcal{H}^{n-1}(y),$$

which suggests that the fractional perimeter is a weighted measure of the variation of the normal vector around the boundary of a set. As a matter of fact, as $s \nearrow 1/2$, the s-perimeter recovers the classical perimeter from many point of views (a sketchy discussion about this will be given in Appendix A).

Also, in Appendix C, we briefly discuss the second variation of the s-perimeter on surfaces of vanishing nonlocal mean curvature and we show that graphs with vanishing nonlocal mean curvature cannot have horizontal normals.

4.3 We will often use the Divergence Theorem here in a rather formal way, by neglecting the possible singularity of the kernel – for a rigorous formulation one has to check that the possible singular contributions average out, at least for smooth sets.

Let us now recall (among the others) an elementary, but useful, application of this notion of fractional perimeter in the framework of digital image reconstruction. Suppose that we have a black and white digitalized image, say a bitmap, in which each pixel is either colored in black or in white. We call E the "black set" and we are interested in measuring its perimeter (the reason for that may be, for instance, that noises or impurities could be distinguished by having "more perimeter" than the "real" picture, since they may present irregular or fractal boundaries). In doing that, we need to be able to compute such perimeter with a very good precision. Of course, numerical errors could affect the computation, since the digital process replaced the real picture by a pixel representation of it, but we would like that our computation becomes more and more reliable if the resolution of the image is sufficiently high, i.e. if the size of the pixels is sufficiently small.

Unfortunately, we see that, in general, an accurate computation of the perimeter is not possible, not even for simple sets, since the numerical error produced by the pixel may not become negligible, even when the pixels are small. To observe this phenomenon (see e.g. [12]) we can consider a grid of square pixels of small side ε and a black square E of side 1, with the black square rotated by 45 degrees with respect to the orientation of the pixels. Now, the digitalization of the square will produce a numerical error, since, say, the pixels that intersect the square are taken as black, and so each side of the square is replaced by a "sawtooth" curve (see Figure 4.1).

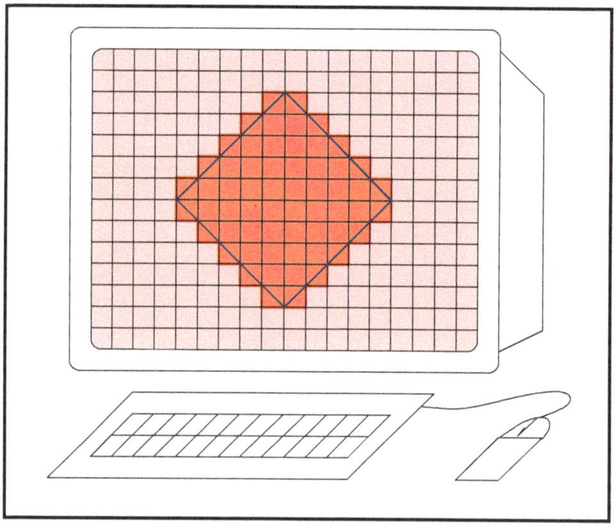

Fig. 4.1: Numerical error in computing the perimeter.

Notice that the length of each of these sawtooth curves is $\sqrt{2}$ (independently on how small each teeth is, that is independently on the size of ε). As a consequence, the perimeter of the digitalized image is $4\sqrt{2}$, instead of 4, which was the original perimeter of the square.

This shows the rather unpleasant fact that the perimeter may be poorly approximated numerically, even in case of high precision digitalization processes. It is a rather remarkable fact that fractional perimeters do not present the same inconvenience and indeed the numerical error in computing the fractional perimeter becomes small when the pixels are small enough. Indeed, the number of pixels which intersect the sides of the original square is $O(\varepsilon^{-1})$ (recall that the side of the square is 1 and the side of each pixel is of size ε). Also, the s-perimeter of each pixel is $O(\varepsilon^{2-2s})$ (since this is the natural scale factor of the interaction in (4.1.11), with $n = 2$). Then, the numerical error in the fractional perimeter comes from the contributions of all these pixels[4.4] and it is therefore $O(\varepsilon^{-1}) \cdot O(\varepsilon^{2-2s}) = O(\varepsilon^{1-2s})$, which tends to zero for small ε, thus showing that the nonlocal perimeters are more efficient than classical ones in this type of digitalization process.

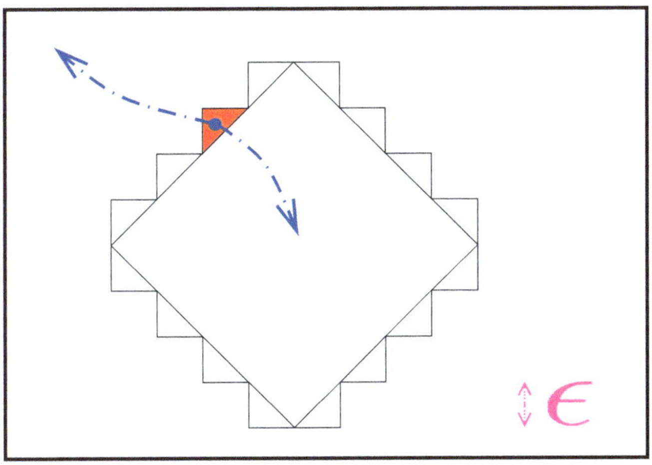

Fig. 4.2: Pixel interactions and numerical errors.

Thus, given its mathematical interest and its importance in concrete applications, it is desirable to reach a better understanding of the surfaces which minimize the s-

4.4 More precisely, when the computer changes the "real" square with the discretized one and produces a staircase border, the only interactions changed are the ones affecting the union of the triangles (that are "half pixels") that are added to the square in this procedure. In the "real" picture, these triangles interact with the square, while in the digitalized picture they interact with the exterior. To compute the error obtained one takes the signed superposition of these effects, therefore, to estimate the error in absolute value, one can just sum up these contributions, which in turn are bounded by the sum of the interactions of each triangle with its complement, see Figure 4.2.

perimeter (that one can call s-minimal surfaces). To start with, let us remark that an analogue of (4.1.9) holds true, in the sense that s-minimal surfaces have vanishing s-mean curvature in a sense that we now briefly describe. Given a set E with smooth boundary and $p \in \partial E$, we define

$$H_E^s(p) := \int_{\mathbb{R}^n} \frac{\chi_{E^c}(x) - \chi_E(x)}{|x-p|^{n+2s}}\,dx. \tag{4.1.14}$$

The expression[4.5] in (4.1.14) is intended in the principal value sense, namely the singularity is taken in an averaged limit, such as

$$H_E^s(p) = \lim_{\rho \searrow 0} \int_{\mathbb{R}^n \setminus B_\rho(p)} \frac{\chi_{E^c}(x) - \chi_E(x)}{|x-p|^{n+2s}}\,dx.$$

For simplicity, we omit the principal value from the notation. It is also useful to recall (4.1.12) and to remark that H_E^s can be computed as a weighted boundary integral of the normal, namely

$$\begin{aligned} H_E^s(p) &= -\frac{1}{2s}\int_{\mathbb{R}^n} (\chi_{E^c}(x) - \chi_E(x))\,\mathrm{div}\,\frac{x-p}{|x-p|^{n+2s}}\,dx \\ &= -\frac{1}{s}\int_{\partial E} \frac{\nu(x)\cdot(p-x)}{|p-x|^{n+2s}}\,d\mathcal{H}^{n-1}(x). \end{aligned} \tag{4.1.15}$$

This quantity H_E^s is what we call the nonlocal mean curvature of E at the point p, and the name is justified by the following observation:

Lemma 4.1.1. *If E is a set with smooth boundary that minimizes the s-perimeter in Ω, then $H_E^s(p) = 0$ for any $p \in (\partial E) \cap \Omega$.*

The proof of Lemma 4.1.1 will be given in Section 4.2. We refer to [8] for a version of Lemma 4.1.1 that holds true (in the viscosity sense) without assuming that the set has

---

4.5 The definition of fractional mean curvature in (4.1.14) may look a bit awkward at a first glance. To make it appear more friendly, we point out that the classical mean curvature of ∂E at a point $x \in \partial E$, up to normalizing constants, can be computed via the average procedure

$$\lim_{r \searrow 0} \frac{1}{r^{n+1}} \int_{B_r(x)} \chi_{E^c}(y) - \chi_E(y)\,dy.$$

To see this, up to rigid motions, one can assume that x is the origin and E, in a small neighborhood of the origin, is the subgraph of a function $u : \mathbb{R}^{n-1} \to \mathbb{R}$ with $u(0) = 0$ and $\nabla u(0) = 0$. Then, we have that

$$\begin{aligned} \lim_{r \searrow 0} \frac{1}{r^{n+1}} \int_{B_r} \chi_{E^c}(y) - \chi_E(y)\,dy &= \lim_{r \searrow 0} \frac{1}{r^{n+1}} \int_{\substack{\{|y'|\leqslant r\} \\ \{y_n \in [-r,r]\}}} \chi_{E^c}(y) - \chi_E(y)\,dy \\ &= -\lim_{r \searrow 0} \frac{2}{r^{n+1}} \int_{\{|y'|\leqslant r\}} u(y')\,dy' = -\lim_{r \searrow 0} \frac{2}{r^{n+1}} \int_{\{|y'|\leqslant r\}} D^2u(0)y'\cdot y' + o(|y'|^2)\,dy' \\ &= -c\,\Delta u(0). \end{aligned}$$

smooth boundary. See also [1] for further comments on this notion of nonlocal mean curvature.

Let us now briefly discuss the fractional analogue of the regularity results in (4.1.4) and (4.1.5). At the moment, a complete regularity theory in the fractional case is still not available. At best, one can obtain regularity results either in low dimension or when s is sufficiently close to $\frac{1}{2}$ (see [27, 10] and also [3] for higher regularity results): namely, the analogue of (4.1.4) is:

Theorem 4.1.2 (Interior regularity results for s-minimal surfaces - I). *Let $E \subset \mathbb{R}^n$ be a minimizer for the s-perimeter in Ω. Assume that*

- *either $n = 2$,*
- *or $n \leqslant 7$ and $\frac{1}{2} - s \leqslant \varepsilon_*$, for some $\varepsilon_* > 0$ sufficiently small.*

Then, $(\partial E) \cap \Omega$ is smooth.

Similarly, a fractional analogue of (4.1.5) is known, by now, only when s is sufficiently close to $\frac{1}{2}$:

Theorem 4.1.3 (Interior regularity results for s-minimal surfaces - II). *Let $E \subset \mathbb{R}^n$ be a minimizer for the s-perimeter in Ω. Assume that $n \geqslant 8$ and $\frac{1}{2} - s \leqslant \varepsilon_n$, for some $\varepsilon_n > 0$ sufficiently small. Then, $((\partial E) \cap \Omega)\setminus\Sigma$ is smooth, being Σ a closed set of Hausdorff dimension at most $n - 8$.*

In contrast with the statement in (4.1.6), it is not known if Theorems 4.1.2 and 4.1.3 are sharp, and in fact there are no known examples of s-minimal surfaces with singular sets: and, as a matter of fact, in dimension $n \leqslant 6$, these pathological examples – if they exist – cannot be built by symmetric cones (which means that they either do not exist or are pretty hard to find!), see [15].

In [12], several quantitative regularity estimates for local minimizers are given (as a matter of fact, these estimates are valid in a much more general setting, but, for simplicity, we focus here on the most basic statements and proofs). For instance, minimizers of the s-perimeter have locally finite perimeter (that is, classical perimeter, not only fractional perimeter), as stated in the next result:

Theorem 4.1.4. *Let $E \subset \mathbb{R}^n$ be a minimizer for the s-perimeter in B_R. Then*

$$\operatorname{Per}(E, B_{1/2}) \leqslant CR^{n-1},$$

for a suitable constant $C > 0$.

We stress that Theorem 4.1.4 presents several novelties with respect to the existing literature. First of all, it provides a scaling invariant regularity estimate that goes beyond the natural scaling of the s-perimeter, that is valid in any dimension and without any topological restriction on the s-minimal surface (analogous results for the classical perimeter are not known in this generality). Also, in spite of the fact that, for the sake of simplicity, we state and prove Theorem 4.1.4 only in the case of minimizers of the s-perimeter, more general versions of this result hold true for stable solutions and for more general interaction kernels (even for kernels without any regularizing effect). This type of results also leads to new compactness and existence theorems, see [12] for full details on this topic.

As a matter of fact, we stress that the analogue of Theorem 4.1.4 for stable surfaces which are critical points of the classical perimeter is only known, up to now, for two-dimensional surfaces that are simply connected and immersed in $\mathbb{R}^3$ (hence, this is a case in which the nonlocal theory can go beyond the local one).

Now, we briefly discuss the fractional analogue of the Bernstein's problem. Let us start by pointing out that, by combining (4.1.4) and (4.1.7), we have an "abstract" version of the Bernstein's problem, which states that *if E is a minimal graph in $\mathbb{R}^{n+1}$ and the minimal surfaces in $\mathbb{R}^n$ are smooth, then E is a halfspace.*

Of course, for the way we have written (4.1.4) and (4.1.7), this abstract statement seems only to say that $8 = 7 + 1$: nevertheless this abstract version of the Bernstein's problem is very useful in the classical case, since it admits a nice fractional counterpart, which is:

Theorem 4.1.5 (Bernstein result for s-minimal surfaces - I). *If E is an s-minimal graph in $\mathbb{R}^{n+1}$ and the s-minimal surfaces in $\mathbb{R}^n$ are smooth, then E is a halfspace.*

This result was proved in [20]. By combining it with Theorem 4.1.2 (using the notation $N := n + 1$), we obtain:

Theorem 4.1.6 (Bernstein result for s-minimal surfaces - II). *Let $E \subset \mathbb{R}^N$ be an s-minimal graph. Assume that*

- *either $N = 3$,*
- *or $N \leqslant 8$ and $\frac{1}{2} - s \leqslant \varepsilon_*$, for some $\varepsilon_* > 0$ sufficiently small.*

Then, E is a halfspace.

This is, at the moment, the fractional counterpart of (4.1.7) (we stress, however, that any improvement in the fractional regularity theory would give for free an improvement in the fractional Bernstein's problem, via Theorem 4.1.5).

We remark again that, differently from the claim in (4.1.8), it is not known if the statement in Theorem 4.1.6 is sharp, since there are no known examples of s-minimal graphs other than the hyperplanes.

It is worth recalling that, by a blow-down procedure, one can deduce from Theorem 4.1.2 that global s-minimal surfaces are hyperplanes, as stated in the following result:

Theorem 4.1.7 (Flatness of s-minimal surfaces). *Let $E \subset \mathbb{R}^n$ be a minimizer for the s-perimeter in any domain of $\mathbb{R}^n$. Assume that*
- *either $n = 2$,*
- *or $n \leqslant 7$ and $\frac{1}{2} - s \leqslant \varepsilon_*$, for some $\varepsilon_* > 0$ sufficiently small.*

Then, E is a halfspace.

Of course, a very interesting spin-off of the regularity theory in Theorem 4.1.7 lies in finding quantitative flatness estimates: namely, if we know that a set E is an s-minimizer in a large domain, can we say that it is sufficiently close to be a halfspace, and if so, how close, and in which sense?

This question has been recently addressed in [12]. As a matter of fact, the results in [12] are richer than the ones we present here, and they are valid for a very general class of interaction kernels and of perimeters of nonlocal type. Nevertheless we think it is interesting to give a flavor of them even in their simpler form, to underline their connection with the regularity theory that we discussed till now.

In this setting, we present here the following result when $n = 2$ (see indeed [12] for more general statements):

Theorem 4.1.8. *Let $R \geqslant 2$. Let $E \subset \mathbb{R}^2$ be a minimizer for the s-perimeter in B_R. Then there exists a halfplane h such that*

$$\left|(E\Delta h) \cap B_1\right| \leqslant \frac{C}{R^s}, \tag{4.1.16}$$

where Δ is here the symmetric difference of the two sets (i.e. $E\Delta h := (E\setminus h) \cup (h\setminus E)$) and $C > 0$ is a constant.

We stress that Theorem 4.1.8 may be seen as a quantitative version of Theorem 4.1.7 when $n = 2$: indeed if $E \subset \mathbb{R}^n$ is a minimizer for the s-perimeter in any domain of $\mathbb{R}^n$ we can send $R \nearrow +\infty$ in (4.1.16) and obtain that E is a halfplane.

We observe that, untill now, we have presented and discussed a series of results which are somehow in accordance, as much as possible, with the classical case. Now we present something that differs strikingly from the classical case. The minimizers of the classical perimeter in a convex domain reach continuously the boundary data (see e.g. Theorem 15.9 in [21]). Quite surprisingly, the minimizers of the fractional perimeter have the tendency to stick at the boundary. This phenomenon has been discovered in [19], where several explicit stickiness examples have been given (see also [5] for other examples in more general settings).

Roughly speaking, the stickiness phenomenon may be described as follows. We know from Lemma 4.1.1 that nonlocal minimal surfaces in a domain Ω need to adjust their shape in order to make the nonlocal minimal curvature vanish inside Ω. This is a rather strong condition, since the nonlocal minimal curvature "sees" the set all over the space. As a consequence, in many cases in which the boundary data are "not favorable" for this condition to hold, the nonlocal minimal surfaces may prefer to modify their shape by sticking at the boundary, where the condition is not prescribed, in order to compensate the values of the nonlocal mean curvature inside Ω.

In many cases, for instance, the nonlocal minimal set may even prefer to "disappear", i.e. its contribution inside Ω becomes empty and its boundary sticks completely to the boundary of Ω. In concrete cases, the fact that the nonlocal minimal set disappears may be induced by a suitable choice of the data outside Ω or by an appropriate choice of the fractional parameter. As a prototype example of these two phenomena, we recall here the following results given in [19]:

Theorem 4.1.9 (Stickiness for small data). *For any $\delta > 0$, let*

$$K_\delta := (B_{1+\delta} \setminus B_1) \cap \{x_n < 0\}.$$

Let E_δ be an s-minimal set in B_1 among all the sets E such that $E \setminus B_1 = K_\delta$.

Then, there exists $\delta_o > 0$, depending on s and n, such that for any $\delta \in (0, \delta_o]$ we have that

$$E_\delta = K_\delta.$$

Theorem 4.1.10 (Stickiness for small s). *As $s \to 0^+$, the s-minimal set in $B_1 \subset \mathbb{R}^2$ that agrees with a sector outside B_1 sticks to the sector.*

More precisely: let E_s be the s-minimizer among all the sets E such that

$$E \setminus B_1 = \Sigma := \{(x, y) \in \mathbb{R}^2 \setminus B_1 \text{ s.t. } x > 0 \text{ and } y > 0\}.$$

Then, there exists $s_o > 0$ such that for any $s \in (0, s_o]$ we have that $E_s = \Sigma$.

We stress the sharp difference between the local and the nonlocal cases exposed in Theorems 4.1.9 and 4.1.10: indeed, in the local framework, in both cases the minimal surface is a segment inside the ball B_1, while in the nonlocal case it coincides with a piece of the circumference ∂B_1.

The stickiness phenomenon of nonlocal minimal surfaces may also be caused by a sufficiently high oscillation of the data outside Ω. This concept is exposed in the following result:

Theorem 4.1.11 (Stickiness coming from large oscillations of the data). *Let $M > 1$ and let $E_M \subset \mathbb{R}^2$ be s-minimal in $(-1, 1) \times \mathbb{R}$ with datum outside $(-1, 1) \times \mathbb{R}$ given* *by* $J_M := J_M^- \cup J_M^+$, *where*

$$J_M^- := (-\infty, -1] \times (-\infty, -M) \quad \text{and} \quad J_M^+ := [1, +\infty) \times (-\infty, M).$$

Then, if M is large enough, E_M *sticks at the boundary. Moreover, the stickiness region gets close to the origin, up to a power of M.*

More precisely: there exist $M_o > 0$ *and* $C_o \geqslant C_o' > 0$, *depending on s, such that if* $M \geqslant M_o$ *then*

$$\begin{aligned} &[-1, 1) \times [C_o M^{\frac{1+2s}{2+2s}}, M] \subseteq E_M^c \\ \text{and} \quad &(-1, 1] \times [-M, -C_o M^{\frac{1+2s}{2+2s}}] \subseteq E_M. \end{aligned} \tag{4.1.17}$$

It is worth remarking that the stickiness phenomenon in Theorem 4.1.11 becomes "more and more visible" as the oscillation of the data increase, since, referring to (4.1.17), we have that

$$\lim_{M \to +\infty} \frac{M^{\frac{1+2s}{2+2s}}}{M} = 0,$$

hence the sticked portion of E_M on $\partial\Omega$ becomes, proportionally to M, larger and larger when $M \to +\infty$.

Also, the exponent $\frac{1+2s}{2+2s}$ in (4.1.17) is optimal, see again [19]. The stickiness phenomenon detected in Theorem 4.1.11 is described in Figure 4.3.

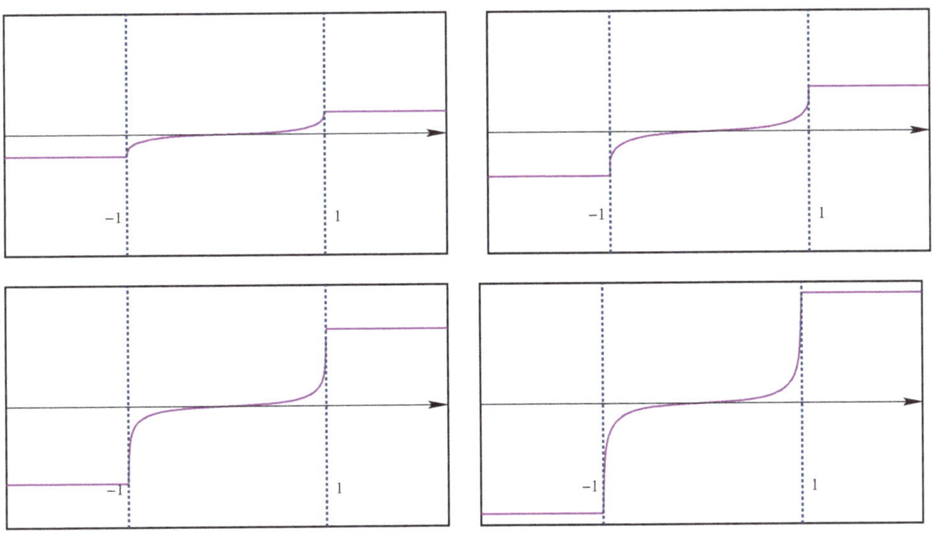

Fig. 4.3: Stickiness coming from large oscillations of the data with the oscillation progressively larger.

We believe that the stickiness phenomenon is rather common among nonlocal minimal surfaces. Indeed, it may occur even under small modifications of boundary data for which the nonlocal minimal surfaces cut the boundary in a transversal way.

A typical, and rather striking, example of this situation happens for perturbation of halfplanes in $\mathbb{R}^2$. That is, an arbitrarily small perturbation of the data corresponding to halfplanes is sufficient for the stickiness phenomenon to occur. Of course, the smaller the perturbation, the smaller the stickiness: nevertheless, small perturbations are enough to cause the fact that the boundary data of nonlocal minimal surfaces are not attained in a continuous way, and indeed they may exhibit jumps (notice that this lack of boundary regularity for s-minimal surfaces is rather surprising, especially after the interior regularity results discussed in Theorem 4.1.2 and 4.1.3 and it shows that the boundary behavior of the halfplanes is rather unstable).

A detailed result goes as follows:

Theorem 4.1.12 (Stickiness arising from perturbation of halfplanes). *There exists $\delta_0 > 0$ such that for any δ in $(0, \delta_0]$ the following statement holds true.*

Let $\Omega := (-1, 1) \times \mathbb{R}$. Let also

$$F_- := [-3, -2] \times [0, \delta], \qquad F_+ := [2, 3] \times [0, \delta], \qquad H := \mathbb{R} \times (-\infty, 0).$$

Assume that $F \subseteq \mathbb{R}^2$, with

$$F \supseteq H \cup F_- \cup F_+.$$

Let E be an s-minimal set in Ω among all the sets which coincide with F outside Ω. Then,

$$E \supseteq (-1, 1) \times [0, \delta^\gamma],$$

for a suitable $\gamma > 1$.

The result of Theorem 4.1.12 is depicted in Figure 4.4.

Let us briefly give some further comments on the stickiness phenomena discussed above. First of all, we would like to convince the reader (as well as ourselves) that these type of behaviors indeed occurs in the nonlocal case.

To this end, let us make an investigation to find how the s-minimal set E_α in $\Omega := (-1, 1) \times \mathbb{R} \subset \mathbb{R}^2$ with datum

$$C_\alpha := \{(x, y) \in \mathbb{R}^2 \text{ s.t. } y < \alpha|x|\}$$

looks like.

When $\alpha = 0$, then $E_\alpha = C_\alpha$ is the halfplane, so the interesting case is when $\alpha \neq 0$; say, up to symmetries, $\alpha > 0$. Now, we know how an investigation works: we need to place all the usual suspects in a row and try to find the culprit.

The line of suspect is on Figure 4.5 (remember that we have to find the s-minimal set among them). Some of the suspects resemble our prejudices on how the culprit should look like. For instance, for what we saw on TV, we have the prejudice that serial

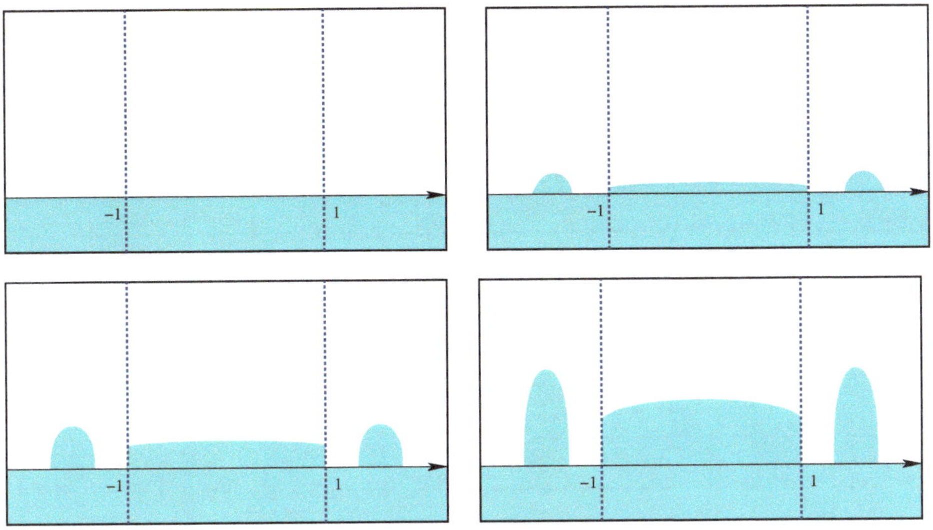

Fig. 4.4: Stickiness arising from perturbation of halfplanes, with the perturbation progressively larger.

killers always wear black gloves and raincoats. Similarly, for what we learnt from the hyperplanes, we may have the prejudice that s-minimal surfaces meet the boundary data in a smooth fashion (this prejudice will turn out to be wrong, as we will see). In this sense, the usual suspects number 1 and 2 in Figure 4.5 are the ones who look like the serial killers.

Then, we have the regular guys with some strange hobbies, we know from TV that they are also quite plausible candidates for being guilty; in our analogy, these are the usual suspects number 3 and 4, which meet the boundary data in a Lipschitz or Hölder fashion (and one may also observe that number 3 is the minimal set in the local case).

Then, we have the candidates which look above suspicion, the ones to which nobody ever consider to be guilty, usually the postman or the butler. In our analogy, these are the suspects number 5 and 6, which are discontinuous at the boundary.

Now, we know from TV how we should proceed: if a suspect has a strong and verified alibi, we can rule him or her out of the list. In our case, an alibi can be offered by the necessary condition for s-minimality given in Lemma 4.1.1. Indeed, if one of our suspects E does not satisfy that $H_E^s = 0$ along $(\partial E) \cap \Omega$, then E cannot be s-minimal and we can cross out E from our list of suspects (E has an alibi!).

Now, it is easily seen that all the suspects number 1, 2, 3, 4 and 5 have an alibi: indeed, from Figure 4.6 we see that $H_E^s(p) \neq 0$, since the set E occupies (in mea-

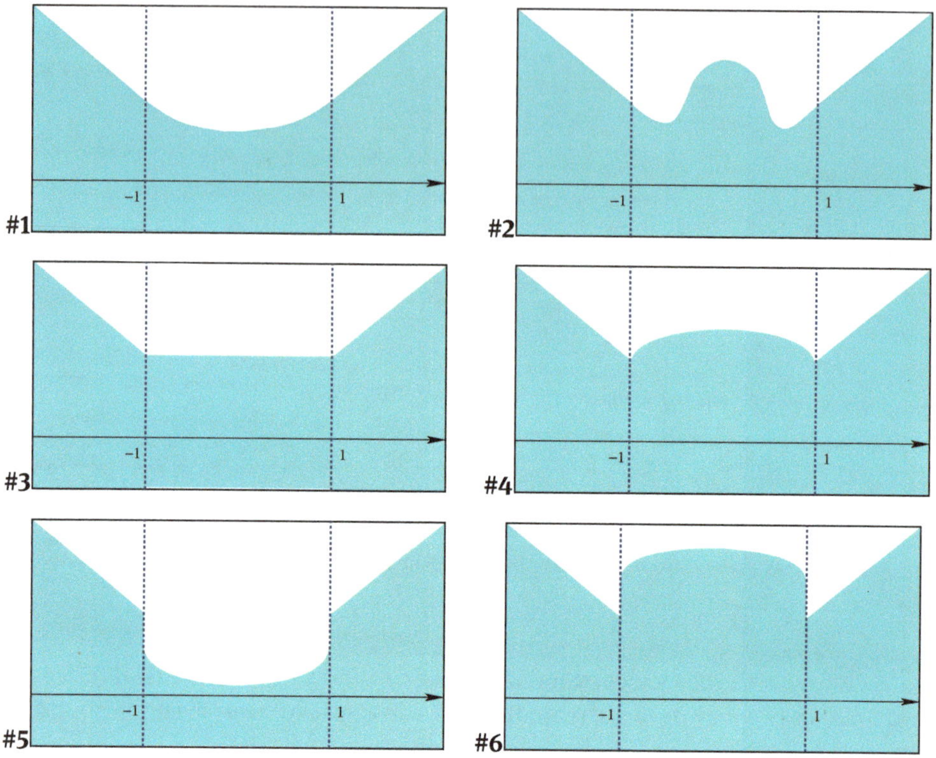

Fig. 4.5: Confrontation between the suspects.

sure, weighted by the kernel in (4.1.14)) more than a halfplane[4.6] passing through p: in Figure 4.6 the point p is the big dot and the halfplane is marked by the line passing through it, so a quick inspection confirms that the alibis of number 1, 2, 3, 4 and 5 check out, hence their nonlocal mean curvature does not vanish at p and consequently they are not s-minimal sets.

On the other hand, the alibi of number 6 doesn't hold water. Indeed, near p, the set E is confined below the horizontal line, but at infinity the set E go well beyond such line: these effects might compensate each other and produce a vanishing mean curvature.

4.6 Indeed, in view of (4.1.14), we know that an s-minimal set, seen from any point of the boundary, satisfies a perfect balance between the weighted measure of the set itself and the weighted measure of its complement (here, weighted is intended with respect to the kernel in (4.1.14)). Since a halfplane also satisfies such perfect balance when seen from any point of its boundary (due to odd symmetry), one can say that a set is s-minimal when, at any point of its boundary, the weighted contributions of the set and its complement produce the same result as the ones of a hyperplane passing through such point. This geometric trick often allows us to "subtract the tangent halfplane" from a set without modifying its fractional curvature (and this is often convenient to observe cancellations).

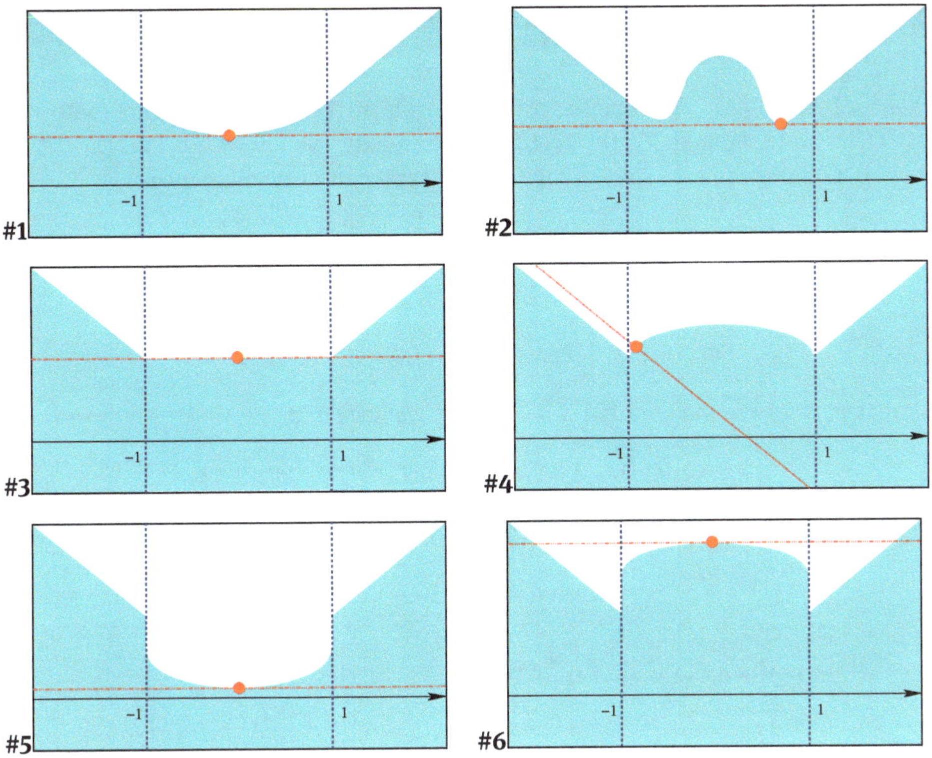

Fig. 4.6: The alibis of the suspects.

So, having ruled out all the suspects but number 6, we have only to remember what the old investigators have taught us (e.g., “When you have eliminated the impossible, whatever remains, however improbable, must be the truth”), to find that the only possible (though, in principle, rather improbable) culprit is number 6.

Of course, once that we know that the butler did it, i.e. that number 6 is s-minimal, it is our duty to prove it beyond any reasonable doubt. Many pieces of evidence, and a complete proof, is given in [19] (where indeed the more general version given in Theorem 4.1.12 is established). Here, we provide some ideas towards the proof of Theorem 4.1.12 in Section 4.5.

This set of notes is organized as follows. In Section 4.2 we present the proof of Lemma 4.1.1. Sections 4.3 and 4.4 are devoted to the proofs of the quantitative estimates in Theorems 4.1.4 and 4.1.8, respectively. Then, Section 4.5 is dedicated to a sketch of the proof of Theorem 4.1.12. We also provide Appendix A to discuss briefly the asymptotics of the s-perimeter as $s \nearrow 1/2$ and as $s \searrow 0$ and Appendix B to discuss the asymptotic expansion of the nonlocal mean curvature as $s \searrow 0$. Finally, in Appendix C we discuss the second variation of the fractional perimeter functional.

4.2 Proof of Lemma 4.1.1

Proof of Lemma 4.1.1. We consider a diffeomorphism $T_\varepsilon(x) := x + \varepsilon v(x)$, with $v \in C_0^\infty(\Omega, \mathbb{R}^n)$ and we take $E_\varepsilon := T_\varepsilon(E)$. By minimality, we know that $\mathrm{Per}_s(E_\varepsilon, \Omega) \geqslant \mathrm{Per}_s(E, \Omega)$ for every $\varepsilon \in (-\varepsilon_0, \varepsilon_0)$, with $\varepsilon_0 > 0$ sufficiently small, hence

$$\mathrm{Per}_s(E_\varepsilon, \Omega) - \mathrm{Per}_s(E, \Omega) = o(\varepsilon). \tag{4.2.1}$$

Suppose, for simplicity, that $I(E\setminus\Omega, E^c\setminus\Omega) < +\infty$, so that we can write

$$\mathrm{Per}_s(E_\varepsilon, \Omega) - \mathrm{Per}_s(E, \Omega) = I(E_\varepsilon, E_\varepsilon^c) - I(E, E^c).$$

Moreover, if we use the notation $X := T_\varepsilon^{-1}(x)$, we have that

$$dx = |\det DT_\varepsilon(X)|\, dX = \big(1 + \varepsilon \operatorname{div} v(X) + o(\varepsilon)\big)\, dX.$$

Similarly, if $Y := T_\varepsilon^{-1}(y)$, we find that

$$\begin{aligned}
&|x-y|^{-n-2s}\\
=\;& |T_\varepsilon(X) - T_\varepsilon(Y)|^{-n-2s}\\
=\;& \big|X - Y + \varepsilon\big(v(X) - v(Y)\big)\big|^{-n-2s}\\
=\;& |X-Y|^{-n-2s} - (n+2s)\,\varepsilon\,|X-Y|^{-n-2s-2}(X-Y)\cdot\big(v(X)-v(Y)\big) + o(\varepsilon).
\end{aligned}$$

As a consequence,

$$\begin{aligned}
&\mathrm{Per}_s(E_\varepsilon, \Omega) - \mathrm{Per}_s(E, \Omega)\\
=\;& \iint_{E_\varepsilon \times E_\varepsilon^c} \frac{dx\,dy}{|x-y|^{n+2s}} - \iint_{E \times E^c} \frac{dx\,dy}{|x-y|^{n+2s}}\\
=\;& \iint_{E \times E^c} \Big[|X-Y|^{-n-2s} - (n+2s)\,\varepsilon\,|X-Y|^{-n-2s-2}(X-Y)\cdot\big(v(X)-v(Y)\big)\Big]\\
&\quad \cdot\big(1+\varepsilon \operatorname{div} v(X)\big)\big(1+\varepsilon \operatorname{div} v(Y)\big)\, dX\, dY\\
&\quad - \iint_{E \times E^c} \frac{dx\,dy}{|x-y|^{n+2s}} + o(\varepsilon)\\
=\;& -(n+2s)\,\varepsilon \iint_{E \times E^c} \frac{(x-y)\cdot\big(v(x)-v(y)\big)}{|x-y|^{n+2s+2}}\, dx\, dy\\
&\quad + \varepsilon \iint_{E \times E^c} \frac{\operatorname{div} v(x) + \operatorname{div} v(y)}{|x-y|^{n+2s}}\, dx\, dy + o(\varepsilon).
\end{aligned}$$

Now we point out that

$$\operatorname{div}_x \frac{v(x)}{|x-y|^{n+2s}} = -(n+2s)\frac{v(x)\cdot(x-y)}{|x-y|^{n+2s+2}} + \frac{\operatorname{div}_x v(x)}{|x-y|^{n+2s}}$$

and so, interchanging the names of the variables,

$$\operatorname{div}_y \frac{v(y)}{|x-y|^{n+2s}} = (n+2s)\frac{v(y)\cdot(x-y)}{|x-y|^{n+2s+2}} + \frac{\operatorname{div}_y v(y)}{|x-y|^{n+2s}}.$$

Consequently,

$$\begin{aligned} &\operatorname{Per}_s(E_\varepsilon,\Omega) - \operatorname{Per}_s(E,\Omega) \\ = \;& \varepsilon \iint\limits_{E\times E^c} \left[\operatorname{div}_x \frac{v(x)}{|x-y|^{n+2s}} + \operatorname{div}_y \frac{v(y)}{|x-y|^{n+2s}}\right] dx\,dy + o(\varepsilon). \end{aligned}$$

Now, using the Divergence Theorem and changing the names of the variables we have that

$$\begin{aligned} \iint\limits_{E\times E^c} \operatorname{div}_x \frac{v(x)}{|x-y|^{n+2s}}\,dx\,dy &= \int_{E^c} dy \left[\int_{\partial E} \frac{v(x)\cdot\nu(x)}{|x-y|^{n+2s}}\,d\mathcal{H}^{n-1}(x)\right] \\ &= \int_{E^c} dx \left[\int_{\partial E} \frac{v(y)\cdot\nu(y)}{|x-y|^{n+2s}}\,d\mathcal{H}^{n-1}(y)\right] \end{aligned}$$

and

$$\iint\limits_{E\times E^c} \operatorname{div}_y \frac{v(y)}{|x-y|^{n+2s}}\,dx\,dy = -\int_E dx \left[\int_{\partial E} \frac{v(y)\cdot\nu(y)}{|x-y|^{n+2s}}\,d\mathcal{H}^{n-1}(y)\right].$$

Accordingly, we find that

$$\begin{aligned} &\operatorname{Per}_s(E_\varepsilon,\Omega) - \operatorname{Per}_s(E,\Omega) \\ = \;& \varepsilon \int_{\partial E} d\mathcal{H}^{n-1}(y)\,v(y)\cdot\nu(y) \left[\int_{E^c} \frac{dx}{|x-y|^{n+2s}} - \int_E \frac{dx}{|x-y|^{n+2s}}\right] + o(\varepsilon) \\ = \;& \varepsilon \int_{\partial E} v(y)\cdot\nu(y)\,H_E^s(y)\,d\mathcal{H}^{n-1}(y) + o(\varepsilon). \end{aligned}$$

Comparing with (4.2.1), we see that

$$\int_{\partial E} v(y)\cdot\nu(y)\,H_E^s(y)\,d\mathcal{H}^{n-1}(y) = 0$$

and so, since v is an arbitrary vector field supported in Ω, the desired result follows. □

4.3 Proof of Theorem 4.1.4

The basic idea goes as follows. One uses the appropriate combination of two general facts: on the one hand, one can perturb a given set by a smooth flow and compare the energy at time t with the one at time $-t$, thus obtaining a second order estimate; on the other hand, the nonlocal interaction always charges a mass on points that are sufficiently close, thus providing a natural measure for the discrepancy between the original set and its flow. One can appropriately combine these two facts with the minimality (or more generally, the stability) property of a set. Indeed, by choosing as smooth flow a translation near the origin, the above arguments lead to an integral estimate of the discrepancy between the set and its translations, which in turn implies a perimeter estimate.

We now give the details of the proof of Theorem 4.1.4. To do this, we fix $R \geqslant 1$, a direction $v \in S^{n-1}$, a function $\varphi \in C_0^\infty(B_{9/10})$ with $\varphi = 1$ in $B_{3/4}$, and a small scalar quantity $t \in \left(-\frac{1}{100}, \frac{1}{100}\right)$, and we consider the diffeomorphism $\Phi^t \in C_0^\infty(B_{9/10})$ given by $\Phi^t(x) := x + t\varphi(x/R)\,v$. Notice that

$$\Phi^t(x) = x + tv \text{ for any } x \in B_{3R/4}. \tag{4.3.1}$$

We also define $E_t := \Phi^t(E)$. We have the following useful auxiliary estimates (that will be used in the proofs of both Theorem 4.1.4 and Theorem 4.1.8):

Lemma 4.3.1. *Let E be a minimizer for the s-perimeter in B_R. Then*

$$\operatorname{Per}_s(E_t, B_R) + \operatorname{Per}_s(E_{-t}, B_R) - 2\operatorname{Per}_s(E, B_R) \leqslant CR^{n-2s-2}\,t^2, \tag{4.3.2}$$

$$2I(E_t \setminus E, E \setminus E_t) \leqslant CR^{n-2s-2}\,t^2, \tag{4.3.3}$$

$$\min\left\{ \left|((E+tv)\setminus E) \cap B_{R/2}\right|, \left|(E \setminus (E+tv)) \cap B_{R/2}\right| \right\} \leqslant C R^{\frac{n-2s-2}{2}}\,|t|, \tag{4.3.4}$$

and

$$\min\left\{ \int_{B_{R/2}} (\chi_E(x+tv) - \chi_E(x))_+\,dx, \int_{B_{R/2}} (\chi_E(x+tv) - \chi_E(x))_-\,dx \right\} \leqslant C R^{\frac{n-2s-2}{2}}\,|t|, \tag{4.3.5}$$

for some $C > 0$.

Proof. First we observe that

$$\operatorname{Per}_s(E_t, B_R) + \operatorname{Per}_s(E_{-t}, B_R) - 2\operatorname{Per}_s(E, B_R) \leqslant \frac{Ct^2}{R^2}\operatorname{Per}_s(E, B_R), \tag{4.3.6}$$

for some $C > 0$. This is indeed a general estimate, which does not use minimality, and which follows by changing variable in the integrals of the fractional perimeter (and noticing that the linear term in t simplifies). We provide some details of the proof of (4.3.6) for the facility of the reader. To this aim, we observe that

$$|\det D\Phi^t(X)| = |\det(\mathbf{1} + tR^{-1}\nabla\varphi(X/R) \otimes v)| = 1 + tR^{-1}\nabla\varphi(X/R)\cdot v + O(t^2R^{-2}).$$

Moreover, if, for any $\xi, \eta \in \mathbb{R}^n$, we set

$$g(\xi, \eta) := \frac{(\varphi(\xi) - \varphi(\eta))\,v}{|\xi - \eta|},$$

we have that g is bounded and

$$\begin{aligned}|\Phi^t(X)-\Phi^t(Y)| &= |X-Y+t(\varphi(X/R)-\varphi(Y/R))\,v|\\ &= |X-Y|\left|\frac{X-Y}{|X-Y|}+tR^{-1}\frac{(\varphi(X/R)-\varphi(Y/R))\,v}{|(X/R)-(Y/R)|}\right|\\ &= |X-Y|\left|\frac{X-Y}{|X-Y|}+tR^{-1}g(X/R,Y/R)\right|.\end{aligned}$$

Therefore

$$|\Phi^t(X)-\Phi^t(Y)|^{-n-2s}=|X-Y|^{-n-2s}\left|\frac{X-Y}{|X-Y|}+tR^{-1}g(X/R,Y/R)\right|^{-n-2s}$$
$$=|X-Y|^{-n-2s}\left(1-(n+2s)tR^{-1}\frac{X-Y}{|X-Y|}\cdot g(X/R,Y/R)+O(t^2R^{-2})\right).$$

Now we observe that Φ^t is the identity outside B_R and therefore if $A\in\{B_R,B_R^c,\mathbb{R}^n\}$ then $E_t\cap A=\Phi^t(E\cap A)$. Accordingly, for any $A,B\in\{B_R,B_R^c,\mathbb{R}^n\}$, a change of variables $x:=\Phi^t(X)$ and $y:=\Phi^t(Y)$ gives that

$$\begin{aligned}& I(E_t\cap A,E_t^c\cap B)\\ =\;& \int_{\Phi^t(E\cap A)}\int_{\Phi^t(E\cap B)}|x-y|^{-n-2s}\,dx\,dy\\ =\;& \int_{E\cap A}\int_{E\cap B}|\Phi^t(X)-\Phi^t(Y)|^{-n-2s}\,|\det D\Phi^t(X)|\,|\det D\Phi^t(Y)|\,dX\,dY\\ =\;& \int_{E\cap A}\int_{E\cap B}|X-Y|^{-n-2s}\left(1-(n+2s)tR^{-1}\frac{X-Y}{|X-Y|}\cdot g(X/R,Y/R)+O(t^2R^{-2})\right)\\ & \cdot\left(1+tR^{-1}\nabla\varphi(X/R)\cdot v+O(t^2R^{-2})\right)\left(1+tR^{-1}\nabla\varphi(Y/R)\cdot v+O(t^2R^{-2})\right)\,dX\,dY\\ =\;& \int_{E\cap A}\int_{E\cap B}|X-Y|^{-n-2s}\left(1-(n+2s)tR^{-1}\tilde g(X/R,Y/R)+O(t^2R^{-2})\right)\,dX\,dY,\end{aligned}$$

for a suitable scalar function $\tilde g$.

Then, replacing t with $-t$ and summing up, the linear term in t simplifies and we obtain

$$I(E_t\cap A,E_t^c\cap B)+I(E_{-t}\cap A,E_{-t}^c\cap B)=(2+O(t^2R^{-2}))\int_{E\cap A}\int_{E\cap B}\frac{dX\,dY}{|X-Y|^{n+2s}}.$$

This, choosing A and B appropriately, establishes (4.3.6).

On the other hand, the s-minimality of E gives that $\mathrm{Per}_s(E,B_R)\leqslant\mathrm{Per}_s(E\cup B_R,B_R)$, which, in turn, is bounded from above by the interaction between B_R and B_R^c, namely $I(B_R,B_R^c)$, which is a constant (only depending on n and s) times R^{n-2s}, due to scale invariance of the fractional perimeter. That is, we have that $\mathrm{Per}_s(E,B_R)\leqslant CR^{n-2s}$, for some $C>0$, and then we can make the right hand side of (4.3.6) uniform in E and obtain (4.3.2), up to renaming $C>0$.

The next step is to charge mass in a ball. Namely, one defines $E_t^{\cup}:=E\cup E_t$ and $E_t^{\cap}:=E\cap E_t$. By counting the interactions of the different sets, one sees that

$$\mathrm{Per}_s(E,B_R)+\mathrm{Per}_s(E_t,B_R)-\mathrm{Per}_s(E_t^{\cup},B_R)-\mathrm{Per}_s(E_t^{\cap},B_R)=2I(E_t\setminus E,E\setminus E_t).\quad(4.3.7)$$

To check this, one observes indeed that the set $E_t \backslash E$ interacts with $E \backslash E_t$ in the computations of $\mathrm{Per}_s(E, B_R)$ and $\mathrm{Per}_s(E_t, B_R)$, while these two sets do not interact in the computations of $\mathrm{Per}_s(E_t^{\cup}, B_R)$ and $\mathrm{Per}_s(E_t^{\cap}, B_R)$ (the interactions of the other sets simplify). This proves (4.3.7). We remark that, again, formula (4.3.7) is a general fact and is not based on minimality. Changing t with $-t$, we also obtain from (4.3.7) that

$$\mathrm{Per}_s(E, B_R) + \mathrm{Per}_s(E_{-t}, B_R) - \mathrm{Per}_s(E_{-t}^{\cup}, B_R) - \mathrm{Per}_s(E_{-t}^{\cap}, B_R) = 2I(E_{-t} \backslash E, E \backslash E_{-t}).$$

This and (4.3.7) give that

$$\begin{aligned} & \mathrm{Per}_s(E_t, B_R) + \mathrm{Per}_s(E_{-t}, B_R) - 2\mathrm{Per}_s(E, B_R) \\ = \; & \mathrm{Per}_s(E_t^{\cup}, B_R) + \mathrm{Per}_s(E_t^{\cap}, B_R) + \mathrm{Per}_s(E_{-t}^{\cup}, B_R) + \mathrm{Per}_s(E_{-t}^{\cap}, B_R) \\ & \quad -4\mathrm{Per}_s(E, B_R) + 2I(E_t \backslash E, E \backslash E_t) + 2I(E_{-t} \backslash E, E \backslash E_{-t}) \\ \geqslant \; & 2I(E_t \backslash E, E \backslash E_t) + 2I(E_{-t} \backslash E, E \backslash E_{-t}), \end{aligned}$$

thanks to the s-minimality of E. In particular,

$$\mathrm{Per}_s(E_t, B_R) + \mathrm{Per}_s(E_{-t}, B_R) - 2\mathrm{Per}_s(E, B_R) \geqslant 2I(E_t \backslash E, E \backslash E_t).$$

This and (4.3.2) imply (4.3.3).

Now, the interaction kernel is bounded away from zero in $B_{R/2}$, and so

$$I(E_t \backslash E, E \backslash E_t) \geqslant |(E_t \backslash E) \cap B_{R/2}| \cdot |(E \backslash E_t) \cap B_{R/2}|.$$

This is again a general fact, not depending on minimality. By plugging this into (4.3.3), we conclude that

$$\begin{aligned} CR^{n-2s-2}\, t^2 \quad & \geqslant \quad |(E_t \backslash E) \cap B_{R/2}| \cdot |(E \backslash E_t) \cap B_{R/2}| \\ & \geqslant \quad \min \left\{ |(E_t \backslash E) \cap B_{R/2}|^2, \; |(E \backslash E_t) \cap B_{R/2}|^2 \right\} \end{aligned}$$

and so, again up to renaming C,

$$\min \left\{ |(E_t \backslash E) \cap B_{R/2}|, \; |(E \backslash E_t) \cap B_{R/2}| \right\} \leqslant CR^{\frac{n-2s-2}{2}}\, t. \tag{4.3.8}$$

Now, we recall (4.3.1) and we observe that $E_t \cap B_{R/2} = (E + tv) \cap B_{R/2}$. Hence, the estimate in (4.3.8) becomes

$$\min \left\{ |((E + tv) \backslash E) \cap B_{R/2}|, \; |(E \backslash (E + tv)) \cap B_{R/2}| \right\} \leqslant CR^{\frac{n-2s-2}{2}}\, t. \tag{4.3.9}$$

Since this is valid for any $v \in S^{n-1}$, we may also switch the sign of v and obtain that

$$\min \left\{ |((E - tv) \backslash E) \cap B_{R/2}|, \; |(E \backslash (E - tv)) \cap B_{R/2}| \right\} \leqslant CR^{\frac{n-2s-2}{2}}\, t. \tag{4.3.10}$$

From (4.3.9) and (4.3.10) we obtain (4.3.4).

Now we observe that, for any sets A and B,

$$\chi_{A \backslash B}(x) \geqslant \chi_A(x) - \chi_B(x). \tag{4.3.11}$$

Indeed, this formula is clearly true if $x \in B$, since in this case the right hand side is nonpositive. The formula is also true if $x \in A \setminus B$, since in this case the left hand side is 1 and the right hand side is less or equal than 1. It remains to consider the case in which $x \notin A \cup B$. In this case, $\chi_A(x) = 0$, hence the right hand side is nonpositive, which gives that (4.3.11) holds true.

By (4.3.11),

$$\chi_{A\setminus B}(x) \geqslant \big(\chi_A(x) - \chi_B(x)\big)_+ .$$

As a consequence,

$$\begin{aligned} \big|((E-tv)\setminus E) \cap B_{R/2}\big| &= \int_{B_{R/2}} \chi_{(E-tv)\setminus E}(x)\,dx \\ &\geqslant \int_{B_{R/2}} \big(\chi_{E-tv}(x) - \chi_E(x)\big)_+\,dx = \int_{B_{R/2}} \big(\chi_E(x+tv) - \chi_E(x)\big)_+\,dx \end{aligned}$$

$$\text{and}\quad \begin{aligned} \big|(E\setminus(E-tv)) \cap B_{R/2}\big| &= \int_{B_{R/2}} \chi_{E\setminus(E-tv)}(x)\,dx \\ &\geqslant \int_{B_{R/2}} \big(\chi_E(x) - \chi_{E-tv}(x)\big)_+\,dx = \int_{B_{R/2}} \big(\chi_E(x) - \chi_E(x+tv)\big)_+\,dx \\ &= \int_{B_{R/2}} \big(\chi_E(x+tv) - \chi_E(x)\big)_-\,dx. \end{aligned}$$

This and (4.3.10) give that

$$CR^{\frac{n-2s-2}{2}}\,t \geqslant \min\left\{ \int_{B_{R/2}} \big(\chi_E(x+tv) - \chi_E(x)\big)_+\,dx,\ \int_{B_{R/2}} \big(\chi_E(x+tv) - \chi_E(x)\big)_-\,dx \right\},$$

which is (4.3.5). This ends the proof of Lemma 4.3.1. □

With the preliminary work done in Lemma 4.3.1 (to be used here with $R = 1$), we can now complete the proof of Theorem 4.1.4. To this end, we observe that

$$\begin{aligned} &\left| \int_{B_{1/2}} \big(\chi_E(x+tv) - \chi_E(x)\big)_+\,dx - \int_{B_{1/2}} \big(\chi_E(x+tv) - \chi_E(x)\big)_-\,dx \right| \\ &= \left| \int_{B_{1/2}} \big(\chi_E(x+tv) - \chi_E(x)\big)\,dx \right| \\ &= \left| \int_{B_{1/2}-tv} \chi_E(x)\,dx - \int_{B_{1/2}} \chi_E(x)\,dx \right| \\ &\leqslant \big|(B_{1/2} - tv)\Delta B_{1/2}\big| \\ &\leqslant C\,t, \end{aligned} \tag{4.3.12}$$

for some $C > 0$.

Also, we observe that, for any $a, b \in \mathbb{R}$,

$$a + b \leqslant |a - b| + 2\min\{a, b\}. \tag{4.3.13}$$

Indeed, up to exchanging a and b, we may suppose that $a \geqslant b$; thus

$$a + b = a - b + 2b = |a - b| + 2\min\{a, b\},$$

which proves (4.3.13).

Using (4.3.5), (4.3.12) and (4.3.13), we obtain that

$$\begin{aligned}
&\int_{B_{1/2}} |\chi_E(x+tv) - \chi_E(x)|\,dx \\
= &\int_{B_{1/2}} (\chi_E(x+tv) - \chi_E(x))_+\,dx + \int_{B_{1/2}} (\chi_E(x+tv) - \chi_E(x))_-\,dx \\
\leqslant &\left|\int_{B_{1/2}} (\chi_E(x+tv) - \chi_E(x))_+\,dx - \int_{B_{1/2}} (\chi_E(x+tv) - \chi_E(x))_-\,dx\right| \\
&+2\min\left\{\int_{B_{1/2}} (\chi_E(x+tv) - \chi_E(x))_+\,dx,\ \int_{B_{1/2}} (\chi_E(x+tv) - \chi_E(x))_-\,dx\right\} \\
\leqslant &\ Ct,
\end{aligned}$$

up to renaming C. Dividing by t and sending $t \searrow 0$ (up to subsequences), one finds that

$$\int_{B_{1/2}} |\partial_v \chi_E(x)|\,dx \leqslant C,$$

for any $v \in S^{n-1}$, in the bounded variation sense. Since the direction v is arbitrary, this proves that

$$\mathrm{Per}\,(E, B_{1/2}) = \int_{B_{1/2}} |\nabla \chi_E(x)|\,dx \leqslant C.$$

This proves Theorem 4.1.4 with $R = 1$, and the general case follows from scaling.

4.4 Proof of Theorem 4.1.8

In this part, we will make use of some integral geometric formulas which compute the perimeter of a set by averaging the number of intersections of straight lines with the boundary of a set.

For this, we recall the notation of the positive and negative part of a function u, namely

$$u_+(x) := \max\{u(x),\ 0\} \qquad \text{and} \qquad u_-(x) := \max\{-u(x),\ 0\}.$$

Notice that $u_\pm \geqslant 0$, that $|u| = u_+ + u_-$ and that $u = u_+ - u_-$.

Also, if $v \in \partial B_1$ and $p \in \mathbb{R}^n$, we define

$$\begin{aligned}
& v^\perp := \{y \in \mathbb{R}^n \text{ s.t. } y \cdot v = 0\} \\
\text{and} \qquad & p + \mathbb{R}v := \{p + tv \text{ s.t. } t \in \mathbb{R}\}.
\end{aligned}$$

That is, $v^\perp$ is the orthogonal linear space to v and $p + \mathbb{R}v$ is the line passing through p with direction v.

Now, given a Caccioppoli set $E \subseteq \mathbb{R}^n$ with exterior normal ν (and reduced boundary denoted by $\partial^* E$), and $v \in \partial B_1$, we set

$$I_{v,\pm}(y) := \sup \mp \int_{y+\mathbb{R}v} \chi_E(x)\, \varphi'(x)\, d\mathcal{H}^1(x), \tag{4.4.1}$$

with the sup taken over all smooth φ supported in the segment $B_1 \cap (y + \mathbb{R}v)$ with image in $[0, 1]$. We have (see e.g. Proposition 4.4 in [12]) that one can compute the directional derivative in the sense of bounded variation by the formula

$$\int_{B_1} (\partial_v \chi_E)_\pm(x)\, dx = \int_{y \in v^\perp} I_{v,\pm}(y)\, d\mathcal{H}^{n-1}(y) \tag{4.4.2}$$

and we also have that $I_{v,\pm}(y)$ is the number of points x that lie in $B_1 \cap (\partial^* E) \cap (y + \mathbb{R}v)$ and such that $\mp v \cdot \nu(x) > 0$. That is, the quantity $I_{v,+}(y)$ (resp., $I_{v,-}(y)$) counts the number of intersections in the ball B_1 between the line $y + \mathbb{R}v$ and the (reduced) boundary of E that occur at points x in which $v \cdot \nu(x)$ is negative (resp., positive). In particular,

$$I_{v,\pm}(y) \in \mathbb{Z} \cap [0, +\infty) = \{0, 1, 2, 3, \dots\}. \tag{4.4.3}$$

Furthermore, the vanishing of $I_{v,+}(y)$ (resp., $I_{v,-}(y)$) is related to the fact that, moving along the segment $B_1 \cap (y + \mathbb{R}v)$, one can only exit (resp., enter) the set E, according to the following result:

Lemma 4.4.1. *If $I_{v,+}(y) = 0$, then the map $B_1 \cap (y + \mathbb{R}v) \ni x \mapsto \chi_E(x)$ is nonincreasing.*

Proof. For any smooth φ supported in the segment $B_1 \cap (y + \mathbb{R}v)$ with image in $[0, 1]$,

$$0 = I_{v,+}(y) \geqslant - \int_{y+\mathbb{R}v} \chi_E(x)\, \varphi'(x)\, d\mathcal{H}^1(x),$$

that is

$$\int_{y+\mathbb{R}v} \chi_E(x)\, \varphi'(x)\, d\mathcal{H}^1(x) \geqslant 0,$$

which gives the desired result. □

Now we define

$$\Phi_\pm(v) := \int_{y \in v^\perp} I_{v,\pm}(y)\, d\mathcal{H}^{n-1}(y). \tag{4.4.4}$$

By (4.4.2),

$$\Phi_\pm(v) = \int_{B_1} (\partial_v \chi_E)_\pm(x)\, dx. \tag{4.4.5}$$

We observe that

Lemma 4.4.2. *Let* $\mathrm{Per}\,(E, B_1) < +\infty$ *and* $n \geqslant 2$. *Then the functions* $\Phi_{\pm}$ *are continuous on* S^{n-1}. *Moreover, there exists* $v_\star$ *such that*

$$\Phi_+(v_\star) = \Phi_-(v_\star). \tag{4.4.6}$$

Proof. Let $v, w \in S^{n-1}$. By (4.4.5),

$$\begin{aligned}|\Phi_+(v) - \Phi_+(w)| &\leqslant \int_{B_1} |(\partial_v \chi_E)_+(x) - (\partial_w \chi_E)_+(x)|\, dx \\ &\leqslant \int_{B_1} |\partial_v \chi_E(x) - \partial_w \chi_E(x)|\, dx \leqslant |v - w| \int_{B_1} |\nabla \chi_E(x)|\, dx = |v - w|\, \mathrm{Per}\,(E, B_1).\end{aligned}$$

This shows that Φ_+ is continuous. Similarly, one sees that Φ_- is continuous.

Now we prove (4.4.6). For this, let $\Psi(v) := \Phi_+(v) - \Phi_-(v)$. By (4.4.5),

$$\Phi_{\pm}(-v) = \Phi_{\mp}(v).$$

Therefore

$$\Psi(-v) = \Phi_+(-v) - \Phi_-(-v) = \Phi_-(v) - \Phi_+(v) = -\Psi(v). \tag{4.4.7}$$

Now, if $\Psi(e_1) = 0$, we can take $v_\star := e_1$ and (4.4.6) is proved. So we can assume that $\Psi(e_1) > 0$ (the case $\Psi(e_1) < 0$ is analogous). By (4.4.7), we obtain that $\Psi(-e_1) < 0$. Hence, since Ψ is continuous, it must have a zero on any path joining e_1 to $-e_1$, and this proves (4.4.6). □

A control on the function $\Phi_{\pm}$ implies a quantitative flatness bound on the set E, as stated here below:

Lemma 4.4.3. *Let* $n = 2$. *There exists* $\mu_o > 0$ *such that for any* $\mu \in (0, \mu_o]$ *the following statement holds.*

Assume that

$$\Phi_-(e_2) \leqslant \mu \tag{4.4.8}$$

and that

$$\max\{\Phi_+(e_1),\ \Phi_-(e_1)\} \leqslant \mu. \tag{4.4.9}$$

Then, there exists a horizontal halfplane $h \subset \mathbb{R}^2$ *such that*

$$|(E \setminus h) \cap B_1| + |(h \setminus E) \cap B_1| \leqslant C\mu, \tag{4.4.10}$$

for some $C > 0$.

Proof. Given $v \in \partial B_1$, we take into account the sets of $y \in v^\perp$ which give a positive contribution to $I_{v,\pm}(y)$. For this, we define

$$\mathcal{B}_{\pm}(v) := \{y \in v^\perp \text{ s.t. } I_{v,\pm}(y) \neq 0\}.$$

From (4.4.3), we know that if $y \in \mathcal{B}_\pm(v)$, then $I_{v,\pm}(y) \geqslant 1$. As a consequence of this and of (4.4.4), we have that

$$\Phi_\pm(v) \geqslant \int_{\mathcal{B}_\pm(v)} I_{v,\pm}(y)\, d\mathcal{H}^1(y) \geqslant \mathcal{H}^1(\mathcal{B}_\pm(v)).$$

Accordingly, by (4.4.8) and (4.4.9), we see that

$$\mathcal{H}^1(\mathcal{B}_-(e_2)) \leqslant \mu \tag{4.4.11}$$

and

$$\mathcal{H}^1(\mathcal{B}_\pm(e_1)) \leqslant \mu. \tag{4.4.12}$$

Furthermore, for any $y \in v^\perp \setminus \mathcal{B}_+(v)$ (resp. $y \in v^\perp \setminus \mathcal{B}_-(v)$), we have that $I_{v,+}(y) = 0$ (resp., $I_{v,-}(y) = 0$) and thus, by Lemma 4.4.1, the map $B_1 \cap (y + \mathbb{R}v) \ni x \mapsto \chi_E(x)$ is nonincreasing (resp., nondecreasing).

Therefore, by (4.4.12), we have that for any vertical coordinate $y \in e_1^\perp$ outside the small set $\mathcal{B}_-(e_1) \cup \mathcal{B}_+(e_1)$ (which has total length of size 2μ), the vertical line $y + \mathbb{R}e_1$ is either all contained in E or in its complement (see Figure 4.7).

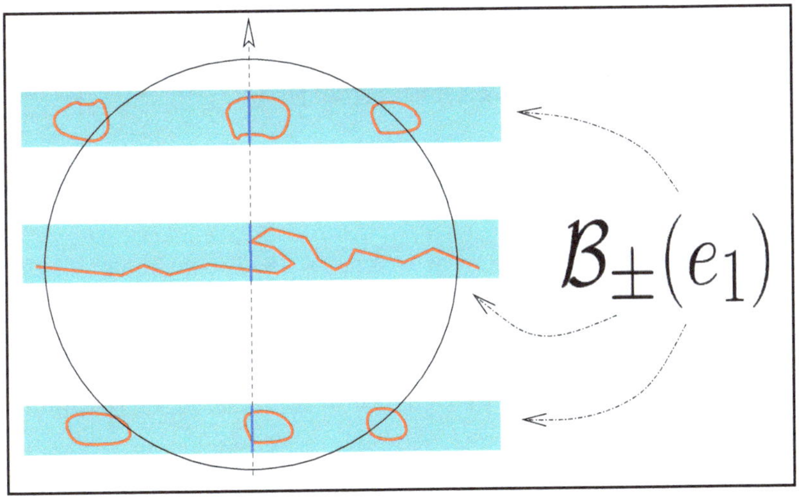

Fig. 4.7: Horizontal lines do not meet the boundary of E, with the exception of a small set $\mathcal{B}_\pm(e_1)$.

That is, we can denote by $\mathcal{G}_E$ the set of vertical coordinates y for which the portion in B_1 of the horizontal line passing through y lies in E and, similarly, by $\mathcal{G}_{E^c}$ the set of vertical coordinates y for which the portion in B_1 of the horizontal line passing through y lies in E^c and we obtain that $\mathcal{G}_E \cup \mathcal{G}_{E^c}$ exhaust the whole of $(-1, 1)$, up to a set of size at most 2μ.

We also remark that $\mathcal{G}_E$ lies below $\mathcal{G}_{E^c}$: indeed, by (4.4.11), we have that vertical lines can only exit the set E (possibly with the exception of a small set of size μ). The situation is depicted in Figure 4.8.

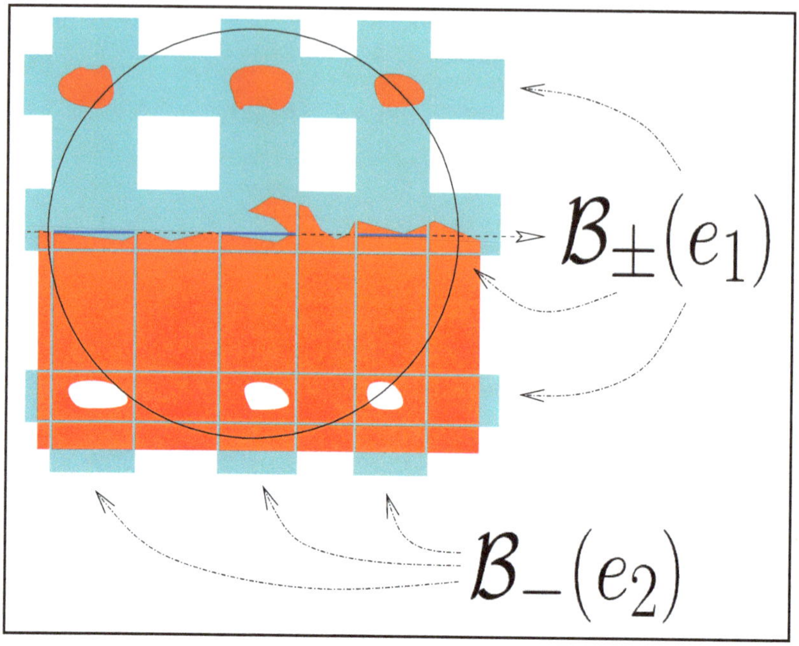

Fig. 4.8: Vertical lines do not meet the boundary of E, with the exception of a small set $\mathcal{B}_-(e_2)$.

Hence, if we take h to be a horizontal halfplane which separates $\mathcal{G}_E$ and $\mathcal{G}_{E^c}$, we obtain (4.4.10). □

With this, we can now complete the proof of Theorem 4.1.8. The main tool for this goal is Lemma 4.4.3. In order to apply it, we need to check that (4.4.8) and (4.4.9) are satisfied. To this end, we argue as follows. First of all, fixed a large $R > 2$, we consider, as in Section 4.3, a diffeomorphism Φ^t such that $\Phi^t(x) = x$ for any $x \in \mathbb{R}^n \setminus B_{9R/10}$, and $\Phi^t(x) = x + tv$ for any $x \in B_{3R/4}$, and we set $E_t := \Phi^t(E)$. From (4.3.5) (recall that here $n = 2$), we have that

$$\min\left\{\int_{B_{R/2}} (\chi_E(x+tv) - \chi_E(x))_+ \, dx, \int_{B_{R/2}} (\chi_E(x+tv) - \chi_E(x))_- \, dx\right\} \leqslant \frac{Ct}{R^s},$$

for some $C > 0$. Thus, dividing by t and sending $t \searrow 0$,

$$\min\left\{\int_{B_{R/2}} (\partial_v \chi_E(x))_+ \, dx, \int_{B_{R/2}} (\partial_v \chi_E(x))_- \, dx\right\} \leqslant \frac{C}{R^s}.$$

That is, recalling (4.4.5),

$$\min\{\Phi_+(v), \ \Phi_-(v)\} \leqslant \frac{C}{R^s}. \tag{4.4.13}$$

We also observe that E has finite perimeter in B_1, thanks to Theorem 4.1.4, and so we can make use of Lemma 4.4.2. In particular, by (4.4.6), after a rotation of coordinates, we may assume that $\Phi_+(e_1) = \Phi_-(e_1)$. Hence (4.4.13) says that

$$\max\{\Phi_+(e_1),\ \Phi_-(e_1)\} = \min\{\Phi_+(e_1),\ \Phi_-(e_1)\} \leqslant \frac{C}{R^s}. \tag{4.4.14}$$

Also, up to a change of orientation, we may suppose that $\Phi_-(e_2) \leqslant \Phi_+(e_2)$, hence in this case (4.4.13) says that

$$\Phi_-(e_2) \leqslant \frac{C}{R^s}.$$

From this and (4.4.14), we see that (4.4.8) and (4.4.9) are satisfied (with $\mu = C/R^s$) and so by Lemma 4.4.3 we conclude that

$$|(E\setminus h) \cap B_1| + |(h\setminus E) \cap B_1| \leqslant \frac{C}{R^s},$$

for some halfplane h. This completes the proof of Theorem 4.1.8: as a matter of fact, the result proven is even stronger, since it says that, after removing horizontal and vertical slabs of size C/R^s, we have that ∂E in B_1 is a graph of oscillation bounded by C/R^s, see Figure 4.8 (in fact, more general statements and proofs can be find in [12]).

4.5 Sketch of the Proof of Theorem 4.1.12

The core of the proof of Theorem 4.1.12 consists in constructing a suitable barrier that can be slided "from below" and which exhibits the desired stickiness phenomenon: if this is possible, since the s-minimal surface cannot touch the barrier, it has to stay above the barrier and stick at the boundary as well.

So, the barrier we are looking for should have negative fractional mean curvature, coincide with F outside $(-1, 1) \times \mathbb{R}$ and contain $(-1, 1) \times (-\infty, \delta^\gamma)$.

Such barrier is constructed in [19] in an iterative way, that we now try to describe.

Step 1. Let us start by looking at the subgraph of the function $y = \frac{x_+}{\ell}$, given $\ell \geqslant 0$. Then, at all the boundary points $X = (x, y)$ with positive abscissa $x > 0$, the fractional mean curvature is at most

$$-\frac{c}{\max\{1, \ell\}\,|X|^{2s}}, \tag{4.5.1}$$

for some $c > 0$. The full computation is given in Lemma 5.1 of [19], but we can give a heuristic justification of it, by saying that for small X the boundary point gets close to the origin, where there is a corner and the curvature blows up (with a negative sign, since there is "more than a hyperplane" contained in the set), see Figure 4.9. Also, the power $2s$ in (4.5.1) follows by scaling.

In addition, if ℓ is close to 0, this first barrier is close to a ninety degree angle, while if ℓ is large it is close to a flat line, and these considerations are also in agreement with (4.5.1).

Step 2. Having understood in Step 1 what happens for the "angles", now we would like to "shift iteratively in a smooth way from one slope to another", see Figure 4.10.

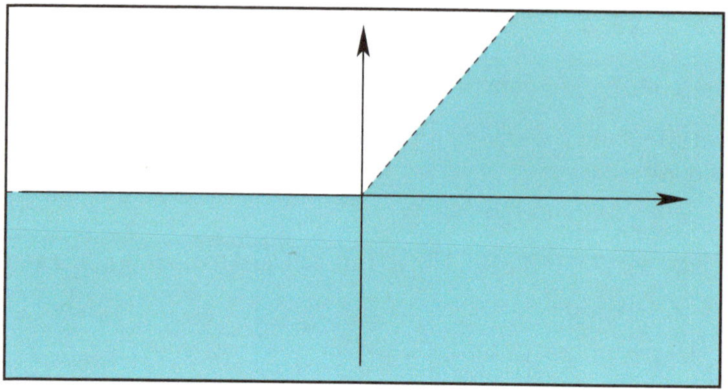

Fig. 4.9: Description of Step 1.

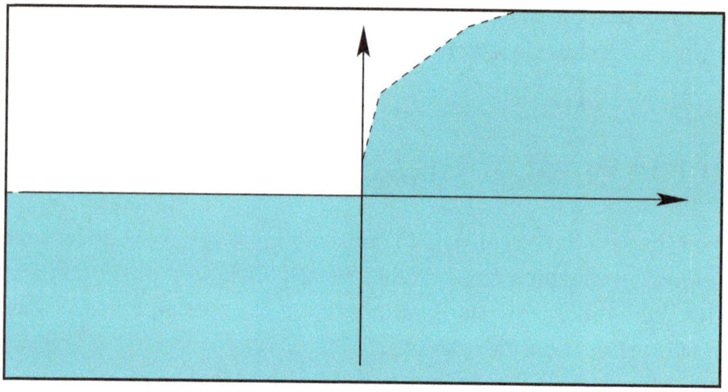

Fig. 4.10: Description of Step 2.

The detailed statement is given in Proposition 5.3 in [19], but the idea is as follows. For any $K \in \mathbf{N}$, $K \geqslant 1$, one looks at the subgraph of a nonnegative function v_K such that

- $v_K(x) = 0$ if $x < 0$,
- $v_K(x) \geqslant a_K$ if $x > 0$, for some $a_K > 0$,
- $v_K(x) = \frac{x+q_K}{\ell_K}$ for any $x \geqslant \ell_K - q_K$, for some $\ell_K \geqslant K$ and $q_K \in \left[0, \frac{1}{K}\right]$,
- at all the boundary points $X = (x, y)$ with positive abscissa $x > 0$, the fractional mean curvature is at most $-\frac{c}{\ell_K |X|^{2s}}$, for some $c > 0$.

Step 3. If K is sufficiently large in Step 2, the final slope is almost horizontal. In this case, one can smoothly glue such barrier with a power like function like $x^{\frac{1}{2}+s+\varepsilon_0}$. Here, ε_0 is any fixed positive exponent (the power γ in the statement of Theorem 4.1.12 is related to ε_0, since $\gamma := \frac{2+\varepsilon_0}{1-2s}$). The details of the barrier constructed in this way are given in Proposition 6.3 of [19]. In this case, one can still control the fractional mean curvature at all the boundary points $X = (x, y)$ with positive abscissa $x > 0$, but the estimate is of the type either $|X|^{-2s}$, for small $|X|$, or $|X|^{-\frac{1}{2}-s+\varepsilon_0}$, for large $|X|$. A sketch of such barrier is given in Figure 4.11.

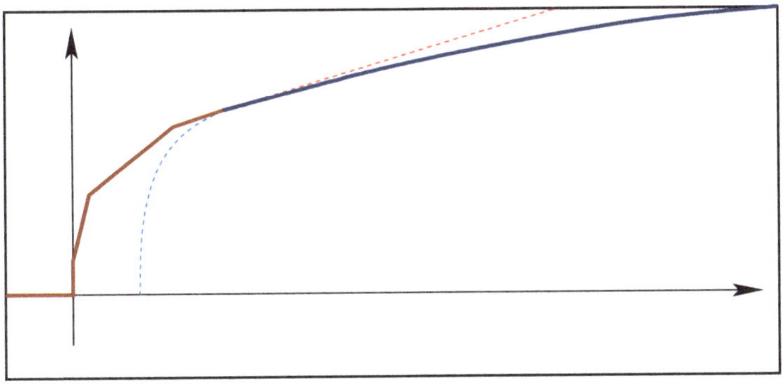

Fig. 4.11: Description of Step 3.

Step 4. Now we use the barrier of Step 3 to construct a compactly supported object. The idea is to take such barrier, to reflect it and to glue it at a “horizontal level”, see Figure 4.12.

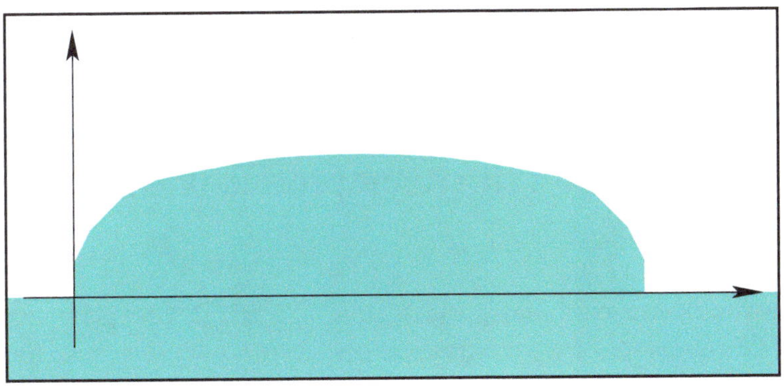

Fig. 4.12: Description of Step 4.

We remark that such barrier has a vertical portion at the origin and one can control its fractional mean curvature from above with a negative quantity for the boundary points $X = (x, y)$ with positive, but not too large, abscissa.

Of course, this type of estimate cannot hold at the maximal point of the barrier, where “more than a hyperplane” is contained in the complement of the set, and therefore the fractional mean curvature is positive (the precise quantitative estimate is given in Proposition 7.1. of [19]).

Step 5. Nevertheless, we can now compensate this error in the fractional mean curvature near the maximal point of the barrier by adding two suitably large domains on the sides of the barriers, see Figure 4.13.

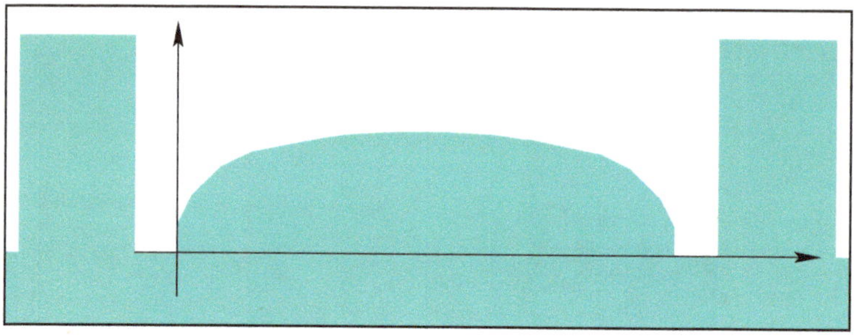

Fig. 4.13: Description of Step 5.

The barrier constructed in this way is described in details in Proposition 7.3 of [19] and its basic feature is to possess a vertical portion near the origin and to possess negative fractional mean curvature.

By keeping good track of the quantitative estimates on the bumps of the barriers and on their fractional mean curvatures, one can now scale the latter barrier and slide it from below, in order to prove Theorem 4.1.12. The full details are given in Section 8 of [19].

A A Short Discussion on the Asymptotics of the *s*-perimeter

In this appendix, we would like to emphasize the fact that, as $s \nearrow 1/2$, the s-perimeter recovers (under different perspectives) the classical perimeter, while, as $s \searrow 0$, the nonlocal features become predominant and the problem produces the Lebesgue measure – or, better to say, convex combinations of Lebesgue measures by interpolation parameters of nonlocal type.

First of all, we show that if E is a bounded set with smooth boundary, then

$$\lim_{s \nearrow 1/2} (1-2s) \operatorname{Per}_s(E, \mathbb{R}^n) = \kappa_{n-1} \operatorname{Per}(E, \mathbb{R}^n), \tag{A.1}$$

where we denoted by κ_n the n-dimensional volume of the n-dimensional unit ball.

For further convenience, we also use the notation

$$\varpi_n := \mathcal{H}^{n-1}(S^{n-1}).$$

Notice that, by polar coordinates,

$$\kappa_n = \int_{S^{n-1}} \left[\int_0^1 \rho^{n-1}\, d\rho \right] d\mathcal{H}^{n-1}(x) = \frac{\varpi_n}{n}. \tag{A.2}$$

We point out that formula (A.1) is indeed a simple version of more general approximation results, for which we refer to [4, 14, 2, 25, 9] and to [10] for the regularity results that can be achieved by approximation methods. See also [17] for further comments and examples.

The proof of (A.1) can be performed by different methods; here we give a simple argument which uses formula (4.1.13). To this aim, we fix $x \in \partial E$ and $\delta > 0$. If $y \in (\partial E) \cap B_\delta(x)$ and δ is sufficiently small, then $\nu(y) = \nu(x) + O(\delta)$. Moreover, for any $\varrho \in (0, \delta]$, the $(n-2)$-dimensional contribution of ∂E in $\partial B_\varrho(x)$ coincides, up to higher orders in δ, with the one of the $(n-2)$-dimensional sphere, that is $\varpi_{n-1}\, \varrho^{n-2}$, see Figure 4.14.

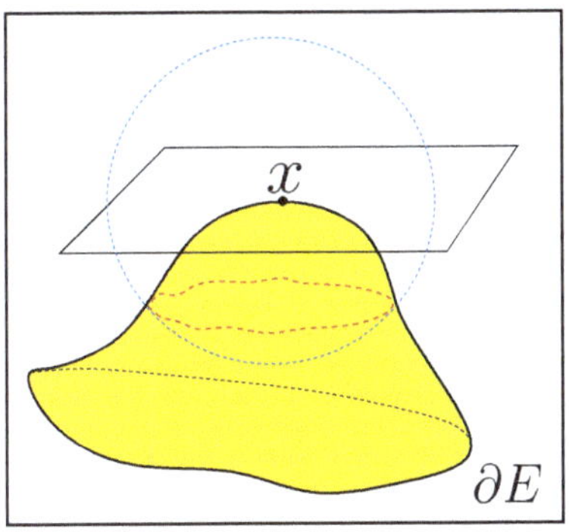

Fig. 4.14: $\mathcal{H}^{n-2}((\partial E) \cap \partial B_\varrho(x))$ (in the picture, $n = 3$).

As a consequence of these observations, we have that

$$\begin{aligned}
\int_{(\partial E)\cap B_\delta(x)} \frac{\nu(x)\cdot\nu(y)}{|x-y|^{n+2s-2}}\, d\mathcal{H}^{n-1}(y) &= \int_{(\partial E)\cap B_\delta(x)} \frac{1+O(\delta)}{|x-y|^{n+2s-2}}\, d\mathcal{H}^{n-1}(y) \\
&= (1+O(\delta)) \int_0^\delta \frac{\mathcal{H}^{n-2}((\partial E)\cap(\partial B_\rho))}{\varrho^{n+2s-2}}\, d\varrho \\
&= (1+O(\delta))\, \varpi_{n-1} \int_0^\delta \frac{\varrho^{n-2}}{\varrho^{n+2s-2}}\, d\varrho \\
&= \frac{(1+O(\delta))\, \varpi_{n-1}\, \delta^{1-2s}}{1-2s}.
\end{aligned}$$

On the other hand,

$$\int_{(\partial E)\setminus B_\delta(x)} \frac{\nu(x)\cdot\nu(y)}{|x-y|^{n+2s-2}}\,d\mathcal{H}^{n-1}(y) \leqslant \frac{\mathcal{H}^{n-1}(\partial E)}{\delta^{n+2s-2}}.$$

Therefore

$$\int_{\partial E} \frac{\nu(x)\cdot\nu(y)}{|x-y|^{n+2s-2}}\,d\mathcal{H}^{n-1}(y) = \frac{(1+O(\delta))\,\varpi_{n-1}\,\delta^{1-2s}}{1-2s} + O(\delta^{-n-2s+2}).$$

Accordingly, recalling (4.1.13),

$$\begin{aligned}
&\lim_{s\nearrow 1/2} (1-2s)\,\mathrm{Per}_s\,(E,\mathbb{R}^n)\\
&= \lim_{s\nearrow 1/2} \frac{1-2s}{2s\,(n+2s-2)} \int_{\partial E}\left[\int_{\partial E} \frac{\nu(x)\cdot\nu(y)}{|x-y|^{n+2s-2}}\,d\mathcal{H}^{n-1}(y)\right] d\mathcal{H}^{n-1}(x)\\
&= \lim_{s\nearrow 1/2} \frac{1-2s}{n-1}\int_{\partial E}\left[\frac{(1+O(\delta))\,\varpi_{n-1}\,\delta^{1-2s}}{1-2s} + O(\delta^{-n-2s+2})\right] d\mathcal{H}^{n-1}(x)\\
&= \lim_{s\nearrow 1/2} \frac{(1+O(\delta))\,\varpi_{n-1}\,\delta^{1-2s} + (1-2s)\,O(\delta^{-n-2s+2})}{n-1}\,\mathcal{H}^{n-1}(\partial E)\\
&= \frac{(1+O(\delta))\,\varpi_{n-1}}{n-1}\,\mathcal{H}^{n-1}(\partial E).
\end{aligned}$$

Hence, by taking δ arbitrarily small,

$$\lim_{s\nearrow 1/2} (1-2s)\,\mathrm{Per}_s\,(E,\mathbb{R}^n) = \frac{\varpi_{n-1}}{n-1}\,\mathcal{H}^{n-1}(\partial E),$$

which gives (A.1), in view of (A.2).

Now we show that, if $n \geqslant 3$ and E is a bounded set with smooth boundary,

$$\lim_{s\searrow 0} s\,\mathrm{Per}_s\,(E,\mathbb{R}^n) = \frac{\varpi_n}{2}\,|E|. \tag{A.3}$$

Once again, more general (and subtle) statements hold true, see [23, 18] for details.

To prove (A.3), we denote by

$$\Gamma(x) := \frac{1}{(n-2)\,\varpi_n\,|x|^{n-2}}$$

the fundamental solution[4.7] of the Laplace operator when $n \geqslant 3$, that is

$$-\Delta\Gamma(x) = \delta_0(x),$$

4.7 It is interesting to understand how the fundamental solution of the Laplacian also occurs when $n = 2$. In this case, we observe that if $c_E := \int_{\partial E} \nu(y)\,d\mathcal{H}^{n-1}(y)$, then of course

$$\iint_{(\partial E)\times(\partial E)} \nu(x)\cdot\nu(y)\,d\mathcal{H}^{n-1}(x)\,d\mathcal{H}^{n-1}(y) = \int_{\partial E} \nu(x)\cdot c_E\,d\mathcal{H}^{n-1}(x) = \int_E \mathrm{div}_x c_E\,dx = \int_E 0\,dx = 0.$$

where δ_0 is the Dirac's Delta centered at the origin. Then, from (4.1.13),

$$\begin{aligned}
\lim_{s\searrow 0} s\,\mathrm{Per}_s(E,\mathbb{R}^n) &= \frac{1}{2(n-2)}\int_{\partial E}\left[\int_{\partial E}\frac{\nu(x)\cdot\nu(y)}{|x-y|^{n-2}}\,d\mathcal{H}^{n-1}(y)\right]d\mathcal{H}^{n-1}(x)\\
&= \frac{\varpi_n}{2}\int_{\partial E}\left[\int_{\partial E}\nu(y)\cdot(\nu(x)\Gamma(x-y))\,d\mathcal{H}^{n-1}(y)\right]d\mathcal{H}^{n-1}(x)\\
&= \frac{\varpi_n}{2}\int_{\partial E}\left[\int_{E}\mathrm{div}_y(\nu(x)\Gamma(x-y))\,dy\right]d\mathcal{H}^{n-1}(x)\\
&= \frac{\varpi_n}{2}\int_{\partial E}\left[\int_{E}\nu(x)\cdot\nabla_y\Gamma(x-y)\,dy\right]d\mathcal{H}^{n-1}(x)\\
&= \frac{\varpi_n}{2}\int_{E}\left[\int_{\partial E}\nu(x)\cdot\nabla_y\Gamma(x-y)\,d\mathcal{H}^{n-1}(x)\right]dy\\
&= \frac{\varpi_n}{2}\int_{E}\left[\int_{E}\mathrm{div}_x(\nabla_y\Gamma(x-y))\,dx\right]dy\\
&= -\frac{\varpi_n}{2}\iint\limits_{E\times E}\Delta\Gamma(x-y)\,dx\,dy\\
&= \frac{\varpi_n}{2}\iint\limits_{E\times E}\delta_0(x-y)\,dx\,dy\\
&= \frac{\varpi_n}{2}\int_E 1\,dy\\
&= \frac{\varpi_n}{2}|E|,
\end{aligned}$$

that is (A.3).

We remark that formula (A.3) is actually a particular case of a more general phenomenon, described in [18]. For instance, if the following limit exists

$$a(E) := \lim_{s\searrow 0}\frac{2s}{\varpi_n}\int_{E\setminus B_1}\frac{dx}{|x|^{n+2s}},$$

then

$$\lim_{s\searrow 0}\frac{2s}{\varpi_n}\mathrm{Per}_s(E,\Omega) = (1-a(E))\,|E\cap\Omega| + a(E)\,|\Omega\setminus E|. \tag{A.4}$$

Hence, we write

$$\frac{1}{|x-y|^{2s}} = \exp(-2s\log|x-y|) = 1 - 2s\log|x-y| + O(s^2),$$

thus

$$\begin{aligned}
&\frac{1}{2s}\iint\limits_{(\partial E)\times(\partial E)}\frac{\nu(x)\cdot\nu(y)}{|x-y|^{2s}}\,d\mathcal{H}^{n-1}(x)\,d\mathcal{H}^{n-1}(y)\\
&= -\iint\limits_{(\partial E)\times(\partial E)}\nu(x)\cdot\nu(y)\,\log|x-y|\,d\mathcal{H}^{n-1}(x)\,d\mathcal{H}^{n-1}(y) + O(s)
\end{aligned}$$

and one can use the same fundamental solution trick as in the case $n \geqslant 3$.

Notice indeed that (A.3) is a particular case of (A.4), since when E is bounded, then $a(E) = 0$. Equation (A.4) has also a suggestive interpretation, since it says that, in a sense, as $s \searrow 0$, the fractional perimeter is a convex interpolation of measure contributions inside the reference set Ω: namely it weights the measures of two contributions of E and the complement of E inside Ω by a convex parameter $a(E) \in [0, 1]$ which in turn takes into account the behavior of E at infinity.

B A Short Discussion on the Asymptotics of the s-mean Curvature

As $s \nearrow 1/2$, the s-mean curvature recovers the classical mean curvature (see [1] for details).

A very natural question raised to us by Jun-Cheng Wei dealt with the asymptotics as $s \searrow 0$ of the s-mean curvature. Notice that, by (A.3), we know that $2s$ times the s-perimeter approaches ϖ_n times the volume. Since the variation of the volume along normal deformations is 1, if one is allowed to "exchange the limits" (i.e. to identify the limit of the variation with the variation of the limit), then she or he may guess that $2s$ times the s-mean curvature should approach ϖ_n.

This is indeed the case, and higher orders can be computed as well, according to the following observation: if E has smooth boundary, $p \in \partial E$ and $E \subseteq B_R(p)$ for some $R > 0$, then

$$2s\, H_E^s(p) = \varpi_n + 2s \left(\int_{B_R(p)} \frac{\chi_{E^c}(x) - \chi_E(x)}{|x-p|^n}\, dx - \varpi_n \log R \right) + o(s), \tag{B.1}$$

as $s \searrow 0$. To prove this, we first observe that, up to a translation, we can take $p = 0$. Moreover, since E lies inside B_R,

$$\begin{aligned} \int_{\mathbb{R}^n \setminus B_R} \frac{\chi_{E^c}(x) - \chi_E(x)}{|x|^{n+2s}}\, dx &= \int_{\mathbb{R}^n \setminus B_R} \frac{dx}{|x|^{n+2s}} \\ &= \frac{\varpi_n}{2s\, R^{2s}} = \frac{\varpi_n}{2s} \exp(-2s \log R) \\ &= \frac{\varpi_n}{2s} (1 - 2s \log R + o(s)). \end{aligned} \tag{B.2}$$

In addition, since ∂E is smooth, we have that (possibly after a rotation) there exists $\delta_o \in (0, \min\{1, R\})$ such that, for any $\delta \in (0, \delta_o]$, $E \cap B_\delta$ contains $\{x_n \leqslant -M|x'|^2\}$ and is contained in $\{x_n \leqslant M|x'|^2\}$. Here, $M > 0$ only depends on the curvatures of E and we are using the notation $x = (x', x_n) \in \mathbb{R}^{n-1} \times \mathbb{R}$ (notice also that since we took $p = 0$, the ball B_δ is actually centered at p).

Therefore, we have that $\chi_{E^c}(x) - \chi_E(x) = -1$ for any $x \in B_\delta \cap \{x_n \leqslant -M|x'|^2\}$ and $\chi_{E^c}(x) - \chi_E(x) = 1$ for any $x \in B_\delta \cap \{x_n \geqslant M|x'|^2\}$. In this way, a cancellation gives that

$$\int_{B_\delta \cap \{|x_n| \geqslant M|x'|^2\}} \frac{\chi_{E^c}(x) - \chi_E(x)}{|x|^{n+2s}}\, dx = 0.$$

As a consequence, for any $\sigma \in [0, s]$, if $s \in (0, 1/4)$,

$$\begin{aligned}
\left| \int_{B_\delta} \frac{\chi_{E^c}(x) - \chi_E(x)}{|x|^{n+2\sigma}}\, dx \right| &\leqslant \int_{\{|x'| \leqslant \delta\}} dx' \int_{\{|x_n| \leqslant M\,|x'|^2\}} dx_n \frac{1}{|x|^{n+2\sigma}} \\
&\leqslant 2M \int_{\{|x'| \leqslant \delta\}} \frac{|x'|^2}{|x'|^{n+2\sigma}}\, dx' \leqslant \frac{2M\,\varpi_n\,\delta^{1-2\sigma}}{1-2\sigma} \leqslant 4M\,\varpi_n\,\delta^{1/2}.
\end{aligned}$$

Therefore, we use this inequality with $\sigma := 0$ and $\sigma := s$ and the Dominated Convergence Theorem, to find that

$$\begin{aligned}
& \lim_{s \searrow 0} \left| \int_{B_R} \frac{\chi_{E^c}(x) - \chi_E(x)}{|x|^n}\, dx - \int_{B_R} \frac{\chi_{E^c}(x) - \chi_E(x)}{|x|^{n+2s}}\, dx \right| \\
\leqslant\ & \lim_{s \searrow 0} \left| \int_{B_R \setminus B_\delta} \frac{\chi_{E^c}(x) - \chi_E(x)}{|x|^n}\, dx - \int_{B_R \setminus B_\delta} \frac{\chi_{E^c}(x) - \chi_E(x)}{|x|^{n+2s}}\, dx \right| + 8M\,\varpi_n\,\delta^{1/2} \\
=\ & 8M\,\varpi_n\,\delta^{1/2}.
\end{aligned}$$

Hence, since we can now take δ arbitrarily small, we conclude that

$$\lim_{s \searrow 0} \left| \int_{B_R} \frac{\chi_{E^c}(x) - \chi_E(x)}{|x|^n}\, dx - \int_{B_R} \frac{\chi_{E^c}(x) - \chi_E(x)}{|x|^{n+2s}}\, dx \right| = 0.$$

In view of this, and recalling (4.1.14) and (B.2), we find that

$$\begin{aligned}
& \lim_{s \searrow 0} \frac{1}{s} \left| 2s\, H_E^s(0) - \varpi_n - 2s \left(\int_{B_R} \frac{\chi_{E^c}(x) - \chi_E(x)}{|x|^n}\, dx - \varpi_n \log R \right) \right| \\
& \leqslant \lim_{s \searrow 0} \frac{1}{s} \left| 2s \int_{\mathbb{R}^n} \frac{\chi_{E^c}(x) - \chi_E(x)}{|x|^{n+2s}}\, dx - \varpi_n - 2s \left(\int_{B_R} \frac{\chi_{E^c}(x) - \chi_E(x)}{|x|^{n+2s}}\, dx - \varpi_n \log R \right) \right| \\
& \quad + 2 \left| \int_{B_R} \frac{\chi_{E^c}(x) - \chi_E(x)}{|x|^n}\, dx - \int_{B_R} \frac{\chi_{E^c}(x) - \chi_E(x)}{|x|^{n+2s}}\, dx \right| \\
& = \lim_{s \searrow 0} \frac{1}{s} \left| \varpi_n \left(1 - 2s \log R + o(s)\right) - \varpi_n + 2s\varpi_n \log R \right| \\
& = 0.
\end{aligned}$$

This proves (B.1).

C Second Variation Formulas and Graphs of Zero Nonlocal Mean Curvature

In this appendix, we show that the second variation (say, with respect to a normal perturbation η) of the fractional perimeter of surfaces with vanishing mean curvature is given by

$$-2\int_{\partial E}\frac{\eta(y)-\eta(x)}{|x-y|^{n+2s}}\,d\mathcal{H}^{n-1}(y)+\int_{\partial E}\frac{\eta(x)\left[1-\nu(x)\cdot\nu(y)\right]}{|x-y|^{n+2s}}\,d\mathcal{H}^{n-1}(y).$$

A rigorous statement for this claim will be given in the forthcoming Lemma C.1: for the moment, we remark that the expression above is related with the Jacobi field along surfaces of vanishing nonlocal mean curvature. We refer to [15] for full details about this type of formulas. See in particular formula (1.6) there, which gives the details of this formula, Lemma A.2 there, which shows that, as $s \nearrow 1/2$, the first integral approaches the Laplace-Beltrami operator and Lemma A.4 there, which shows that the latter integral produces, as $s \nearrow 1/2$, the norm squared of the second fundamental form, in agreement with the classical case.

Here, for simplicity, we reduce to the case in which E is a graph and we consider a small normal deformation of its boundary, plus an additional small translation, and we write the resulting manifold as an appropriate normal deformation. The details go as follows:

Lemma C.1. *Let $\Sigma \subset \mathbb{R}^n$ be a graph of class C^2, and let E be the corresponding epigraph. Let $\nu = (\nu_1, \dots, \nu_n)$ be the exterior normal of $\Sigma = \partial E$.*

Given $\varepsilon > 0$ and $\overline{x} \in \Sigma$, we set

$$\Sigma_\varepsilon^* := \{x + \varepsilon\eta(x)\,\nu(x) - \varepsilon\eta(\overline{x})\,\nu(\overline{x}),\ x \in \Sigma\}. \tag{C.1}$$

Then, if ε is sufficiently small, Σ_ε^* is a graph, with epigraph a suitable E_ε^*, with $\overline{x} \in \partial E_\varepsilon^*$, and

$$\begin{aligned}\lim_{\varepsilon\to 0}\frac{1}{2\varepsilon}\left(H_E^s(\overline{x}) - H_{E_\varepsilon^*}^s(\overline{x})\right) &= \int_\Sigma \frac{\eta(y)-\eta(\overline{x})\,\nu(\overline{x})\cdot\nu(y)}{|\overline{x}-y|^{n+2s}}\,d\mathcal{H}^{n-1}(y)\\ &= \int_\Sigma \frac{\eta(y)-\eta(\overline{x})}{|\overline{x}-y|^{n+2s}}\,d\mathcal{H}^{n-1}(y)\\ &\quad+\int_\Sigma \frac{\eta(\overline{x})\left[1-\nu(\overline{x})\cdot\nu(y)\right]}{|\overline{x}-y|^{n+2s}}\,d\mathcal{H}^{n-1}(y).\end{aligned}$$

Proof. We denote by $\gamma : \mathbb{R}^{n-1} \to \mathbb{R}$ the graph of class C^2 that describes Σ. In this way, we can write $E = \{x_n < \gamma(x')\}$ and

$$\nu(x) = \nu(x', \gamma(x')) = \frac{(-\nabla\gamma(x'), 1)}{\sqrt{1+|\nabla\gamma(x')|^2}}.$$

We also write $\kappa = (\kappa', \kappa_n) := \eta(\overline{x})\,\nu(\overline{x})$. Then

$$\begin{aligned}\Sigma_\varepsilon^* &= \left\{(x', \gamma(x')) + \varepsilon\eta(x', \gamma(x'))\,\frac{(-\nabla\gamma(x'), 1)}{\sqrt{1+|\nabla\gamma(x')|^2}} - \varepsilon\kappa,\ x' \in \mathbb{R}^{n-1}\right\}\\ &= \left\{\left(x' - \varepsilon\kappa' - \frac{\varepsilon\eta(x', \gamma(x'))\nabla\gamma(x')}{\sqrt{1+|\nabla\gamma(x')|^2}},\ \gamma(x') - \varepsilon\kappa_n + \frac{\varepsilon\eta(x', \gamma(x'))}{\sqrt{1+|\nabla\gamma(x')|^2}}\right),\ x' \in \mathbb{R}^{n-1}\right\}.\end{aligned}$$

So we define

$$y' = y'(x') := x' - \varepsilon\kappa' - \frac{\varepsilon\eta(x', \gamma(x'))\nabla\gamma(x')}{\sqrt{1 + |\nabla\gamma(x')|^2}}. \tag{C.2}$$

Notice that, if ε is sufficiently small

$$\det \frac{\partial y'(x')}{\partial x'} \neq 0.$$

Moreover, $|\nabla\gamma(x')| \leqslant 1 + |\nabla\gamma(x')|^2$ and therefore

$$|y'(x)| \geqslant |x'| - \varepsilon\kappa' - \varepsilon \to +\infty \text{ as } |x| \to +\infty.$$

Hence, by the Global Inverse Function Theorem (see e.g. Corollary 4.3 in [24]), we have that y' is a global diffeomorphism of class C^2 of $\mathbb{R}^{n-1}$, with inverse diffeomorphism $x' = x'(y')$. Thus, we obtain

$$\Sigma_\varepsilon^* = \left\{ \left(y',\ \gamma(x'(y')) - \varepsilon\kappa_n + \frac{\varepsilon\eta(x'(y'), \gamma(x'(y')))}{\sqrt{1 + \left|\nabla\gamma(x'(y'))\right|^2}} \right),\ y' \in \mathbb{R}^{n-1} \right\}.$$

This is clearly a graph, whose corresponding epigraph can be written as $E_\varepsilon^* = \{y_n < \gamma_\varepsilon^*(y')\}$, with

$$\gamma_\varepsilon^*(y') := \gamma(x'(y')) - \varepsilon\kappa_n + \frac{\varepsilon\eta(x'(y'), \gamma(x'(y')))}{\sqrt{1 + \left|\nabla\gamma(x'(y'))\right|^2}}.$$

By (C.2), we have that $y'(x_0') = x_0'$, therefore $\gamma_\varepsilon^*(\overline{x}') = \gamma(\overline{x}')$ and so $\overline{x} \in \partial E_\varepsilon^*$. We also notice that

$$\begin{aligned}
\gamma_\varepsilon^*(y') &= \gamma(y') + \nabla\gamma(y')\cdot(x'(y) - y') - \varepsilon\kappa_n + \frac{\varepsilon\eta(y', \gamma(y'))}{\sqrt{1 + |\nabla\gamma(y')|^2}} + \varepsilon^2 R(y') \\
&= \gamma(y') + \nabla\gamma(y')\cdot\left(\varepsilon\kappa' + \frac{\varepsilon\eta(y', \gamma(y'))\nabla\gamma(y')}{\sqrt{1 + |\nabla\gamma(y')|^2}}\right) - \varepsilon\kappa_n + \frac{\varepsilon\eta(y', \gamma(y'))}{\sqrt{1 + |\nabla\gamma(y')|^2}} + \varepsilon^2 R(y') \\
&= \gamma(y') + \varepsilon\sqrt{1 + |\nabla\gamma(y')|^2}\left(\eta(y', \gamma(y')) - \kappa\cdot\frac{(\nabla\gamma(y'), -1)}{\sqrt{1 + |\nabla\gamma(y')|^2}}\right) + \varepsilon^2 R(y')
\end{aligned}$$

for a suitable remainder functions R (possibly varying from line to line), that are bounded if so is $|D^2\gamma|$.

Accordingly,

$$\begin{aligned}
E_\varepsilon^* \setminus E &= \{\gamma(y') \leqslant y_n < \gamma_\varepsilon^*(y')\} \\
&= \left\{\gamma(y') \leqslant y_n < \gamma(y') + \varepsilon\left(\Xi(y') + \varepsilon^2 R(y')\right)^+\right\},
\end{aligned}$$

where

$$\Xi(y') := \sqrt{1 + |\nabla\gamma(y')|^2}\ \left(\eta(y', \gamma(y')) - \kappa\cdot\tilde{\nu}(y')\right)$$

and $$\tilde{\nu}(y') := \frac{(\nabla\gamma(y'), -1)}{\sqrt{1 + |\nabla\gamma(y')|^2}}.$$

Notice that $\tilde{\nu}(y') = \nu(y', \gamma(y'))$. Similarly,

$$E \setminus E^*_\varepsilon \subseteq \left\{\gamma(y') - \varepsilon\big(\Xi(y') + \varepsilon^2 R(y')\big)^- \leqslant y_n < \gamma(y')\right\}.$$

Therefore

$$\begin{aligned}\lim_{\varepsilon\to 0} \frac{1}{\varepsilon} \int_{E^*_\varepsilon \setminus E} \frac{dy}{|\overline{x} - y|^{n+2s}} &= \lim_{\varepsilon\to 0} \frac{1}{\varepsilon} \int_{\mathbb{R}^{n-1}} \left[\int_{\gamma(y')}^{\gamma(y')+\varepsilon\,(\Xi(y')+\varepsilon R(y'))^+} \frac{dy_n}{|\overline{x} - y|^{n+2s}}\right] dy' \\ &= \int_{\mathbb{R}^{n-1}} \frac{\Xi^+(y')}{(|\overline{x}' - y'|^2 + |\overline{x}_n - \gamma(y')|^2)^{\frac{n+2s}{2}}}\, dy'\end{aligned}$$

and, similarly

$$\lim_{\varepsilon\to 0} \frac{1}{\varepsilon} \int_{E \setminus E^*_\varepsilon} \frac{dy}{|\overline{x} - y|^{n+2s}} = \int_{\mathbb{R}^{n-1}} \frac{\Xi^-(y')}{(|\overline{x}' - y'|^2 + |\overline{x}_n - \gamma(y')|^2)^{\frac{n+2s}{2}}}\, dy'.$$

As a consequence,

$$\begin{aligned}&\lim_{\varepsilon\to 0} \frac{1}{2\varepsilon} \left(H^s_E(\overline{x}) - H^s_{E^*_\varepsilon}(\overline{x})\right) \\ =\ &\lim_{\varepsilon\to 0} \frac{1}{\varepsilon} \left[\int_{E^*_\varepsilon \setminus E} \frac{dy}{|\overline{x} - y|^{n+2s}} - \int_{E \setminus E^*_\varepsilon} \frac{dy}{|\overline{x} - y|^{n+2s}}\right] \\ =\ &\int_{\mathbb{R}^{n-1}} \frac{\Xi(y')}{(|\overline{x}' - y'|^2 + |\overline{x}_n - \gamma(y')|^2)^{\frac{n+2s}{2}}}\, dy' \\ =\ &\int_{\mathbb{R}^{n-1}} \sqrt{1 + |\nabla\gamma(y')|^2}\, \frac{\eta(y', \gamma(y')) - \kappa \cdot \tilde{\nu}(y')}{(|\overline{x}' - y'|^2 + |\overline{x}_n - \gamma(y')|^2)^{\frac{n+2s}{2}}}\, dy' \\ =\ &\int_\Sigma \frac{\eta(y) - \kappa \cdot \nu(y)}{|\overline{x} - y|^{n+2s}}\, d\mathcal{H}^{n-1}(y),\end{aligned}$$

that is the desired result. □

An interesting consequence of Lemma C.1 is that graphs with vanishing nonlocal mean curvature cannot have horizontal normals, as given by the following result:

Theorem C.2. *Let $E \subset \mathbb{R}^n$. Suppose that ∂E is globally of class C^2 and that $H^s_E(x) = 0$ for any $x \in \partial E$.*

Let $\nu = (\nu_1(x), \dots, \nu_n(x))$ be the exterior normal of E at $x \in \partial E$.

Then $\nu_n(x) \neq 0$, for any $x \in \partial E$.

To prove Theorem C.2, we first compare deformations and translations of a graph. Namely, we show that a normal deformation of size $\varepsilon\nu_n$ of a graph with normal $\nu = (\nu_1, \dots, \nu_n)$ coincides with a vertical translation of the graph itself, up to order of ε^2. The precise result goes as follows:

Lemma C.3. *Let $\Sigma \subset \mathbb{R}^n$ be a graph of class C^2 globally, and let E be the corresponding epigraph. Let $\nu = (\nu_1, \dots, \nu_n)$ be the exterior normal of $\Sigma = \partial E$.*

Given $\varepsilon > 0$, let

$$\Sigma_\varepsilon := \{x + \varepsilon \nu_n(x)\, \nu(x),\ \ x \in \Sigma\}. \tag{C.3}$$

Then, if ε is sufficiently small, Σ_ε is a graph, for some epigraph E_ε, and there exists a C^2-diffeomorphism Ψ of $\mathbb{R}^n$ that is $C\varepsilon^2$-close to the identity in $C^2(\mathbb{R}^n)$, for some $C > 0$, such that

$$\Psi(E_\varepsilon) = E + \varepsilon e_n.$$

Proof. We denote by $\gamma : \mathbb{R}^{n-1} \to \mathbb{R}$ the graph that describes Σ. In this way, we can write $E = \{x_n < \gamma(x')\}$ and

$$\nu(x) = \nu(x', \gamma(x')) = \frac{(-\nabla\gamma(x'), 1)}{\sqrt{1 + |\nabla\gamma(x')|^2}}.$$

Accordingly,

$$\begin{aligned} \Sigma_\varepsilon &= \left\{(x', \gamma(x')) + \varepsilon \frac{(-\nabla\gamma(x'), 1)}{1 + |\nabla\gamma(x')|^2},\ x' \in \mathbb{R}^{n-1}\right\} \\ &= \left\{\left(x' - \varepsilon \frac{\nabla\gamma(x')}{1 + |\nabla\gamma(x')|^2},\ \gamma(x') + \frac{\varepsilon}{1 + |\nabla\gamma(x')|^2}\right),\ x' \in \mathbb{R}^{n-1}\right\}. \end{aligned}$$

To write Σ_ε as a graph, we take as new coordinate

$$y' = y'(x') := x' - \varepsilon \frac{\nabla\gamma(x')}{1 + |\nabla\gamma(x')|^2}. \tag{C.4}$$

Notice that, if ε is sufficiently small

$$\det \frac{\partial y'(x')}{\partial x'} \neq 0.$$

Moreover, $|\nabla\gamma(x')| \leqslant 1 + |\nabla\gamma(x')|^2$ and therefore

$$|y'(x)| \geqslant |x'| - \varepsilon \to +\infty \text{ as } |x'| \to +\infty.$$

As a consequence, by the Global Inverse Function Theorem (see e.g. Corollary 4.3 in [24]), we have that y' is a global diffeomorphism of class C^2 of $\mathbb{R}^{n-1}$, we write $x' = x'(y')$ the inverse diffeomorphism and we have that

$$\Sigma_\varepsilon = \left\{\left(y',\ \gamma(x'(y')) + \frac{\varepsilon}{1 + \left|\nabla\gamma(x'(y'))\right|^2}\right),\ y' \in \mathbb{R}^{n-1}\right\}.$$

So we can write the epigraph of Σ_ε as

$$E_\varepsilon = \left\{y_n < \gamma(x'(y')) + \frac{\varepsilon}{1 + \left|\nabla\gamma(x'(y'))\right|^2}\right\}.$$

Now we define

$$\Phi(y') := \gamma(y') - \gamma(x'(y')) + \varepsilon - \frac{\varepsilon}{1 + \left|\nabla\gamma(x'(y'))\right|^2} \tag{C.5}$$

and $z = \Psi(y) = \Psi(y', y_n) := y + \Phi(y')e_n$. By construction, we have that

$$\Psi(E_\varepsilon) = \left\{z_n < \gamma(z') + \varepsilon\right\} = E + \varepsilon e_n.$$

To complete the proof of Lemma C.3, we need to show that

$$\|\Phi\|_{C^2(\mathbb{R}^n)} \leqslant C\varepsilon^2, \tag{C.6}$$

for some $C > 0$. To this aim, we use (C.4) to see that

$$x' = y' + \varepsilon\frac{\nabla\gamma(y')}{1 + |\nabla\gamma(y')|^2} + \varphi_1(y'),$$

with $\|\varphi_1\|_{C^2(\mathbb{R}^n)} \leqslant C\varepsilon^2$. Accordingly, by (C.5), we have that

$$\begin{aligned}\Phi(y') &= \gamma(y') - \gamma\left(y' + \varepsilon\frac{\nabla\gamma(y')}{1 + |\nabla\gamma(y')|^2} + \varphi_1(y')\right)\\ &\quad + \varepsilon - \frac{\varepsilon}{1 + \left|\nabla\gamma\left(y' + \varepsilon\frac{\nabla\gamma(y')}{1+|\nabla\gamma(y')|^2} + \varphi_1(y')\right)\right|^2}\\ &= \gamma(y') - \gamma(y') - \varepsilon\frac{|\nabla\gamma(y')|^2}{1 + |\nabla\gamma(y')|^2} + \varepsilon - \frac{\varepsilon}{1 + |\nabla\gamma(y')|^2} + \varphi_2(y')\\ &= \varphi_2(y'),\end{aligned}$$

with $\|\varphi_2\|_{C^2(\mathbb{R}^n)} \leqslant C\varepsilon^2$. This proves (C.6), as desired. □

From Lemma C.3 here and Theorem 1.1 in [13], we obtain:

Corollary C.4. *In the setting of Lemma C.3, for any $p \in \Sigma_\varepsilon = \partial E_\varepsilon$ we have that*

$$\left|H^s_{E_\varepsilon}(p) - H^s_{E+\varepsilon e_n}(\Psi(p))\right| \leqslant C\varepsilon^2,$$

for some $C > 0$.

Now we complete the proof of Theorem C.2. To this aim, we observe that

$$\nu_n(x) \geqslant 0 \text{ for any } x \in \partial E, \tag{C.7}$$

since E is a graph. Suppose that, by contradiction,

$$\nu_n(\overline{x}) = 0 \text{ for some } \overline{x} \in \partial E. \tag{C.8}$$

We use this and Lemma C.1 with $\eta := \nu_n$ and we find that

$$\lim_{\varepsilon\to 0}\frac{1}{2\varepsilon}\left(H^s_E(\overline{x}) - H^s_{E^*_\varepsilon}(\overline{x})\right) = \int_\Sigma \frac{\nu_n(y)}{|\overline{x} - y|^{n+2s}}\,d\mathcal{H}^{n-1}(y). \tag{C.9}$$

Also, comparing (C.1) (with $\eta := \nu_n$) and (C.3), and using again (C.8), we see that $E_\varepsilon^* = E_\varepsilon$ and so Corollary C.4 gives that

$$H^s_{E^*_\varepsilon}(\overline{x}) = H^s_{E+\varepsilon e_n}(\overline{y}) + O(\varepsilon^2),$$

for some $\overline{y} \in \partial E + \varepsilon e_n$. Since H^s_E vanishes, we can use the translation invariance to see that also $H^s_{E+\varepsilon e_n}$ vanishes. So we conclude that

$$H^s_{E^*_\varepsilon}(\overline{x}) = O(\varepsilon^2).$$

These observations and (C.9) imply that

$$\int_\Sigma \frac{\nu_n(y)}{|\overline{x} - y|^{n+2s}}\, d\mathcal{H}^{n-1}(y) = 0.$$

Hence, in view of (C.7), we see that ν_n must vanish identically along Σ. This says that Σ is a vertical hyperplane, in contradiction with the graph assumption. This ends the proof of Theorem C.2.

Acknowledgment: The first author has been supported by the Alexander von Humboldt Foundation. The second author has been supported by the ERC grant 277749 *E.P.S.I.L.O.N. Elliptic Pde's and Symmetry of Interfaces and Layers for Odd Nonlinearities.* Authors have been supported by the Australian Research Council Discovery Project Grant N.E.W. Nonlocal Equations at Work.

Online lectures

There are a few videotaped lectures online which collect some of the material presented in this set of notes. The interest reader may look at

- http://www.birs.ca/events/2014/5-day-workshops/14w5017/videos/watch/201405271048-Valdinoci.html
- https://www.youtube.com/watch?v=2j2r1ykoyuE
- https://www.youtube.com/watch?v=EDJ8uBpYpB4
- https://www.youtube.com/watch?v=s_RRzgZ7VcM&list=PLj6jTBBj-5B_Vx5qA-HelhGUnGrCu7SdW&index=7
- https://www.youtube.com/watch?v=okXncmRbCZc&index=14&list=PLj6jTBBj-5B_Vx5qA-HelhGUnGrCu7SdW

– http://www.fields.utoronto.ca/video-archive/2016/06/2022-15336

– http://www.mathtube.org/lecture/video/nonlocal-equations-various-perspectives-lecture-1

– http://www.mathtube.org/lecture/video/nonlocal-equations-various-perspectives-lecture-2

– http://www.mathtube.org/lecture/video/nonlocal-equations-various-perspectives-lecture-3

– http://www.birs.ca/events/2016/5-day-workshops/16w5065/videos/watch/201609291100-Dipierro.html

Bibliography

[1] Nicola Abatangelo and Enrico Valdinoci. A notion of nonlocal curvature. *Numer. Funct. Anal. Optim.*, 35(7-9):793–815, 2014.

[2] Luigi Ambrosio, Guido De Philippis, and Luca Martinazzi. Gamma-convergence of nonlocal perimeter functionals. *Manuscripta Math.*, 134(3-4):377–403, 2011.

[3] Begoña Barrios, Alessio Figalli, and Enrico Valdinoci. Bootstrap regularity for integro-differential operators and its application to nonlocal minimal surfaces. *Ann. Sc. Norm. Super. Pisa Cl. Sci. (5)*, 13(3):609–639, 2014.

[4] Jean Bourgain, Haïm Brezis, and Petru Mironescu. Limiting embedding theorems for $W^{s,p}$ when $s \uparrow 1$ and applications. *J. Anal. Math.*, 87:77–101, 2002. Dedicated to the memory of Thomas H. Wolff.

[5] Claudia Bucur, Luca Lombardini, and Enrico Valdinoci. Complete stickiness of nonlocal minimal surfaces for small values of the fractional parameter. *Preprint*, 2016.

[6] Claudia Bucur and Enrico Valdinoci. *Nonlocal diffusion and applications*, volume 20 of *Lecture Notes of the Unione Matematica Italiana*. Springer, [Cham]; Unione Matematica Italiana, Bologna, 2016.

[7] Renato Caccioppoli. Sulle quadratura delle superficie piane e curve. *Atti Accad. Naz. Lincei, Rend., VI. Ser.*, 6:142–146, 1927.

[8] L. Caffarelli, J.-M. Roquejoffre, and O. Savin. Nonlocal minimal surfaces. *Comm. Pure Appl. Math.*, 63(9):1111–1144, 2010.

[9] Luis Caffarelli and Enrico Valdinoci. Uniform estimates and limiting arguments for nonlocal minimal surfaces. *Calc. Var. Partial Differential Equations*, 41(1-2):203–240, 2011.

[10] Luis Caffarelli and Enrico Valdinoci. Regularity properties of nonlocal minimal surfaces via limiting arguments. *Adv. Math.*, 248:843–871, 2013.

[11] Luis A. Caffarelli and Panagiotis E. Souganidis. Convergence of nonlocal threshold dynamics approximations to front propagation. *Arch. Ration. Mech. Anal.*, 195(1):1–23, 2010.

[12] Eleonora Cinti, Joaquim Serra, and Enrico Valdinoci. Quantitative flatness results and BV-estimates for stable nonlocal minimal surfaces. J. Differential Geom.

[13] Matteo Cozzi. On the variation of the fractional mean curvature under the effect of $C^{1,\alpha}$ perturbations. *Discrete Contin. Dyn. Syst.*, 35(12):5769–5786, 2015.

[14] J. Dávila. On an open question about functions of bounded variation. *Calc. Var. Partial Differential Equations*, 15(4):519–527, 2002.

[15] Juan Dávila, Manuel del Pino, and Juncheng Wei. Nonlocal s-minimal surfaces and Lawson cones. *J. Differential Geom.*

[16] Eleonora Di Nezza, Giampiero Palatucci, and Enrico Valdinoci. Hitchhiker's guide to the fractional Sobolev spaces. *Bull. Sci. Math.*, 136(5):521–573, 2012.

[17] S. Dipierro. Asymptotics of fractional perimeter functionals and related problems. *Rend. Semin. Mat. Univ. Politec. Torino*, 72(1-2):3–16, 2014.

[18] Serena Dipierro, Alessio Figalli, Giampiero Palatucci, and Enrico Valdinoci. Asymptotics of the s-perimeter as $s \searrow 0$. *Discrete Contin. Dyn. Syst.*, 33(7):2777–2790, 2013.

[19] Serena Dipierro, Ovidiu Savin, and Enrico Valdinoci. Boundary behavior of nonlocal minimal surfaces. *J. Funct. Anal.*, 272(5):1791–1851, 2017.

[20] Alessio Figalli and Enrico Valdinoci. Regularity and Bernstein-type results for nonlocal minimal surfaces. *J. Reine Angew. Math.* 729:263–273, 2017.

[21] Enrico Giusti. *Minimal surfaces and functions of bounded variation*. Department of Pure Mathematics, Australian National University, Canberra, 1977. With notes by Graham H. Williams; Notes on Pure Mathematics, 10.

[22] Cyril Imbert. Level set approach for fractional mean curvature flows. *Interfaces Free Bound.*, 11(1):153–176, 2009.

[23] V. Mazya and T. Shaposhnikova. On the Bourgain, Brezis, and Mironescu theorem concerning limiting embeddings of fractional Sobolev spaces. *J. Funct. Anal.*, 195(2):230–238, 2002.

[24] Richard S. Palais. Natural operations on differential forms. *Trans. Amer. Math. Soc.*, 92:125–141, 1959.

[25] Augusto C. Ponce. A new approach to Sobolev spaces and connections to Γ-convergence. *Calc. Var. Partial Differential Equations*, 19(3):229–255, 2004.

[26] Ovidiu Savin and Enrico Valdinoci. Γ-convergence for nonlocal phase transitions. *Ann. Inst. H. Poincaré Anal. Non Linéaire*, 29(4):479–500, 2012.

[27] Ovidiu Savin and Enrico Valdinoci. Regularity of nonlocal minimal cones in dimension 2. *Calc. Var. Partial Differential Equations*, 48(1-2):33–39, 2013.

[28] Augusto Visintin. Generalized coarea formula and fractal sets. *Japan J. Indust. Appl. Math.*, 8(2):175–201, 1991.

Rupert L. Frank

Eigenvalue Bounds for the Fractional Laplacian: A Review

Abstract: We review some recent results on eigenvalues of fractional Laplacians and fractional Schrödinger operators. We discuss, in particular, Lieb–Thirring inequalities and their generalizations, as well as semi-classical asymptotics.

5.1 Introduction

An attempt is made, at the request of the editors of this volume to whom the author is grateful, to review some recent developments concerning eigenvalues of fractional Laplacians and fractional Schrödinger operators. Such review is necessarily incomplete and biased towards the author's interests. It is hoped, however, that this collection of results will provide a useful snapshot of a certain line of research and that the open questions mentioned here stimulate some further research.

As is well known, the fractional Laplacian appears in many different areas in connection with non-local phenomena. Here we are particularly interested in problems related to quantum mechanics, where the square root of the Laplacian is used to model relativistic effects. Early works on the one-body and many-body theory include [64, 30] and [36, 83, 84, 44, 85], respectively, and we refer to these for further physical motivations.

Let us define the operators in question. For an open set $\Omega \subset \mathbb{R}^d$ we denote by $\mathring{H}^s(\Omega)$ the set of all functions in the Sobolev space $H^s(\mathbb{R}^d)$ which vanish almost everywhere in $\mathbb{R}^d \setminus \Omega$. We denote the Fourier transform of ψ by

$$\hat{\psi}(p) := \frac{1}{(2\pi)^{d/2}} \int_{\mathbb{R}^d} e^{-ip\cdot x} \psi(x)\, dx\,.$$

The non-negative quadratic form

$$\int_{\mathbb{R}^d} |p|^{2s} |\hat{\psi}(p)|^2\, dp\,, \qquad \psi \in \mathring{H}^s(\Omega)\,,$$

(note that ψ is zero almost everywhere on $\mathbb{R}^d \setminus \Omega$) is closed in the Hilbert space $L^2(\Omega)$ and therefore generates a self-adjoint, non-negative operator

$$H_\Omega^{(s)} \qquad \text{in } L^2(\Omega)\,.$$

Rupert L. Frank, Rupert L. Frank, Ludwig-Maximilans Universität München, Theresienstr. 39, 80333 München, Germany, and Mathematics 253-37, Caltech, Pasadena, CA 91125, USA, E-mail: rlfrank@caltech.edu

https://doi.org/10.1515/9783110571561-007

For $0 < s < 1$ we call $H_\Omega^{(s)}$ the *fractional Laplacian* in Ω. When $s = 1$, this construction gives the usual Dirichlet Laplacian, which we denote by $-\Delta_\Omega = H_\Omega^{(1)}$. When $\Omega = \mathbb{R}^d$, then $H_{\mathbb{R}^d}^{(s)}$ coincides with the fractional power s (in the sense of the functional calculus) of the operator $-\Delta_{\mathbb{R}^d}$ and we will simplify notation by writing $(-\Delta)^s = H_{\mathbb{R}^d}^{(s)}$. It is important to note that, if $\Omega \neq \mathbb{R}^d$ (up to sets of capacity zero), then $H_\Omega^{(s)}$ does *not* coincide with the fractional power of the operator $-\Delta_\Omega$ and, in fact, the comparison of these two operators is one of the recurring themes in this review.

There is a useful alternative expression for the fractional Laplacian, namely,

$$\int_{\mathbb{R}^d} |p|^{2s} |\hat{\psi}(p)|^2 \, dp = a_{d,s} \iint_{\mathbb{R}^d \times \mathbb{R}^d} \frac{|\psi(x) - \psi(y)|^2}{|x - y|^{d+2s}} \, dx \, dy$$

for all $\psi \in H^s(\mathbb{R}^d)$ with

$$a_{d,s} = 2^{2s-1} \pi^{-d/2} \frac{\Gamma(\frac{d+2s}{2})}{|\Gamma(-s)|} . \tag{5.1.1}$$

This is a classical computation, which we recall in Appendix A.

Besides the fractional Laplacian on an open set we will also be interested in the fractional Schrödinger operator $(-\Delta)^s + V$. Heuristically, the connection between the two operators is that the fractional Laplacian in Ω is the special case of the fractional Schrödinger operator with the potential V which equals 0 on Ω and $+\infty$ on its complement. This intuition can be made precise as a limiting theorem, at least in the case of a not too irregular boundary, but we will not make use of this here. Nevertheless, it is useful to keep this connection in mind when comparing the results for both operators.

As we said, our main concern here are eigenvalue bounds for $H_\Omega^{(s)}$ and $(-\Delta)^s + V$. It is technically convenient to consider, instead of eigenvalues, the numbers given by the variational principle. Namely, for a general self-adjoint operator A with quadratic form a in a Hilbert space and for $n \in \mathbf{N}$ we define

$$E_n(A) := \sup_{\psi_1,\dots,\psi_{n-1}} \left(\inf_{0 \neq \psi \perp \psi_1,\dots,\psi_{n-1}} \frac{a[\psi]}{\|\psi\|^2} \right) .$$

According to the variational principle (see, e.g., [93, Theorem XIII.1]), if $E_n(A) < \inf \operatorname{ess-spec}(A)$, then $E_n(A)$ is the n-th eigenvalue of A, counting multiplicities. In general, however, $E_n(A)$ need not be an eigenvalue. Since our tools in this paper are of variational nature, they lead naturally to inequalities for $E_n(A)$, independently of whether or not it actually is an eigenvalue.

Let us briefly outline the structure this review. In Section 5.2 we begin with lower bounds on the ground state energies $E_1(H_\Omega^{(s)})$ and $E_1((-\Delta)^s + V)$. These lower bounds come naturally from the shape optimization problems of minimizing $E_1(H_\Omega^{(s)})$ among all Ω with given measure and minimizing $E_1((-\Delta)^s + V)$ among all V with given L^p norm. The (classical) answers are given in Theorems 5.2.1 and 5.2.2. We then turn to comparing the eigenvalues of the operators $H_\Omega^{(s)}$ and $(-\Delta_\Omega)^s$ and recall a theorem from [32].

In Section 5.3 we discuss the asymptotics of $E_n(H_\Omega^{(s)})$ as $n \to \infty$ and of $\#\{n : E_n((-\Delta)^s + \alpha V) < 0\}$ as $\alpha \to \infty$. Both questions are closely related, because studying the asymptotics of $E_n(H_\Omega^{(s)})$ as $n \to \infty$ is the same as studying $\#\{n : E_n(H_\Omega^{(s)}) < \mu\}$ as $\mu \to \infty$, which is the same as studying $\#\{n : E_n(h^{2s}H_\Omega^{(s)} - 1) < 0\}$ as $h \to 0$. Clearly, studying $\#\{n : E_n((-\Delta)^s + \alpha V) < 0\}$ as $\alpha \to \infty$ is the same as studying $\#\{n : E_n(h^{2s}H_\Omega^{(s)} - V) < 0\}$ as $h \to 0$, so both questions correspond to a semi-classical limit with an effective Planck constant h tending to zero. While the leading term in the asymptotics is well known and given by a Weyl-type formula, there are still open questions corresponding to subleading corrections.

In Section 5.4 we supplement the asymptotic results on the number and sums of eigenvalues by 'uniform' inequalities which hold not only in the asymptotic regimes considered in the previous section. The important feature of these inequalities is, however, that they have a form reminiscent of the asymptotics. We present such eigenvalue bounds not only for $H_\Omega^{(s)}$ and $(-\Delta)^s + V$, but also for operators of the form $(-\Delta)^s - W + V$, where W is an explicit 'Hardy weight'.

We conclude with a short Section 5.5 on (some of) the topics that we have not treated in this paper.

5.2 Bounds on Single Eigenvalues

5.2.1 The fractional Faber–Krahn inequality

We recall that $E_1(H_\Omega^{(s)})$ denotes the ground state energy of the fractional Laplacian on an open set $\Omega \subset \mathbb{R}^d$. Using Sobolev interpolation inequalities on $\mathbb{R}^d$ (see, for instance, (5.2.4) below) and Hölder's inequality it is easy to prove that

$$E_1(H_\Omega^{(s)}) \geqslant C_{d,s}|\Omega|^{-2s/d}$$

for some positive constant $C_{d,s}$ depending only on d and s. The fractional Faber–Krahn inequality in the following theorem says that the optimal value of the constant $C_{d,s}$ is attained when Ω is a ball. We recall that for any measurable set $E \subset \mathbb{R}^d$ of finite measure, E^* denotes the centered, open ball with radius determined such that $|E^*| = |E|$.

Theorem 5.2.1. *Let $\Omega \subset \mathbb{R}^d$ be open with finite measure. Then*

$$E_1(H_\Omega^{(s)}) \geqslant E_1(H_{\Omega*}^{(s)})$$

with equality if and only if Ω is a ball.

This theorem follows easily using symmetric decreasing rearrangement (see, e.g., [80] for a textbook presentation). We know [6] that

$$\|(-\Delta)^{s/2}\psi\|^2 \geqslant \|(-\Delta)^{s/2}\psi^*\|^2\,, \tag{5.2.1}$$

where ψ^* denotes the symmetric decreasing rearrangement of ψ, and since $\|\psi^*\|^2 = \|\psi\|^2$ and ψ^* is supported in $\overline{\Omega^*}$, we obtain the inequality in the theorem. The uniqueness of the ball follows from the strictness statement for (5.2.1), see [27, 56]. (We also mention that a version of (5.2.1) for functions on a interval appears in [59].) An alternative proof of Theorem 5.2.1, based on a comparison result for the corresponding heat equations, can be found in [98]. For results related to and generalizing Theorem 5.2.1, see [15].

It would be interesting to supplement Theorem 5.2.1 with a stability result analogous to [58, 24], namely to show that $E_1(H_\Omega^{(s)}) - E_1(H_{\Omega^*}^{(s)})$ is bounded from below by a constant (depending only on s and d) times $|\Omega|^{-2s/d-2}\inf\{|B\Delta\Omega|^2 : B \text{ ball with } |B| = |\Omega|\}$.

Theorem 5.2.1 corresponds to minimizing $E_1(H_\Omega^{(s)})$ among all sets Ω with given measure. Another interesting problem is to minimize $E_1(H_\Omega^{(s)})$ among all *convex* sets Ω with given inner radius $r_{\text{in}}(\Omega) := \sup_{x\in\Omega}\operatorname{dist}(x, \Omega^c)$. Optimal results for this question appear in [8, 87].

5.2.2 The fractional Keller inequality

We recall that $E_1((-\Delta)^s + V)$ denotes the ground state energy of the fractional Schrödinger operator. In [70] Keller asked for $s = 1$ how small the ground state energy can be for a given L^p norm of the potential; see also [82]. The following theorem generalizes this result to the fractional case.

Theorem 5.2.2. *Let $d \geqslant 1$, $0 < s < 1$ and $\gamma > 0$. If $d = 1$ and $s > 1/2$ we assume in addition that $\gamma \geqslant 1 - 1/(2s)$. Then*

$$K_{\gamma,d,s} := -\inf_V \frac{E_1((-\Delta)^s + V)}{\|V\|_{\gamma+d/(2s)}^{1+d/(2s\gamma)}} < \infty\,.$$

Moreover, there is a positive, radial, symmetric decreasing function W such that the inequality

$$E_1((-\Delta)^s + V) \geqslant -K_{\gamma,d,s}\left(\int_{\mathbb{R}^d}|V|^{\gamma+d/(2s)}\,dx\right)^{1/\gamma} \tag{5.2.2}$$

is strict unless $V = -b^{-2s}W((x-a)/b)$ for some $a \in \mathbb{R}^d$ and $b > 0$.

Let us briefly sketch the proof. The key idea (essentially contained in [82] for $s = 1$) is that the inequality $K_{\gamma,d,s} < \infty$ is equivalent to a Sobolev interpolation inequality.

According to the variational definition of $E_1((-\Delta)^s + V)$ we have

$$-K_{\gamma,d,s} = \inf_V \inf_\psi \frac{\|(-\Delta)^{s/2}\psi\|^2 + \int_{\mathbb{R}^d} V|\psi|^2\,dx}{\|\psi\|^2\,\|V\|_{\gamma+d/(2s)}^{1+d/(2s\gamma)}}$$

Since the quotient in this formula remains invariant if we replace both $V(x)$ by $b^2V(bx)$ and $\psi(x)$ by $c\psi(bx)$ for arbitrary $b, c > 0$, we can restrict the infimum to potentials V with $\|V\|_{\gamma+d/(2s)} = 1$ and to functions ψ with $\|\psi\| = 1$. Moreover, since the quotient does not increase if we replace V by $-|V|$ we can restrict the infimum to potentials $V \leqslant 0$. We summarize these findings as

$$-K_{\gamma,d,s} = \inf\left\{\mathcal{E}_q[\psi] + \mathcal{H}_q[\psi, U] :\ U \geqslant 0\,,\ \|\psi\| = \|U\|_{q/(q-2)} = 1\right\}$$

with $q \geqslant 2$ such that $1/(\gamma + d/(2s)) + 2/q = 1$,

$$\mathcal{E}_q[\psi] = \|(-\Delta)^{s/2}\psi\|^2 - \|\psi\|_q^2$$

and

$$\mathcal{H}_q[\psi, U] = \|\psi\|_q^2 - \int_{\mathbb{R}^d} U|\psi|^2\,dx\,.$$

By Hölder's inequality we have $\mathcal{H}_q[\psi, U] \geqslant 0$ for $U \geqslant 0$ with $\|U\|_{q/(q-2)} = 1$, and equality holds if and only if $U = (|\psi|/\|\psi\|_q)^{q-2}$. Thus, $K_{\gamma,d,s} < \infty$ is equivalent to

$$\inf\{\mathcal{E}_q[\psi] :\ \|\psi\| = 1\} > -\infty\,, \tag{5.2.3}$$

and there is a bijective correspondence between V's realizing equality in (5.2.2) and ψ's realizing the infimum in (5.2.3). The statement (5.2.3) is, by scaling, equivalent to the Sobolev interpolation inequality

$$\|(-\Delta)^{s/2}\psi\|^{2\vartheta}\|\psi\|^{2(1-\vartheta)} \geqslant \mathcal{S}_{d,q,s}\|\psi\|_q^2 \tag{5.2.4}$$

with a constant $\mathcal{S}_{d,q,s} > 0$ (and some $\vartheta \in (0, 1)$ uniquely determined by scaling). This inequality is well known to hold for $2 \leqslant q \leqslant 2d/(d-2s)$ if $d > 2s$, for $2 \leqslant q < \infty$ if $d = 2s$ and for $2 \leqslant q \leqslant \infty$ if $d < 2s$. Therefore, we deduce that $K_{\gamma,d,s} < \infty$ under the assumptions on γ in the theorem. Moreover, if $\mathcal{S}_{d,q,s}$ denotes the optimal constant in (5.2.4), then it is also well-known that there is a minimizer ψ for which equality holds (see, for instance, [29] for a proof for $s = 1$; the necessary modifications for $s < 1$ are, for instance, in [19]). By the rearrangement inequality (5.2.1), this ψ can be chosen positive, radial and symmetric decreasing. It was recently proved in [49, 50] that there is a unique function Q such that any function achieving equality in (5.2.4) coincides with Q after translation, dilation and multiplication by a constant, which leads to the uniqueness statement in Theorem 5.2.2. This completes our sketch of the proof of the theorem.

We expect that the method from [29], together with the non-degeneracy results from [49, 50], leads to a stability version of (5.2.2).

5.2.3 Comparing eigenvalues of $H_\Omega^{(s)}$ and $(-\Delta_\Omega)^s$

It is important to distinguish between $H_\Omega^{(s)}$, the fractional Laplacian on Ω, and the fractional power $(-\Delta_\Omega)^s$ of the Dirichlet Laplacian. These two operators are different, but, as shown in the following theorem, the first one is always less or equal than the second one. We recall that for two operators A, B, which are bounded from below, we write $A \leqslant B$ if their quadratic forms a, b with form domains $\mathcal{D}[a]$, $\mathcal{D}[b]$ satisfy $\mathcal{D}[a] \supset \mathcal{D}[b]$ and $a[u] \leqslant b[u]$ for every $u \in \mathcal{D}[b]$. Note that $A \leqslant B$ implies $E_n(A) \leqslant E_n(B)$ for all $n \in \mathbb{N}$.

Theorem 5.2.3. *Let $\Omega \subset \mathbb{R}^d$ be open and $0 < s < 1$. Then*

$$H_\Omega^{(s)} \leqslant (-\Delta_\Omega)^s \tag{5.2.5}$$

In particular,

$$E_n(H_\Omega^{(s)}) \leqslant E_n((-\Delta_\Omega)^s) = (E_n(-\Delta_\Omega))^s \qquad \text{for all } n \in \mathbb{N}\,. \tag{5.2.6}$$

Moreover, unless $\mathbb{R}^d \setminus \Omega$ has zero capacity, the operators $H_\Omega^{(s)}$ and $(-\Delta_\Omega)^s$ do not coincide.

The first part of the theorem is due to Chen–Song [32] (see also [39] and its generalization in [33]), which extends earlier results in [9] for $s = 1/2$ and in [38] for s irrational. The second part concerning strictness is from [47], where it is also shown that, in a certain sense, $E_n(H_\Omega^{(s)})$ and $E_n((-\Delta_\Omega)^s)$ have the same leading term as $n \to \infty$, but a different subleading term; see Corollary 5.3.4 below for a precise statement. An alternative proof of Theorem 5.2.3, which yields strict inequality in (5.2.6) for any n for bounded Ω, is in [88] and is based on the Caffarelli–Silvestre extension technique [31]. In fact, it was recently shown in [75], using Jensen's inequality, that $(0,1) \ni s \mapsto E_n(H_\Omega^{(s)})^{1/s}$ is *strictly* increasing for any n if Ω is bounded.

Let us sketch the idea of the proof of Theorem 5.2.3 in [47]. It is based on the observation that, if A is a non-negative operator in a Hilbert space with trivial kernel, P an orthogonal projection and φ an operator monotone function on $(0, \infty)$, then

$$P\varphi(PAP)P \geqslant P\varphi(A)P\,. \tag{5.2.7}$$

(This is closely related to the Sherman–Davis inequality, see, e.g., [28, Theorem 4.19].) Using Loewner's integral representation of operator monotone functions (see, for instance, [18] or [97, Theorem 1.6]), (5.2.7) follows from

$$P(PAP)^{-1}P \leqslant PA^{-1}P\,, \tag{5.2.8}$$

which, in turn, can be proved using a variational characterization of the inverse operator in the spirit of Dirichlet's principle. Inequality (5.2.5) follows immediately from

(5.2.7) with the choice $A = -\Delta$ in $L^2(\mathbb{R}^d)$, P = multiplication by the characteristic function of Ω (note that $H_\Omega^{(s)} = PA^sP$ in the quadratic form sense) and $\varphi(E) = E^s$ (which is operator monotone for $0 < s < 1$). Note that this argument gives an analogue of (5.2.5) for any operator monotone function. If we only want (5.2.5), we do not need Loewner's theorem, but only the integral representation

$$E^s = \frac{\sin(\pi s)}{\pi} \int_0^\infty t^s \left(\frac{1}{t} - \frac{1}{t+E} \right) dt \qquad \text{if } 0 < s < 1 .$$

Analyzing the cases of equality in (5.2.8) shows that, under the assumption that φ is not affine linear, equality in (5.2.7) holds iff $\operatorname{ran} P$ is a reducing subspace of A. (This is stated in [47] only for positive definite A, which is needed for (5.2.8), but when passing from (5.2.8) to (5.2.7) one always has a positive definite operator.) Since $L^2(\Omega)$ is not a reducing subspace for $-\Delta$ in $L^2(\mathbb{R}^d)$ unless $\mathbb{R}^d \backslash \Omega$ has capacity zero, we obtain the second part of the theorem.

While Theorem 5.2.3 gives an upper bound on $E_n(H_\Omega^{(s)})$ in terms of $E_n(-\Delta_\Omega^s)$, the following theorem, also due to Chen–Song [32], yields a lower bound.

Theorem 5.2.4. *Let $\Omega \subset \mathbb{R}^d$ be bounded and satisfy the exterior cone condition and let $0 < s < 1$. Then there is a $c_{\Omega,s} > 0$ such that*

$$E_n(H_\Omega^{(s)}) \geqslant c_{\Omega,s}\, E_n((-\Delta_\Omega)^s) = c_{\Omega,s}\, (E_n(-\Delta_\Omega))^s \qquad \textit{for all } n \in \mathbf{N}. \tag{5.2.9}$$

If Ω is convex, (5.2.9) *holds with $c_{\Omega,s} = 1/2$.*

We note that Theorem 5.2.4 allows one to obtain lower bounds on $E_n(H_\Omega^{(s)})$ from lower bounds on $E_n(-\Delta_\Omega)$. For instance, one can show that for convex domains $E_1(H_\Omega^{(s)})$ is bounded from below by a constant times $r_{\text{in}}(\Omega)^{-2s}$ [32]. This gives weaker inequalities, however, than the direct approach in [8, 87].

5.3 Eigenvalue Asymptotics

5.3.1 Eigenvalue asymptotics for the fractional Laplacian

From a (fractional analogue) of Rellich's compactness lemma we know that $H_\Omega^{(s)}$ has purely discrete spectrum when $\Omega \subset \mathbb{R}^d$ has finite measure. In this subsection we discuss the asymptotics of the eigenvalues $E_n(H_\Omega^{(s)})$ as $n \to \infty$. The basic result is due to Blumenthal and Getour [22] (see also [12, Rem. 2.2] and [61]).

Theorem 5.3.1. *Let $\Omega \subset \mathbb{R}^d$ be open with finite measure. Then*

$$\lim_{n\to\infty} \frac{E_n(H_\Omega^{(s)})}{n^{2s/d}} = (2\pi)^{2s} \omega_d^{-2s/d} |\Omega|^{-2s/d} \tag{5.3.1}$$

with $\omega_d = |\{\xi \in \mathbb{R}^d : |\xi| < 1\}|$.

Alternatively, one can write (5.3.1) as

$$\lim_{\mu\to\infty} \mu^{-d/(2s)} N(\mu, H_\Omega^{(s)}) = (2\pi)^{-d}\omega_d|\Omega|\,, \tag{5.3.2}$$

where for an arbitrary self-adjoint operator A, which is bounded from below, we set $N(\mu, A) = \#\{n : E_n(A) < \mu\}$. If A has discrete spectrum in $(-\infty, \mu)$, then $N(\mu, A)$ denotes the total number of eigenvalues below μ, counting multiplicities.

For later purposes we record that (5.3.1) implies

$$\lim_{N\to\infty} N^{-1-2s/d} \sum_{n=1}^{N} E_n(H_\Omega^{(s)}) = \frac{d}{d+2s}(2\pi)^{2s}\omega_d^{-2s/d}|\Omega|^{-2s/d}\,. \tag{5.3.3}$$

We also note that (5.3.2) and integration in μ shows that, for any $\gamma > 0$,

$$\lim_{\mu\to\infty} \mu^{-\gamma-d/(2s)} \operatorname{Tr}\left(H_\Omega^{(s)} - \mu\right)_-^\gamma = L_{\gamma,d,s}^{\mathrm{cl}}|\Omega|\,, \tag{5.3.4}$$

where

$$\operatorname{Tr}(H_\Omega^{(s)} - \mu)_-^\gamma = \sum_n \left(E_n(H_\Omega^{(s)}) - \mu\right)_-^\gamma = \gamma\int_0^\infty N(\mu, H_\Omega^{(s)})\mu^{\gamma-1}\,d\mu$$

and

$$L_{\gamma,d,s}^{\mathrm{cl}} := \frac{1}{(2\pi)^d}\int_{\mathbb{R}^d}\left(|p|^{2s} - 1\right)_-^\gamma dp = \frac{\omega_d}{(2\pi)^d}\frac{\Gamma(\gamma+1)\,\Gamma(\frac{d}{2s}+1)}{\Gamma(\gamma+\frac{d}{2s}+1)}\,. \tag{5.3.5}$$

A classical result of Weyl states that

$$\lim_{\mu\to\infty} \mu^{-d/2} N(\mu, -\Delta_\Omega) = (2\pi)^{-d}\omega_d\,|\Omega|\,,$$

and therefore, by the spectral theorem,

$$\lim_{\mu\to\infty} \mu^{-d/(2s)} N(\mu, (-\Delta_\Omega)^s) = \lim_{\mu'\to\infty} (\mu')^{-d/2} N(\mu', -\Delta_\Omega) = (2\pi)^{-d}\omega_d\,|\Omega|\,.$$

Comparing this with (5.3.1) we see that $E_n(H_\Omega^{(s)})$ and $E_n(-\Delta_\Omega^s) = (E_n(-\Delta_\Omega))^s$ coincide to leading order as $n \to \infty$. In the following we will be interested in subleading corrections to the asymptotics in Theorem 5.3.1.

We begin with the case $d = 1$. After a translation and a dilation we can assume without loss of generality that $\Omega = (-1, 1)$.

Theorem 5.3.2. *Let* $\Omega = (-1, 1) \subset \mathbb{R}$. *Then*

$$E_n(H_\Omega^{(s)}) = \left(\frac{n\pi}{2} - \frac{(1-s)\pi}{4}\right)^{2s} + O(n^{-1}) \tag{5.3.6}$$

This theorem is due to Kwaśnicki [74] and generalizes an earlier result [72] for $s = 1/2$. A key role is played by a detailed analysis of the half line problem [73].

Asymptotics (5.3.6) are remarkably precise. For $s > 1/2$ they give the first three terms as $n \to \infty$. We also see that the lim inf and the lim sup of $N(\mu, H_\Omega^{(s)}) - \pi^{-1}|\Omega|\mu^{1/(2s)}$ as $\mu \to \infty$ are finite, but do not coincide. (In fact, the lim sup is positive for $s \in (0,1)$, which shows that the analogue of Pólya's conjecture fails in the fractional case. This was first observed in [75].)

We now turn to the higher-dimensional case. The authors of [14] posed the problem to prove that, under suitable assumptions on Ω, the quantity

$$n^{-(2s-1)/d}\left(E_n(H_\Omega^{(s)}) - n^{2s/d}(2\pi)^{2s}\omega_d^{-2s/d}|\Omega|^{-2s/d}\right)$$

has a limit. For $s = 1$ this is a celebrated result by Ivrii [66] which holds under the assumption that the set of periodic billiards has measure zero. In fact, after the first version of this review was submitted, Ivrii [67] announced a solution of the above problem for $s \in (0,1)$ under the same assumption.

The following theorem from [47] verifies the existence of a limit in the Cesàro sense, that is, the quantity

$$N^{-(2s-1)/d}\left(N^{-1}\sum_{n=1}^{N} E_n(H_\Omega^{(s)}) - \frac{d}{d+2s}N^{2s/d}(2\pi)^{2s}\omega_d^{-2s/d}|\Omega|^{-2s/d}\right) \tag{5.3.7}$$

has a limit. Just like (5.3.1) is equivalent to (5.3.2) and (5.3.3) is equivalent to (5.3.4) with $\gamma = 1$, the existence of the limit of (5.3.7) is equivalent to the existence of the limit of

$$\mu^{-1-(d-1)/(2s)}\left(\operatorname{Tr}(H_\Omega^{(s)} - \mu)_- - \mu^{1+d/(2s)}L_{1,d,s}^{\mathrm{cl}}|\Omega|\right). \tag{5.3.8}$$

(These equivalences are elementary facts about sequences; see, e.g., [47, Lemma 21].) The advantage of (5.3.2), (5.3.4) and (5.3.8) over (5.3.1), (5.3.3) and (5.3.7), respectively, is that disjoint subsets of Ω have asymptotically an additive influence on the asymptotics, which allows for localization techniques.

The main result from [47] is

Theorem 5.3.3. *For any $d \geqslant 1$ and $0 < s < 1$ there is a constant $B_{d,s}^{\mathrm{cl}} > 0$ such that for any bounded domain $\Omega \subset \mathbb{R}^d$ with C^1 boundary,*

$$\lim_{\mu\to\infty} \mu^{-1-(d-1)/(2s)}\left(\operatorname{Tr}(H_\Omega^{(s)} - \mu)_- - \mu^{1+d/(2s)}L_{1,d,s}^{\mathrm{cl}}|\Omega|\right) = -B_{d,s}^{\mathrm{cl}}\,\sigma(\partial\Omega). \tag{5.3.9}$$

Here $\sigma(\partial\Omega)$ denotes the surface measure of $\partial\Omega$. In [47] this is stated for domains with $C^{1,\alpha}$ boundary, $0 < \alpha \leqslant 1$, and a quantitative remainder whose order depends on α. The same argument as in [48], however, yields the result for C^1 boundaries with a $o(1)$ remainder.

In [47] we obtain an expression for $B_{d,s}^{\mathrm{cl}}$ which is explicit enough to deduce that it is different (in fact, smaller) than the corresponding expression for the fractional power of the Dirichlet Laplacian. In order to state this precisely, we recall that there is a constant $\tilde{B}^{\mathrm{cl}}_{d,s} > 0$ such that for any bounded domain $\Omega \subset \mathbb{R}^d$ with C^1 boundary,

$$\lim_{\mu\to\infty} \mu^{-1-(d-1)/(2s)} \left(\operatorname{Tr}((-\Delta_\Omega)^s - \mu)_- - \mu^{1+d/(2s)} L^{\mathrm{cl}}_{1,d,s} |\Omega| \right) = -\tilde{B}^{\mathrm{cl}}_{d,s}\, \sigma(\partial\Omega)\,;$$

see, e.g., [46] for a proof for domains with $C^{1,\alpha}$ boundary, $0 < \alpha \leqslant 1$, which again can be modified to yield the result for C^1 boundaries. We prove [47, Sec. 6.4]

$$B^{\mathrm{cl}}_{d,s} < \tilde{B}^{\mathrm{cl}}_{d,s}$$

and deduce

Corollary 5.3.4. *For any bounded domain $\Omega \subset \mathbb{R}^d$ with C^1 boundary,*

$$\begin{aligned} &\lim_{\mu\to\infty} \mu^{-1-(d-1)/(2s)} \left(\operatorname{Tr}(H^{(s)}_\Omega - \mu)_- - \operatorname{Tr}((-\Delta_\Omega)^s - \mu)_- \right) \\ &\quad = -\left(B^{\mathrm{cl}}_{d,s} - \tilde{B}^{\mathrm{cl}}_{d,s} \right) \sigma(\partial\Omega) > 0\,. \end{aligned}$$

Theorem 5.3.3 implies via integration that

$$\lim_{t\to 0} t^{(d-1)/(2s)} \left(\operatorname{Tr} e^{-tH^{(s)}_\Omega} - t^{-d/(2s)} \frac{\omega_d\, \Gamma(1+\frac{d}{2s})}{(2\pi)^d} |\Omega| \right) = -\Gamma(2+\tfrac{d-1}{2s})\, B^{\mathrm{cl}}_{d,s}\, \sigma(\partial\Omega)\,. \tag{5.3.10}$$

(This is essentially the argument that convergence in Cesàro sense implies convergence in Abel sense.) Asymptotics (5.3.10) are, in fact, even true for Ω with Lipschitz boundary, as had earlier been shown in [14]. This extends the result from [26] for $s = 1$ to the fractional case. See also [12] for remainder terms in (5.3.10) under stronger regularity assumptions on the boundary.

It seems to be unknown whether Theorem 5.3.3 remains true for Lipschitz domains.

Asymptotics like (5.3.10) have been shown for more general non-local operators, see, e.g., [16, 92, 23, 62].

5.3.2 Eigenvalue asymptotics for fractional Schrödinger operators

The analogue of Theorem 5.3.1 for fractional Schrödinger operators is

Theorem 5.3.5. *Let $0 < s < 1$ and let V be a continuous function on $\mathbb{R}^d$ with compact support. Then*

$$\lim_{\alpha\to\infty} \alpha^{-d/(2s)} N((-\Delta)^s + \alpha V) = (2\pi)^{-d} \omega_d \int_{\mathbb{R}^d} V_-^{d/(2s)}\, dx\,. \tag{5.3.11}$$

Similarly to (5.3.11) one can show that for any $\gamma > 0$,

$$\lim_{\alpha\to\infty} \alpha^{-\gamma-d/(2s)} \operatorname{Tr}((-\Delta)^s + \alpha V)_-^\gamma = L^{\mathrm{cl}}_{\gamma,d,s} \int_{\mathbb{R}^d} V_-^{\gamma+d/(2s)}\,dx \tag{5.3.12}$$

with $L^{\mathrm{cl}}_{\gamma,d,s}$ from (5.3.5). The assumptions on V for (5.3.11) and (5.3.12) to hold can be relaxed. In particular, for $d \geqslant 2$, as well as for $d = 1$ and $0 < s < 1/2$, one can show that the asymptotics hold under the sole assumption $V_- \in L^{\gamma+d/(2s)}$. This will be explained after Theorem 5.4.2. The case $d = 1$ and $1/2 \leqslant s < 1$ is more subtle. In analogy with [21, 89] one might wonder whether there are $V \in L^{1/(2s)}$ for which $N((-\Delta)^s + \alpha V)$ grows faster than $\alpha^{1/(2s)}$ or like $\alpha^{1/(2s)}$ but with an asymptotic constant strictly larger than $\int_{\mathbb{R}} V_-^{1/(2s)}\,dx$. Apparently this question has not been studied.

We are also not aware of sharp remainder estimates or subleading terms in (5.3.11) and (5.3.12). Note that, due to the non-smoothness of $p \mapsto |p|^{2s}$ at $p = 0$, the operator $(-h^2\Delta)^s + V$ is not an admissible operator in the sense of [63]. For a remainder bound for the massive analogue of (5.3.12) with $\gamma = 1/2$ we refer to [99].

5.4 Bounds on Sums of Eigenvalues

5.4.1 Berezin–Li–Yau inequalities

In this subsection we discuss bounds on sums of eigenvalues of $H_\Omega^{(s)}$. The bounds in the following theorem are called Berezin–Li–Yau inequalities since they generalize the corresponding bounds for $s = 1$ [20, 79] to the fractional case.

Theorem 5.4.1. *Let $\Omega \subset \mathbb{R}^d$ be an open set of finite measure. Then for any $\mu > 0$,*

$$\sum_n \left(E_n(H_\Omega^{(s)}) - \mu\right)_- \leqslant \mu^{1+d/(2s)} L^{\mathrm{cl}}_{1,d,s}\,|\Omega| \tag{5.4.1}$$

and, equivalently, for any $N \in \mathbf{N}$,

$$\sum_{n=1}^N E_n(H_\Omega^{(s)}) \geqslant \frac{d}{d+2s}(2\pi)^{2s}\omega_d^{-2s/d}|\Omega|^{-2s/d} N^{1+2s/d}\,. \tag{5.4.2}$$

Inequality (5.4.1) is a special case of a result in [76]. To see that (5.4.1) and (5.4.2) are equivalent, denote the left and right side of (5.4.1) by $f_l(\mu)$ and $f_r(\mu)$, respectively, by $g_l(\nu)$ the piecewise linear function which coincides with the left side of (5.4.2) for $\nu = N \in \mathbf{N}$ and by $g_r(\nu)$ the right side of (5.4.2) with N replaced by a continuous variable ν. Note that (5.4.2) is equivalent to $g_l(\nu) \geqslant g_r(\nu)$ for *all* $\nu > 0$. We have defined four convex functions and we note that $f_\#$ and $g_\#$ are Legendre transforms of each other with $\# = l, r$. Thus, the equivalence follows from the fact that the Legendre transform reverses inequalities.

The important feature of (5.4.1) and (5.4.2) is that the constant on the right side coincides with the asymptotic value as μ or N tend to infinity; see (5.3.3) and (5.3.4). For remainder terms in (5.4.2) we refer to [101].

Bounding the left side of (5.4.2) from above by $NE_N(H_\Omega^{(s)})$ or the left side of (5.4.1) from below by $(\Lambda-\mu)_- N(\Lambda, H_\Omega^{(s)})$ and optimizing in $\Lambda<\mu$ we obtain

$$N(\Lambda, H_\Omega^{(s)}) \leqslant \left(\frac{d+2s}{d}\right)^{\frac{d}{2s}} \frac{\omega_d}{(2\pi)^d} |\Omega| \Lambda^{\frac{d}{2s}}, \qquad E_N(H_\Omega^{(s)}) \geqslant \frac{d}{d+2s} \frac{(2\pi)^{2s}}{\omega_d^{\frac{2s}{d}}} |\Omega|^{-\frac{2s}{d}} N^{\frac{2s}{d}}. \tag{5.4.3}$$

It is a challenging open question (the fractional analogue of Pólya's conjecture) whether the factors $((d+2s)/d)^{d/(2s)}$ and $d/(d+2s)$ can be removed in these bounds. It was recently shown [75] that Pólya's conjecture fails in $d=1$ for $s\in(0,1)$ and in $d=2$ at least for all sufficiently small values of s. (As an aside, we mention that Pólya's conjecture also fails for the Laplacian in two-dimensions with a constant magnetic field and that in this case the factors $((d+2s)/d)^{d/(2s)}=2$ and $d/(d+2s)=2$ are optimal [55].)

We finally mention a well known inequality for the heat kernel. From the maximum principle for the heat equation we know that the heat kernel $k_t(x,x')$ for $H_\Omega^{(s)}$ satisfies

$$0 \leqslant k_t(x,x') \leqslant \int_{\mathbb{R}^d} e^{-t|p|^{2s}} e^{ip\cdot(x-x')} \frac{dp}{(2\pi)^d} \qquad \text{for all } x, x' \in \Omega.$$

(The right side is the heat kernel of $(-\Delta)^s$.) We evaluate this inequality for $x=x'$. If $H_\Omega^{(s)}$ has discrete spectrum (which is the case, for instance, if $|\Omega|<\infty$) and ψ_n denote the normalized eigenfunctions corresponding to the $E_n(H_\Omega^{(s)})$, then we obtain

$$\sum_n e^{-tE_n(H_\Omega^{(s)})} |\psi_n(x)|^2 \leqslant \frac{\omega_d \Gamma(1+d/(2s))}{(2\pi)^d} t^{-d/(2s)} \qquad \text{for all } x\in\Omega. \tag{5.4.4}$$

By integration over $x\in\Omega$ we obtain

$$\sum_n e^{-tE_n(H_\Omega^{(s)})} \leqslant \frac{\omega_d \Gamma(1+d/(2s))}{(2\pi)^d} |\Omega|\, t^{-d/(2s)},$$

which, in turn, could have been obtained directly by integrating (5.4.1) against $t^2 e^{-t\mu}$ over $\mu\in\mathbb{R}_+$. However, in some applications the local information in (5.4.4) is crucial. For example, one useful consequence of (5.4.4) comes by bounded the left side from below by $e^{-t\mu}\sum_{E_n(H_\Omega^{(s)})<\mu} |\psi_n(x)|^2$. Optimizing the resulting inequality over $t>0$ yields

$$\sum_{E_n(H_\Omega^{(s)})<\mu} |\psi_n(x)|^2 \leqslant \frac{\omega_d \Gamma(1+d/(2s))}{(2\pi)^d} \left(\frac{2se}{d}\right)^{d/(2s)} \mu^{d/(2s)}. \tag{5.4.5}$$

While yielding a worse constant than (5.4.3) when integrated over $x\in\Omega$, this a-priori bound on the 'local number of eigenvalues' is useful when proving $\mu\to\infty$ asymptotics.

5.4.2 Lieb–Thirring inequalities

Lieb–Thirring inequalities [82] provide bounds of sums of powers of negative eigenvalues of Schrödinger operators in terms of integrals of the potential. They play an important role in the proof of stability of matter by Lieb and Thirring; see [81] for a textbook presentation. For further background and references about Lieb–Thirring inequalities we also refer to the reviews [77, 65].

The following theorem summarizes Lieb–Thirring inequalities for fractional Schrödinger operators.

Theorem 5.4.2. *Let $d \geqslant 1$, $0 < s < 1$ and*

$$\begin{cases} \gamma \geqslant 1 - 1/(2s) & \text{if } d = 1 \text{ and } s > 1/2\,, \\ \gamma > 0 & \text{if } d = 1 \text{ and } s = 1/2\,, \\ \gamma \geqslant 0 & \text{if } d \geqslant 2 \text{ or } d = 1 \text{ and } s < 1/2\,. \end{cases}$$

Then there is a $L_{\gamma,d,s}$ such that for all V,

$$\operatorname{Tr}\left((-\Delta)^s + V\right)_-^\gamma \leqslant L_{\gamma,d,s} \int_{\mathbb{R}^d} V_-^{\gamma+d/(2s)}\, dx\,. \tag{5.4.6}$$

This theorem, with the additional assumption $\gamma > 1 - 1/(2s)$ if $d = 1$, $s > 1/2$, appears in [36], which also has explicit values for $L_{\gamma,d,s}$ in the physically most relevant cases. Since we have not found the case $\gamma = 1 - 1/(2s)$ if $d = 1$, $s > 1/2$, in the literature, we provide a proof in Appendix B.

To appreciate the strength of Theorem 5.4.2, we note that by bounding the sum over all eigenvalues by a single one, we deduce from (5.4.6) that

$$E_1((-\Delta)^s + V) \geqslant -\left(L_{\gamma,d,s} \int_{\mathbb{R}^d} V_-^{\gamma+d/(2s)}\, dx\right)^{1/\gamma},$$

which is the bound from Theorem 5.2.2 and which we have seen to be equivalent to the Sobolev inequality (5.2.4). Moreover, replacing V by αV and comparing with Theorem 5.3.5 we see that the right side of (5.4.6) has the correct order of growth in the large coupling limit $\alpha \to \infty$. Thus, Theorem 5.4.2 shows that the semi-classical approximation is, up to a multiplicative constant, a uniform upper bound. This observation and a density argument based on Ky-Fan's eigenvalue inequality (see, e.g., [96, Theorem 1.7]) can be used to show that for γ as in Theorem 5.4.2 the asymptotics (5.3.11) and (5.3.12) hold for all V with $V_- \in L^{\gamma+d/(2s)}(\mathbb{R}^d)$.

Let us comment on the case $\gamma = 0$ if $d = 1$ and $s = 1/2$. In this case it is easy to see that

$$\inf_{\|\psi\|=1} \left(\|(-\Delta)^{1/4}\psi\|^2 + \int_{\mathbb{R}} V|\psi|^2\, dx \right) < 0 \qquad \text{if } \int_{\mathbb{R}} V\, dx < 0\,,$$

and so inequality (5.4.6) necessarily fails for $\gamma = 0$. Remarkably, in this case one can show a *reverse bound*,

$$\operatorname{Tr}\left((-\Delta)^{1/2} + V\right)_-^0 \geqslant c \int_{\mathbb{R}} V_- \, dx \qquad \text{if } V \leqslant 0 \,. \tag{5.4.7}$$

(This is contained in [94] up to a conformal transformation.)

While there has been substantial progress concerning the sharp constants in the $s = 1$ analogue of Theorem 5.4.2, no sharp constant seems to be known in the case $s < 1$.

Our final topic are Hardy–Lieb–Thirring inequalities. We recall [64] that Hardy's inequality states that for $0 < s < d/2$ and $\psi \in \dot{H}^s(\mathbb{R}^d)$, the homogeneous Sobolev space,

$$\int_{\mathbb{R}^d} |p|^{2s} |\hat{\psi}(p)|^2 \, dp \geqslant \mathcal{C}_{s,d} \int_{\mathbb{R}^d} |x|^{-2s} |\psi(x)|^2 \, dx$$

with the sharp constant

$$\mathcal{C}_{s,d} = 2^{2s} \frac{\Gamma((d+2s)/4)^2}{\Gamma((d-2s)/4)^2} \,.$$

As a consequence, $(-\Delta)^s - \mathcal{C}_{s,d}|x|^{-2s}$ is a non-negative operator. The following theorem says that, up to avoiding the endpoint $\gamma = 0$ and modifying the constant, the Lieb–Thirring inequalities from Theorem 5.4.2 remain valid when $(-\Delta)^s$ is replaced by $(-\Delta)^s - \mathcal{C}_{s,d}|x|^{-2s}$.

Theorem 5.4.3. *Let $d \geqslant 1$, $0 < s < d/2$ and $\gamma > 0$. Then there is a constant $L^{\mathrm{HLT}}_{\gamma,d,s}$ such that*

$$\operatorname{Tr}\left((-\Delta)^s - \mathcal{C}_{s,d}|x|^{-2s} + V\right)_-^\gamma \leqslant L^{\mathrm{HLT}}_{\gamma,d,s} \int_{\mathbb{R}^d} V_-^{\gamma+d/(2s)} \, dx \,. \tag{5.4.8}$$

We emphasize that the assumption $s \leqslant 1$ is *not* needed here. Moreover, arguing as before (5.4.7) one can show that the inequality does not hold for $\gamma = 0$.

Theorem 5.4.3 was initially proved for $s = 1$ in [42] and then extended in [52] to $0 < s < 1$ (with $0 < s < 1/2$ if $d = 1$). The full result is from [45] and uses an idea from [99].

The proof in [52] (for $0 < s \leqslant 1$) allows for the inclusion of a magnetic field. This leads to the proof of stability of relativistic matter with magnetic fields for nuclear charges up to and including the critical value; see also [51].

Let us briefly comment on the proof of Theorem 5.4.3 in [52], since this will also be relevant in the following. Similarly as after Theorem 5.4.2 we observe that by bounding the sum over all eigenvalues by a single one, we deduce from (5.4.8) that

$$E_1((-\Delta)^s - \mathcal{C}_{s,d}|x|^{-2s} + V) \geqslant -\left(L^{\mathrm{HLT}}_{\gamma,d,s} \int_{\mathbb{R}^d} V_-^{\gamma+d/(2s)} \, dx\right)^{1/\gamma} ,$$

which in turn, by the argument in the proof of Theorem 5.2.2, is equivalent to the Hardy–Sobolev inequality

$$\left(\|(-\Delta)^{s/2}\psi\|^2 - \mathcal{C}_{s,d}\||x|^{-s}\psi\|^2\right)^\vartheta \|\psi\|^{2(1-\vartheta)} \geqslant C_{d,q,s}\|\psi\|_q^2 \tag{5.4.9}$$

with $1/(\gamma + d/(2s)) + 2/q = 1$ (and some $\vartheta \in (0,1)$ uniquely determined by scaling). The proof in [52] proceeds by first showing the latter inequality (at this point the assumption $s \leqslant 1$ enters through the use of the rearrangement inequality (5.2.1) for $\|(-\Delta)^{s/2}\psi\|^2$) and then by proving, in an abstract set-up (see also [53]), that a Sobolev inequality, in fact, *implies* a Lieb–Thirring inequality. (To be more precise, there is an arbitrarily small loss in the exponent. For instance, (5.4.9) for a given q implies (5.4.8) for any γ with $1/(\gamma + d/(2s)) + 2/q < 1$. But since we want to prove (5.4.8) for an open set of exponents γ, this loss is irrelevant for us.) This concludes our discussion of the proof of Theorem 5.4.3.

The Hardy inequalities discussed so far involve the function $|x|^{-2s}$ with a singularity at the origin. For convex domains there are also Hardy inequalities with the function $\operatorname{dist}(x, \Omega^c)^{-2s}$, or more generally, for arbitrary domains with the function

$$m_{2s}(x) := \left(\frac{2\pi^{\frac{d-1}{2}} \Gamma(\frac{1+2s}{2})}{\Gamma(\frac{d+2s}{2})} \right)^{\frac{1}{2s}} \left(\int_{\mathbb{S}^{d-1}} \frac{d\omega}{d_\omega(x)^{2s}} \right)^{-\frac{1}{2s}},$$

where $d_\omega(x) := \inf\{|t| : x + t\omega \notin \Omega\}$. (We say 'more generally' since one can show that $m_{2s}(x) \leqslant \operatorname{dist}(x, \Omega^c)$ for convex Ω; see [86].) The sharp Hardy inequality of Loss and Sloane [86] states that for $d \geqslant 2$, $1/2 < s < 1$, any open $\Omega \subset \mathbb{R}^d$ and any $\psi \in C_c^1(\Omega)$,

$$\int_{\mathbb{R}^d} |p|^{2s} |\hat{\psi}(p)|^2 \, dp \geqslant \mathcal{C}_s' \int_\Omega m_{2s}(x)^{-2s} |\psi(x)|^2 \, dx$$

with the sharp constant

$$\mathcal{C}_s' = \frac{\Gamma(\frac{1+2s}{2})}{|\Gamma(-s)|} \, \frac{B(\frac{1+2s}{2}, 1-s) - 2^{2s}}{2s\sqrt{\pi}} .$$

This inequality is the fractional analogue of Davies' inequality [37]. The fractional inequality in the special case of a half space is due to [25].

The analogue of Theorem 5.4.3 is

Theorem 5.4.4. *Let $d \geqslant 2$, $1/2 < s < 1$ and $\gamma \geqslant 0$. Then there is a constant $L^{\mathrm{HLT}'}_{\gamma,d,s}$ such that for all open $\Omega \subset \mathbb{R}^d$ and all V,*

$$\operatorname{Tr} \left(H_\Omega^{(s)} - \mathcal{C}_s' m_{2s}^{-2s} + V \right)_-^\gamma \leqslant L^{\mathrm{HLT}'}_{\gamma,d,s} \int_\Omega V_-^{\gamma + d/(2s)} \, dx . \tag{5.4.10}$$

We emphasize that, in contrast to Theorem 5.4.3, now $\gamma = 0$ is allowed.

Theorem 5.4.4 is the analogue of a result for $s = 1$, $d \geqslant 3$ in [54]. Since it appears here for the first time, we comment briefly on its proof. Adapting an argument of Aizenman and Lieb [5] to our setting we see that it suffices to prove the inequality for $\gamma = 0$. As in the proof of Theorem 5.4.3 from [52] the first step is the 'single function result', that is, the analogue of (5.4.9), which reads

$$\|(-\Delta)^{s/2}\psi\|^2 - \mathcal{C}_s' \|m_{2s}^{-s}\psi\|^2 \geqslant C_{d,s} \|\psi\|^2_{2d/(d-2s)} \tag{5.4.11}$$

for $\psi \in C_c^1(\Omega)$. This inequality is proved in [40]. With (5.4.11) at hand one can apply the abstract machinery from [53] in the same way as in [54] to obtain the theorem.

5.5 Some Further Topics

We conclude with some brief comments on further topics in the spectral theory of fractional Laplacians which are not included in the main part of this text.

(1) Positivity and uniqueness of the ground state. This is a classical result which can be derived using Perron–Frobenius arguments and the positivity of the heat kernel or by the maximum principle.

(2) Simplicity of excited states for radial fractional Schrödinger operators operators. This question has some relevance in non-linear problems and has recently been investigated in [49, 50] for Schrödinger operators with radially increasing potentials.

(3) Decay of eigenfunctions. In contrast to the local case $s = 1$, the decay of eigenfunctions of Schrödinger operators with potentials tending to zero at infinity is only algebraic; see [30]. (Earlier bounds in the massive case are in [90, 91].) For bounds for growing potentials see, e.g., [69].

(4) Shape of the ground state and of some excited states for the fractional Laplacian on a (convex) set. See [9, 13] for some results in $d = 1$ and [71] for a related result in $d = 2$. For superharmonicity in any d for some s, see [7]. For antisymmetry of the first excited state on a ball, see [41]. (This has also numerical methods for upper and lower bounds on the eigenvalues on a ball).

(5) Number of nodal domains. Is Sturm's bound in $d = 1$ valid? Is Courant's bound in $d \geqslant 2$ valid? For some partial results, see [9, 49, 50].

(6) Regularity of eigenfunctions. Despite the non-locality of the fractional Laplacian, eigenfunctions of $(-\Delta)^s + V$ can be shown to be regular where V is regular [34, 35]. For improved Hölder continuity results for radial potentials, see [78].

(7) Bounds on the gap $E_2(H_\Omega^{(s)}) - E_1(H_\Omega^{(s)})$ for convex Ω. See [10, 11, 68]; there are some conjectures in [10].

(8) Heat trace asymptotics for fractional Schrödinger operators and heat content asymptotics. See [17, 2, 3, 4].

(9) Many-body Coulomb systems. Stability of matter [36, 83, 84, 44, 85, 52, 51]. Proof of the Scott correction without [57, 99] and with (self-generated) magnetic field [43].

A Proof of (5.1.1)

The following computation of $a_{s,d}$ is a slight simplification of [52, Lemma 3.1]. It follows from Plancherel's theorem that

$$\iint_{\mathbb{R}^d\times\mathbb{R}^d} \frac{|\psi(x)-\psi(y)|^2}{|x-y|^{d+2s}}\,dx\,dy = \iint_{\mathbb{R}^d\times\mathbb{R}^d} \frac{|\psi(x)-\psi(x+h)|^2}{|h|^{d+2s}}\,dx\,dh = \int_{\mathbb{R}^d} t(p)|\hat{\psi}(p)|^2\,dp$$

with

$$t(p) = \int_{\mathbb{R}^d} \frac{|1 - e^{ip\cdot h}|^2}{|h|^{d+2s}}\, dh = 2 \int_{\mathbb{R}^d} \frac{1 - \cos(p \cdot h)}{|h|^{d+2s}}\, dh\,.$$

By homogeneity and rotation invariance we have

$$t(p) = a_{d,s}^{-1}\, |p|^{2s}$$

with

$$a_{d,s}^{-1} = 2 \int_{\mathbb{R}^d} \frac{1 - \cos(h_d)}{|h|^{d+2s}}\, dh\,.$$

It remains to compute this integral. We begin with the case $d = 1$, which is an exercise in complex analysis. First let $0 < s < 1/2$, so

$$a_{1,s}^{-1} = 4\,\mathrm{Re} \int_0^\infty \frac{1 - e^{ih}}{h^{1+2s}}\, dh\,.$$

Since $(1-e^{iz})/z^{1+2s}$ is analytic in the upper right quadrant and sufficiently fast decaying as $|z| \to \infty$, we can move the integration from the positive real axis to the positive imaginary axis and obtain

$$\int_0^\infty \frac{1 - e^{ih}}{h^{1+2s}}\, dh = -i \int_0^\infty \frac{1 - e^{-t}}{(it)^{1+2s}}\, dt = -i^{-2s} \int_0^\infty \frac{1 - e^{-t}}{t^{1+2s}}\, dt\,.$$

The integral here can be recognized as a gamma function. Indeed, we have if $\mathrm{Re}\, z > 0$,

$$\Gamma(z) - \frac{1}{z} = -\int_0^1 \left(1 - e^{-t}\right) t^{z-1}\, dt + \int_1^\infty e^{-t} t^{z-1}\, dt\,.$$

Since the right side is analytic in $\{\mathrm{Re}\, z > -1\}$, the formula extends to this region and, in particular,

$$\Gamma(z) = -\int_0^\infty \left(1 - e^{-t}\right) t^{z-1}\, dt \qquad \text{if } -1 < \mathrm{Re}\, z < 0\,.$$

Thus, we have shown that

$$a_{1,s}^{-1} = 4\,\mathrm{Re}\, i^{-2s}\Gamma(-2s) = 4\cos(\pi s)\Gamma(-2s)\,.$$

Using the duplication formula $\Gamma(-2s) = \pi^{-1/2} 2^{2s-1} \Gamma(-s)\Gamma((1-2s)/2)$ and the reflection formula $\Gamma((1+2s)/2)\Gamma((1-2s)/2) = -\pi/\cos(\pi s)$ we obtain the claimed formula for $a_{1,s}$. When $1/2 < s < 1$, we start from

$$a_{1,s}^{-1} = 4\,\mathrm{Re} \int_0^\infty \frac{1 + ih - e^{ih}}{h^{1+2s}}\, dh$$

and argue similarly using

$$\Gamma(z) = -\int_0^\infty \left(1 - t - e^{-t}\right) t^{z-1}\, dt \qquad \text{if } -2 < \mathrm{Re}\, z < -1\,.$$

Finally, the formula for $s = 1/2$ follows by continuity. This concludes the proof of (5.1.1) for $d = 1$.

Now let $d \geqslant 2$ and write $h = (h', h_d) \in \mathbb{R}^{d-1} \times \mathbb{R}$ and compute for fixed $h_d \in \mathbb{R}$

$$\int_{\mathbb{R}^{d-1}} \frac{dh'}{(h'^2 + h_d^2)^{(d+2s)/2}} = \frac{b_{d,s}}{|h_d|^{1+2s}}$$

with

$$b_{d,s} = \int_{\mathbb{R}^{d-1}} \frac{d\eta}{(1+\eta^2)^{(d+2s)/2}} .$$

Thus,

$$a_{d,s}^{-1} = 2 b_{d,s} \int_{\mathbb{R}} \frac{1 - \cos h_d}{|h_d|^{1+2s}} \, dh_d = b_{d,s} a_{1,s}^{-1} ,$$

and it remains to compute $b_{d,s}$. To do so we use [1, (6.2.1), (6.2.2)] and obtain

$$\begin{aligned} b_{d,s} &= |\mathbb{S}^{d-2}| \int_0^\infty \frac{r^{d-2}\, dr}{(1+r^2)^{(d+2s)/2}} = \frac{|\mathbb{S}^{d-2}|}{2} \int_0^\infty \frac{t^{(d-3)/2}\, dt}{(1+t)^{(d+2s)/2}} \\ &= \frac{|\mathbb{S}^{d-2}|}{2} \frac{\Gamma((d-1)/2)\,\Gamma((1+2s)/2)}{\Gamma((d+2s)/2)} = \pi^{(d-1)/2} \frac{\Gamma((1+2s)/2)}{\Gamma((d+2s)/2)} . \end{aligned}$$

This concludes the proof of (5.1.1) for $d \geqslant 2$.

B Lieb–Thirring Inequality in the Critical Case

Our goal in this appendix is to prove Theorem 5.4.2 in the critical case $d = 1$, $1/2 < s < 1$ and $\gamma = 1 - 1/(2s)$. Our argument will be a modification of Weidl's argument [100] in the $s = 1$ case (see also the unpublished manuscript [95]).

For $1/2 < s < 1$, any bounded interval $Q \subset \mathbb{R}$ and any $\psi \in H^s(Q)$, we define

$$t_Q^{(s)}[\psi] := a_{1,s} \iint\limits_{Q\times Q} \frac{|\psi(x) - \psi(y)|^2}{|x-y|^{1+2s}} \, dx\, dy ,$$

where $a_{1,s}$ is the constant from (5.1.1). We shall need the following Poincaré–Sobolev inequality for this quadratic form.

Lemma B.1. *Let $d = 1$ and $1/2 < s < 1$. Then there is a constant C_s such that for any bounded interval $Q \subset \mathbb{R}$ and any $\psi \in H^s(Q)$ with $\int_Q \psi \, dx = 0$,*

$$\sup_Q |\psi|^2 \leqslant C_s |Q|^{2s-1} t_Q^{(s)}[\psi] .$$

Proof. By a density argument we may assume that ψ is continuous. We know from [60] (with $\Psi(x) = x^2$ and $p(x) = |x|^{s+1/2}$) and a simple scaling argument that for any $a < b$ and any continuous function φ on $[a, b]$,

$$\frac{|\varphi(a) - \varphi(b)|^2}{(b-a)^{2s-1}} \leqslant D_s \int_a^b \int_a^b \frac{|\varphi(x) - \varphi(y)|^2}{|x-y|^{1+2s}} \, dx\, dy$$

with $D_s = (16(2s+1)/(2s-1))^2$. Since $\int_Q \psi\,dx = 0$ there is a $c \in Q$ such that $\psi(c) = 0$. Moreover, let $d \in Q$ be such that $|\psi(d)| = \sup|\psi|$. We apply the above inequality with $a = \min\{c, d\}$ and $b = \max\{c, d\}$ and note that $b - a \leqslant |Q|$ to obtain the lemma. □

The quadratic form $t_Q^{(s)}[\psi]$ is non-negative and closed in $L^2(Q)$ and therefore generates a self-adjoint operator, which we denote by $T_Q^{(s)}$. In some sense this corresponds to imposing Neumann boundary conditions on ∂Q.

Lemma B.2. *Let $d = 1$, $1/2 < s < 1$ and let C_s be the constant from Lemma B.1. Let $Q \subset \mathbb{R}$ be a bounded interval and assume that $V \in L^1(Q)$ satisfies*

$$\alpha := |Q|^{2s-1} \int_Q V_-\,dx < C_s^{-1}.$$

Then $T_Q^{(s)} + V$ has at most one negative eigenvalue E and this eigenvalue satisfies, if it exists,

$$E \geqslant -\alpha^{-1/(2s-1)}(1 - C_s\alpha)^{-1} \left(\int_Q V_-\,dx\right)^{2s/(2s-1)}.$$

Proof. If $\psi \in H^s(Q)$ satisfies $\int_Q \psi\,dx = 0$, then by Lemma B.1

$$\begin{aligned} t_Q^{(s)}[\psi] + \int_Q V|\psi|^2\,dx &\geqslant t_Q^{(s)}[\psi] - \int_Q V_-\,dx \sup_Q |\psi|^2 \geqslant t_Q^{(s)}[\psi]\left(1 - C_s|Q|^{2s-1}\int_Q V_-\,dx\right) \\ &= t_Q^{(s)}[\psi]\,(1 - C_s\alpha) \geqslant 0. \end{aligned}$$

Thus, $E_2(T_Q^{(s)} + V) \geqslant 0$.

For general $\psi \in H^s(Q)$ we set $\psi_Q := |Q|^{-1}\int_Q \psi\,dx$ and bound similarly, for any $\beta > 0$,

$$\begin{aligned} t_Q^{(s)}[\psi] + \int_Q V|\psi|^2\,dx &\geqslant t_Q^{(s)}[\psi] - \int_Q V_-\,dx \left(\sup_Q |\psi - \psi_Q| + |\psi_Q|\right)^2 \\ &\geqslant t_Q^{(s)}[\psi] - \int_Q V_-\,dx \left(\left(C_s|Q|^{2s-1} t_Q^{(s)}[\psi]\right)^{1/2} + |Q|^{-1/2}\|\psi\|\right)^2 \\ &\geqslant t_Q^{(s)}[\psi]\left(1 - (1+\beta)C_s|Q|^{2s-1}\int_Q V_-\,dx\right) \\ &\qquad - (1+\beta^{-1})|Q|^{-1}\int_Q V_-\,dx\,\|\psi\|^2 \\ &= t_Q^{(s)}[\psi]\,(1 - (1+\beta)C_s\alpha) \\ &\qquad - (1+\beta^{-1})\alpha^{-1/(2s-1)}\left(\int_Q V_-\,dx\right)^{2s/(2s-1)}\|\psi\|^2. \end{aligned}$$

With the choice $\beta = (1 - C_s\alpha)/(C_s\alpha)$ we finally obtain

$$t_Q^{(s)}[\psi] + \int_Q V|\psi|^2\,dx \geqslant -\frac{1}{1 - C_s\alpha}\alpha^{-1/(2s-1)}\left(\int_Q V_-\,dx\right)^{2s/(2s-1)}\|\psi\|^2,$$

which implies the lower bound on $E_1(T_Q^{(s)} + V)$ in the lemma. □

Proof of Theorem 5.4.2 for $d = 1$, $1/2 < s < 1$, $\gamma = 1 - 1/(2s)$. Let C_s be the constant from Lemma B.1 and fix $0 < \alpha < C_s$ to be chosen later. We claim that there are disjoint open intervals Q_n whose closed union covers $\operatorname{supp} V_-$ and such that

$$|Q_n|^{2s-1} \int_{Q_n} V_- \, dx = \alpha \qquad \text{for all } n\,.$$

In fact, pick $x_0 \in \mathbb{R}$ arbitrary and define x_{k+1} inductively, given x_k, as follows: If $V_- \equiv 0$ on (x_k, ∞) we stop the procedure. Otherwise, since $\ell \mapsto \ell^{2s-1} \int_{x_k}^{x_k+\ell} V_- \, dx$ is non-decreasing and unbounded, we can find $x_{k+1} > x_k$ such that

$$(x_{k+1} - x_k)^{2s-1} \int_{x_k}^{x_{k+1}} V_- \, dx = \alpha\,.$$

Since $(x_{k+1} - x_k)^{2s-1} \geqslant \alpha / \int_{x_0}^{\infty} V_- \, dx$, we will eventually cover $\operatorname{supp} V_- \cap [x_0, \infty)$. Now we repeat the same argument to the left of x_0. The Q_n's are all the intervals (x_k, x_{k+1}).

We have

$$\|(-\Delta)^{s/2}\psi\|^2 = a_{1,s} \iint\limits_{\mathbb{R}\times\mathbb{R}} \frac{|\psi(x) - \psi(y)|^2}{|x-y|^{1+2s}} \, dx\, dy \geqslant \sum_n t_{Q_n}^{(s)}[\psi]\,,$$

which, by the variational principle, implies that

$$(-\Delta)^s + V \geqslant \sum_n \left(T_{Q_n}^{(s)} + V_{Q_n} \right),$$

where V_{Q_n} denotes the restriction of V to Q_n, and therefore

$$\operatorname{Tr}\left((-\Delta)^s + V\right)_-^{\frac{2s-1}{2s}} \leqslant \operatorname{Tr}\left(\sum_n \left(T_{Q_n}^{(s)} + V_{Q_n} \right) \right)_-^{\frac{2s-1}{2s}} = \sum_n \operatorname{Tr}\left(T_{Q_n}^{(s)} + V_{Q_n} \right)_-^{\frac{2s-1}{2s}}.$$

According to Lemma B.2,

$$\operatorname{Tr}\left(T_{Q_n}^{(s)} + V_{Q_n} \right)_-^{\frac{2s-1}{2s}} \leqslant \alpha^{-\frac{1}{2s}} (1 - C_s\alpha)^{-\frac{2s-1}{2s}} \int_{Q_n} V_- \, dx\,.$$

Summing over n, we obtain

$$\operatorname{Tr}\left((-\Delta)^s + V\right)_-^{\frac{2s-1}{2s}} \leqslant \alpha^{-\frac{1}{2s}} (1 - C_s\alpha)^{-\frac{2s-1}{2s}} \int_{\mathbb{R}} V_- \, dx\,.$$

We can optimize this in α by choosing $\alpha = 1/(2sC_s)$ and obtain

$$\operatorname{Tr}\left((-\Delta)^s + V\right)_-^{\frac{2s-1}{2s}} \leqslant C_s^{\frac{1}{2s}} \frac{2s}{(2s-1)^{\frac{2s-1}{2s}}} \int_{\mathbb{R}} V_- \, dx.$$

This proves the theorem. □

Acknowledgment: The author would like to thank B. Dyda, L. Geisinger, E. Lenzmann, E. Lieb, R. Seiringer and L. Silvestre for collaborations involving the fractional Laplacian. He thanks R. Bañuelos, M. Kwaśnicki, F. Maggi and R. Song for helpful comments. Partial support by U.S. National Science Foundation DMS-1363432 is acknowledged.

Bibliography

[1] M. Abramowitz, I. A. Stegun, *Handbook of mathematical functions with formulas, graphs, and mathematical tables*. National Bureau of Standards Applied Mathematics Series **55**, Washington, D.C. 1964.

[2] L. Acuña Valverde, *Trace asymptotics for fractional Schrödinger operators*. J. Funct. Anal. **266** (2014), no. 2, 514–559.

[3] L. Acuña Valverde, *Heat Content for Stable Processes in Domains of* $\mathbb{R}^d$. J. Geometric Anal., to appear.

[4] L. Acuña Valverde, R. Bañuelos, *Heat content and small time asymptotics for Schrödinger operators on* $\mathbb{R}^d$. Potential Anal. **42** (2015), no. 2, 457–482.

[5] M. Aizenman, E. H. Lieb, *On semiclassical bounds for eigenvalues of Schrödinger operators*. Phys. Lett. A **66** (1978), no. 6, 427–429.

[6] F. J. Almgren Jr., E. H. Lieb, *Symmetric decreasing rearrangement is sometimes continuous*. J. Amer. Math. Soc. **2** (1989), no. 4, 683–773.

[7] R. Bañuelos, D. DeBlassie, *On the first eigenfunction of the symmetric stable process in a bounded Lipschitz domain*. Potential Anal. **42** (2015), no. 2, 573–583.

[8] R. Bañuelos, R. Latała, P. J. Méndez-Hernández, *A Brascamp–Lieb–Luttinger-type inequality and applications to symmetric stable processes*. Proc. Amer. Math. Soc. **129** (2001), 2997–3008.

[9] R. Bañuelos, T. Kulczycki, *The Cauchy process and the Steklov problem*. J. Funct. Anal. **211** (2004), 355–423.

[10] R. Bañuelos, T. Kulczycki, *Eigenvalue gaps for the Cauchy process and a Poincaré inequality*. J. Funct. Anal. **234** (2006), no. 1, 199–225.

[11] R. Bañuelos, T. Kulczycki, *Spectral gap for the Cauchy process on convex, symmetric domains*. Comm. Partial Differential Equations **31** (2006), no. 10-12, 1841–1878.

[12] R. Bañuelos, T. Kulczycki, *Trace estimates for stable processes*. Probab. Theory Related Fields **142** (2008), no. 3-4, 313–338.

[13] R. Bañuelos, T. Kulczycki, P. J. Méndez-Hernández, *On the shape of the ground state eigenfunction for stable processes*. Potential Anal. **24** (2006), no. 3, 205–221.

[14] R. Bañuelos, T. Kulczycki, B. Siudeja, *On the trace of symmetric stable processes on Lipschitz domains*. J. Funct. Anal. **257** (2009), no. 10, 3329–3352.

[15] R. Bañuelos, P. J. Méndez-Hernández, *Symmetrization of Lévy processes and applications*. J. Funct. Anal. **258** (2010), no. 12, 4026–4051.

[16] R. Bañuelos, J. B. Mijena, E. Nane, *Two-term trace estimates for relativistic stable processes*. J. Math. Anal. Appl. **410** (2014), no. 2, 837–846.

[17] R. Bañuelos, S. Y. Yolcu, *Heat trace of non-local operators*. J. Lond. Math. Soc. (2) **87** (2013), no. 1, 304–318.

[18] R. Bhatia, *Matrix analysis*. Graduate Texts in Mathematics **169**. Springer-Verlag, New York, 1997.

[19] J. Bellazzini, R. L. Frank, N. Visciglia, *Maximizers for Gagliardo–Nirenberg inequalities and related non-local problems*. Math. Ann. **360** (2014), no. 3-4, 653–673.

[20] F. A. Berezin, *Covariant and contravariant symbols of operators*. Math. USSR-Izv. **6** (1972), 1117–1151.

[21] M. Sh. Birman, A. Laptev, *The negative discrete spectrum of a two-dimensional Schrödinger operator*. Comm. Pure Appl. Math. **49** (1996), no. 9, 967–997.

[22] R. M. Blumenthal, R. K. Getoor, *The asymptotic distribution of the eigenvalues for a class of Markov operators*. Pacific J. Math. **9** (1959), 399–408.

[23] K. Bogdan, B. A. Siudeja, *Trace estimates for unimodal Lévy processes*. J. Evol. Equ. **16** (2016), no. 4, 857–876.

[24] L. Brasco, G. De Philippis, B. Velichkov, *Faber–Krahn inequalities in sharp quantitative form*. Duke Math. J. **164** (2015), no. 9, 1777–1831.

[25] K. Bogdan, B. Dyda, *The best constant in a fractional Hardy inequality*. Math. Nachr. **284** (2011), no. 5-6, 629–638.

[26] R. M. Brown, *The trace of the heat kernel in Lipschitz domains*. Trans. Amer. Math. Soc. **339** (1993), no. 2, 889–900.

[27] A. Burchard, H. Hajaiej, *Rearrangement inequalities for functionals with monotone integrands*. J. Funct. Anal. **233** (2006), no. 2, 561–582.

[28] E. A. Carlen, *Trace inequalities and quantum entropy: an introductory course*. In: Entropy and the quantum, 73–140, Contemp. Math. **529**, Amer. Math. Soc., Providence, RI, 2010.

[29] E. A. Carlen, R. L. Frank, E. H. Lieb, *Stability estimates for the lowest eigenvalue of a Schrödinger operator*. Geom. Funct. Anal. **24** (2014), no. 1, 63–84.

[30] R. Carmona, W. C. Masters, B. Simon, *Relativistic Schrödinger operators: asymptotic behavior of the eigenfunctions*. J. Funct. Anal. **91** (1990), no. 1, 117–142.

[31] L. Caffarelli, L. Silvestre, *An extension problem related to the fractional Laplacian*. Comm. Partial Differential Equations **32** (2007), no. 7-9, 1245–1260.

[32] Z.-Q. Chen, R. Song, *Two-sided eigenvalue estimates for subordinate processes in domains*. J. Funct. Anal. **226** (2005), no. 1, 90–113.

[33] Z.-Q. Chen, R. Song, *Continuity of eigenvalues of subordinate processes in domains*. Math. Z. **252** (2006), no. 1, 71–89.

[34] A. Dall'Acqua, S. Fournais, T. Ø. Sørensen, E. Stockmeyer, *Real analyticity away*

from the nucleus of pseudorelativistic Hartree-Fock orbitals. Anal. PDE **5** (2012), no. 3, 657–691.

[35] A. Dall'Acqua, S. Fournais, T. Ø. Sørensen, E. Stockmeyer, *Real analyticity of solutions to Schrödinger equations involving a fractional Laplacian and other Fourier multipliers*. In: XVIIth International Congress on Mathematical Physics, 600–609, World Sci. Publ., Hackensack, NJ, 2014.

[36] I. Daubechies, *An uncertainty principle for fermions with generalized kinetic energy*. Comm. Math. Phys. **90** (1983), no. 4, 511–520.

[37] E. B. Davies, *Some norm bounds and quadratic form inequalities for Schrödinger operators. II*. J. Operator Theory **12** (1984), no. 1, 177–196.

[38] R. D. DeBlassie, *Higher order PDEs and symmetric stable processes*. Probab. Theory Related Fields **129** (2004), no. 4, 495–536. Correction: ibid., **133** (2005), no. 1, 141–143.

[39] R. D. Deblassie, P. J. Méndez-Hernández, *α-continuity properties of the symmetric α-stable process*. Trans. Amer. Math. Soc. **359** (2007), no. 5, 2343–2359.

[40] B. Dyda, R. L. Frank, *Fractional Hardy–Sobolev–Maz'ya inequality for domains*. Studia Math. **208** (2012), no. 2, 151–166.

[41] B. Dyda, A. Kuznetsov, M. Kwaśnicki, *Eigenvalues of the fractional Laplace operator in the unit ball*. Preprint, arXiv:1509.08533.

[42] T. Ekholm, R. L. Frank, *On Lieb–Thirring inequalities for Schrödinger operators with virtual level*. Comm. Math. Phys. **264** (2006), no. 3, 725–740.

[43] L. Erdos, S. Fournais, J. P. Solovej, *Relativistic Scott correction in self-generated magnetic fields*. J. Math. Phys. **53** (2012), no. 9, 095202, 26 pp.

[44] C. Fefferman, R. de la Llave, *Relativistic stability of matter. I*. Rev. Mat. Iberoamericana **2** (1986), no. 1-2, 119–213.

[45] R. L. Frank, *A simple proof of Hardy–Lieb–Thirring inequalities*. Comm. Math. Phys. **290** (2009), no. 2, 789–800.

[46] R. L. Frank, L. Geisinger, *Two-term spectral asymptotics for the Dirichlet Laplacian on a bounded domain*. In: Mathematical results in quantum physics, P. Exner (ed.), 138–147, World Sci. Publ., Hackensack, NJ, 2011.

[47] R. L. Frank, L. Geisinger, *Refined semiclassical asymptotics for fractional powers of the Laplace operator*. J. Reine Angew. Math. **712** (2016), 1–37.

[48] R. L. Frank, L. Geisinger, *Semi-classical analysis of the Laplace operator with Robin boundary conditions*. Bull. Math. Sci. **2** (2012), no. 2, 281–319.

[49] R. L. Frank, E. Lenzmann, *Uniqueness of non-linear ground states for fractional Laplacians in $\mathbb{R}$*. Acta Math. **210** (2013), no. 2, 261–318.

[50] R. L. Frank, E. Lenzmann, L. Silvestre, *Uniqueness of radial solutions for the fractional Laplacian*. Comm. Pure Appl. Math. **69** (2016), no. 9, 1671–1726.

[51] R. L. Frank, E. H. Lieb, R. Seiringer, *Stability of relativistic matter with magnetic fields for nuclear charges up to the critical value*. Comm. Math. Phys. **275** (2007), no. 2, 479–489.

[52] R. L. Frank, E. H. Lieb, R. Seiringer, *Hardy–Lieb–Thirring inequalities for fractional Schrödinger operators*. J. Amer. Math. Soc. **21** (2008), no. 4, 925–950.
[53] R. L. Frank, E. H. Lieb, R. Seiringer, *Equivalence of Sobolev inequalities and Lieb–Thirring inequalities*. In: XVIth International Congress on Mathematical Physics, 523–535, World Sci. Publ., Hackensack, NJ, 2010.
[54] R. L. Frank, M. Loss, *Hardy–Sobolev–Maz'ya inequalities for arbitrary domains*. J. Math. Pures Appl. (9) **97** (2012), no. 1, 39–54.
[55] R. L. Frank, M. Loss, T. Weidl, *Pólya's conjecture in the presence of a constant magnetic field*. J. Eur. Math. Soc. (JEMS) **11** (2009), no. 6, 1365–1383.
[56] R. L. Frank, R. Seiringer, *Non-linear ground state representations and sharp Hardy inequalities*. J. Funct. Anal. **255** (2008), no. 12, 3407–3430.
[57] R. L. Frank, H. Siedentop, S. Warzel, *The ground state energy of heavy atoms: Relativistic lowering of the leading energy correction*. Comm. Math. Phys. **278** (2008), no. 2, 549–566.
[58] N. Fusco, F. Maggi, A. Pratelli, *Stability estimates for certain Faber–Krahn, isocapacitary and Cheeger inequalities*. Ann. Sc. Norm. Super. Pisa Cl. Sci. (5) **8** (2009), no. 1, 51–71.
[59] A. M. Garsia, E. Rodemich, *Monotonicity of certain functionals under rearrangement*. Ann. Inst. Fourier **24** (1974), No. 2, 67–116.
[60] A. M. Garsia, E. Rodemich, H. Rumsey Jr., *A real variable lemma and the continuity of paths of some Gaussian processes*. Indiana Univ. Math. J. **20** (1970/1971) 565–578.
[61] L. Geisinger, *A short proof of Weyl's law for fractional differential operators*. J. Math. Phys. **55** (2014), no. 1, 011504, 7 pp.
[62] S. Gottwald, *Two-term spectral asymptotics for the Dirichlet pseudo-relativistic kinetic energy operator on a bounded domain*. PhD thesis (2016), LMU Munich, https://edoc.ub.uni-muenchen.de/view/subjects/fak16.html
[63] B. Helffer, D. Robert, *Calcul fonctionnel par la transformation de Mellin et opérateurs admissibles*. J. Funct. Anal. **53** (1983), no. 3, 246–268.
[64] I. W. Herbst, *Spectral theory of the operator* $(p^2+m^2)^{1/2}-Ze^2/r$. Comm. Math. Phys. **53** (1977), no. 3, 285–294.
[65] D. Hundertmark, *Some bound state problems in quantum mechanics*. In: Spectral theory and mathematical physics: a Festschrift in honor of Barry Simon's 60th birthday, 463–496, Proc. Sympos. Pure Math., **76**, Part 1, Amer. Math. Soc., Providence, RI, 2007.
[66] V. Ja. Ivriĭ, *The second term of the spectral asymptotics for a Laplace-Beltrami operator on manifolds with boundary*. Functional Anal. Appl. **14** (1980), no. 2, 98–106.
[67] V. Ja. Ivriĭ, *Spectral asymptotics for fractional Laplacians*. Preprint (2016), arXiv:1603.06364.
[68] K. Kaleta, *Spectral gap lower bound for the one-dimensional fractional Schrödinger operator in the interval*. Studia Math. **209** (2012), no. 3, 267–287.

[69] K. Kaleta, T. Kulczycki, *Intrinsic ultracontractivity for Schrödinger operators based on fractional Laplacians*. Potential Anal. **33** (2010), no. 4, 313–339.
[70] J. B. Keller, *Lower bounds and isoperimetric inequalities for eigenvalues of the Schrödinger equation*. J. Mathematical Phys. **2** (1961), 262–266.
[71] T. Kulczycki, *On concavity of solutions of the Dirichlet problem for the equation* $(-\Delta)^{1/2}\varphi = 1$ *in convex planar regions*. J. Eur. Math. Soc. (JEMS) **19** (2017), no. 5, 1361–4120.
[72] T. Kulczycki, M. Kwaśnicki, J. Małecki, A. Stos, *Spectral properties of the Cauchy process on half-line and interval*. Proc. Lond. Math. Soc. (3) **101** (2010), no. 2, 589–622.
[73] M. Kwaśnicki, *Spectral analysis of subordinate Brownian motions on the half-line*. Studia Math. **206** (2011), no. 3, 211–271.
[74] M. Kwaśnicki, *Eigenvalues of the fractional Laplace operator in the interval*. J. Funct. Anal. **262** (2012), no. 5, 2379–2402.
[75] M. Kwaśnicki, R. Laugesen, B. Siudeja, *Pólya's conjecture fails for the fractional Laplacian*. Preprint (2016), arXiv:1608.06905.
[76] A. Laptev, *Dirichlet and Neumann eigenvalue problems on domains in Euclidean spaces*. J. Funct. Anal. **151** (1997), no. 2, 531–545.
[77] A. Laptev, T. Weidl, *Recent results on Lieb–Thirring inequalities*. Journées "Équations aux Dérivées Partielles" (La Chapelle sur Erdre, 2000), Exp. No. XX, 14 pp., Univ. Nantes, Nantes, 2000.
[78] M. Lemm, *On the Hölder regularity for the fractional Schrödinger equation and its improvement for radial data*. Comm. Part. Diff. Eq. **41** (2016), no. 11, 1761–1792.
[79] P. Li, S. T. Yau, *On the Schrödinger equation and the eigenvalue problem*. Comm. Math. Phys. **88** (1983), 309–318.
[80] E. H. Lieb, M. Loss, *Analysis*. Second edition. Graduate Studies in Mathematics **14**. American Mathematical Society, Providence, RI, 2001.
[81] E. H. Lieb, R. Seiringer, *The stability of matter in quantum mechanics*. Cambridge University Press, Cambridge, 2010.
[82] E. H. Lieb, W. Thirring, *Inequalities for the moments of the eigenvalues of the Schrödinger Hamiltonian and their relation to Sobolev inequalities*. In: Stud. in Math. Phys., Princeton Univ. Press, 1976, pp. 269–303.
[83] E. H. Lieb, W. Thirring, *Gravitational collapse in quantum mechanics with relativistic kinetic energy*. Ann. Phys. **155** (1984), no. 2, 494–512.
[84] E. H. Lieb, H.-T. Yau, *The Chandrasekhar theory of stellar collapse as the limit of quantum mechanics*. Comm. Math. Phys. **112** (1987), no. 1, 147–174.
[85] E. H. Lieb, H.-T. Yau, *The stability and instability of relativistic matter*. Comm. Math. Phys. **118** (1988), no. 2, 177–213.
[86] M. Loss, C. Sloane, *Hardy inequalities for fractional integrals on general domains*. J. Funct. Anal. **259** (2010), no. 6, 1369–1379.
[87] P. J. Méndez-Hernández, *Brascamp–Lieb–Luttinger inequalities for convex domains of finite inradius*. Duke Math. J. **113** (2002), no. 1, 93–131.

[88] R. Musina, A. I. Nazarov, *On fractional Laplacians*. Comm. Partial Differential Equations **39** (2014), no. 9, 1780–1790.
[89] K. Naimark, M. Solomyak, *Regular and pathological eigenvalue behavior for the equation* $-\lambda u'' = Vu$ *on the semiaxis*. J. Funct. Anal. **151** (1997), no. 2, 504–530.
[90] F. Nardini, *Exponential decay for the eigenfunctions of the two-body relativistic Hamiltonian*. J. Analyse Math. **47** (1986), 87–109.
[91] F. Nardini, *On the asymptotic behaviour of the eigenfunctions of the relativistic N-body Schrödinger operator*. Boll. Un. Mat. Ital. A (7) **2** (1988), no. 3, 365–369.
[92] H. Park, R. Song, *Trace estimates for relativistic stable processes*. Potential Anal. **41** (2014), no. 4, 1273–1291.
[93] M. Reed, B. Simon *Methods of modern mathematical physics. IV. Analysis of operators*. Academic Press, New York - London, 1978.
[94] E. Shargorodsky, *An estimate for the Morse index of a Stokes wave*. Arch. Ration. Mech. Anal. **209** (2013), no. 1, 41–59.
[95] B. Simon, *Critical Lieb–Thirring bounds for one-dimensional Schrödinger operators and Jacobi matrices with regular ground states*. arXiv:0705.3640.
[96] B. Simon, *Trace ideals and their applications*. Second edition. Mathematical Surveys and Monographs **120**. American Mathematical Society, Providence, RI, 2005.
[97] B. Simon, *Loewner's theorem on monotone matrix functions*. In preparation.
[98] Y. Sire, J. L. Vãzquez, B. Volzone, *Symmetrization for fractional elliptic and parabolic equations and an isoperimetric application*. Preprint (2015), arXiv:1506.07199.
[99] J. P. Solovej, T. Ø. Sørensen, W. L. Spitzer, *Relativistic Scott correction for atoms and molecules*. Comm. Pure Appl. Math. **63** (2010), no. 1, 39–118.
[100] T. Weidl, *On the Lieb–Thirring constants* $L_{\gamma,1}$ *for* $\gamma \geqslant 1/2$. Comm. Math. Phys. **178** (1996), no. 1, 135–146.
[101] S. Yıldırım Yolcu, T. Yolcu, *Estimates for the sums of eigenvalues of the fractional Laplacian on a bounded domain*. Common. Contemp. Math. **15** (2013), no. 3, 1250048, 15 pp.

María del Mar González

Recent Progress on the Fractional Laplacian in Conformal Geometry

6.1 Introduction

The aim of this paper is to report on recent development on the conformal fractional Laplacian, both from the analytic and geometric points of view, but especially towards the PDE community.

The basic tool in conformal geometry is to understand the geometry of a space by studying the transformations on such space. More precisely, let M^n be a smooth Riemannian manifold of dimension $n \geqslant 2$ with a Riemannian metric h. A *conformal* change of metric is such that it preserves angles so, mathematically, two metrics h, $\tilde{h}$ are conformally related if

$$\tilde{h} = fh \quad \text{for some function } f > 0.$$

We say that an operator $A (= A_h)$ on M is *conformally covariant* if under the change of metric $h_w = e^{2w}h$, then A satisfies the transformation (sometimes called intertwining) law

$$A_{h_w}\varphi = e^{-bw}A_h(e^{aw}\varphi) \quad \text{for all} \quad \varphi \in \mathcal{C}^\infty(M), \tag{6.1.1}$$

for some $a, b \in \mathbb{R}$. One may associate to such A a notion of curvature with interesting conformal properties defined by

$$Q_A^h := A_h(1).$$

The intertwining rule (6.1.1) then yields the Q_A-curvature equation

$$A_h(e^{aw}) = e^{bw}Q_A^{h_w}.$$

The most well known example of a conformally covariant operator is the conformal Laplacian

$$L_h := -\Delta_h + \tfrac{n-2}{4(n-1)}R_h, \tag{6.1.2}$$

and its associated curvature is precisely the scalar curvature R_h (modulo a multiplicative constant). The conformal transformation law is usually written as

$$L_{h_u}(\cdot) = u^{-\frac{n+2}{n-2}}L_h(u\,\cdot) \tag{6.1.3}$$

María del Mar González, Universidad Autónoma de Madrid, Spain, E-mail: mariamar.gonzalezn@uam.es

https://doi.org/10.1515/9783110571561-008

for a change of metric

$$h_u = u^{\frac{4}{n-2}} h, \quad u > 0,$$

and it gives rise to interesting semilinear equations with reaction term of power type, such as the constant scalar curvature (or Yamabe) equation

$$-\Delta_h u + \tfrac{n-2}{4(n-1)} R_h = \tfrac{n-2}{4(n-1)} c u^{\frac{n+2}{n-2}}. \tag{6.1.4}$$

The Yamabe problem (see [92, 68] for a general background) has been one of the multiple examples of the interaction between analysis and geometry.

A higher order example of a conformally covariant operator is the Paneitz operator ([87]), which is defined as the bi-Laplacian $(-\Delta_h)^2$ plus lower order curvature terms. Its associated curvature, known as Q-curvature, a is fourth-order geometric object that has received a lot of attention (see [21] and the references therein). The generalizaton to all even orders $2k$ was investigated by Graham, Jenne, Mason and Sparling (*GJMS*) in [50] and is based on the ambient metric construction of [35].

These operators belong to a general framework in which on the manifold (M, h) there exists a meromorphic family of conformally covariant pseudodifferential operators of fractional order

$$P_s^h = (-\Delta_h)^s + \dots \qquad \text{for any} \quad s \in (0, n/2).$$

P_s^h will be called the *conformal fractional Laplacian*. The main goal of this discussion is to describe and to give some examples, applications and open problems for this non-local object. The uniqueness issue will be postponed to Section 6.7.

To each of these operators there exists an associated curvature Q_s^h, that generalizes the scalar curvature, the Q-curvature and the mean curvature. The Q_s^h constitute a one-parameter family of non-local curvatures on M; the objective is to understand the geometric and topological information they contain, together with the study of the new non-local fractional order PDE that arise.

The conformal fractional Laplacian is defined on the boundary of a Poincaré-Einstein manifold in terms of scattering theory (all the necessary background will be explained in Section 6.2). Research on Poincaré-Einstein metrics has its origins in the work of Newman, Penrose and LeBrun [66] on four dimensional space-time Physics. The subsequent work of Fefferman and Graham [35] provided the mathematical framework for the study of conformally invariant (o covariant) operators on the boundary (denoted as the conformal infinity) of a Poincaré-Einstein manifold (the ambient) through this approach, through the study of the asymptotics of an eigenvalue problem in the spirit of Maldacena's AdS/CFT correspondence.

The celebrated AdS/CFT correspondence in string theory [73, 2, 96] establishes a connection between conformal field theories in n dimensions and gravity fields on a $(n+1)$-dimensional spacetime of anti-de Sitter type, to the effect that correlation functions in conformal field theory are given by the asymptotic behavior at infinity of the supergravity action. Mathematically, this involves describing the solution to the gravitational field equations in $(n+1)$ dimensions (which, in the simplest case of a scalar field reduces to the Klein–Gordon equation) in terms of a conformal field, which plays the role of the boundary data imposed on the (timelike) conformal boundary.

An equivalent construction for P_s^h has been recently proposed [20] in the setting of metric measure spaces in relation to Perelman's W-functional [88]. This point of view has the advantage that it justifies the notion of a harmonic function in fractional dimensions that was sketched in the classical paper of Caffarelli and Silvestre for the usual fractional Laplacian [17].

In Section 6.3 we will explain the construction of the conformal fractional Laplacian from a purely analytical point of view. Caffarelli and Silvestre [17] gave a construction for the standard fractional Laplacian $(-\Delta_{\mathbb{R}^n})^s$ as a Dirichlet-to-Neumann operator of a uniformly degenerate elliptic boundary value problem. In the manifold case, Chang and the author [22] related the original definition of the conformal fractional Laplacian coming from scattering theory to a Dirichlet-to-Neumann operator for a related elliptic extension problem, thus allowing for an analytic treatment of Yamabe-type problems in the non-local setting ([41]).

The fractional Yamabe problem, proposed in [41] poses the question of finding a constant fractional curvature metric in a given conformal class. In the simplest case, the resulting (non-local) PDE is

$$(-\Delta)^s u = c u^{\frac{n+2s}{n-2s}} \quad \text{in } \mathbb{R}^n, \quad u > 0. \tag{6.1.5}$$

The underlying idea is to pass to the extension, looking for a solution of a (possibly degenerate) elliptic equation with a nonlinear boundary reaction term, which can be handled through a variational argument where the main difficulty is the lack of compactness. As in the usual Yamabe problem, the proof is divided into several cases; some of them still remain open.

From the geometric point of view, the fractional Yamabe problem is a generalization of Escobar's classical problem [32] on the construction of a constant mean curvature metric on the boundary of a given manifold, and in the particular case $s = 1/2$ it reduces to it modulo some lower order error terms.

We turn to examples in Sections 6.4 and 6.5. As the standard fractional Laplacian $(-\Delta_{\mathbb{R}^n})^s$, that can be characterized in terms of a Fourier symbol or, equivalently, as a singular integral with a convolution kernel, the conformal fractional Laplacian on the sphere $\mathbb{S}^n$ and on the cylinder $\mathbb{R} \times \mathbb{S}^{n-1}$ may be defined in both ways.

Thus we first review the classical construction for the conformal fractional Laplacian on the sphere coming from representation theory, which yields its Fourier symbol, and then prove some new results on the characterization of this operator using only stereographic projection from $\mathbb{R}^n$. We show, in particular, a singular integral formulation for $P_s^{\mathbb{S}^n}$ that resembles the classical formula for the standard fractional Laplacian.

Next we follow a parallel construction for the cylinder, recalling the results of [27, 28]. More precisely, we give the explicit formula for the conformal fractional Laplacian on the cylinder in terms of its Fourier symbol, and then, a singular integral formula for a convolution kernel.

This second example is interesting because it is the natural geometric characterization of an isolated singularity for the fractional Laplacian

$$\begin{cases} (-\Delta)^s u = c\, u^{\frac{n+2s}{n-2s}} \text{ in } \mathbb{R}^n \backslash \{0\}, \quad u > 0, \\ u(x) \to \infty \text{ as } |x| \to 0. \end{cases} \tag{6.1.6}$$

Radially symmetric solutions for (6.1.6) have been constructed in [28] and are known as Delaunay solutions for the fractional curvature, since they generalize the classical construction of radially symmetric constant mean curvature surfaces [29, 30], or radially symmetric constant scalar curvature surfaces [64, 90].

In addition, the cylinder is the simplest example of a non-compact manifold where the conformal fractional Laplacian may be constructed. However, being a non-local operator, it may not be well defined in the presence of general singularities. In Section 6.6 we give the latest development on this issue. This raises challenging questions in the area of nonlocal PDE and removability of singularities, with implications both in harmonic analysis and pseudo-differential operators.

Then, in Section 6.7, we consider the issue of uniqueness. Since the conformal fractional Laplacian is defined on the boundary of a Poincaré-Einstein manifold, it will depend on this filling. We will review here the well known construction of two different Poincaré-Einstein fillings for the same boundary manifold [56]. Unfortunately, we have not been able to find an explicit expression for the corresponding operators P_s^i, $i = 1, 2$.

Our last Section 6.8 is of independent interest. It is motivated by the following question: given a smooth domain Ω in $\mathbb{R}^{n+1}$, is there a canonical way to define the conformal fractional Laplacian on $M = \partial\Omega$ using only the information on the Euclidean metric in Ω? More generally, what are the (extrinsic) conformal invariants for a hypersurface M^n of X^{n+1}? Some invariants have been very recently constructed in [49, 44, 45]; these resemble the conformal non-local quantities we have defined on the boundary of a Poincaré-Einstein manifold but the new approach is much more general and applies to any embedded hypersurface.

In particular, when M is a surface in Euclidean 3-space, one recovers the Willmore invariant with this construction (the interested reader may look at the survey [76] for the latest development on the Willmore conjecture), so an interesting consequence of this approach is that one produces new (extrinsic) conformal invariants for hypersurfaces in higher dimensions that generalize the Willmore invariant of a two-dimensional surface.

We conclude this introduction with some remarks on further generalizations to other geometries. First, note that the formulas for the conformal fractional Laplacian in the sphere case as the boundary of the Poincaré ball have been long known in the representation theory community. They arise from the theory of joint eigenspaces in symmetric spaces, since the Poincaré model for hyperbolic space is the simplest example of a non-compact symmetric space of rank one. But there are other examples of rank one symmetric spaces: the complex hyperbolic space, that yields the CR fractional Laplacian on the Heisenberg group or the quaternionic hyperbolic space [10, 37].

On the contrary, the picture is more complex in the higher rank case, and it is related to the theory of quantum N-body scattering (see [84] and related references). In the paper [42] we aim to provide an analytical formulation for this problem without the represention theory machinery, when possible. The idea is to construct conformally covariant non-local operators on the boundary M of a higher rank symmetric space X^{n+k}, which is a submanifold of codimension $k > 1$. Analytically, the difficulties come from considering boundary value problems for systems of (possibly degenerate) linear partial differential equations with regular singularities ([60]).

Many of the ideas above still hold if one switches from Riemannian to Lorentzian geometry. In particular, the conformal fractional Laplacian becomes the conformal wave operator, and one needs to move from elliptic to dispersive machinery. The papers [95, 31] provide a first approach to this setting, but many open questions still remain.

6.2 Scattering Theory and the Conformal Fractional Laplacian

We first provide the general geometric setting for our construction and, in particular, the definition of a Poincaré-Einstein filling.

Let X^{n+1} be (the interior of) a smooth Riemannian manifold of dimension $n+1$ with compact boundary $\partial X = M^n$. A function ρ is a *defining function* of M in X if

$$\rho > 0 \text{ in } X, \quad \rho = 0 \text{ on } \partial X, \quad d\rho \neq 0 \text{ on } \partial X.$$

We say that a metric g^+ is *conformally compact* if the new metric

$$\bar{g} := \rho^2 g^+$$

extends smoothly to $\bar{X}$ for a defining function ρ so that $(\bar{X}, \bar{g})$ is a compact Riemannian manifold. This induces a conformal class of metrics $[h]$ on M for $h = \bar{g}|_{TM}$ as the defining function varies. $(M^n, [h])$ is called the *conformal infinity* and (X^{n+1}, g^+) the ambient manifold or filling.

A metric g^+ is said to be *asymptotically hyperbolic* if it is conformally compact and the sectional curvature approaches to -1 at the conformal infinity, which is equivalent to saying that $|d\rho|_{\bar{g}} \to 1$. A more restrictive condition is to demand that g^+ is conformally compact Einstein (*Poincaré-Einstein*), i.e., it is conformally compact and its Ricci tensor satisfies

$$Ric_{g^+} = -ng^+.$$

Given a representative h of the conformal infinity $(M, [h])$, there is a unique geodesic defining function ρ such that, on a neighborhood $M \times (0, \delta)$ in X, g^+ has the normal form

$$g^+ = \rho^{-2}(d\rho^2 + h_\rho) \tag{6.2.1}$$

where h_ρ is a one parameter family of metrics on M such that $h_\rho|_{\rho=0} = h$ ([46]). In the following, we will always assume that the defining function for the problem is chosen so that the metric g^+ is written in normal form once the representative h of the conformal infinity is fixed. $\bar{g}$ will be always defined with respect to this defining function.

Let (X, g^+) be Poincaré-Einstein manifold with conformal infinity $(M, [h])$. The *conformal fractional Laplacian* P_s^h is a nonlocal operator on M which is constructed as the Dirichlet-to-Neumann operator for a generalized eigenvalue problem on (X, g^+), that we describe next. Classical references are [81, 51, 59, 52], for instance.

The spectrum of the Laplacian $-\Delta_{g^+}$ of an asymptotically hyperbolic manifold is well known ([81, 80]). More precisely, it consists of the interval $[n^2/4, \infty)$ and a finite set of L^2-eigenvalues contained in $(0, n^2/4)$. Traditionally one writes the spectral parameter as $\sigma(n-\sigma)$; in the rest of the paper we will always assume that this value it is not an L^2-eigenvalue. Then, for $\sigma \in \mathbb{C}$ with $\mathrm{Re}(\sigma) > n/2$ and such that $\sigma \notin n/2 + \mathbb{N}$, for each Dirichlet-type data $u \in \mathcal{C}^\infty(M)$, the generalized eigenvalue problem

$$-\Delta_{g^+} w - \sigma(n-\sigma)w = 0 \quad \text{in } X \tag{6.2.2}$$

has a solution of the form

$$w = \mathcal{U}\rho^{n-\sigma} + \tilde{\mathcal{U}}\rho^{\sigma}, \quad \mathcal{U}, \tilde{\mathcal{U}} \in \mathcal{C}^\infty(\overline{X}), \quad \mathcal{U}|_{\rho=0} = u. \tag{6.2.3}$$

Fixed $s \in (0, n/2)$, $s \notin \mathbb{N}$, and $\sigma = n/2 + s$ as above, the *conformal fractional Laplacian* on M with respect to the metric h is defined as the normalized scattering operator

$$P_s^h u = d_s \tilde{\mathcal{U}}|_{\rho=0}, \tag{6.2.4}$$

for the constant

$$d_s = 2^{2s}\frac{\Gamma(s)}{\Gamma(-s)}, \tag{6.2.5}$$

where Γ is the ordinary Gamma function.

Remark here that the operator P_s^h is non-local, since it depends on the extension metric g^+ even if we do not indicate it explicitly. For the rest of this paper we will always assume that a background metric g^+ has been fixed.

The main properties of the conformal fractional Laplacian are summarized in the following:

i. P_s^h is a self-adjoint pseudo-differential operator on M with principal symbol the same as $(-\Delta_h)^s$, i.e.,

$$P_s^h \in (-\Delta_h)^s + \Psi_{s-1},$$

where Ψ_l is the set of pseudo-differential operators of loss l.

ii. In the case that $M = \mathbb{R}^n$ with the Euclidean metric $|dx|^2$ and its canonical extension to $\mathbb{R}^{n+1}_+$, all the curvature terms vanish and

$$P_s^{\mathbb{R}^n} = (-\Delta_{\mathbb{R}^n})^s,$$

i.e., we recover the classical fractional Laplacian.

iii. P_s^h is a conformally covariant operator, in the sense that under the conformal change of metric

$$h_u = u^{\frac{4}{n-2s}} h, \quad u > 0,$$

it satisfies the transformation law

$$P_s^{h_u}(\cdot) = u^{-\frac{n+2s}{n-2s}} P_s^h(u\,\cdot). \tag{6.2.6}$$

The *fractional order curvature* of the metric h on M associated to the conformal fractional Laplacian P_s^h is defined as

$$Q_s^h = P_s^h(1),$$

although note that other authors use a different normalization constant. From the above relation (6.2.6) we obtain the curvature equation

$$P_s^h(u) = Q_s^{h_u}\, u^{\frac{n+2s}{n-2s}} \quad \text{in } M^n, \tag{6.2.7}$$

which is a non-local semilinear equation with critical power nonlinearity generalizing (6.1.5) to the curved case.

One of the main observations is that the P_s^h constitute a one-parameter meromorphic family of conformally covariant operators on M, for $s \in (0, n/2)$, $s \notin \mathbb{N}$. At the integer powers, the conformal s-Laplacian can be constructed by a residue formula thanks to the normalization constant (6.2.5) (see [51]). In addition, when s is a positive integer, P_s^h is a local operator that coincides with the classical GJMS operator from [50, 35]. In particular:

– For $s = 1$, P_1 is precisely the conformal Laplacian defined in (6.1.2), i.e.,

$$P_1^h = L_h = -\Delta_h + \tfrac{n-2}{4(n-1)} R_h, \tag{6.2.8}$$

and the associated curvature is a multiple of the scalar curvature

$$Q_1^h = \tfrac{n-2}{4(n-1)} R_h.$$

– For $s = 2$, the conformal fractional Laplacian coincides with the well known Paneitz operator ([87])

$$P_2^h = (-\Delta_h)^2 + \delta(a_n R_h + b_n Ric_h)d + \tfrac{n-4}{2} Q_2^h,$$

and Q_2 is (up to multiplicative constant) the so-called fourth order Q-curvature.

For any other powers $s \notin \mathbb{N}$, P_s^h is a non-local operator on M and reflects the geometry of the filling (X, g^+). Some explicit examples will be considered in Sections 6.4 and 6.5.

We also remark that the same construction is true for a general asymptotically hyperbolic manifold, except for values $s \in \mathbb{N}/2$, unless the expansion of the term h_ρ in the normal form (6.2.1) is even up to a suitable order [51, 52]. In this exposition we will explain in detail only the case $s \in (0, 1)$, where we will explain the role played by mean curvature (see Theorem 6.3.2 below).

6.3 The Extension and the *s*-Yamabe Problem

It was observed in [22] (see also [20] for the most recent development) that the generalized eigenvalue problem (6.2.2)-(6.2.3) on (X, g^+) is equivalent to a linear degenerate elliptic problem on the compactified manifold $(\overline{X}, \overline{g})$. Hence they reconciled the definition of the conformal fractional Laplacian P_s^h given in the previous section as the normalized scattering operator and the one given in the spirit of the Dirichlet-to-Neumann operators by Caffarelli and Silvestre in [17].

In this section we will assume that $s \in (0, 1)$. For higher powers $s > 1$ we refer to [20], [97] and [26]. As in the introduction chapter, we set $a = 1 - 2s$. $W^{1,2}(X, \rho^a)$ will denote the weighted Sobolev space $W^{1,2}$ on X with weight ρ^a.

Theorem 6.3.1 ([22]). *Let (X, g^+) be a Poincaré-Einstein manifold with conformal infinity $(M, [h])$. Then, given $u \in \mathcal{C}^\infty(M)$, the generalized eigenvalue problem* (6.2.2)-(6.2.3) *is equivalent to the degenerate elliptic equation*

$$\begin{cases} -\operatorname{div}(\rho^a \nabla U) + E(\rho)U = 0 & \text{in } (X, \overline{g}), \\ U|_{\rho=0} = u & \text{on } M, \end{cases} \tag{6.3.1}$$

where the derivatives are taken with respect to the original metric $\overline{g}$*, and* $U = \rho^{n/2-s}w$*. The zero-th order term is*

$$E(\rho) = \rho^{-1-\sigma}\left(-\Delta_{g^+} - \sigma(n-\sigma)\right)\rho^{n-\sigma}.$$

Notice that, in a neighborhood $M \times (0, \delta)$ *where the metric* g^+ *is in normal form (6.2.1), this expression simplifies to*

$$E(\rho) = \tfrac{n-1+a}{4n} R_{\overline{g}}\rho^a. \tag{6.3.2}$$

Such U *is the unique minimizer of the energy*

$$F[V] = \int_X \rho^a |\nabla V|^2_{\overline{g}}\, dv_{\overline{g}} + \int_X E(\rho)|V|^2\, dv_{\overline{g}}$$

among all the functions $V \in W^{1,2}(X, \rho^a)$ *with fixed trace* $V|_{\rho=0} = u$*. Moreover, we recover the conformal fractional Laplacian on* M *as*

$$P_s^h u = -d_s^* \lim_{\rho\to 0} \rho^a \partial_\rho U,$$

where

$$d_s^* = -\frac{2^{2s-1}\Gamma(s)}{s\Gamma(-s)}.$$

Before we continue with the exposition, let us illustrate these concepts with the simplest example of a Poincaré-Einstein manifold: the hyperbolic space $\mathbb{H}^{n+1}$. It can be characterized as the upper half-space $\mathbb{R}^{n+1}_+$ (with coordinates $x \in \mathbb{R}^n$, $y \in \mathbb{R}_+$), endowed with the metric

$$g^+ = \frac{dy^2 + |dx|^2}{y^2}.$$

In this case, y is a defining function and the conformal infinity $\{y = 0\}$ is just the Euclidean space $\mathbb{R}^n$ with its flat metric $|dx|^2$. Then problem (6.3.1) with Dirichlet condition u reduces to

$$\begin{cases} -\operatorname{div}(y^a \nabla U) = 0 & \text{in } \mathbb{R}^{n+1}_+, \\ U|_{y=0} = u & \text{on } \mathbb{R}^n, \end{cases} \tag{6.3.3}$$

and the fractional Laplacian at the boundary $\mathbb{R}^n$ is just

$$P_s^{\mathbb{R}^n} u = (-\Delta_{\mathbb{R}^n})^s u = -d_s^* \lim_{y\to 0}\left(y^a \partial_y U\right),$$

which is precisely the usual construction for the fractional Laplacian as a Dirichlet-to-Neumann operator from [17]. We note that it is possible to write $U = P *_x u$, where P is the Poisson kernel for this extension problem as given in section 0.2.1 of chapter 0.

If the background manifold (X, g^+) is not Poincaré-Einstein, but only asymptotically hyperbolic, we have a similar extension problem but here the mean curvature of the boundary M respect to the metric $\overline{g}$ in $\overline{X}$, denoted by H, plays an essential role:

Theorem 6.3.2 ([22]). *Let $(X^{n+1}, \bar{g}^+)$ be an asymptotically hyperbolic manifold with a geodesic defining function ρ and conformal infinity $(M^n, [h])$, with the metric written in normal form (6.2.1). Then the conformal fractional Laplacian can be constructed through the following extension problem: for each given smooth function u on M, consider*

$$\begin{cases} -\operatorname{div}(\rho^a \nabla U) + E(\rho)U = 0 & \text{in } (\overline{X}, \bar{g}), \\ U = u & \text{on } M, \end{cases}$$

where

$$E(\rho) = \tfrac{n-1-a}{4n}\left[R_{\bar{g}} - \{n(n+1) + R_{g^+}\}\rho^{-2}\right]\rho^a. \tag{6.3.4}$$

Then there exists a unique solution U and moreover,

1. *For $s \in (0, 1/2)$,*

$$P_s^h u = -d_s^* \lim_{\rho \to 0} \rho^a \partial_\rho U, \tag{6.3.5}$$

2. *For $s = 1/2$, we have an extra term*

$$P_{\frac{1}{2}}^h u = \lim_{\rho \to 0} \partial_\rho U + \tfrac{n-1}{2} H u.$$

3. *If $s \in (1/2, 1)$, the limit in the right hand side of (6.3.5) exists if and only if the the mean curvature H vanishes identically, in which case, (6.3.5) holds too.*

Remark 6.1. *Note that, in the particular case that $s = 1/2$, the fractional curvature $Q_{1/2}$ reduces to the mean curvature of M (up to multiplicative constant).*

This dichotomy in Theorem 6.3.2 due to the presence of mean curvature also appears in many other non-local problems such as [18, 39, 89]. The underlying idea is that, for $s \in (0, 1/2)$, non-local curvature is the essential term, but for $s \in (1/2, 1)$, mean curvature takes over.

Our next objective is, in the Poincaré-Einstein case, to compare the geometric extension (6.3.1) to the Euclidean one (6.3.3). It was observed in [22] that it is possible to find a special defining function ρ^* such that when we rewrite the scattering equation (6.3.1) for the new metric $\bar{g}^* = (\rho^*)^2 g^+$, the lower order term $E(\rho^*)$ vanishes; thus making the extension as close as possible to the Euclidean one. This construction was inspired in the special defining function of [67] and is also an essential ingredient in the formulation of the scattering problem in the metric measure space setting of [20], which is a very interesting development for which we do not have space here. It is also crucially used in the construction of a Hamiltonian quantity for a non-local ODE [**?**].

As it was pointed out in [20], it is necessary to assume, here and in the rest of this exposition, that the first eigenvalue for $-\Delta_{g^+}$ satisfies $\lambda_1(-\Delta_{g^+}) > \frac{n^2}{4} - s^2$. We arrive at the following improvement of Theorem 6.3.1:

Proposition 6.3.3 ([22, 20]). *Let w^* be the solution to (6.2.2)-(6.2.3) with Dirichlet data $u \equiv 1$, and set $\rho^* = (w^*)^{\frac{1}{n/2-s}}$. The function ρ^* is a defining function of M in X such that $E(\rho^*) \equiv 0$. Moreover, it has the asymptotic expansion near the conformal infinity*

$$\rho^*(\rho) = \rho \left[1 + \frac{Q_s^h}{(n/2-s)d_s}\rho^{2s} + O(\rho^2) \right].$$

By construction, if U^* is the solution to

$$\begin{cases} -\operatorname{div}\left((\rho^*)^a \nabla U^*\right) = 0 & \text{in } (X, \bar{g}^*), \\ U^* = u & \text{on } M, \end{cases}$$

with respect to the metric $\bar{g}^* = (\rho^*)^2 g^+$, then

$$P_s^h u = -d_s^* \lim_{\rho^* \to 0} (\rho^*)^a \, \partial_{\rho^*} U^* + u\, Q_s^h.$$

Next, we give an interpretation of the fractional curvature as a variation of weighted volume, in analogy to the usual mean curvature situation. The notion of renormalized volume was first investigated by the physicists in relation to the AdS/CFT correspondence, and was considered by [36, 24]. Given (X^{n+1}, g^+) a Poincaré-Einstein manifold with boundary M^n and defining function ρ, one may compute the asymptotic expansion of the volume of the region $\{\rho > \varepsilon\}$; the renormalized volume is defined as one very specific term in this asymptotic expansion. When the dimension n is odd, the renormalized volume is a conformal invariant of the conformally compact structure, and it can be calculated as the conformal primitive of the Q-curvature coming from the scattering operator (this is the case $s = n/2$). In that case that n is even, the picture is more complex, and one can show that the renormalized volume is one term of the Chern-Gauss-Bonnet formula in higher dimensions.

When $s \in (0, 1)$ one can also give a weighted version for volume (see [38]), and to obtain the fractional curvature Q_s as its first variation. More precisely, for each $\varepsilon > 0$ we set

$$vol_{g^+,s}(\{\rho > \varepsilon\}) := \int_{\{\rho > \varepsilon\}} (\rho^*)^{\frac{n}{2}-s} \, dv_{g^+}, \tag{6.3.6}$$

where ρ^* is the special defining function from Proposition 6.3.3.

Proposition 6.3.4 ([38]). *Let (X, g^+) be a Poincaré-Einstein manifold with conformal infinity $(M, [h])$. The weighted volume (6.3.6) has an asymptotic expansion in ε when $\varepsilon \to 0$ given by*

$$vol_{g^+,s}(\{\rho > \varepsilon\}) = \varepsilon^{-\frac{n}{2}-s} \left[\left(\tfrac{n}{2}+s\right)^{-1} vol(M) + \varepsilon^{2s} V_s^h + \textit{higher order terms} \right]$$

where

$$V_s^h := \frac{1}{d_s} \frac{1}{n/2-s} \int_M Q_s^h \, dv_h.$$

Finally, and as an application of the extension Theorems 6.3.1 and 6.3.2, we give a summary of the recent development on the fractional Yamabe problem.

The resolution of the classical Yamabe problem by Aubin, Schoen, Trudinger, has been one of the most significant advances in geometric analysis (see [68, 92], and the references therein). Given a smooth background metric, the problem is to find a conformal one that has constant scalar curvature. In PDE language, this is (6.1.4).

One may pose then the analogous question of finding a constant Q_s-curvature in the same conformal class as a given one. This study was initiated by the author and Qing in [41], and it amounts to, given a background metric (M^n, h), solve the following non-local semilinear geometric equation with critical exponent (recall (6.2.7)),

$$P_s^h(u) = cu^{\frac{n+2s}{n-2s}}, \quad u > 0, \tag{6.3.7}$$

for some constant c on M.

Theorem 6.3.1 allows to write (6.3.7) as a local elliptic equation in the extension with a non-linear boundary reaction term:

$$\begin{cases} -\operatorname{div}(\rho^a U) + E(\rho)U = 0 & \text{in } (X^{n+1}, \bar{g}), \\ -d_s^* \lim_{\rho\to 0} \rho^a \partial_\rho U = cu^{\frac{n+1}{n-1}} & \text{on } M^n, \quad u > 0, \end{cases} \tag{6.3.8}$$

where we have written $u = U(\cdot, 0)$.

Even though (6.3.7) is a non-local equation, the resolution to the fractional Yamabe problem follows the same scheme as in the original Yamabe problem for the scalar curvature, using a variational method. In addition, the $s = 1/2$ case is deeply related to the prescribing constant mean curvature problem (also known as the boundary Yamabe problem) considered by Escobar [32], Brendle-Chen [13], Li-Zhu [69], Marques [74, 75], Almaraz [3], Mayer-Ndiaye [77] and others, and which corresponds to the following Dirichlet-to-Neumann operator

$$\begin{cases} -\Delta_{\bar{g}} u + \frac{n-1}{4n} R_{\bar{g}} u = 0 & \text{in } (X^{n+1}, \bar{g}), \\ \partial_\nu u + \frac{n-1}{2} Hu = cu^{\frac{n+1}{n-1}} & \text{on } M^n. \end{cases} \tag{6.3.9}$$

This connection will be made precise right below (6.3.14). However, there is a subtle issue: in the proof one will need to find particular background metric $(X, \bar{g})$ with very precise asymptotic behavior near a point $p \in M$ in a good coordinate system. However, in contrast to the study of (6.3.9), where they are free to choose conformal Fermi coordinates on the filling $(\bar{X}, \bar{g})$, our freedom of choice of metrics for (6.3.8) is restricted to the boundary. Once a metric $h_1 \in [h]$ is chosen, the corresponding defining function ρ_1 is determined and the extension metric $\bar{g}_1$, written in normal form (6.2.1) for $\bar{g}_1 = (\rho_1)^2 g^+$, is unique and cannot be simplified.

Let us set up the notation for (6.3.7). We consider a scale-free functional on metrics in the class $[h]$ on M given by

$$I_s[h] = \frac{\int_M Q_s^h \, dv_h}{\left(\int_M dv_h\right)^{\frac{n-2s}{n}}}.$$

Or, if we set a base metric h and write a conformal metric $h_u = u^{\frac{4}{n-2s}} h$, then

$$I_s[u,h] = \frac{\int_M u P_s^h(u)\, dv_h}{\left(\int_M u^{2^*}\, dv_h\right)^{\frac{2}{2^*}}},$$

where

$$2^* = \frac{2n}{n-2s}.$$

We will call I_s the s-Yamabe functional.

Our objective is to find a metric in the conformal class $[h]$ that minimizes the functional I_s. It is clear that a metric h_u, where u is a minimizer of $I_s[u,h]$, is an admissible solution for (6.3.7) (positivity will be guaranteed by an application of a suitable maximum principle). This suggests that we define the *s-Yamabe constant*

$$\Lambda_s(M,[h]) = \inf\{I_s[h_1] : h_1 \in [h]\}.$$

It is then apparent that $\Lambda_s(M,[h])$ is an invariant on the conformal class $[h]$ when g^+ is fixed. In addition, it is proved in [41] that the sign of $\Lambda_s(M,[h])$ governs the sign of the possible constants c in (6.3.7), and the sign of the first eigenvalue for P_s^h.

In the mean time, based on Theorem 6.3.1, we set

$$\bar{I}_s[U,\bar{g}] = \frac{\int_X \rho^a |\nabla U|_{\bar{g}}^2\, dv_{\bar{g}} + \int_X E(\rho)\, U^2\, dv_{\bar{g}}}{\left(\int_M |U|^{2^*}\, dv_h\right)^{\frac{2}{2^*}}}. \tag{6.3.10}$$

Note then

$$\Lambda_s(X,[h]) = \inf\left\{\bar{I}_s[U,\bar{g}] : U \in W^{1,2}(X,\rho^a)\right\}. \tag{6.3.11}$$

As a consequence, fixing the integral $\int_M u^{2^*}\, dv_h = 1$, if U is a minimizer of the functional $\bar{I}_s[\,\cdot\,,\bar{g}]$, then its trace $u = U(\cdot,0)$ is a solution for (6.3.7).

This minimization procedure is related to the trace Sobolev embedding

$$W^{1,2}(X,\rho^a) \hookrightarrow H^s(M) \hookrightarrow L^{2^*}(M),$$

which is continuous, but not compact. Hence the difficulty comes from this lack of compactness, which is well understood in the Euclidean case below:

Theorem 6.3.5 ([70]). *There exists a positive constant $C_{n,s}$ such that for every function U in $W^{1,2}(\mathbb{R}_+^{n+1}, y^a)$ we have that*

$$\|u\|_{L^{2^*}(\mathbb{R}^n)}^2 \leqslant C_{n,s} \int_{\mathbb{R}_+^{n+1}} y^a |\nabla U|^2\, dx\, dy,$$

where u is the trace $u := U(\cdot,0)$. Moreover equality holds if and only if u is a "bubble", i.e.,

$$u(x) = C\left(\frac{\mu}{|x-x_0|^2+\mu^2}\right)^{\frac{n-2s}{2}}, \quad x \in \mathbb{R}^n, \tag{6.3.12}$$

for $C \in \mathbb{R}$, $\mu > 0$ and $x_0 \in \mathbb{R}^n$ fixed, and $U = P *_x u$ its Poisson extension.

We also remark that all entire positive solutions to

$$(-\Delta)^s u = u^{\frac{n+2s}{n-2s}}, \quad u > 0,$$

have been completely classified (see [58], for instance, for an account of references). In particular, they must be the standard "bubbles" (6.3.12). Other non-linearities for fractional Laplacian equations have been considered, for instance in [15, 14, 94, 8], although by no means this list is exhaustive.

Going back to the minimization problem (6.3.11), we observe that the variational method that was used in the resolution of the classical Yamabe problem can still be applied, but the difficulty comes from the specific structure of metric in the filling $(X, \bar{g})$. In any case, one starts by comparing the Yamabe constant on M to the Yamabe constant on the sphere.

Using stereographic projection, from Theorem 6.3.5 it is easily seen that

$$\Lambda_s(\mathbb{S}^n, [h_{\mathbb{S}^n}]) = \frac{1}{C_{n,s}},$$

where $[h_{\mathbb{S}^n}]$ is the canonical conformal class of metrics on the sphere $\mathbb{S}^n$ understood as the conformal infinity of the Poincaré ball.

Suppose that (X^{n+1}, g^+) is an asymptotically hyperbolic manifold with a geodesic defining function ρ and set $\bar{g} = \rho^2 g^+$. Let $(M^n, [h])$ be its conformal infinity. One can show that ([41, 19]) the fractional Yamabe constant satisfies

$$-\infty < \Lambda_s(M, [h]) \leqslant \Lambda_s(\mathbb{S}^n, [h_{\mathbb{S}^n}]).$$

Theorem 6.3.6 ([41]). *In the setting above, if*

$$\Lambda_s(M, [h]) < \Lambda_s(\mathbb{S}^n, [h_{\mathbb{S}^n}]), \tag{6.3.13}$$

then the s-Yamabe problem is solvable for $s \in (0, 1)$.

Therefore, it suffices to find a suitable test function in the functional (6.3.10) that attains this strict inequality. As we have mentioned, one needs to construct suitable conformal normal coordinates on M by conformal change, and then deal with the corresponding extension metric. Hence one needs to make some assumptions on the behavior of the asymptotically hyperbolic manifold g^+. The underlying idea here is to have g^+ as close as possible as a Poincaré-Einstein manifold. The first one of these assumptions is

$$R_{g^+} + n(n+1) = o(\rho^2) \quad \text{as} \quad \rho \to 0,$$

which looks very reasonable in the light of (6.3.4). In particular, under this condition one has that

$$E(\rho) = \tfrac{n-1+a}{4n} R_{\bar{g}} \rho^a + o(\rho^2) \quad \text{as} \quad \rho \to 0. \tag{6.3.14}$$

(compare to (6.3.2)). Another consequence of this expression is that the $1/2$-Yamabe problem coincides to the prescribing constant mean curvature problem (6.3.9), up to a small error. In general one needs a higher order of vanishing for g^+ (see [61] for the precise statements), which is automatically true if g^+ is Poincaré-Einstein and not just asymptotically hyperbolic. This also shows that the natural geometric setting for an asymptotically hyperbolic g^+ is to demand that g^+ has constant scalar curvature $R_{g^+} = -n(n+1)$.

The first attempt to prove (6.3.13) was [41] in the non-umbilic case, where the authors use a bubble as a test function. The umbilic, non-locally conformally flat case in high dimensions was considered in [43]. Finally, Kim, Musso and Wei [61] have provided an unified development, covering all the cases that do not need a positive mass theorem for the conformal fractional Laplacian. Their test function is not a "bubble" but instead it has a more complicated geometry. Summarizing, some hypothesis under which the fractional Yamabe problem for $s \in (0,1)$ is solvable (in addition to those on g^+ above) are:

- $n \geqslant 2$, $s \in (0, 1/2)$, M has a point of negative mean curvature.
- $n \geqslant 4$, $s \in (0, 1)$, M is not umbilic.
- $n > 4 + 2s$, M is umbilic but not locally conformally flat.
- M is locally conformally flat or $n = 2$, and the fractional positive mass theorem holds.

However, we see from this last point that to cover all the cases with this method one still needs to develop a positive mass theorem for the Green's function of the conformal fractional Laplacian, which is at this time a puzzling open question. From another point of view, we mention the work [78], where they use the the barycenter technique of Bahri-Coron to bypass the positive mass issue for the locally flat and umbilic conformal infinity.

Finally, one may look at the lack of compactness phenomenon. In general, Palais-Smale sequences can be decomposed into the solution of the limit equation plus a finite number of bubbles. Moreover, the multi-bubbles are non-interfering even though the operator is non-local (see, for instance, [34, 86, 62, 63]).

6.4 The Conformal Fractional Laplacian on the Sphere

In this section we look at the sphere $\mathbb{S}^n$ with the round metric $h_{\mathbb{S}^n}$, understood as the conformal infinity of the Poincaré ball model for hyperbolic space $\mathbb{H}^{n+1}$. Note that hyperbolic space is the simplest example of a Poincaré-Einstein manifold, and the model for the general development.

On $\mathbb{S}^n$ one explicitly knows ([11], see also the lecture notes [12], for instance) that the conformal fractional Laplacian (or intertwining operator) has the explicit expression

$$P_s^{\mathbb{S}^n} = \frac{\Gamma\left(A_{1/2}+s+\frac{1}{2}\right)}{\Gamma\left(A_{1/2}-s+\frac{1}{2}\right)}, \quad A_{1/2}=\sqrt{-\Delta_{\mathbb{S}^n}+\left(\frac{n-1}{2}\right)^2}, \tag{6.4.1}$$

for all $s \in (0, n/2)$. From here one easily calculates that the fractional curvature of the sphere is a positive constant

$$Q_s^{\mathbb{S}^n} = P_s^{\mathbb{S}^n}(1) = \frac{\Gamma\left(\frac{n}{2}+s\right)}{\Gamma\left(\frac{n}{2}-s\right)}. \tag{6.4.2}$$

Formula (6.4.1) may be easily derived from the scattering problem (6.2.2)-(6.2.3). A proof can be found in the book [9], which also makes the link to the representation theory community. Note, however, a different factor of 2, which is always an issue when passing from representation theory to geometry. For convenience of the reader not familiar with this subject we provide a direct proof below.

Consider the Poincaré metric for hyperbolic space $\mathbb{H}^{n+1}$, written in normal form (6.2.1) as

$$g^+ = \rho^{-2}\left(d\rho^2 + \left(1-\tfrac{\rho^2}{4}\right)^2 h_{\mathbb{S}^n}\right),$$

for $\rho \in (0, 2]$. Remark that $\rho = 2$ corresponds to the origin of the Poincaré ball and thus the apparent singularity is just a consequence of the expression for the metric in polar-like coordinates.

Calculating the Laplace-Beltrami operator with respect to g^+ we obtain, recalling that $\sigma = \frac{n}{2} + s$, that the eigenvalue equation (6.2.2) is equivalent to the following:

$$\rho^{n+1}\left(1-\tfrac{\rho^2}{4}\right)^{-n}\partial_\rho\left[\rho^{-n+1}\left(1-\tfrac{\rho^2}{4}\right)^{n}\partial_\rho w\right]+\rho^2\left(1-\tfrac{\rho^2}{4}\right)^{-2}\Delta_{\mathbb{S}^n}w+\left(\tfrac{n^2}{4}-s^2\right)w=0. \tag{6.4.3}$$

We will show that the operator $P_s^{\mathbb{S}^n}$ diagonalizes in the spherical harmonic decomposition for $\mathbb{S}^n$. With some abuse of notation, let $\mu_m = m(m+n-1)$, $m = 0, 1, 2, \ldots$ be the eigenvalues of $-\Delta_{\mathbb{S}^n}$, repeated according to multiplicity, and $\{E_m\}$ be the corresponding basis of eigenfunctions. The projection of (6.4.3) onto each eigenspace $\langle E_m \rangle$ yields

$$\rho^{n+1}\left(1-\tfrac{\rho^2}{4}\right)^{-n}\partial_\rho\left[\rho^{-n+1}\left(1-\tfrac{\rho^2}{4}\right)^{n}\partial_\rho w_m\right]-\rho^2\left(1-\tfrac{\rho^2}{4}\right)^{-2}\mu_m w_m + \left(\tfrac{n^2}{4}-s^2\right)w_m=0.$$

This is a hypergeometric ODE with general solution

$$w_m(\rho) = c_1\rho^{\frac{n}{2}-s}\varphi_1(\rho) + c_2\rho^{\frac{n}{2}+s}\varphi_2(\rho), \quad c_1, c_2 \in \mathbb{R}, \tag{6.4.4}$$

for

$$\varphi_1(\rho) := (\rho^2-4)^{\frac{-n-\beta+1}{2}}\,{}_2F_1\left(\tfrac{-\beta+1}{2}, \tfrac{-\beta+1}{2}-s, 1-s, \tfrac{\rho^2}{4}\right),$$

$$\varphi_2(\rho) := (\rho^2-4)^{\frac{-n-\beta+1}{2}}\,{}_2F_1\left(\tfrac{-\beta+1}{2}, \tfrac{-\beta+1}{2}+s, 1+s, \tfrac{\rho^2}{4}\right),$$

where we have defined

$$\beta := \sqrt{(n-1)^2 + 4\mu_m}$$

and ${}_2F_1$ is the usual Hypergeometric function.

In order to calculate the conformal fractional Laplacian, first one needs to obtain an asymptotic expansion of the form (6.2.3) for $\mathcal{U}$, $\tilde{\mathcal{U}}$ smooth up to $\overline{X}$. Since w must be smooth at the central point $\rho = 2$, one should choose the constants c_1, c_2 such that in (6.4.4) the singularities of φ_1 and φ_2 at $\rho = 2$ cancel out. This is,

$$c_1 2^{\frac{n}{2}-s}\,{}_2F_1\left(\tfrac{-\beta+1}{2}, \tfrac{-\beta+1}{2} - s, 1-s, 1\right) + c_2 2^{\frac{n}{2}+s}\,{}_2F_1\left(\tfrac{-\beta+1}{2}, \tfrac{-\beta+1}{2} + s, 1+s, 1\right) = 0. \tag{6.4.5}$$

In order to simplify this expression, recall the following property of the Hypergeometric function from [1]: if $a + b < c$, then

$${}_2F_1(a, b, c, 1) = \frac{\Gamma(c)\Gamma(c-a-b)}{\Gamma(c-a)\Gamma(c-b)}.$$

After some calculation, (6.4.5) yields

$$\frac{c_2}{c_1} = 2^{-2s}\frac{\Gamma(\frac{1}{2}+s+\frac{\beta}{2})\Gamma(-s)}{\Gamma(\frac{1}{2}-s+\frac{\beta}{2})\Gamma(s)}. \tag{6.4.6}$$

Next, looking at the definition of the conformal fractional Laplacian from (6.2.4), and noting that both φ_1, φ_2 are smooth at $\rho = 0$, we conclude from (6.4.6) that

$$P_s^{\mathbb{S}^n}|_{\langle E_m\rangle} u_m = d_s \frac{c_2}{c_1} u_m = \frac{\Gamma(\frac{1}{2}+s+\frac{\beta}{2})}{\Gamma(\frac{1}{2}-s+\frac{\beta}{2})} u_m.$$

This concludes the proof of (6.4.1) when $s \in (0, n/2)$ is not an integer.

For integer powers $k \in \mathbb{N}$, it can be shown that (6.4.1) also yields the factorization formula for the GJMS operators on the sphere

$$P_k^{\mathbb{S}^n} = \prod_{j=1}^{k} \left\{-\Delta_{\mathbb{S}^n} + \left(\tfrac{n}{2}+j-1\right)\left(\tfrac{n}{2}-j\right)\right\}. \tag{6.4.7}$$

The paper [47] by Graham independently derives this expression just by using the formula for the corresponding operator on Euclidean space $\mathbb{R}^n$ and then stereographic projection to translate it back to the sphere $\mathbb{S}^n$.

Here we show that Graham's method using stereographic projection also works for non-integer s, yielding a factorization formula in the spirit of (6.4.7). The advantage of this formulation is that it does not require the extension, but only the conformal property (6.2.6) from Euclidean space to the sphere.

Proposition 6.4.1. *Let $s_0 \in (0,1)$, $k \in \mathbb{N}$. Then*

$$P^{\mathbb{S}^n}_{k+s_0} = \prod_{j=1}^{k} \left(P^{\mathbb{S}^n}_1 + c_j\right) P^{\mathbb{S}^n}_{s_0}.$$

for $c_j = -(s_0 + j - 1)(s_0 + j)$.

Here $P^{\mathbb{S}^n}_1$ is the usual conformal Laplacian (6.2.8) on $\mathbb{S}^n$, i.e.,

$$P^{\mathbb{S}^n}_1 = -\Delta_{\mathbb{S}^n} + \tfrac{n(n-2)}{4}. \tag{6.4.8}$$

Before we give a proof of this result we set up we set up the notation for the stereographic projection from South pole. $\mathbb{S}^n$ is parameterized by coordinates $z' = (z_1, \dots, z_n)$, z_{n+1} such that $|z'|^2 + z^2_{n+1} = 1$, and $\mathbb{R}^n$ by coordinates $x \in \mathbb{R}^n$. Let $\varphi : \mathbb{S}^n \backslash \{S\} \to \mathbb{R}^n$ be the map given by

$$x := \varphi(z', z_{n+1}) = \frac{z'}{1 + z_{n+1}}.$$

The push forward map is just

$$\varphi^* \left(\frac{2}{1 + |x|^2}\right) = 1 + z_{n+1}$$

and, by conformality, it transforms the metric as

$$\varphi^* h_{eq} = (1 + z_{n+1})^{-2} h_{\mathbb{S}^n}, \tag{6.4.9}$$

where $h_{eq} = |dx|^2$ is the Euclidean metric. For simplicity, we denote the conformal factor as

$$B^p f = (1 + z_{n+1})^p f, \quad B_p f = 2^p (1 + |x|^2)^{-p} f,$$

and note that the change of variable between them is simply

$$B^p \varphi^* = \varphi^* B_p,$$

which will be used repeatedly in the following.

Let $-\Delta$ be the standard Laplacian on $\mathbb{R}^n$. It is related to the conformal Laplacian on the sphere (6.4.8) by the transformation law (6.1.3), written as

$$P^{\mathbb{S}^n}_1 B^{1-n/2} \varphi^* = B^{-1-n/2} \varphi^* (-\Delta). \tag{6.4.10}$$

The conformal fractional Laplacian also satisfies the conformal covariance property (6.2.6), which is

$$P^{\mathbb{S}^n}_s B^{s-n/2} \varphi^* = B^{-s-n/2} \varphi^* (-\Delta)^s, \tag{6.4.11}$$

where $(-\Delta)^s$ is the standard fractional Laplacian on $\mathbb{R}^n$ with respect to the Euclidean metric.

We show some preliminary commutator identities on $\mathbb{R}^n$:

Lemma 6.4.2. *Let $X = \sum x_i \partial_{x_i}$. Then*

$$\begin{aligned}
&[-\Delta, X] = 2(-\Delta),\\
&[X, B_p] = -p\,|x|^2\, B_{p+1},\\
&[-\Delta, B_p] = pB_p(2X + n - (p-1)B_1\,|x|^2)B_1,\\
&[(-\Delta)^s, B_{-1}] = -s\,(2X + n + 2(s-1))\,(-\Delta)^{s-1}.
\end{aligned} \tag{6.4.12}$$

Proof. The first three are direct calculations and can be found in [47], while the last one is proved by Fourier transform. Indeed, compute

$$[(-\Delta)^s, B_{-1}]u(x) = \tfrac{1}{2}\left[(-\Delta)^s\{|x|^2 u(x)\} - |x|^2(-\Delta)^s u(x)\right]. \tag{6.4.13}$$

Fourier transform, with the multiplicative constants normalized to one, yields

$$\begin{aligned}
\mathcal{F}\{|x|^2(-\Delta_x)^s u(x)\} &= -\Delta_\xi\{|\xi|^{2s}\hat{u}(\xi)\}\\
&= -2s(n+2(s-1))|\xi|^{2(s-1)}\hat{u}(\xi) - 4s|\xi|^{2(s-1)}\sum_{i=1}^n \xi_i\partial_{\xi_i}\hat{u}(\xi) - |\xi|^{2s}\Delta_\xi\hat{u}(\xi)\\
&= 2s(n+2(s-1))|\xi|^{2(s-1)}\hat{u}(\xi) - 4s\sum_{i=1}^n \partial_{\xi_i}\left\{\xi_i|\xi|^{2(s-1)}\hat{u}(\xi)\right\} - |\xi|^{2s}\Delta_\xi\hat{u}(\xi).
\end{aligned}$$

Taking the inverse Fourier transform we obtain

$$|x|^2(-\Delta_x)^s u(x) = 2s(n+2(s-1))(-\Delta_x)^{s-1}u(x) + 4sXu(x) + (-\Delta_x)^s\{|x|^2u(x)\}$$

which, in view of (6.4.13), immediately yields the fourth identity in (6.4.12). □

Proof of theorem 6.4.1: By induction, it is clear that it is enough to show that

$$P_{1+s}^{\mathbb{S}^n} = \left(P_1^{\mathbb{S}^n} + c_s\right)P_s^{\mathbb{S}^n}, \quad \text{for} \quad c_s = -s(s+1).$$

Let $\mathcal{P} := \left(P_1^{\mathbb{S}^n} + c_s\right)P_s^{\mathbb{S}^n}$. We claim that $\mathcal{P}$ is conformally covariant of order $1+s$ in the sense of (6.2.6), which, by uniqueness, will imply the proof of the theorem. Thus it is enough to show that $\mathcal{P}$ satisfies the conformal covariance identity

$$\mathcal{P}B^{s+1-n/2}\varphi^* = B^{-s-1-n/2}\varphi^*(-\Delta)^{s+1}. \tag{6.4.14}$$

For this, we first expand the left hand side of (6.4.14). The idea is to use both the conformal invariance for the fractional Laplacian of exponent s (6.4.11) and for the standard conformal Laplacian (6.4.10), in order to relate the operator on $\mathbb{S}^n$ to the equivalent one on $\mathbb{R}^n$. We have

$$\begin{aligned}
\text{(LHS)} &= \left(P_1^{\mathbb{S}^n} + c_s\right)P_s^{\mathbb{S}^n}B^{s+1-n/2}\varphi^*\\
&= \left(P_1^{\mathbb{S}^n} + c_s\right)P_s^{\mathbb{S}^n}B^{s-n/2}\varphi^*B_1\\
&= \left(P_1^{\mathbb{S}^n} + c_s\right)B^{-s-n/2}\varphi^*(-\Delta)^sB_1\\
&= \left(P_1^{\mathbb{S}^n} + c_s\right)B^{1-n/2}\varphi^*B_{-1-s}(-\Delta)^sB_1.
\end{aligned} \tag{6.4.15}$$

Recalling again the conformal invariance for the conformal Laplacian $P_1^{\mathbb{S}^n}$ from (6.4.10),

$$\begin{aligned}(\text{LHS}) &= \left[B^{-1-n/2}\varphi^*(-\Delta) + c_s B^{1-n/2}\varphi^*\right] B_{-1-s}(-\Delta)^s B_1 \\ &= \varphi^* B_{-1-n/2}\left[(-\Delta) + c_s B_2\right] B_{-1-s}(-\Delta)^s B_1. \end{aligned} \tag{6.4.16}$$

We next claim that

$$\left[(-\Delta) + c_s B_2\right] B_{-1-s}(-\Delta)^s B_1 = B_{-s}(-\Delta)^{s+1}, \tag{6.4.17}$$

whose proof is presented below. Therefore, when we substitute the previous expression into (6.4.16) we obtain

$$(\text{LHS}) = \varphi^* B_{-1-n/2-s}(-\Delta)^{s+1}, \tag{6.4.18}$$

which indeed implies (6.4.14) as we wished.

Now we give a proof for (6.4.17). First note that

$$\begin{aligned}(-\Delta)B_{-1-s}(-\Delta)^s B_1 &= \{(-\Delta)B_{-s}\} B_{-1}(-\Delta)^s B_1 \\ &= B_{-s}(-\Delta)B_{-1}(-\Delta)^s B_1 + [(-\Delta), B_{-s}]B_{-1}(-\Delta)^s B_1\end{aligned}$$

and

$$\begin{aligned}(-\Delta)B_{-1}(-\Delta)^s &= (-\Delta)\left\{(-\Delta)^s B_{-1} + [B_{-1}, (-\Delta)^s]\right\} \\ &= (-\Delta)^{1+s} B_{-1} + (-\Delta)[B_{-1}, (-\Delta)^s],\end{aligned}$$

so putting both expressions together yields

$$\left[(-\Delta) + c_s B_2\right] B_{-1-s}(-\Delta)^s B_1 = B_{-s}(-\Delta)^{s+1} + F_s,$$

where

$$F_s := B_{-s}(-\Delta)[B_{-1}, (-\Delta)^s]B_1 + [(-\Delta), B_{-s}]B_{-1}(-\Delta)^s B_1 + c_s B_{1-s}(-\Delta)^s B_1.$$

A straightforward computation using the properties of the commutator from Lemma 6.4.2 gives that $F_s \equiv 0$, and thus (6.4.17) is proved.

This concludes the proof of the Theorem.

□

From another point of view, on $\mathbb{R}^n$ with the Euclidean metric, the fractional Laplacian for $s \in (0, 1)$ can be computed as the principal value of the integral

$$(-\Delta)^s u(x) = C(n, s) \int_{\mathbb{R}^n} \frac{u(x) - u(\xi)}{|x - \xi|^{n+2s}}\, d\xi. \tag{6.4.19}$$

Our next objective is to give an analogous expression for $P_s^{\mathbb{S}^n}$ in terms of a singular integral operator, using stereographic projection in expression (6.4.19):

Proposition 6.4.3. *Let $s \in (0,1)$. Given $u(z)$ in $\mathcal{C}^\infty(\mathbb{S}^n)$, it holds*

$$P_s^{\mathbb{S}^n} u(z) = \int_{\mathbb{S}^n} [u(z) - u(\zeta)]\, K_s(z,\zeta)\, d\zeta + A_{n,s} u(z),$$

where the kernel K_s is given by

$$K_s(z,\zeta) = 2^{s+n/2} C(n,s) \left(\frac{1-z_{n+1}}{1+z_{n+1}}\right)^{s+n/2} \left(\frac{1-\zeta_{n+1}}{1+\zeta_{n+1}}\right)^{s+n/2} \frac{1}{(1-z\cdot\zeta)^{s+n/2}}.$$

and the (positive) constant

$$A_{n,s} = \frac{\Gamma(\frac{n}{2}+s)}{\Gamma(\frac{n}{2}-s)}.$$

Proof. We recall the conformal covariance property for $P_s^{\mathbb{S}^n}$ from (6.2.6)

$$P_s^{\mathbb{S}^n} B^{s-n/2} \varphi^* = B^{-s-n/2} \varphi^* (-\Delta)^s,$$

that for $u \in \mathcal{C}^\infty(\mathbb{S}^n)$ is equivalent to

$$P_s^{\mathbb{S}^n} u = B^{-s-n/2} \varphi^* (-\Delta)^s \left[\varphi_* B^{-s+n/2} u\right].$$

From (6.4.19) we have

$$\begin{aligned}(-\Delta)^s &\left[B_{-s+n/2} \varphi_* u\right] \\ &= 2^{-s+\frac{n}{2}} C(n,s) \int_{\mathbb{R}^n} \frac{(1+|x|^2)^{s-n/2} u(\varphi^{-1}(x)) - (1+|\xi|^2)^{s-n/2} u(\varphi^{-1}(\xi))}{|x-\xi|^{n+2s}}\, d\xi.\end{aligned}$$

We pull back to $\mathbb{S}^n$, with coordinates

$$x = \varphi(z), \quad \zeta = \varphi(\xi),$$

recalling the Jacobian of the transformation from (6.4.9). Also note that

$$|x-\xi|^2 = 2\frac{1-z\cdot\zeta}{(1-z_{n+1})(1-\zeta_{n+1})}.$$

Therefore

$$\begin{aligned} P_s^{\mathbb{S}^n} u(z) = & C(n,s) 2^{s+n/2} (1+z_{n+1})^{-s-n/2} \\ & \cdot \int_{\mathbb{S}^n} \left[(1+z_{n+1})^{-s+\frac{n}{2}} u(z) - (1+\zeta_{n+1})^{-s+\frac{n}{2}} u(\zeta)\right] \\ & \cdot \frac{(1-z_{n+1})^{s+n/2} (1-\zeta_{n+1})^{s+n/2}}{(1-z\cdot\zeta)^{s+n/2}} (1+\zeta_{n+1})^{-n}\, d\zeta. \end{aligned}$$

Writing

$$u(z) = u(z)\frac{(1+\zeta_{n+1})^{-s+n/2}}{(1+z_{n+1})^{-s+n/2}} + u(z)\left[1 - \frac{(1+\zeta_{n+1})^{-s+n/2}}{(1+z_{n+1})^{-s+n/2}}\right],$$

we can arrive at

$$P_s^{\mathbb{S}^n} u(z) = \int_{\mathbb{S}^n} [u(z) - u(\zeta)]\, K_s(z,\zeta)\, d\zeta + u(z)\tilde{K}_s(z), \tag{6.4.20}$$

for

$$\begin{aligned}\tilde{K}_s(z) =& 2^{\frac{n}{2}+s} C(n,s) \\ &\cdot \int_{\mathbb{S}^n} \frac{(1+z_{n+1})^{-s+\frac{n}{2}} - (1+\zeta_{n+1})^{-s+\frac{n}{2}}}{(1+z\cdot\zeta)^{s+\frac{n}{2}}} \frac{(1-z_{n+1})^{s+\frac{n}{2}}}{(1+z_{n+1})^n} \frac{(1-\zeta_{n+1})^{s+\frac{n}{2}}}{(1+\zeta_{n+1})^n}\, d\zeta.\end{aligned}$$

On the other hand, it is possible to show that $\tilde{K}_s$ is constant in z. We have not attempted a direct proof; instead, we compare (6.4.2) and (6.4.20) applied to $u \equiv 1$. As a consequence,

$$\tilde{K}_s(z) \equiv \frac{\Gamma(\frac{n}{2}+s)}{\Gamma(\frac{n}{2}-s)}.$$

This yields the proof of the Proposition. □

6.5 The Conformal Fractional Laplacian on the Cylinder

Up to now, we have just considered conformally compact manifolds, for which the conformal infinity $(M,[h])$ is compact. But one could also look at the non-compact case. This is, perhaps, one of the most interesting issues since the definition on of the fractional conformal Laplacian, being a non-local operator, is not clear when M has singularities. In this section we consider the particular case when M is a cylinder.

Let $M = \mathbb{R}^n \setminus \{0\}$ with the cylindrical metric given by

$$h_0 := \frac{1}{r^2}|dx|^2$$

for $r = |x|$. Use the Emden-Fowler change of variable $r = e^{-t}$, $t \in \mathbb{R}$, and remark that the Euclidean metric may be written as

$$|dx|^2 = dr^2 + r^2 h_{\mathbb{S}^{n-1}} = e^{-2t}[dt^2 + h_{\mathbb{S}^{n-1}}] =: e^{-2t} h_0. \tag{6.5.1}$$

Thus, in these new coordinates, M may be identified with the cylinder $\mathbb{R}\times\mathbb{S}^{n-1}$ with the metric $h_0 = dt^2 + h_{\mathbb{S}^{n-1}}$.

The conformal covariance property (6.2.6) allows to formally write the conformal fractional Laplacian on the cylinder from the standard fractional Laplacian on Euclidean space. Indeed,

$$P_s^{h_0}(v) = r^{\frac{n+2s}{2}} P_s^{|dx|^2}(r^{-\frac{n-2s}{2}} v) = r^{\frac{n+2s}{2}}(-\Delta)^s u,$$

where we have set

$$u = r^{-\frac{n-2s}{2}} v. \tag{6.5.2}$$

This relation also allows to calculate the fractional curvature of a cylinder. It is the (positive) constant

$$c_{n,s} := Q_s^{h_0} = P_s^{h_0}(1) = r^{\frac{n+2s}{2}}(-\Delta)^s(r^{-\frac{n-2s}{2}}) = 2^{2s}\left(\frac{\Gamma(\frac{1}{2}(\frac{n}{2}+s))}{\Gamma(\frac{1}{2}(\frac{n}{2}-s))}\right)^2, \tag{6.5.3}$$

where the last equality is shown taking into account the Fourier transform of a homogeneous distribution.

In [27] the authors compute the principal symbol of the operator $P_s^{h_0}$ on $\mathbb{R}\times\mathbb{S}^{n-1}$ using the spherical harmonic decomposition for $\mathbb{S}^{n-1}$. This proof is close in spirit to the calculation we presented in the previous section for the sphere case, once we understand the underlying geometry. In fact, the standard cylinder $(\mathbb{R}\times\mathbb{S}^{n-1}, h_0)$ is the conformal infinity of the Riemannian AdS space, which is another simple example of a Poincaré-Einstein manifold. AdS space may be described as the $(n+1)$-dimensional manifold with metric

$$g^+ = \rho^{-2}\left(d\rho^2 + \left(1+\tfrac{\rho^2}{4}\right)^2 dt^2 + \left(1-\tfrac{\rho^2}{4}\right)^2 h_{\mathbb{S}^{n-1}}\right),$$

where $\rho\in(0,2]$ and $t\in\mathbb{R}$. As in the sphere case, the calculation of the Fourier symbol of $P_s^{h_0}$ goes by reducing the scattering problem (6.2.2)-(6.2.3) to an ODE in the variable ρ and then looking at its asymptotic behavior at $\rho=0$ and $\rho=2$. We will not present the proof of Theorem 6.5.1 below but refer to the original paper [27], since the new difficulties are of technical nature only.

With some abuse of notation, let $\mu_m = m(m+n-2)$, $m=0,1,2,\ldots$ be the eigenvalues of $-\Delta_{\mathbb{S}^{n-1}}$, repeated according to multiplicity. Then, any function v on $\mathbb{R}\times\mathbb{S}^{n-1}$ may be decomposed as $\sum_m v_m(t)E_m$, where $\{E_m\}$ is a basis of eigenfunctions. Let

$$\hat{v}(\xi) = \frac{1}{\sqrt{2\pi}}\int_{\mathbb{R}} e^{-i\xi\cdot t} v(t)\,dt$$

be our normalization for the one-dimensional Fourier transform. Then the operator $P_s^{h_0}$ diagonalizes under such eigenspace decomposition, and moreover, it is possible to calculate the Fourier symbol of each projection. More precisely:

Theorem 6.5.1 ([27]). *Fix $s\in(0,\frac{n}{2})$ and let P_s^m be the projection of the operator $P_s^{h_0}$ over each eigenspace $\langle E_m\rangle$. Then*

$$\widehat{P_s^m(v_m)} = \Theta_s^m(\xi)\,\widehat{v_m},$$

and this Fourier symbol is given by

$$\Theta_s^m(\xi) = 2^{2s}\frac{\left|\Gamma\left(\frac{1}{2}+\frac{s}{2}+\frac{\sqrt{(\frac{n}{2}-1)^2+\mu_m}}{2}+\frac{\xi}{2}i\right)\right|^2}{\left|\Gamma\left(\frac{1}{2}-\frac{s}{2}+\frac{\sqrt{(\frac{n}{2}-1)^2+\mu_m}}{2}+\frac{\xi}{2}i\right)\right|^2}. \tag{6.5.4}$$

Let us restrict to the space of radial functions $v = v(t)$, which corresponds to the eigenspace with $m = 0$, and denote $\mathcal{L}_s := P_s^0$. Then the Fourier symbol of $\mathcal{L}_s$ is given by

$$\Theta_s^0(\xi) = 2^{2s}\frac{\left|\Gamma(\frac{n}{4}+\frac{s}{2}+\frac{\xi}{2}i)\right|^2}{\left|\Gamma(\frac{n}{4}-\frac{s}{2}+\frac{\xi}{2}i)\right|^2}.$$

Again, in parallel to Proposition 6.4.3 in the sphere case, it is possible to give a singular integral formulation for the pseudodifferential operator $\mathcal{L}_s$ acting on $\langle E_0\rangle$:

Proposition 6.5.2 ([28]). *Given $v = v(t)$ smooth, $t \in \mathbb{R}$, we have for $\mathcal{L}_s$:*

$$\mathcal{L}_s v(t) = C(n,s)P.V.\int_{-\infty}^{\infty}(v(t)-v(\tau))K(t-\tau)\,d\tau + c_{n,s}v(t), \tag{6.5.5}$$

for the kernel

$$K(t) = \kappa_{n,s}e^{-\frac{n+2s}{2}t}\,{}_2F_1\left(\tfrac{n+2s}{2}, 1+s; \tfrac{n}{2}; e^{-2t}\right),$$

for $t > 0$, extended evenly for $t < 0$. $\kappa_{n,s}$ is a constant and the value of $c_{n,s}$ is given in (6.5.3).

It can be shown ([28]) that the asymptotic behavior of this kernel K is

$$\begin{aligned} K(\xi) &\sim |\xi|^{-1-2\gamma} \quad \text{as} \quad |\xi| \to 0,\\ K(\xi) &\sim e^{-|\xi|\frac{n+2\gamma}{2}} \quad \text{as} \quad |\xi| \to \infty. \end{aligned}$$

This shows that, at the origin, its singular behavior corresponds to that of the one-dimensional fractional Laplacian but, at infinity, it has a much faster decay. Levy processes arising from this type of generators are known as tempered stable processes; they combine both α-stable (in the short range) and Gaussian (in the long range) trends. In addition, invertibility properties of the operator with symbol $\Theta_m(\xi)-\lambda$ have been considered in [?]. In particular, they construct the Green's function for a Hardy type operator with fractional Laplacian.

Before we continue with this exposition let us mention a related problem: to construct solutions to the fractional Yamabe problem on $\mathbb{R}^n$, $s \in (0,1)$, with an isolated singularity at the origin. This means that one seeks positive solutions of

$$(-\Delta)^s u = c_{n,s}u^{\frac{n+2s}{n-2s}} \text{ in } \mathbb{R}^n\setminus\{0\}, \tag{6.5.6}$$

where $c_{n,s}$ is any positive constant that will be normalized as in (6.5.3), and such that

$$u(x) \to \infty \quad \text{as} \quad |x| \to 0.$$

For technical reasons, one needs to assume here that $n > 2+2s$. Because of the well known extension theorem for the fractional Laplacian (6.5.6) is equivalent to the boundary reaction problem

$$\begin{cases} -\operatorname{div}(y^a \nabla U) = 0 \text{ in } \mathbb{R}^{n+1}_+, \\ U = u \text{ on } \mathbb{R}^n \backslash \{0\}, \\ -d_s^* \lim_{y\to 0} y^a \partial_y u = c_{n,s} u^{\frac{n+2s}{n-2s}} \text{ on } \mathbb{R}^n \backslash \{0\}. \end{cases} \tag{6.5.7}$$

Our model for an isolated singularity is the cylindrical solution, given by $U_1 = P *_x u_1$ with $u_1(r) = r^{-\frac{n-2s}{2}}$. In the recent paper [16] the authors characterize all the nonnegative solutions to (6.5.7). Indeed, if the origin is not a removable singularity, then $u(x)$ is radial in the x variable and, if $u = U(\cdot, 0)$, then near the origin one must have that

$$c_1 r^{-\frac{n-2s}{2}} \leqslant u(x) \leqslant c_2 r^{-\frac{n-2s}{2}},$$

where c_1, c_2 are positive constants.

Positive radial solutions for (6.5.6) have been studied in the papers [27, 28]. It is clear from the above that one should look for solutions of the form (6.5.2) for some function $v = v(r)$ satisfying $0 < c_1 \leqslant v \leqslant c_2$. In the classical case $s = 1$, equation (6.5.6) reduces to a standard second order ODE. However, in the fractional case it becomes a fractional order ODE and, as a consequence, classical methods cannot be directly applied here.

One possible point of view is to rewrite (6.5.6) in the new metric h_0. Since the metrics $|dx|^2$ and h_0 are conformally related by (6.5.1), we prefer to use h_0 as a background metric and thus any conformal change may be rewritten as

$$h_v = u^{\frac{4}{n-2s}} |dx|^2 = v^{\frac{4}{n-2s}} h_0,$$

where u and v are related by (6.5.2). Then the original problem (6.5.6) is equivalent to the fractional Yamabe problem on $\mathbb{R} \times \mathbb{S}^{n-1}$: fixed h_0 as a background metric on $\mathbb{R} \times \mathbb{S}^{n-1}$, find a conformal metric h_v of positive constant fractional curvature $Q_s^{h_v}$, i.e., find a positive smooth solution v for

$$P_s^{h_0}(v) = c_{n,s} v^{\frac{n+2s}{n-2s}} \quad \text{on} \quad \mathbb{R} \times \mathbb{S}^{n-1}. \tag{6.5.8}$$

A complete study of radial solutions $v = v(t)$, $0 < c_1 \leqslant v \leqslant c_2$, for this equation is not available since is not an ODE. The local case $s = 1$, however, is well known since it reduces to understanding the phase-portrait of a Hamiltonian ODE (see the lecture notes [90], for instance), and periodic solutions constructed in this way are known as Delaunay solutions for the scalar curvature.

Fractional Delaunay solutions v_L to equation (6.5.8), i.e., radially symmetric periodic solutions in the variable $t \in \mathbb{R}$ for a given period L, are constructed in [28] using a variational method that we sketch here: if v is radial, then (6.5.8) is equivalent to

$$\mathcal{L}_s v = c_{n,s} v^{\frac{n+2s}{n-2s}}, \quad t \in \mathbb{R},$$

where $\mathcal{L}_s$ is given in (6.5.5). For periodic functions $v(t+L) = v(t)$ it can be rewritten as

$$\mathcal{L}_s^L v = c_{n,s} v^{\frac{n+2s}{n-2s}}, \quad t \in [0, L], \tag{6.5.9}$$

for

$$\mathcal{L}_s^L v(t) = C(n,s) P.V. \int_0^L (v(t) - v(\tau)) K_L(t-\tau)\, d\tau + c_{n,s} v(t),$$

and K_L is a periodic singular kernel given by $K_L(\xi) = \sum_{j\in\mathbb{Z}} K(\xi - jL)$.

We find a bifurcation behavior at the value of L, denoted by L_0, where the first eigenvalue for the linearization of problem (6.5.9) crosses zero. Moreover, for each period $L > L_0$ there exists a periodic solution v_L:

Theorem 6.5.3 ([28]). *Consider the variational formulation for equation* (6.5.9), *written as*

$$b(L) = \inf\{\mathcal{F}_L(v) \,:\, v \in H^s(0,L),\ v \text{ is } L\text{-periodic}\}$$

where

$$\mathcal{F}_L(v) = \frac{C(n,s) \int_0^L \int_0^L (v(t) - v(\tau))^2 K_L(t-\tau)\, dt\, d\tau + c_{n,s} \int_0^L v(t)^2\, dt}{\left(\int_0^L v(t)^{2*}\, dt\right)^{2/2*}}.$$

Then there is a unique $L_0 > 0$ such that $b(L)$ is attained by a nonconstant (positive) minimizer v_L when $L > L_0$ and when $L < L_0$ $b(L)$ is attained by the constant only.

Such v_L for $L > L_0$ are known as the Delaunay solutions for the fractional curvature, and they can be characterized almost explicitly. We remark that, geometrically, the constant solution v_{L_0} corresponds to the standard cylinder, while $v_L \to v_\infty$ as $L \to \infty$, where $v_\infty(t) = c(\cosh(t))^{-\frac{n-2s}{2}}$ corresponds to a standard sphere (i.e., the bubble solution (6.3.12), normalized accordingly). For other values of L we have a characterization as a bubble tower; in fact:

Proposition 6.5.4 ([5]). *We have that*

$$v_L = \sum_{j\in\mathbb{Z}} v_\infty(\cdot - jL) + \varphi_L,$$

where

$$\|\varphi_L\|_{H^s(0,L)} \to 0 \quad \text{as} \quad L \to \infty. \tag{6.5.10}$$

Moreover, for L large we have the following Holder estimates on φ_L:

$$\|\varphi_L\|_{\mathcal{C}^\alpha([0,L])} \leqslant C e^{-\frac{(n-2s)L}{4}(1+\xi)}$$

for some $\alpha \in (0,1)$ and $\xi > 0$ independent of L.

Finally, Delaunay-type solutions can be used in gluing problems since they model isolated singularities. In [5] the authors modify the methods in [91, 83] to construct complete metrics on the sphere of constant scalar curvature with a finite number of isolated singularities:

Theorem 6.5.5 ([5]). *Let $\Lambda = \{p_1, \dots, p_N\}$ be a set of N points in $\mathbb{R}^n$. There exists a smooth positive solution u for the problem*

$$\begin{cases} (-\Delta)^s u = u^{\frac{n+2s}{n-2s}} & \text{in } \mathbb{R}^n \setminus \Lambda, \\ u(x) \to \infty & \text{as } x \to \Lambda. \end{cases} \tag{6.5.11}$$

The proof of this theorem consists of a Lyapunov-Schmidt reduction involving a different perturbation of each bubble in the bubble tower (6.5.4), in the spirit of [72]. Note that the compatibility conditions that arise come from an infinite-dimensional Toda-type system; in addition, they do not impose any restriction on the location of the singular points, only on the neck sizes L at each singularity.

6.6 The Non-compact Case

Once the isolated singularities case has been reasonably well understood, we turn to the study of higher dimensional singularities. From the analysis point of view, one wishes to understand the semilinear problem

$$\begin{cases} (-\Delta)^s u = c\, u^{\frac{n+2s}{n-2s}} \text{ in } \mathbb{R}^n \setminus \Lambda, \\ u(x) \to \infty \text{ as } x \to \Lambda, \end{cases} \tag{6.6.1}$$

where Λ is a closed set of Hausdorff dimension $0 < k < n$ and $c \in \mathbb{R}$. The first difficulty one encounters is precisely how to define the fractional Laplacian $(-\Delta)^s$ on $\Omega := \mathbb{R}^n \setminus \Lambda$ since it is a non-local operator. Nevertheless, as in the cylinder case, this is better understood from the conformal geometry point of view.

In order to put (6.6.1) into a broader context, let us give a brief review of the classical singular Yamabe problem ($s = 1$). Let (M, h) be a compact n-dimensional Riemannian manifold, $n \geqslant 3$, and $\Lambda \subset M$ is any closed set as above. We are concerned with the existence and geometric properties of complete (non-compact) metrics of the form $h_u = u^{\frac{4}{n-2}} h$ with constant scalar curvature. This corresponds to solving the partial differential equation (recall (6.1.4))

$$-\Delta_h u + \tfrac{n-2}{4(n-1)} R_h u = \tfrac{n-2}{4(n-1)} R\, u^{\frac{n+2}{n-2}}, \quad u > 0,$$

where $R_{h_u} \equiv R$ is constant and with a boundary condition that $u \to \infty$ sufficiently quickly at Λ so that h_u is complete. It is known that solutions with $R < 0$ exist quite generally if Λ is large in a capacitary sense ([71, 65]), whereas for $R > 0$ existence is only known when Λ is a smooth submanifold (possibly with boundary) of dimension $k < (n-2)/2$ ([82, 33]).

There are both analytic and geometric motivations for studying this problem. For example, in the positive case ($R > 0$), solutions to this problem are actually weak solutions across the singular set ([93]), so these results fit into the broader investigation of possible singular sets of weak solutions of semilinear elliptic equations.

On the geometric side, a well-known theorem by Schoen and Yau ([93, 92]) states that if (M, h) is a compact manifold with a locally conformally flat metric h of positive scalar curvature, then the developing map D from the universal cover $\widetilde{M}$ to $\mathbb{S}^n$, which by definition is conformal, is injective, and moreover, $\Lambda := \mathbb{S}^n \backslash D(\widetilde{M})$ has Hausdorff dimension less than or equal to $(n-2)/2$. Regarding the lifted metric $\tilde{h}$ on $\widetilde{M}$ as a metric on Ω, this provides an interesting class of solutions of the singular Yamabe problem which are periodic with respect to a Kleinian group, and for which the singular set Λ is typically nonrectifiable. More generally, they also show that if $h_{\mathbb{S}^n}$ is the canonical metric on the sphere and if $h = u^{\frac{4}{n-2}} h_{\mathbb{S}^n}$ is a complete metric with positive scalar curvature and bounded Ricci curvature on a domain $\Omega = \mathbb{S}^n \backslash \Lambda$, then $\dim \Lambda \leqslant (n-2)/2$.

Going back to the non-local case, although it is not at all clear how to define $P_s^{\tilde{h}}$ and $Q_s^{\tilde{h}}$ on a general complete (non-compact) manifold $(\Omega, \tilde{h})$, in the paper [40] the authors give a reasonable definition when Ω is an open dense set in a compact manifold M and the metric $\tilde{h}$ is conformally related to a smooth metric h on M. Namely, one can define them by demanding that the conformal property (6.2.6) holds (as usual, we assume that a Poincaré-Einstein filling (X, g^+) has been fixed). Note, however, that this is not as simple as it first appears since, because of the nonlocal character of $P_s^{\tilde{h}}$, we must extend u as a distribution on all of M. There is no difficulty in using the relationship (6.2.6) to define $P_s^{\tilde{h}}\varphi$ when $\varphi \in \mathcal{C}_0^\infty(\Omega)$. From here one can use an abstract functional analytic argument to extend $P_s^{\tilde{h}}$ to act on any $\varphi \in L^2(\Omega, dv_{\tilde{h}})$. Indeed, it is straightforward to check that the operator $P_s^{\tilde{h}}$ defined in this way is essentially self-adjoint on $L^2(\Omega, dv_{\tilde{h}})$ when s is real. However, observe that $P_s^{\tilde{h}} = (-\Delta_{\tilde{h}})^s + \mathcal{K}$, where $\mathcal{K}$ is a pseudo-differential operator of order $2s-1$. Furthermore, $(-\Delta_{\tilde{h}})^s$ is self-adjoint. Since $\mathcal{K}$ is a lower order symmetric perturbation, then $P_s^{\tilde{h}}$ is also essentially self-adjoint.

Another interesting development is [54], where they give a sharp spectral characterization of Poincaré-Einstein manifolds with conformal infinity of positive Yamabe type.

The singular fractional Yamabe problem on $(M, [h])$ is then formulated as

$$\begin{cases} P_s^h u = c u^{\frac{n+2s}{n-2s}} & \text{in } M \backslash \Lambda, \\ u(x) \to \infty & \text{as } x \to \Lambda, \end{cases} \tag{6.6.2}$$

for $c \equiv Q_s^{\tilde{h}}$ constant. A separate, but also very interesting issue, is whether $c > 0$ implies that the conformal factor u is actually a weak solution of (6.6.2) on all of M.

The first result in [40] partially generalizes Schoen-Yau's theorem:

Theorem 6.6.1 ([40]). *Suppose that (M^n, h) is compact and $h_u = u^{\frac{4}{n-2s}} h$ is a complete metric on $\Omega = M \backslash \Lambda$, where Λ is a smooth k-dimensional submanifold in M. Assume furthermore that u is polyhomogeneous along Λ with leading exponent $-n/2 + s$. If $s \in (0, \frac{n}{2})$, and if $Q_s^h > 0$ everywhere for any choice of asymptotically Poincaré-Einstein extension (X, g^+) then n, k and s are restricted by the inequality*

$$\Gamma\left(\frac{n}{4} - \frac{k}{2} + \frac{s}{2}\right) \Big/ \Gamma\left(\frac{n}{4} - \frac{k}{2} - \frac{s}{2}\right) > 0. \tag{6.6.3}$$

This inequality holds in particular when

$$k < \frac{n - 2s}{2}, \tag{6.6.4}$$

and in this case then there is a unique distributional extension of u on all of M which is still a solution of (6.6.2) *on all of M.*

As we have noted, inequality (6.6.3) is satisfied whenever $k < (n - 2s)/2$, and in fact is equivalent to this simpler inequality when $s = 1$. When $s = 2$, i.e. for the standard Q–curvature, this result is already known: [23] shows that complete metrics with $Q_2 > 0$ and positive scalar curvature must have singular set with dimension less than $(n - 4)/2$, which again agrees with (6.6.3).

Of course, the main open question is to remove the smoothness assumption on the singular set Λ. Recent results of [98] show that, under a positive scalar curvature assumption, if $Q_s > 0$ for $s \in (1, 2)$, then (6.6.4) holds for any Λ.

We also remark that a dimension estimate of the type (6.6.3) implies some topological restrictions on M: on the homotopy ([92], chapter VI), on the cohomology ([85]), or even classification results in the low dimensional case ([57]).

On the contrary, to give conditions for sufficience is a delicate issue, and only partial results exist when the singular set is a smooth submanifold of dimension $k < (n - 2s)/2$ [6, **?**].

6.7 Uniqueness

One of the main questions that arises is, given a manifold (M^n, h), is there a canonical way to define the conformal fractional Laplacian on M? this question is equivalent to ask how many Poincaré-Einstein fillings (X^{n+1}, g^+) one can find. The answer is, in general, not unique, unless the conformal infinity is the round sphere (or equivalently, $\mathbb{R}^n$) (see the survey [25], for instance).

In this section we would like to describe two Poincaré-Einstein fillings on the topologically same 4-manifold with the same conformal infinity. This construction comes from the study of thermodynamics of black holes in Anti-de Sitter Space [56], is well known and it is explained in [25] and [48], for instance, but we repeat it here for completeness.

The AdS-Schwarzchild manifold is an Einstein 4-manifold described as $\mathbb{R}^2 \times \mathbb{S}^2$ with the metric [56] (see also the survey [25])

$$g_m^+ = Vdt^2 + V^{-1}dr^2 + r^2 h_{\mathbb{S}^2} \quad \text{for} \quad V = 1 + r^2 - \frac{2m}{r}.$$

We call r_m the positive root for $1 + r^2 - \frac{2m}{r} = 0$, and we restrict $r \in [r_m, +\infty)$. $m > 0$ is known as the mass parameter, to be chosen later.

Even though this metric seems singular ar r_m, we will prove that this is not the case if we make the t variable periodic, i.e., $t \in \mathbb{S}^1(L)$. To see this, define a function $\rho : (r_h, \infty) \to (0, \infty)$ by

$$\rho(r) = \int_{r_m}^{r} V^{-1/2}.$$

One can check that for r near r_m,

$$g_{+1}^m \sim d\rho^2 + \frac{(V'(r_m))^2}{4}\rho^2 dt^2 + r^2 h_{\mathbb{S}^2},$$

so the singularity at $r = r_m$ is of the same type as the origin in standard polar coordinates. Thus thus we need to make the t variable periodic, i.e, $0 \leqslant t \leqslant 2\pi L$, for

$$L := L(m) = \frac{V'(r_m)}{2} = \frac{2r_m}{3r_m^2 + 1}. \tag{6.7.1}$$

To show that g_m^+ is conformally compact, we change to the defining function $\tilde{r} = \frac{1}{r}$. Since $V(r) \approx \frac{1}{\tilde{r}^2}$ when $\tilde{r} \to 0$, then

$$g_m^+ \sim \frac{1}{\tilde{r}^2}[d\tilde{r}^2 + dt^2 + h_{\mathbb{S}^2}] \quad \text{as} \quad \tilde{r} \to 0.$$

Therefore, for each $m > 0$, g_m^+ is Poincaré-Einstein and its conformal infinity is $\mathbb{S}^1(L) \times \mathbb{S}^2$ with the metric $h_m := dt^2 + h_{\mathbb{S}^2}$.

But we could ask the reverse question of, given L, how many Poincaré-Einstein fillings one can find for $S^1(L) \times \mathbb{S}^2$. Looking at (9.1.2), $r_m = \frac{1}{\sqrt{3}}$ is a critical point for $L(r_m)$, actually a maximum with value

$$L(\tfrac{1}{\sqrt{3}}) = \tfrac{1}{\sqrt{3}}.$$

There holds

– For $0 < L < 1/\sqrt{3}$, we can find two different masses m_1 and m_2 with the same $L(m)$.
– For $L = 1/\sqrt{3}$, there exist only one mass m which gives $L(m)$.
– If $L > 1/\sqrt{3}$, there does not exist any mass which gives $L(m)$.

Thus for the same conformal infinity $\mathbb{S}^1(L) \times \mathbb{S}^2$, when $0 < L < \frac{1}{\sqrt{3}}$ there are two non-isometric AdS-Schwarzschild spaces with metrics $g^+_{m_1}$ and $g^+_{m_2}$. The natural question now is to calculate the symbol of the conformal fractional Laplacian P^m_s on the conformal infinity for each model. This calculation is similar to that of (6.4.1) for the sphere and (6.5.4) for the cylinder. But, unfortunately, the spherical harmonic decomposition yields a more complicated ODE that we have not been able to solve analytically.

6.8 An Introduction to Hypersurface Conformal Geometry

Let $(\overline{X}^{n+1}, \overline{g})$ be any smooth compact manifold with boundary (M^n, h), where $h = \overline{g}|_M$, for instance, a domain in $\mathbb{R}^{n+1}$ with the Euclidean metric. One would like to understand the conformal geometry of M as an embedded hypersurface with respect to the given filling metric $\overline{g}$, and to produce new extrinsic conformal invariants on this hypersurface.

In this discussion we are mostly interested in the construction of non-local objects on M, in particular, the conformal fractional Laplacian, and to understand how this new P^h_s depends on the geometry of the background metric $\overline{g}$. A good starting reference is the recent paper [49], although there the author is more interested in renormalized volume rather than scattering (see also the parallel development by [44, 45] in the language of tractor calculus).

Let ρ be a geodesic defining function for M. This means, in particular, that $\overline{g} = d\rho^2 + h_\rho$, where h_ρ is a one parameter family of metrics on M with $h_\rho|_{\rho=0} = h$. We would like to produce a suitable asymptotically hyperbolic filling metric g^+ for which the scattering problem (6.2.2)-(6.2.3) can be solved in terms of information from $\overline{g}$ only. Looking at (6.3.4), the reasonable assumption is to ask that g^+ has constant scalar curvature

$$R_{g^+} = -n(n+1). \tag{6.8.1}$$

Thus we seek a new defining function $\hat{\rho} = \hat{\rho}(\rho)$ such that if we define

$$g^+ = \frac{\overline{g}}{\hat{\rho}^2},$$

then this g^+ is asymptotically hyperbolic and satisfies (6.8.1). Remark that sometimes it will be more convenient to write $g^+ = u^{\frac{4}{n-1}}\overline{g}$ for $u = \hat{\rho}^{-\frac{n-1}{2}}$.

This problem for g^+ reduces to the singular Yamabe problem of Loewner-Nirenberg ([71]) for constant negative scalar curvature and it has been well studied ([79, 7, 4], for instance). In PDE language, looking at the conformal transformation law for the usual conformal Laplacian (6.1.4), it amounts to find a positive solution u in X to the equation

$$-\Delta_{\overline{g}}u + \tfrac{n-1}{4n}R_{\overline{g}}u = -\tfrac{n^2-1}{4}u^{\frac{n+3}{n-1}}$$

that has the asymptotic behavior

$$u \sim \rho^{-\frac{n-1}{2}} \quad \text{near } \partial X$$

(recall that we are working on an $(n+1)$-dimensional manifold). It has been shown that such solution exists and it has a very specific polyhomogeneous expansion near ∂X, so that

$$g^+ = \frac{\overline{g}(1+\rho\alpha+\rho^{n+1}\beta)}{\rho^2},$$

where $\alpha \in \mathcal{C}^\infty(\overline{X})$ and $\beta \in \mathcal{C}^\infty(X)$ has a polyhomogeneous expansion with log terms. This type of expansions often appears in geometric problems, such as in the related [55], and each of the terms in the expansion has a precise geometric meaning (some are local, others non-local).

In this general setting, scattering for g^+ can be considered ([52]), and one is able to construct the conformal fractional Laplacian on M with respect to the starting $\overline{g}$ once the log terms in the expansion are controlled. For $s \in (0,1)$ these log terms do not affect the asymptotic expansions at the boundary and one has:

Theorem 6.8.1 ([53] for $s = 1/2$, [22] in general). *Fix $s \in (0,1)$. Let $(\overline{X}^{n+1}, \overline{g})$ be a smooth compact manifold with boundary M^n and set $h := \overline{g}|_M$. Let ρ be a geodesic defining function. Then there exists a defining function $\hat{\rho}$ as in the above construction. Moreover, if U is a solution to the following extension problem*

$$\begin{cases} -\operatorname{div}(\hat{\rho}^a \nabla U) + E(\hat{\rho})U = 0 & \text{in } (\overline{X}, \overline{g}), \\ U = u & \text{on } M, \end{cases}$$

for $E(\hat{\rho})$ given in (6.3.2), *then the conformal fractional Laplacian P_s^h on M with respect to the metric h may be constructed as in Theorem 6.3.2.*

One could also look at higher values of $s \in (0, n/2)$. For example, when M is a surface in Euclidean 3-space, one recovers the Willmore invariant with this construction, so an interesting consequence of this approach is that it produces new (extrinsic) conformal invariants for hypersurfaces in higher dimensions that generalize the Willmore invariant for a two-dimensional surface. Many open questions still remain since this is a growing subject.

Acknowledgment: M.d.M. González is supported by the Spanish Government project MTM2014-52402-C3-1-P and the 2016 BBVA Foundation grant for Investigadores y Creadores Culturales, and is part of the Catalan research group 2014SGR1083.

Bibliography

[1] M. Abramowitz, I. A. Stegun. *Handbook of mathematical functions with formulas, graphs, and mathematical tables*, volume 55 of *National Bureau of Standards Applied Mathematics Series*. For sale by the Superintendent of Documents, U.S. Government Printing Office, Washington, D.C., 1964.
[2] O. Aharony, S. S. Gubser, J. Maldacena, H. Ooguri, Y. Oz. Large N Field Theories, String Theory and Gravity. *Phys. Rept.* 323:183–386, 2000.
[3] S. Almaraz. An existence theorem of conformal scalar-flat metrics on manifolds with boundary. *Pacific J. Math.* 248 (2010), no. 1, 1–22.
[4] L. Andersson, P. T. Chruściel, H. Friedrich. On the regularity of solutions to the Yamabe equation and the existence of smooth hyperboloidal initial data for Einstein's field equations. *Comm. Math. Phys.*, 149(3):587–612, 1992.
[5] W. Ao, H. Chan, A. DelaTorre, M. Fontelos. M.d.M. González, J. Wei. On higher dimensional singularities for the fractional Yamabe problem: a non-local Mazzeo-Pacard program. Preprint.
[6] W. Ao, H. Chan, M.d.M. González, J. Wei. Existence of positive weak solutions for fractional Lane-Emden equations with prescribed singular sets. Preprint.
[7] P. Aviles, R. McOwen. Complete conformal metrics with negative scalar curvature in compact Riemannian manifolds. *Duke Math. J.* 56 (1988), no. 2, 395–398.
[8] B. Barrios, E. Colorado, A. De Pablo, U. Sánchez. On some critical problems for the fractional Laplacian operator. *Journal of Differential Equations*, 252 (11): 6133–6162, 2012.
[9] H. Baum, A. Juhl. *Conformal differential geometry. Q-curvature and conformal holonomy*. Oberwolfach Seminars, 40. Birkhäuser Verlag, Basel, 2010.
[10] O. Biquard, *Métriques d'Einstein asymptotiquement symétriques*. Astérisque No. 265 (2000).
[11] T. P. Branson. Sharp inequalities, the functional determinant, and the complementary series. *Trans. Amer. Math. Soc.*, 347(10):3671–3742, 1995.
[12] T. P. Branson. Spectral theory of invariant operators, sharp inequalities, and representation theory. The Proceedings of the 16th Winter School "Geometry and Physics" (Srní, 1996). *Rend. Circ. Mat. Palermo* (2) Suppl. No. 46 (1997), 29–54.
[13] S. Brendle, S.-Y. S. Chen. An existence theorem for the Yamabe problem on manifolds with boundary. *J. Eur. Math. Soc. (JEMS)* 16 (2014), no. 5, 991–1016.
[14] X. Cabré, Y. Sire. Non-linear equations for fractional Laplacians I: regularity, maximum principles and Hamiltoniam estimates. To appear in *Transactions of AMS*.

[15] X. Cabré and J. Tan. Positive solutions of nonlinear problems involving the square root of the Laplacian. *Adv. Math.* 226, Issue 2 (2011), 1410–1432.

[16] L. Caffarelli, T. Jin, Y. Sire, J. Xiong. Local analysis of solutions of fractional semi-linear elliptic equations with isolated singularities. *Arch. Ration. Mech. Anal.*, 213(1):245–268, 2014.

[17] L. Caffarelli, L. Silvestre. An extension problem related to the fractional Laplacian. *Comm. Partial Differential Equations*, 32(7-9):1245–1260, 2007.

[18] L. A. Caffarelli, P. E. Souganidis. Convergence of nonlocal threshold dynamics approximations to front propagation. *Arch. Ration. Mech. Anal.* 195 (2010), no. 1, 1–23.

[19] J. Case. Some energy inequalities involving fractional GJMS operators. Analysis and PDE, Vol. 10 (2017), No. 2, 253–280.

[20] J. Case, S.-Y. A. Chang. On fractional GJMS operators. To appear in *Communications on Pure and Applied Mathematics*.

[21] S.-Y. A. Chang. *Non-linear elliptic equations in conformal geometry*. Zurich Lectures in Advanced Mathematics. European Mathematical Society (EMS), Zürich, 2004.

[22] S.-Y. A. Chang, M. d. M. González. Fractional Laplacian in conformal geometry. *Adv. Math.*, Vol. 226, Issue 2, 30 January 2011, 1410–1432.

[23] S.-Y. A. Chang, F. Hang, P. C. Yang. On a class of locally conformally flat manifolds. *Int. Math. Res. Not.*, (4):185–209, 2004.

[24] S.-Y. A. Chang, J. Qing, P. Yang. Renormalized volumes for conformally compact Einstein manifolds. *Sovrem. Mat. Fundam. Napravl.*, 17:129–142, 2006.

[25] S.-Y. A. Chang, J. Qing, P. Yang. Some progress in conformal geometry. *SIGMA Symmetry Integrability Geom. Methods Appl.* 3 (2007), Paper 122, 17 pp.

[26] S.-Y. A. Chang, R. Yang. On a class of non-local operators in conformal geometry. Chinese Annals of Mathematics, Series B January 2017, Vol. 38, Issue 1, 215–234.

[27] A. DelaTorre, M. González. Isolated singularities for a semilinear equation for the fractional Laplacian arising in conformal geometry. Preprint arXiv:1504.03493, To appear in Revista Matemática Iberoamericana

[28] A. DelaTorre, M. del Pino, M.d.M. González, J. Wei. Radial solutions with an isolated singularity for a semilinear equation with fractional Laplacian. *Mathematische Annalen* 369 (2017) 597–626.

[29] C. Delaunay. Sur la surface de revolution dont la courbure moyenne est constante. *J. de Math. Pure et Appl.* 6:309–314.

[30] J. Eells The surfaces of Delaunay. *The Mathematical Intelligencer*, 9(1):53–57, 1987.

[31] A. Enciso, M.d.M. González, B. Vergara. Fractional powers of the wave operator via Dirichlet-to-Neumann maps in anti-de Sitter spaces. Journal of Functional Analysis 273 (2017), no. 6, 2144–216.

[32] J. F. Escobar. Conformal deformation of a Riemannian metric to a scalar flat metric with constant mean curvature on the boundary. *Ann. of Math. (2)*, 136(1):1–50, 1992.

[33] S. Fakhi. Positive solutions of $\Delta u + u^p = 0$ whose singular set is a manifold with boundary. *Calc. Var. Partial Differential Equations*, 17(2):179–197, 2003.

[34] Y. Fang, M.d.M. González. Asymptotic behavior of Palais-Smale sequences associated with fractional Yamabe type equations. *Pacific J. Math*, 278, No. 2, 369–405, 2015.

[35] C. Fefferman, C. R. Graham. The ambient metric. Annals of Mathematics Studies (Book 178). Princeton University Press, 2011.

[36] C. Fefferman, C. R. Graham. *Q*-curvature and Poincaré metrics. *Math. Res. Lett.*, 9(2-3):139–151, 2002.

[37] R. Frank, M.d.M. González, D. Monticelli, J. Tan. Conformal fractional Laplacians on the Heisenberg group. *Advances in Mathematics*, 270, 97–137, 2015.

[38] M.d.M. González. A weighted notion of renormalized volume related to the fractional Laplacian. *Pacific J. Math.* Vol. 257, no. 2, 379–394, 2012.

[39] M.d.M. González. Gamma convergence of an energy functional related to the fractional Laplacian. *Calc. Var. Partial Differential Equations* 36 (2009), no. 2, 173–210.

[40] M. d. M. González, R. Mazzeo, Y. Sire. Singular solutions of fractional order conformal Laplacians. *J. Geom. Anal.*, 22(3):845–863, 2012.

[41] M. d. M. Gonzalez, J. Qing. Fractional conformal Laplacians and fractional Yamabe problems, Analysis & PDE 6.7 (2013), 1535–1576.

[42] M. d. M. Gonzalez, M. Sáez. Fractional Laplacians and extension problems: the higher rank case. Preprint 2016, To appear in Transactions of the AMS

[43] M. d. M. González, M. Wang. Further results on the fractional yamabe problem: the umbilic case. Journal of Geometric Analysis 28 (2018), 22–60.

[44] A. R. Gover, A. Waldron. Conformal hypersurface geometry via a boundary Loewner-Nirenberg-Yamabe problem. Prepint arXiv:1506.02723.

[45] A. R. Gover, A. Waldron. Renormalized Volume. Preprint arXiv:1603.07367.

[46] C. R. Graham. Volume and Area Renormalizations for Conformally Compact Einstein Metrics. *math.DG: arXiv:9909042*, The Proceedings of the 19th Winter School "Geometry and Physics" (Srn, 1999). *Rend. Circ. Mat. Palermo* (2) Suppl. No. 63 (2000), 31–42.

[47] C. R. Graham. Conformal powers of the Laplacian via stereographic projection. *SIGMA Symmetry Integrability Geom. Methods Appl.*, 3:Paper 121, 4, 2007.

[48] R. Graham. Lecture notes from the Mini-courses on Nonlinear Elliptic Equations. May 13-18, 2013, Department of Mathematics, Rutgers University.

[49] C. R. Graham. Volume renormalization for singular Yamabe metrics. Proceedings of the American Mathematical Society, Volume 145, Number 4, April 2017, 1781–1792.

[50] C. R. Graham, R. Jenne, L. J. Mason, G. A. J. Sparling. Conformally invariant powers of the Laplacian. I. Existence. *J. London Math. Soc. (2)*, 46(3):557–565, 1992.

[51] C. R. Graham, M. Zworski. Scattering matrix in conformal geometry. *Invent. Math.*, 152(1):89–118, 2003.

[52] C. Guillarmou. Meromorphic properties of the resolvent on asymptotically hyperbolic manifolds. *Duke Math. J.*, 129(1):1–37, 2005.
[53] C. Guillarmou, L. Guillopé. The determinant of the Dirichlet-to-Neumann map for surfaces with boundary. *Int. Math. Res. Not. IMRN*, (22):Art. ID rnm099, 26, 2007.
[54] C. Guillarmou, J. Qing. Spectral characterization of Poincaré-Einstein manifolds with infinity of positive Yamabe type. *Int. Math. Res. Not. IMRN*, (9):1720–1740, 2010.
[55] Q. Han, X. Jiang. Boundary expansions for minimal graphs in the hyperbolic space. Preprint
[56] S. Hawking, D. Page. Thermodynamics of black holes in anti-de Sitter space. *Comm. Math. Phys.* 87 (1982/83), no. 4, 577–588.
[57] H. Izeki. Limit sets of Kleinian groups and conformally flat Riemannian manifolds. *Invent. Math.*, 122(3), 603–625 (1995).
[58] T. Jin, Y. Li, J. Xiong. On a fractional Nirenberg problem, part I: blow up analysis and compactness of solutions. *J. Eur. Math. Soc. (JEMS)*, 16(6):1111–1171, 2014.
[59] A. Juhl. *Families of conformally covariant differential operators, Q-curvature and holography*, volume 275 of *Progress in Mathematics*. Birkhäuser Verlag, Basel, 2009.
[60] M. Kashiwara, A. Kowata, K. Minemura, K. Okamoto, T. Ōshima, M. Tanaka. Eigenfunctions of invariant differential operators on a symmetric space. *Ann. of Math. (2)*, 107(1):1–39, 1978.
[61] S. Kim, M. Musso, J. Wei. Existence theorems of the fractional Yamabe problem. Analysis & PDE, Vol. 11 (2018), No. 1, 75–113.
[62] S. Kim, M. Musso, J. Wei. A non-compactness result on the fractional Yamabe problem in large dimensions. Preprint arXiv:1505.06183.
[63] S. Kim, M. Musso, J. Wei. A compactness theorem of the fractional Yamabe problem. Part I: the non-umbilic conformal infinity. Journal of Functional Analysis Vol. 273, Issue 12, 3759–3830.
[64] N. Korevaar, R. Mazzeo, F. Pacard and R. Schoen. Refined asymptotics for constant scalar curvature metrics with isolated singularities. *Invent. Math.* 135(2):233–272, 1999.
[65] D. A. Labutin. Wiener regularity for large solutions of nonlinear equations. *Ark. Mat.*, 41(2):307–339, 2003.
[66] C. LeBrun. H-space with a cosmological constant. *Proc. Roy. Soc. London Ser.* A 380, no. 1778, 171–185 (1982).
[67] J. Lee. The spectrum of an asymptotically hyperbolic Einstein manifold. *Comm. Anal. Geom.* 3 (1995), no. 1-2, 253–271.
[68] J. M. Lee, T. H. Parker. The Yamabe problem. *Bull. Amer. Math. Soc. (N.S.)*, 17(1):37–91, 1987.
[69] Y. Li, M. Zhu. Sharp Sobolev trace inequalities on Riemannian manifolds with boundaries. *Comm. Pure Appl. Math.* 50 (1997), no. 5, 449–487.

[70] E. H. Lieb. Sharp constants in the Hardy-Littlewood-Sobolev and related inequalities. *Ann. of Math. (2)*, 118(2):349–374, 1983.
[71] C. Loewner, L. Nirenberg. Partial differential equations invariant under conformal or projective transformations. *Contributions to analysis (a collection of papers dedicated to Lipman Bers)*, pp. 245–272. Academic Press, New York, 1974.
[72] A. Malchiodi. Some new entire solutions of semilinear elliptic equations on $\mathbb{R}^n$. *Adv. in Math* 221 (2009), 1843–1909.
[73] J. Maldacena. The large *N* limit of superconformal field theories and supergravity. *Adv. Theor. Math. Phys.* 2 (1998) 231–252.
[74] F. C. Marques. Existence results for the Yamabe problem on manifolds with boundary. *Indiana Univ. Math. J.* 54 (2005), no. 6, 1599–1620.
[75] F. C. Marques. Conformal deformations to scalar-flat metrics with constant mean curvature on the boundary. Comm. Anal. Geom. 15 (2007), no. 2, 381–405.
[76] F. C. Marques, A. Neves. The Willmore conjecture. *Jahresber. Dtsch. Math.-Ver.* 116 (2014), no. 4, 201–222.
[77] M. Mayer, C. B. Ndiaye. Barycenter technique and the Riemann mapping problem of Cherrier-Escobar. J. Differential Geom. Vol. 107, Number 3 (2017), 519–560.
[78] M. Mayer, C. B. Ndiaye. Fractional Yamabe problem on locally flat and umbilic conformal infinity. Preprint.
[79] R. Mazzeo. Regularity for the singular Yamabe problem. *Indiana Univ. Math. J.*, 40(4):1277–1299, 1991.
[80] R. Mazzeo. Unique continuation at infinity and embedded eigenvalues for asymptotically hyperbolic manifolds. *Am. J. Math.* 113, 25–45, 1991.
[81] R. R. Mazzeo, R. B. Melrose. Meromorphic extension of the resolvent on complete spaces with asymptotically constant negative curvature. *J. Funct. Anal.*, 75(2):260–310, 1987.
[82] R. Mazzeo, F. Pacard. A construction of singular solutions for a semilinear elliptic equation using asymptotic analysis. *J. Differential Geom.*, 44(2):331–370, 1996.
[83] R. Mazzeo, F. Pacard. Constant scalar curvature metrics with isolated singularities. *Duke Math. J.* 99, no. 3, 353–418, 1999.
[84] R. Mazeo, A. Vasy. Analytic continuation of the resolvent of the Laplacian on symmetric spaces of noncompact type. *J. Funct. Anal.* 228 (2005), no. 2, 311–368.
[85] S. Nayatani. Patterson-Sullivan measure and conformally flat metrics. *Math. Z.* 225, no. 1, 115–131 (1997).
[86] G. Palatucci, A. Pisante. A global compactness type result for Palais-Smale sequences in fractional Sobolev spaces. *Nonlinear Anal.*, 117, 1–7, 2015.
[87] S. M. Paneitz. A quartic conformally covariant differential operator for arbitrary pseudo-Riemannian manifolds (summary). *SIGMA Symmetry Integrability Geom. Methods Appl.*, 4:Paper 036, 3, 2008.
[88] G. Perelman. The entropy formula for the Ricci flow and its geometric applications. Preprint arXiv:0211159.

[89] O. Savin, E. Valdinoci. Γ-convergence for nonlocal phase transitions. *Ann. Inst. H. Poincaré Anal. Non Linéaire* 29 (2012), no. 4, 479–500.
[90] R. M. Schoen. Variational theory for the total scalar curvature functional for Riemannian metrics and related topics. In *Topics in calculus of variations (Montecatini Terme, 1987)*, volume 1365 of *Lecture Notes in Math.*, pages 120–154. Springer, Berlin, 1989.
[91] R. Schoen. The existence of weak solutions with prescribed singular behavior for a conformally invariant scalar equation. *Comm. Pure Appl. Math.* 41 (1988), no. 3, 317–392.
[92] R. Schoen, S.-T. Yau. *Lectures on differential geometry*. Conference Proceedings and Lecture Notes in Geometry and Topology, I. International Press, 1994.
[93] R. Schoen, S.-T. Yau. Conformally flat manifolds, Kleinian groups and scalar curvature. *Invent. Math.*, 92(1):47–71, 1988.
[94] R. Servadei, E. Valdinoci. The Brezis-Nirenberg result for the fractional Laplacian. *Trans. Amer. Math. Soc.* 367 (2015), no. 1, 67–102.
[95] A. Vasy, The wave equation on asymptotically anti de Sitter spaces. *Anal. PDE* 5 (2012) 81–144.
[96] E. Witten. Anti de Sitter space and holography. *Adv. Theor. Math. Phys.* 2(2):253–291, 1998.
[97] R. Yang. On higher order extensions for the fractional laplacian. Preprint arXiv:1302.4413.
[98] R. Zhang. Non-local Curvature and Topology of Locally Conformally Flat Manifolds. Preprint arXiv:1510.00957.

Moritz Kassmann

Jump Processes and Nonlocal Operators

Abstract: The aim of these notes is to present basic material on jump processes and their connection to nonlocal operators. We discuss the martingale problem and show how the existence of Markov jump processes follows from well-posedness of the deterministic Cauchy problem for integrodifferential operators. Furthermore, we explain how to use jump processes for proving regularity results for integrodifferential equations. The notes do not contain any original new result.

Introduction

Many results about harmonic functions can be proved making use of Brownian Motion. The same is true for solutions to linear partial differential equations of second order, if one uses general Markov diffusion processes. In these notes, we discuss several connections between solutions to integrodifferential equations and properties of corresponding Markov jump processes.

The role of the fractional Laplace operator as the generator of the semigroup generated by the rotationally symmetric stable process has been known for a long time. The corresponding connection between the more general nonlocal operators and jump processes has recently led to interesting studies such as regularity results in Hölder spaces. We provide details of the approach to such results which is based on properties of the underlying stochastic process.

In Section 7.1 we review fundamental concepts of probability theory. Section 7.2 is devoted to Lévy processes. Their translation invariant generators are studied in Section 7.3. In Section 7.4 we explain the notion of the martingale problem and discuss its well-posedness. We provide a detailed list of references in this framework. We omit overviews of the literature in the other sections because they can easily be found elsewhere. In Section 7.5 we formulate and prove regularity estimates for solutions to integrodifferential equations under minimal regularity assumptions.

7.1 Prerequisites and Lévy Processes

In this section, we review some basic concepts of probability theory and we define Lévy processes.

Moritz Kassmann, Faculty of Mathematics, Bielefeld University, P.O. Box 10 01 31, D-33501 Bielefeld, Germany, E-mail: moritz.kassmann@uni-bielefeld.de

https://doi.org/10.1515/9783110571561-009

Definition 7.1.1. *Let $(\Omega, \mathcal{F}, \mathbb{P})$ be a probability space.*

(i) A measurable map $X : \Omega \to \mathbb{R}^d$ is called random variable. Measurable subsets $E \in \mathcal{F}$ are sometimes called events. Given $A \in \mathcal{B}(\mathbb{R}^d)$, we write $(X \in A)$ instead of $\{\omega \in \Omega | X(\omega) \in A\}$ or $X^{-1}(A)$. Thus, $\mathbb{P}(X \in A)$ denotes the probability that X takes value in A.

(ii) If $X : \Omega \to \mathbb{R}, X \in L^1(\Omega, \mathrm{d}\mathbb{P})$, then

$$\mathbb{E}[X] = \int_\Omega X(\omega)\,\mathrm{d}\mathbb{P}(\omega)$$

is called expectation of X.

(iii) Every random variable X induces a measure $\mathbb{P}_X$ on $\mathcal{B}(\mathbb{R}^d)$ via

$$\mathbb{P}_X(B) = \int_B 1\,\mathrm{d}\mathbb{P}_X = \mathbb{P}(X \in B).$$

$\mathbb{P}_X$ is called distribution of X.

(iv) A family $\{X_1, \ldots, X_n\}$ of random variables X_i is independent if for all $B_1, \ldots, B_n \in \mathcal{B}(\mathbb{R}^d)$

$$\mathbb{P}(X_1 \in B_1, \ldots, X_n \in B_n) = \mathbb{P}(X_1 \in B_1) \cdot \ldots \cdot \mathbb{P}(X_n \in B_n).$$

The following facts are important properties on distributions.

(i) If X is a random variable and $f \in C_b(\mathbb{R}^d)$, then

$$\mathbb{E}[f(X)] = \int_\Omega f(X(\omega))\,\mathrm{d}\mathbb{P}(\omega) = \int_{\mathbb{R}^d} f(z)\,\mathrm{d}\mathbb{P}_X(z).$$

(ii) If X, Y are independent random variables, then

$$\mathbb{P}_{X+Y} = \mathbb{P}_X * \mathbb{P}_Y.$$

We proceed with the definition of two concepts of convergence for random variables.

Definition 7.1.2. *Let X_n, $n \in \boldsymbol{N}$, be a sequence of random variables. We say X_n converges to a random variable X in probability if for any $\varepsilon > 0$*

$$\lim_{n\to\infty} \mathbb{P}(|X_n - X| > \varepsilon) \to 0.$$

Note, that every sequence of random variables converging in probability possesses a subsequence that convergences almost surely.

Definition 7.1.3. *Let X_n, $n \in \boldsymbol{N}$, be a sequence of random variables. We say that X_n convergences in distribution to a random variable X, if the sequence of distributions $\mathbb{P}_{X_n}$ convergences weakly to $\mathbb{P}_X$, that is*

$$\lim_{n\to\infty} \int_{\mathbb{R}^d} f\,\mathrm{d}\mathbb{P}_{X_n} = \int_{\mathbb{R}^d} f\,\mathrm{d}\mathbb{P}_X \quad \textit{for all } f \in C_b(\mathbb{R}^d).$$

The next proposition relates the two concepts of convergence in the following way.

Proposition 7.1.4. *Assume X_n, $n \in \boldsymbol{N}$ is a sequence of random variables that converges to a random variable X in probability. Then X_n converges to X in distribution.*

Proof. Since X_n convergences to X in probability, there is a subsequence $(n_k)_k$ such that $X_{n_k}(\omega)$ convergences to $X(\omega)$ for almost every $\omega \in \Omega$.

For $f \in C_b(\mathbb{R}^d)$, we obtain by the dominated convergence theorem

$$\mathbb{E}[f(X_{n_k})] \to \mathbb{E}[f(X)]. \tag{7.1.1}$$

The same reasoning can be applied to any subsequence of (X_n). Thus, (7.1.1) holds for the whole sequence, which finishes the proof. □

Let μ_n, $n \in \boldsymbol{N}$ be a sequence of distributions and μ a distribution. For abbreviation, we simply write $\mu_n \to \mu$ for convergence in distribution, when no confusion can arise.

A family $\{X_t : \Omega \to \mathbb{R}^d \,|\, t \geqslant 0\}$ of random variables is called stochastic process and is denoted by $(X_t)_{t\geqslant 0}$, (X_t) or simply by X. It can be interpreted as a time-ordered sequence of random events. Given $\omega \in \Omega$, the map $t \mapsto X_t(\omega)$ is called path. A stochastic process Y is called modification of a given stochastic process X if

$$\mathbb{P}(X_t = Y_t) = 1 \quad \text{for all } t > 0.$$

Definition 7.1.5. *Let X be a stochastic process. Given $t_1, \ldots, t_n \geqslant 0$, the measure $\mathbb{P}_{(X_{t_1},\ldots,X_{t_n})} : \mathcal{B}(\mathbb{R}^d) \to [0,1]$ is the joint (finite dimensional) distribution, defined as follows:*

$$\mathbb{P}_{(X_{t_1},\ldots,X_{t_n})}(B_1 \times B_2 \times \cdots \times B_n) = \mathbb{P}(X_{t_1} \in B_1, X_{t_2} \in B_2, \ldots, X_{t_n} \in B_n).$$

Using the notion of finite dimensional distributions, we can define equality in law for two stochastic processes.

Definition 7.1.6. *Two stochastic processes are said to be equal in law if all of their finite dimensional distributions coincide.*

We now define a class of stochastic processes, which play an important role in many fields like population models or financial stock prices.

Definition 7.1.7. *A stochastic process X is called Lévy process if the following holds.*

(i) $\mathbb{P}(X_0 = 0) = 1$.
(ii) *For every $t > 0$, the increments $X_{s+t} - X_s$ do not depend on $s \geqslant 0$.*
(iii) *For every $n \in \mathbb{N}$ and every choice $0 \leqslant t_1 < t_2 < \cdots < t_n$, the family of random variables $\{X_{t_2} - X_{t_1}, \ldots, X_{t_n} - X_{t_{n-1}}\}$ is independent.*
(iv) *For every $t > 0$, X_s converges to X_t in probability for $s \to t$.*

(v) *There is* $\Omega_0 \in \mathcal{F}$ *with* $\mathbb{P}(\Omega_0) = 1$ *such that for every* $w \in \Omega_0$ *the path* $t \mapsto X_t(w)$ *is continuous from the right and has left limits.*

Let us comment on the foregoing definition.

Remark 7.1. *The first property tells us that for almost all* $\omega \in \Omega$ *the paths* $t \mapsto X_t(\omega)$ *start in zero.*

Condition (ii) is the so-called stationarity of increments and is a homogeneity property on the time.

The third property tells us that all increments of the process are independent.

Condition (iv) describes some continuity in measure but not continuity of the paths. A stochastic process satisfying (i)–(iv) is called Lévy process in law. Several authors call such a process Lévy process.

Property (v) is known as "continue à droite limite à gauche" (càdlàg) and means right continuous, left limits. Sometimes, in English it is abbreviated by RCLL.

Let us give an important example of a Lévy process.

Example 7.1.8. *A Lévy process X with values in* $\mathbb{R}$ *resp. in* $\mathbb{N}$ *is called Poisson process with parameter* $\lambda > 0$ *if for every* $t > 0$

$$\mathbb{P}_{X_t}(\{k\}) = e^{-\lambda t}\frac{(\lambda t)^k}{k!} \quad (k \in \mathbb{N}_0),$$
$$\mathbb{P}_{X_t}(B) = 0 \qquad \text{for } B \cap \mathbb{N}_0 = \varnothing.$$

Given a Poisson process, its distribution can be represented in the following way.

Proposition 7.1.9. *Let X be a Poisson process with parameter* $\lambda > 0$. *Then there are real valued random variables* $W_1, W_2, \ldots$ *such that* $T_n = W_n - W_{n-1}$ *has exponential distribution with parameter* λ, *i.e. for every* $n \in \mathbf{N}$

$$\mathbb{P}(T_n \in B) = \lambda \int_{B \cap (0,\infty)} e^{-\lambda x}\, \mathrm{d}\lambda,$$

and

$$X_t(w) = n \Leftrightarrow W_n(w) \leqslant t < W_{n+1}(w).$$

Let $M(d \times d)$ denote the set of all $d \times d$ matrices with entries in $\mathbb{R}$ and let $A \in M(d \times d)$. To shorten notation, we write $A > 0$ if the matrix is positive definite.

Definition 7.1.10. *A random variable X is said to have a nondegenerate Gaussian distribution with mean* $\gamma \in \mathbb{R}^d$ *and covariance* $A \in M(d \times d)$, $A > 0$, *if for every* $B \in \mathcal{B}(\mathbb{R}^d)$

$$\mathbb{P}_X(B) = \frac{1}{\sqrt{(2\pi)^d (\det A)}} \int_B e^{-\frac{1}{2}(x-\gamma, A^{-1}(x-\gamma))}\, \mathrm{d}x.$$

Definition 7.1.11. *A Lévy process X is called Brownian motion or Wiener process if*

(i) *X_t has Gaussian distribution with $\gamma = 0$, $A = tId$,*

(ii) *there is a $\Omega_0 \in \mathcal{F}$ with $\mathbb{P}(\Omega_0) = 1$ such that the paths $t \mapsto X_t(\omega)$ are continuous for every $\omega \in \Omega_0$.*

7.2 Lévy-Khintchine Representation

Lévy processes have the unique feature that, once you know $\mathbb{P}_{X_1}$, then you know all distributions $\mathbb{P}_{X_t}$. Moreover, every Lévy process can be described by a single function $\mathbb{R}^d \to \mathbb{C}$. The aim of this section is to understand this function. All results and proofs from this section can be found in [45].

Definition 7.2.1. *Let μ be a probability measure on $\mathcal{B}(\mathbb{R}^d)$. Then $\widehat{\mu} : \mathbb{R}^d \to \mathbb{C}$,*

$$\widehat{\mu}(z) = \int_{\mathbb{R}^d} e^{i(x,z)} \mu(\mathrm{d}x)$$

is called the characteristic function of μ.

Remark 7.2. *Note that $\widehat{\mu}$ is just the Fourier transformation of the measure μ.*

For a random variable X, we can assign a characteristic function by the characteristic function of its distribution $\mathbb{P}_X$.

The following proposition gives some properties of the characteristic function.

Proposition 7.2.2. *Let μ_n, $n \in \mathbf{N}$ be a sequence of distributions on $\mathbb{R}^d$ and μ a distribution on $\mathbb{R}^d$. Then*

(i) *$\widehat{\mu}$ is uniformly continuous,*

(ii) *$\widehat{\mu}(0) = 1$ and for all $z \in \mathbb{R}^d$: $|\widehat{\mu}(z)| \leqslant 1$.*

(iii) *If $\widehat{\mu}_1 = \widehat{\mu}_2$, then $\mu_1 = \mu_2$.*

(iv) *If $\mu = \mu_1 * \mu_2$, then $\widehat{\mu} = \widehat{\mu}_1 \cdot \widehat{\mu}_2$.*
If X, Y are independent random variables then

$$\widehat{\mathbb{P}_{X+Y}} = \widehat{\mathbb{P}}_X \cdot \widehat{\mathbb{P}}_Y.$$

(v) *If $X_1, \dots, X_n$ are random variables and $X = (X_1, \dots, X_n)$ is a $\mathbb{R}^{d\times n}$ random variable, then $X_1, \dots, X_n$ are independent iff for every $z = (z_1, \dots, z_n) \in \mathbb{R}^{d\times n}$*

$$\widehat{\mathbb{P}}_X(z) = \widehat{\mathbb{P}}_{X_1}(z_1) \cdot \dots \cdot \widehat{\mathbb{P}}_{X_n}(z_n).$$

(vi) *If $\mu_n \to \mu$, then $\widehat{\mu}_n \to \widehat{\mu}$ uniformly on compact subsets.*
Furthermore, if $\widehat{\mu}_n(z) \to \widehat{\mu}(z)$ for all $z \in \mathbb{R}^d$, then $\mu_n \to \mu$.

(vii) *Let $\varphi : \mathbb{R}^d \to \mathbb{C}$ be continuous at zero. If $\widehat{\mu}_n \to \varphi$, then φ itself is a characteristic function.*

(viii) *If* $\int |\widehat{\mu}(z)| dz < \infty$, *then* μ *is absolutely continuous w.r.t. the Lebesgue measure with density* $g \in C_b(\mathbb{R}^d)$ *satisfying*

$$g(x) = (2\pi)^{-d} \int e^{-i(x,z)} \widehat{\mu}(z) dz.$$

In the following steps, the connection between the characteristic function of a Lévy processes and a special class of probability measures is studied. This is the class of infinitely divisible distributions. By the Lévy-Khintchine formula the characteristic functions of Lévy processes and infinitely divisible distributions will be set into a one-to-one correspondence.

Definition 7.2.3. *A probability measure* μ *on* $\mathcal{B}(\mathbb{R}^d)$ *is infinitely divisible if, for any* $n \in \mathbb{N}$, *there is a probability measure* μ_n *on* $\mathcal{B}(\mathbb{R}^d)$ *with*

$$\mu = \mu_n * \cdots * \mu_n = (\mu_n)^n. \tag{7.2.1}$$

Note, that by 7.2.2 the convolution of two distributions is expressed by the product of the corresponding characteristic functions, which can be characterized again by the sum of independent random variables. Hence a random variable X has an infinitely divisible distribution if for each $n \in \boldsymbol{N}$ there are i.i.d. random variables $X_{1,n}, \ldots, X_{n,n}$ such that

$$X \stackrel{d}{=} X_{1,n} + \cdots + X_{n,n}.$$

We proceed by showing that the characteristic function of an infinitely divisible distribution has no zero. For $z \in \mathbb{C}$, we write $\overline{z}$ for the complex conjugate.

Lemma 7.2.4. *If* μ *is an infinitely divisible distribution, then* $\widehat{\mu} \neq 0$ *on* $\mathbb{R}^d$.

Proof. For each $n \in \boldsymbol{N}$ there is μ_n such that

$$(\widehat{\mu}_n(z))^n = \widehat{\mu}(z) \qquad \text{for all } z \in \mathbb{R}^d. \tag{7.2.2}$$

Define $\tilde{\mu}$ by $\tilde{\mu}(B) = \mu(-B)$ and $\mu^\#$ by $\mu^\# = \mu * \tilde{\mu}$. Then $\widehat{\mu^\#} = |\widehat{\mu}|^2$. Thus $|\widehat{\mu}_n|^2$ is a characteristic function and $|\widehat{\mu}_n|^2 = |\widehat{\mu}|^{2/n}$. By 7.2.2 $\widehat{\mu}(0) = 1$ and $\widehat{\mu}$ is continuous.

Define

$$\varphi(z) = \lim_{n\to\infty} |\widehat{\mu}_n(z)|^2$$

and observe

$$\varphi(z) = \begin{cases} 1 & \text{if } \widehat{\mu}(z) \neq 0 \\ 0 & \text{if } \widehat{\mu}(z) = 0 \end{cases}$$

7.2.2 (vii) ensures that φ is a characteristic function, thus φ is continuous. Since $\widehat{\mu}(0) = 1$ and $\widehat{\mu}$ is continuous, it follows that $\varphi = 1$ on $B_\varepsilon(0)$ for some $\varepsilon > 0$. Since φ is continuous and $\varphi \in \{0, 1\}$, we conclude $\varphi \equiv 1$. Hence $\widehat{\mu} \neq 0$ on $\mathbb{R}^d$. ☐

Remark 7.3. *In general, $\hat{\mu} \neq 0$ on $\mathbb{R}^d$ does not imply that μ is an infinitely divisible distribution.*

Lemma 7.2.5. *Suppose $\varphi : \mathbb{R}^d \to \mathbb{C}$ is continuous with $\varphi(0) = 1$, $\varphi \neq 0$ on $\mathbb{R}^d$. Then there is a unique continuous function $f : \mathbb{R}^d \to \mathbb{C}$, with $f(0) = 0$ and*

$$e^{f(z)} = \varphi(z) \qquad \text{for all } z \in \mathbb{R}^d. \tag{7.2.3}$$

We write $\log \varphi(z)$ instead of $f(z)$. Note that in general $\varphi(z_1) = \varphi(z_2)$ does not imply $\log \varphi(z_1) = \log \varphi(z_2)$. The following results provide useful properties of this new function.

Lemma 7.2.6. *Assume $\varphi : \mathbb{R}^d \to \mathbb{C}$ and for any $n \in \boldsymbol{N}$ let $\varphi_n : \mathbb{R}^d \to \mathbb{C}$ such that φ, φ_n are continuous with $\varphi(0) = \varphi_n(0) = 1$ and $\varphi \neq 0$, $\varphi_n \neq 0$ on $\mathbb{R}^d$ for all $n \in \boldsymbol{N}$. Assume $\varphi_n \to \varphi$ uniformly on compact sets as $n \to \infty$. Then $\log \varphi_n \to \log \varphi$ uniformly on compact sets as $n \to \infty$.*

Corollary 7.2.7. *Let (μ_n) be a sequence of infinitely divisible distributions. Let μ be an arbitrary distribution with $\mu_n \to \mu$. Then μ is an infinitely divisible distribution itself.*

Lemma 7.2.8. *Let X be a Lévy process. Then for every $t > 0$ the measure $\mathbb{P}_{X_t}$ is infinitely divisible. Moreover, $\mathbb{P}_{X_t} = (\mathbb{P}_{X_1})^t$, where the power is defined appropriately.*

Proof. Assume $X = (X_t)$ is a Lévy process. Let $n \in \mathbb{N}$. Set $\mu = \mathbb{P}_{X_t}$ and $\mu_n = \mathbb{P}_{X_{t_k} - X_{t_{k-1}}}$, where $t_k = \frac{t}{n} \cdot k$ for $k \in \{0, 1, \dots, n\}$.

By the stationary of increments of the Lévy process X, the choice of μ_n does not depend on k. Note

$$X_t = (X_{t_n} - X_{t_{n-1}}) + \cdots + (X_{t_1} - X_{t_0}). \tag{7.2.4}$$

Thus, X_t equals the sum of n independent random variables with equal distributions μ_n.

Hence $\mu = (\mu_n)^n$. From $\mathbb{P}_{X_{\frac{1}{n}}} = (\mathbb{P}_{X_1})^{\frac{1}{n}}$, we deduce $\mathbb{P}_{X_{\frac{m}{n}}} = (\mathbb{P}_{X_1})^{\frac{m}{n}}$. If $t \in \mathbb{Q}$, we are done. If $t \in \mathbb{R} \backslash \mathbb{Q}$, choose $(r_n) \subset \mathbb{Q}$ such that $r_n \to t$ for $n \to \infty$.

By definition of Lévy processes, we know $X_{r_n} \to X_t$ in probability, hence $\mathbb{P}_{X_{r_n}} \to \mathbb{P}_{X_t}$ for $n \to \infty$. Set $\nu_n = \mathbb{P}_{X_{r_n}}$ and $\nu = \mathbb{P}_{X_1}$. Then for every $z \in \mathbb{R}^d$ the characteristic function $\hat{\nu}_n(z)$ converges to $e^{t \log \hat{\nu}(z)} = (\hat{\nu}(z))^t$, where $\hat{\nu}$ is continuous. Thus $(\hat{\nu}(\cdot))^t$ is the characteristic function of some distribution, which we define as $(\mathbb{P}_{X_1})^t$. □

Let's summarize:

Theorem 7.2.9. *Let X be a Lévy process on $\mathbb{R}^d$. Then $\mathbb{P}_{X_t}$ is infinitely divisible for every $t > 0$. The corresponding characteristic function satisfies*

$$\widehat{\mathbb{P}}_{X_t}(z) = e^{t \log \widehat{\mathbb{P}}_{X_1}(z)} \tag{7.2.5}$$

The measures $(\mathbb{P}_{X_t})_{t>0}$ *form a convolution semigroup, i.e. for* $\mu_t = \mathbb{P}_{X_t}$ *we have*

- $\mu_t * \mu_s = \mu_{t+s} \qquad \forall t, s > 0$
- $\mu_0 = \delta_0$
- $\mu_t \to \delta_0$ *for* $t \to 0$

Definition 7.2.10. *The function*

$$z \longmapsto \log \widehat{\mathbb{P}}_{X_1}(z) = \psi(z) \tag{7.2.6}$$

is called symbol of the Lévy process X.

Let us look at examples of Lévy processes and symbols.

1. Assume X is a Poisson process, i.e., $\mathbb{P}_{X_1}(\{k\}) = e^{-\lambda}\frac{\lambda^k}{k!}$. Then $\psi(z) = \lambda(e^{iz} - 1) = \log \widehat{\mathbb{P}}_{X_1}(z)$.
2. Assume X is a Gaussian process with parameters $A \in M(d \times d)$, $A > 0$, and $\gamma \in \mathbb{R}^d$. Then $\psi(z) = -\frac{1}{2}(z, Az) + i(\gamma, z)$.

Theorem 7.2.11 (Lévy-Khintchine).

(i) Let μ *be an infinitely divisible probability measure. There exist a symmetric* $A \in M(d \times d)$ *with* $A \geqslant 0$, $\gamma \in \mathbb{R}^d$ *and a measure* ν *on* $\mathcal{B}(\mathbb{R}^d)$ *with* $\nu(\{0\}) = 0$ *and* $\int_{\mathbb{R}^d} \min(1, |h|^2)\nu(\mathrm{d}h) < \infty$, *such that the characteristic function* $\widehat{\mu}$ *of* μ *is given by*

$$\widehat{\mu}(z) = \exp\left[-\frac{1}{2}(z, Az) + i(\gamma, z) + \int_{\mathbb{R}^d} \left(e^{i(x,z)} - 1 - i(x,z)\mathbb{1}_{B_1}(x) \right)\nu(\mathrm{d}x) \right]. \tag{7.2.7}$$

(ii) The representation of $\widehat{\mu}$ *in terms of* (A, γ, ν) *is unique.*

(iii) The reverse direction of (i) is true. Given (A, γ, ν) *and* $\widehat{\mu}$ *with the representation as in (7.2.7), there is an infinitely divisible distribution* μ *with* $\widehat{\mu}$ *as in (7.2.7).*

The matrix A is called Gaussian variance and $\gamma \in \mathbb{R}^d$ is the so-called drift parameter. The measure ν is called the Lévy measure. It is a Radon measure which describes the jump of underlying Lévy process. (A, γ, ν) is called Lévy triplet. Since the characteristic function of any $\mathbb{R}^d$-valued random variable completely defined its probability distribution, the corresponding Lévy process is fully determined by (A, γ, ν).

Example 7.2.12. *Let* $A = 0$, $\gamma = 0$ *and* $\nu(dh) = C_{\alpha,d}|h|^{-d-\alpha}\,\mathrm{d}h$ *for* $\alpha \in (0, 2)$ *and an appropriate constant* $C_{\alpha,d}$, *chosen such that*

$$C_{\alpha,d} \int_{\mathbb{R}^d} \frac{1 - \cos(h_1)}{|h|^{d+\alpha}}\,\mathrm{d}h = 1.$$

By 7.2.11 this yields to

$$\widehat{\mu}(z) = \exp(-|z|^\alpha).$$

The associated Lévy process is the so-called isotropic α*-stable Lévy process.*

Another important class of jump processes is given by the compound Poisson distribution.

Definition 7.2.13. *A distribution on $\mathbb{R}^d$ is called compound Poisson if $\widehat{\mu}(z) = e^{\lambda(\widehat{\sigma}(z)-1)}$ for some $\lambda > 0$ and some distribution σ with $\sigma(\{0\}) = 0$.*

Example 7.2.14. *Let $d = 1$ and $\sigma = \delta_1$. Then*

$$\widehat{\sigma}(z) = \int_{\mathbb{R}} e^{izh} \sigma(dh) = e^{iz}.$$

Hence

$$\log \widehat{\mu}(z) = \lambda(\widehat{\sigma}(z) - 1) = \lambda(e^{iz} - 1).$$

Remark 7.4.

1. *The integral expression in the representation of the characteristic function in the Lévy-Khintchine formula is well-defined by the properties on the Lévy measure ν. Note that*

   $$e^{i(z,h)} - 1 - i(z,h)\mathbb{1}_{B_1}(h) = \mathcal{O}(|h|^2) \text{ at } |h| \to 0$$

   and is bounded for $|h| > 1$.
2. *The concrete form of the „cutoff" term $\mathbb{1}_{B_1}$ in the integral in (7.2.7) is not important. One option to replace it is given by $\chi : \mathbb{R}^d \to \mathbb{R}$ with the following two properties*

   $$\chi(h) = 1 + \sigma(|h|) \text{ for } |h| \to 0,$$
   $$\chi(h) = \mathcal{O}(\frac{1}{|h|}) \text{ for } |h| \to \infty.$$

   If one replaces $\mathbb{1}_{B_1}(h)$ in (7.2.7) by $\chi(h)$, then one also needs to replace the drift term γ by

   $$\gamma_n = \gamma + \int_{\mathbb{R}^d} h(\chi(h) - \mathbb{1}_{B_1}(h))\nu(\mathrm{d}h). \tag{7.2.8}$$

   Some possible examples for χ are:
   - $\chi(h) = \mathbb{1}_{B_{1/93}}(h)$
   - $\chi(h) = \frac{1}{(1+|h|^2)}$
   - $\chi(h) = \mathbb{1}_{B_1}(h) + (2 - |h|)\mathbb{1}_{B_2/B_1}$

Let us comment on the proof of the Lévy-Khintchine theorem. We select some important steps of the proof provided in [45] and refer to this book for the important details.

Idea of the proof of (ii) in 7.2.11: First, one shows

$$\lim_{s\to\infty} s^{-2} \log \widehat{\mu}(sz) = -\frac{1}{2}(z, Az) \tag{7.2.9}$$

so μ determines A. Second, one determines ν. Last, the uniqueness of γ follows.

Proof of (iii) in 7.2.11: Define $\varphi : \mathbb{R}^d \to \mathbb{C}$ and $\varphi_n : \mathbb{R}^d \to \mathbb{C}$, $n \in N$, as follows. Let $\varphi(z)$ be the right-hand side of (7.2.7) and let φ_n equal the right-hand side of (7.2.7), where the area of integration in the integral is not over $\mathbb{R}^d$ but $\mathbb{R}^d \backslash B_{1/n}$.

Then $\varphi_n = \exp(-\frac{1}{2}(z, Az) + \widehat{\sigma}_n(z))$ with some appropriate compound Poisson distribution σ_n. φ_n is a characteristic function of an infinitely divisible distribution and $\varphi_n(z) \to \varphi(z)$. Since φ is continuous, φ itself turns out to be the characteristic function of an infinitely divisible distribution.

Idea of the proof of (i) in 7.2.11. Let μ be an infinitely divisible distribution. For $n \in \mathbb{N}$, let $t_n \in (0, \infty)$ with $\lim_{n\to\infty} t_n = 0$ and define μ_n through

$$
\begin{aligned}
\widehat{\mu}_n(z) &= \exp\left(\frac{(\widehat{\mu}(z))^{t_n} - 1}{t_n}\right) \\
&= \exp\left(t_n^{-1} \int_{\mathbb{R}^d \backslash \{0\}} (e^{i(z,h)} - 1)\mu^{t_n}(dh)\right) \\
&= \exp\left(t_n^{-1}(e^{t_n \log \widehat{\mu}(z)} - 1)\right) \\
&= \exp\left(t_n^{-1}(t_n \log \widehat{\mu}(z) + \mathcal{O}(t_n^2))\right) \\
&\to \exp(\log \widehat{\mu}(z)) = \widehat{\mu}(z).
\end{aligned}
$$

Thus by 7.2.2, $\mu_n \to \mu$. The proof is completed, once one characterizes the convergence of infinitely divisible distributions by the convergence of Lévy triplets as in the following theorem:

Theorem 7.2.15. *Assume χ is a bounded continuous cutoff function. Let (μ_n) be a sequence of infinitely divisible distributions with Lévy triplet $(A_n, \nu_n, \gamma_n)^\chi$. Let μ be any distribution on $\mathbb{R}^d$. Then $\mu_n \to \mu$ holds true if and only if the following two conditions are satisfied:*

(i) μ is an infinitely divisible distribution.

(ii) μ has Lévy triplet $(A, \nu, \gamma)^\chi$ with

$$
\lim_{\varepsilon\to 0} \limsup_{n\to\infty} |(z, A_{n,\varepsilon} z) - (z, Az)| = 0,
$$

$$
\int_{\mathbb{R}^d} f \, d\nu_n \to \int_{\mathbb{R}^d} f \, d\nu \quad \textit{for } f \in C_b(\mathbb{R}^d) \cap \{f = 0 \textit{ in some } B_r(0)\},
$$

and $\gamma_n \to \gamma$.

where $A_{n,\varepsilon}$ is defined as follows:
$(z, A_{n,\varepsilon} z) = (z, A_n z) + \int_{B_\varepsilon(0)} |(z, h)|^2 \nu_n(dh)$.

7.3 Generators of Lévy Processes

In this section, we study infinitesimal generators of strongly continuous semigroups related to Lévy processes. The material is standard and there are several good sources, e.g., [45] or [22]. Suppose $X = (X_t)_{t\geqslant 0}$ is a Lévy process on $\mathbb{R}^d$ with $\mu = \mathbb{P}_{X_1}$. For $t \geqslant 0$ and $B \in \mathcal{B}(\mathbb{R}^d)$ we define the corresponding transition function by

$$P_t(x, B) = \mu^t(B - x).$$

We define $P_t : L^\infty(\mathbb{R}^d) \to L^\infty(\mathbb{R}^d)$ for $t > 0$ by

$$P_tf(x) = \int_{\mathbb{R}^d} f(y)P_t(x, dy) = \int_{\mathbb{R}} f(x + h)\mu^t\,(dh) = \mathbb{E}[f(x + X_t)].$$

Further let $C_\infty(\mathbb{R}^d)$ be the space of all continuous functions vanishing at infinity, that is

$$C_\infty(\mathbb{R}^d) = \{v \in C_b(\mathbb{R}^d) : \lim_{|x|\to\infty} v(x) = 0\} \tag{7.3.1}$$

and

$$C^2_\infty(\mathbb{R}^d = \{v \in C_\infty(\mathbb{R}^d) : \forall|\alpha| \leqslant 2\ \partial^\alpha f \in C_\infty(\mathbb{R}^d)\}. \tag{7.3.2}$$

Proposition 7.3.1. *The family (P_t) forms a strongly continuous semigroup on $C_\infty(\mathbb{R}^d)$, i.e.*

$$P_0 = Id, \quad P_{t+s} = P_tP_s \text{ for all } t, s \geqslant 0 \tag{7.3.3}$$

and

$$\|P_tf - f\| \to 0 \text{ for } t \to 0. \tag{7.3.4}$$

Furthermore, $\|P_t\| = 1$ for all $t \geqslant 0$.

Proof. The properties $P_0 = \mathrm{Id}$ and $P_{t+s} = P_tP_s$ are easy to see.

Let's prove $\|P_tf - f\| \to 0$. Let $f \in C_\infty(\mathbb{R}^d)$. Note, f is uniformly continuous. Given $\varepsilon > 0$, choose $\delta = \delta(\varepsilon)$ with

$$|h| < \delta \Rightarrow |f(x + h) - f(x)| < \varepsilon \text{ for all } x \in \mathbb{R}^d. \tag{7.3.5}$$

For $x \in \mathbb{R}^d$ and $t > 0$

$$\begin{aligned}|P_tf(x) - f(x)| &\leqslant \Big|\int_{B_\delta} (f(x + h) - f(x))\mu^t(dh)\Big| \\ &\quad + \Big|\int_{\mathbb{R}^d\setminus B_\delta} (f(x + h) - f(x))\mu^t(dh)\Big|\end{aligned}$$

$$\leqslant \varepsilon + 2\|f\|_\infty \mu^t(\mathbb{R}^d \backslash B_\delta)$$
$$\leqslant \varepsilon + 2\|f\|_\infty \varepsilon \text{ for } t < t_0(\varepsilon)\,,$$

where we used the stochastic continuity. The estimate $\|P_t\| \leqslant 1$ is evident. To show $\|P_t\| = 1$, let $f_n : \mathbb{R}^d \to \mathbb{R}$ be a sequence in $C_\infty(\mathbb{R}^d)$ with $0 \leqslant f_n \leqslant 1$ and $f_n = 1$ on $\overline{B_n}$. Then $\lim_{n\to\infty} P_t f_n = 1$ and therefore $\|P_t\| = 1$. □

A strongly continuous semigroup is called a strongly continuous contraction semigroup if $\|P_t\| \leqslant 1$ holds for every $t \geqslant 0$.

Given a strongly continuous contraction semigroup, one can define the infinitesimal generator as follows:

Definition 7.3.2. *The infinitesimal generator L of a strongly continuous contraction semigroup (P_t) on a Banach space X is defined by*

$$Lf = \lim_{t\to 0+} \frac{P_t f - f}{t} \tag{7.3.6}$$

for $f \in D(L) = \{f \in X : \lim_{t\downarrow 0} \frac{P_t f - f}{t}$ exists $\}$.

In the following proposition we study the Fourier transform of the semigroup and its generator.

Proposition 7.3.3.
(i) For $f \in L^1(\mathbb{R}^d) \cap L^\infty(\mathbb{R}^d)$, $t \geqslant 0$

$$\widehat{P_t f}(\xi) = e^{t\psi(-\xi)}\widehat{f}(\xi) \tag{7.3.7}$$

(ii) For $f \in \mathcal{D}(L)$, $Lf \in L^1(\mathbb{R}^d)$

$$\widehat{Lf}(\xi) = \psi(-\xi)\widehat{f}(\xi), \tag{7.3.8}$$

where $(L, \mathcal{D}(L))$ is the infinitesimal generator of (P_t).

Proof.

(i)

$$\widehat{P_t f}(\xi) = \int_{\mathbb{R}^d} e^{i(\xi,x)}\mathbb{E}(f(x+X_t))dx = \mathbb{E}[\int_{\mathbb{R}^d} e^{i(\xi,y-X_t)} f(y)dy]$$
$$= \mathbb{E}[e^{i(-\xi,X_t)}]\int e^{i(\xi,y)} f(y)dy$$

(ii) follows from (i).

□

It is a reasonable question to ask about a relation between the Lévy measure ν and ψ. Assume $\nu(dh) = k(h)dh$ with $K(h) = K(-h)$. Assume that for some $R_0 \in (0, +\infty)$ and some $\Lambda \geqslant 1$

$$\Lambda^{-1}\frac{\ell(|h|)}{|h|^d} \leqslant k(h) \leqslant \Lambda\frac{\ell(|h|)}{|h|^d} \quad \text{for } |h| < R_0, \tag{7.3.9}$$

where $\ell : (0, R_0) \to (0, +\infty)$ satisfies

$$\int_0^{R_0} \ell(s)\frac{ds}{s} = +\infty \tag{7.3.10}$$

plus some weak scaling condition.

Proposition 7.3.4 ([26]). *Set $\mathcal{L}(r) = \int_r^{R_0} \ell(s)\frac{ds}{s}$. Then there are $c \geqslant 1$ and $r_0 > 0$ such that*

$$c^{-1}\mathcal{L}(|\xi|^{-1}) \leqslant \psi(\xi) \leqslant c\mathcal{L}(|\xi|^{-1}) \text{ for } \xi \in \mathbb{R}^d, |\xi| \geqslant r_0\,. \tag{7.3.11}$$

Let us give some examples for the function ℓ and $\mathcal{L}$ for 7.3.4.

Example 7.3.5.

(i) $\ell(s) = s^{-\alpha}, \mathcal{L}(s) \asymp s^{-\alpha}$ *for* $0 < \alpha < 2$

(ii) $\ell(s) = \ln(\frac{2}{s}), \mathcal{L}(s) = \ln^2(\frac{2}{s})$

(iii) $\ell(s) = 1, \mathcal{L}(s) = \ln(\frac{1}{s})$

Theorem 7.3.6. *Let $(L, \mathcal{D}(L))$ denote the infinitesimal generator of (P_t) on $C_\infty(\mathbb{R}^d)$. Then $C_c^\infty(\mathbb{R}^d)$ is a core for L. Moreover $C_\infty^2(\mathbb{R}^d) \subset \mathcal{D}(L)$ and for $f \in C_\infty^2(\mathbb{R}^d)$*

$$\begin{aligned} Lf(x) &= \frac{1}{2}\sum_{i,j} a_{ij}\partial_i\partial_j f + \sum_i \gamma_i\partial_i f \\ &\quad + \int_{\mathbb{R}^d} [f(x+h) - f(x) - (h, \nabla f(x))\mathbb{1}_{B_1}(h)]\,\mu(\mathrm{d}h), \end{aligned} \tag{7.3.12}$$

where (A, γ, ν) is the Lévy triplet of X.

If A and γ equal zero, then L becomes an integrodifferential operator or, if $\mu(\mathbb{R}^d)$ is finite, an integral operator. Let us assume that μ is an isotropic α-stable measure for some $\alpha \in (0, 2)$, i.e.,

$$\mu(\mathrm{d}h) = C_{\alpha,d}\,|h|^{-d-\alpha}\,\mathrm{d}h\,,$$

where $C_{\alpha,d}$ is a specific constant, cf. 7.2.12. Then

$$Lf(x) = C_{\alpha,d}\int_{\mathbb{R}^d} [f(x+h) - f(x) - \mathbb{1}_{B_1}(h)(\nabla f(x), h)]|h|^{-d-\alpha}\,\mathrm{d}h$$

$$= C_{\alpha,d} \lim_{\varepsilon\to 0+} \int\limits_{\mathbb{R}^d\setminus B_\varepsilon} [f(x+h) - f(x)]|h|^{-d-\alpha}\,\mathrm{d}h$$

$$= \tfrac{1}{2} C_{\alpha,d} \int\limits_{\mathbb{R}^d} [f(x+h) - 2f(x) + f(x-h)]|h|^{-d-\alpha}\,\mathrm{d}h\,.$$

It is important to note that, again for $f \in C_\infty(\mathbb{R}^d)$, the following identity

$$C_{\alpha,d} \lim_{\varepsilon\to 0+} \int\limits_{\mathbb{R}^d\setminus B_\varepsilon} [f(x+h) - f(x)]|h|^{-d-\alpha}\,\mathrm{d}h = -(-\Delta)^{\alpha/2} f(x)$$

holds true. Here, the operator $(-\Delta)^{\alpha/2} f$ is defined as follows:

$$(-\Delta)^{\alpha/2} f\,(\xi) = |\xi|^\alpha \widehat{f}(\xi) \qquad (\xi \in \mathbb{R}^d, f \in C_c^\infty(\mathbb{R}^d))$$

In this sense, the fractional Laplace operator appears as the generator of the isotropic α-stable process. Of course, the precise value of $C_{\alpha,d}$ is important for the equality above. However, for most applications, the asymptotic behavior $C_{\alpha,d} \asymp \alpha(2-\alpha)$ is sufficient.

7.4 Nonlocal Operators and Jump Processes

In Section 7.3 we have studied generators of strongly continuous semigroups, which correspond to Lévy processes. As shown in 7.3.6, these generators can be represented as translation invariant integrodifferential operators. In this section, we comment on the relation between more general jump processes and more general integrodifferential operators. A possibility to link these objects is given by the martingale problem.

The aim of this section is to explain how solvability of the deterministic Cauchy problem for an integrodifferential operator of the form $\partial_t - L$ implies well-posedness of the martingale problem for L. We restrict ourselves to purely nonlocal operators of the form

$$Lf(x) = \int\limits_{\mathbb{R}^d} \left[f(x+h) - f(x) - (h\nabla f(x))\mathbb{1}_{B_1}(h)\right] n(x,h)\,\mathrm{d}h, \tag{7.4.1}$$

where $n : \mathbb{R}^d \times \mathbb{R}^d\setminus\{0\} \to [0,\infty]$ satisfies

$$\sup_{x\in\mathbb{R}^d} \int (1 \wedge |h|^2) n(x,h)\,\mathrm{d}h < \infty.$$

Solvability of the Cauchy problem requires some additional conditions, which we discuss further below. Note that $Lf(x)$ is well defined for bounded functions $f : \mathbb{R}^d \to \mathbb{R}$, which are sufficiently regular in a neighborhood of $x \in \mathbb{R}^d$. The material of this section closely follows the presentation in the preprint [1]. Note that is has not been published in [2]. Since there seems to be no survey article on the material of this chapter, we give many references.

Let us state the martingale problem. By $\mathcal{D}([0,\infty);\mathbb{R}^d)$ we denote the space of all càdlàg paths. Below we give a precise definition and a short discussion of $\mathcal{D}([0,\infty);\mathbb{R}^d)$.

Definition 7.4.1. *A probability measure $\mathbb{P}^\mu$ on $\mathcal{D}([0,\infty);\mathbb{R}^d)$ is a solution to the martingale problem for $(L, D(L))$ with domain $D(L)$ being contained in the set of bounded functions $f\colon \mathbb{R}^d \to \mathbb{R}$, L defined as in (7.4.1) and μ a probability measure on $\mathbb{R}^d$ if, for every $\varphi \in D(L)$,*

$$\Big(\varphi(\Pi_t) - \varphi(\Pi_0) - \int_0^t (L\varphi)(\Pi_s)\,ds\Big)_{t\geqslant 0}$$

is a $\mathbb{P}^\mu$-martingale with respect to the filtration $(\sigma(\Pi_s; s \leqslant t))_{t\geqslant 0}$ and $\mathbb{P}^\mu(\Pi_0 = \mu) = 1$. Here Π is the usual coordinate process, i.e., $\Pi\colon [0,\infty) \times \mathcal{D}([0,\infty);\mathbb{R}^d) \to \mathbb{R}^d$, $\Pi_t(\omega) = \omega(t)$. If for every μ there is a unique solution $\mathbb{P}^\mu$ of the martingale problem, we say that the martingale problem for $(L, D(L))$ is well-posed.

The well-posedness of the martingale problem is closely related to weak uniqueness for the corresponding stochastic differential equation. In these notes, we mainly omit related questions of stochastic analysis. A very good source for this is [5].

7.4.1 References for the martingale problem for nonlocal operators

Let us mention some important results concerning the martingale problem for nonlocal operators. Note that the case where $n(x,h)$ does not depend on x is very special because, in this case, L is translation invariant and generates a Lévy process, cf. [9], [45]. One could say that L has constant coefficients in this case.

The martingale problem for an operator of the form $A + L$ where A is a non-degenerate elliptic operator and L is an operator of our type has been studied in [31, 48, 37]. Since A is a second order operator, L is a lower order perturbation of A. [32, 33] treat the martingale problem for pure jump processes generated by operators like L, i.e., $A = 0$. See also [42].

Using pseudodifferential operators and anisotropic Sobolev spaces [19] proves the well-posedness of the martingale problem under assumptions like $x \mapsto n(x,h) \in C^{3d}(\mathbb{R}^d)$. [19] allows for a rather general dependence of $n(x,h)$ on h. More general cases are treated in [30]. In the setting of [19], a parametrix for the pseudodifferential operator is constructed in [13]. The parametrix methods is also applied in [29], [36], [35] and [12].

The, by now classical, method to prove uniqueness for the martingale problem was established in [49]. The main idea is to solve the deterministic Cauchy problem for the operator L in sufficiently regular function spaces. This approach has been carried out in [40, 39, 41, 2].

The uniqueness for the martingale problem is closely related to the uniqueness in law for stochastic differential equations, see [51, 4, 6]. Recent results on stochastic differential equations driven by jump processes can be found in [43, 38, 18, 44, 15, 16]. Systems of such stochastic differential equations often lead to nonlocal operators with singular measures. The martingale problem for such operators has been studied in [7]. We also draw the readers attention to the interesting recent work [25] where counterexamples and sufficient criteria for uniqueness of the martingale problem are presented.

A particular case of nonlocal operators arises if $n(x, h)$ is bounded from above and below by $|h|^{-d-\alpha(x)}$, where $\alpha : \mathbb{R}^d \to (0, 2)$ is a function. [21] provides a nice introduction into this framework including existence results. Well-posedness of the martingale problem is proved in one spatial dimension in [3] when $\alpha(\cdot)$ is Dini-continuous. Uniqueness problems for stochastic differential equations in similar situations including higher dimensions and diffusion coefficients are considered in [50]. The techniques of [3] can be extended to higher dimensions and to a larger class of problems. [23], [28], and [34] provide sufficient conditions for L to extend to a generator of a Feller processes in this framework. [20] provides such a result together with well-posedness of the martingale problem when $x \mapsto \alpha(x)$ is smooth.

7.4.2 The path space of càdlàg paths

The standard reference for the martingale problem for diffusion operators is [49]. Since the paths of jump processes are not continuous by nature we have to set up the martingale problem for the path space $\mathcal{D}([0, \infty); \mathbb{R}^d)$ of all càdlàg paths. Good sources for this space are [11], [17], [24], but the first edition [10] is sufficient for many purposes. A good reference for the martingale problem on $\mathcal{D}([0, \infty); \mathbb{R}^d)$ is [17].

We denote by $\mathcal{D}([0, \infty); \mathbb{R}^d)$ the set of all functions $\omega : [0, \infty) \to \mathbb{R}^d$ satisfying for all $t \geqslant 0$

$$\lim_{s \to t+} \omega(s) = \omega(t) , \qquad \exists\, \omega(t-) = \lim_{s \to t-} \omega(s) .$$

A basic fact about $\mathcal{D}([0, \infty); \mathbb{R}^d)$ is that any $\omega \in \mathcal{D}([0, \infty); \mathbb{R}^d)$ has at most countably many points of discontinuity. As on the space of continuous functions the mapping d_{uc} defined by

$$d_{uc}(\omega_1, \omega_2) = \sum_{k \in N} 2^{-k} \min\{1, \sup_{t \leqslant k} |\omega_1(t) - \omega_2(t)|\}$$

defines a metric. The space $(\mathcal{D}([0,\infty);\mathbb{R}^d), d_{uc})$ is a complete metric space but, in contrast to the case of continuous functions, it is not separable. To see this, consider

$$M := \left\{\omega_s \in \mathcal{D}([0,\infty);\mathbb{R}^d); \omega_s(t) = \mathbb{1}_{[s,\infty)}(t), s \in [0,1)\right\}.$$

There cannot exist a countable dense subset A of the uncountable set M since $d_{uc}(\omega_s, \omega_t) = \frac{1}{2}$ as along as $s \neq t$. The set A would need to be uncountable right away.

Nevertheless, there exists a metrizable topology on $\mathcal{D}([0,\infty);\mathbb{R}^d)$ such that it becomes a complete, separable metric space. We summarize the main results on this space in the following theorem. Since the space $\mathcal{D}([0,\infty);\mathbb{R}^d)$ is not too well known among analysts we include many details in this theorem. It is almost identical to Theorem VI.1.14 in [24].

Proposition 7.4.2. *(1) There exists a metrizable topology on $\mathcal{D}([0,\infty);\mathbb{R}^d)$, called the Skohorod topology for which the space is complete and separable. Denote this metric by d. Then $d(\omega_n, \omega) \to 0$ is equivalent to the existence of a sequence of strictly increasing functions $\lambda_n : [0,\infty) \to [0,\infty)$, satisfying $\lambda_n(0) = 0$, $\lambda_n(t) \nearrow \infty$ for $t \to \infty$ and at the same time*

$$\begin{cases} \sup\limits_{s\geqslant 0} |\lambda_n(s) - s| \to 0 \quad \text{as } n \to \infty\,, \\ \left(\sup\limits_{s\leqslant k} |\omega_n(\lambda_n(s)) - \omega(s)| \to 0 \quad \text{as } n \to \infty\right) \quad \forall\, k \in \mathbf{N}\,. \end{cases}$$

(2) A set $M \subset \mathcal{D}([0,\infty);\mathbb{R}^d)$ is relatively compact for the Skohorod topology if and only if

$$\begin{cases} \sup\limits_{\omega\in M} \sup\limits_{s\leqslant k} |\omega(s)| < \infty \quad \forall k \in \mathbf{N}\,, \\ \lim\limits_{\rho\to 0+} \sup\limits_{\omega\in M} \gamma_k(\omega,\rho) = 0 \quad \forall\, k \in \mathbf{N}\,. \end{cases}$$

where $\gamma_k(\omega, t)$ is a generalized modulus of continuity, defined via

$$\gamma_k(\omega,\rho) = \inf\Big\{ \max_{i\leqslant L} \gamma(\omega; [t_{i-1}, t_i)) : 0 \\ = t_0 < \ldots < t_L = k, \inf_{i<L}(t_i - t_{i-1}) \geqslant \rho\Big\}\,,$$

where $\gamma(\omega; I)$ is the usual modulus of continuity for ω on the interval $I \subset \mathbb{R}$.
(3) For given $t \geqslant 0$ let us denote by Π_t the projection $\mathcal{D}([0,\infty);\mathbb{R}^d) \to \mathbb{R}^d$, $\omega \mapsto \omega(t) = \Pi_t(\omega)$. With this notation the Borel σ-field $\mathcal{B}(\mathcal{D}([0,\infty);\mathbb{R}^d), d)$ equals $\sigma(\Pi_t; t \geqslant 0)$.
(4) The vector space $(\mathcal{D}([0,\infty);\mathbb{R}^d), d)$ is not a topological vector space since addition of two elements is not continuous with respect to this topology.

A stochastic process X with paths in $\mathcal{D}([0,\infty);\mathbb{R}^d)$ can be interpreted as a random variable

$$X\colon (\Omega, \mathcal{F}, \mathbb{P}) \to \mathcal{D}([0,\infty);\mathbb{R}^d)$$

with $X_t(\omega) = \omega(t)$ where $(\Omega, \mathcal{F}, \mathbb{P})$ is an abstract probability space. Given a family $(X^\alpha)_{\alpha\in\mathcal{A}}$ of such processes we say that $(X^\alpha)_{\alpha\in\mathcal{A}}$ is relatively compact if the family $(\mathbb{P}_{X^\alpha})_{\alpha\in\mathcal{A}}$ of image measures $\mathbb{P}_{X^\alpha} = \mathbb{P} \circ (X^\alpha)^{-1}$ is relatively compact which, due to Prokhorov's theorem, amounts to saying that $(\mathbb{P}_{X^\alpha})_{\alpha\in\mathcal{A}}$ is tight.

7.4.3 Uniqueness of the martingale problem

As explained above, it is much more difficult to prove the well-posedness of the martingale problem than mere solvability. The following lemma provides an essential tool for proving uniqueness. It says that finite-dimensional distributions form a convergence determining class, see Theorem 3.7.8. in [17].

Lemma 7.4.3. *Suppose that X_0, $(X^n)_{n\in N}$ is a relative compact family of stochastic processes $X^n : (\Omega, \mathcal{F}, \mathbb{P}) \to \mathcal{D}([0,\infty); \mathbb{R}^d)$ and there is a dense subset $J \subset [0,\infty)$ such that*

$$(X^n(t_1), \ldots X^n(t_N)) \overset{d}{\Rightarrow} (X(t_1), \ldots X(t_N))$$

or, equivalently, $$\mathbb{P}_{\left(X^n(t_1),\ldots X^n(t_N)\right)} \to \mathbb{P}_{\left(X(t_1),\ldots X(t_N)\right)} \text{ weakly}$$

for all finite subsets $\{t_1, \ldots, t_N\} \subset J$. Then $X^n \overset{d}{\Rightarrow} X$ or, equivalently $\mathbb{P}_{X^n} \to \mathbb{P}_X$ weakly.

The situation turns out to be even better for solutions to the martingale problem. The following universal result says that even one-dimensional distributions determine the measure provided they agree for all initial distributions μ, see Theorem 4.4.2 in [17].

Lemma 7.4.4. *Consider the linear operator $(L, D(L))$ with L defined as in (7.4.1). Assume that for any initial distribution μ and any two corresponding solutions $\mathbb{P}^\mu$, $\mathbb{Q}^\mu$ to the martingale problem*

$$\mathbb{P}^\mu_{\Pi_t} = \mathbb{Q}^\mu_{\Pi_t} \quad \forall t \geqslant 0\,,$$

then there exists at most one solution to the martingale problem for any initial distribution μ.

The key to the proof of the previous lemma is to show that regular conditional probabilities solve the martingale problem. Finally, we can prove uniqueness for the martingale problem. In order to have well-posedness, one also needs to prove solvability i.e. one needs to prove that, given a distribution μ, there exists a solution $\mathbb{P}^\mu$. This is much easier ans such results have been established under very mild assumptions, cf. Theorem 2.2 in [48], Theorem IX.2.41 in [24] and Theorem 3.2. in [19].

We will now explain how uniqueness for the martingale problem follows from solvability of the corresponding deterministic Cauchy problem. For $s > 0$ we denote by $\mathcal{C}^s(\mathbb{R}^d)$ the Hölder-Zygmund space and by $\mathcal{C}^s_0(\mathbb{R}^d)$ the closure of $C^\infty_c(\mathbb{R}^d)$ with respect to the norm of $\mathcal{C}^s(\mathbb{R}^d)$. Given a Banach space X, we say that f belongs to the space $C^{1,s}([0,T];X)$ if $f:[0,T]\to X$ is continuously differentiable with $\frac{\mathrm{d}}{\mathrm{d}t}f\in C^s([0,T];X)$. Let L be as in Section 7.4.1.

Theorem 7.4.5. *Assume $\alpha\in(0,2)$, $s\in(0,1)$, $\vartheta\in(0,1)$, and $T>0$. Assume that*
(i) L is a bounded operator from $\mathcal{C}_0^{s+\alpha}(\mathbb{R}^d)$ to $\mathcal{C}_0^s(\mathbb{R}^d)$,
(ii) for every $f\in C^\vartheta([0,T];\mathcal{C}_0^s(\mathbb{R}^d))$ there is a unique solution $u\in C^{1,\vartheta}([0,T];\mathcal{C}^s(\mathbb{R}^d))\cap C^\vartheta([0,T];\mathcal{C}_0^{s+\alpha}(\mathbb{R}^d))$
Then the martingale problem for $(L, C_c^\alpha(\mathbb{R}^d))$ is well posed.

Obviously condition (ii) is a very strong assumption. It is a challenging task to find sufficient conditions on $n(x,h)$ such that (ii) holds true, cf. [2].

Proof. Assume that there are two solutions $\mathbb{P}^\mu$, $\mathbb{Q}^\mu$ to the martingale problem for a given distribution μ. A key step is to show that, for any $T>0$ the stochastic process $M=(M_t)_{t\in[0,T]}$ defined via

$$M_t=v(t,\Pi_t)-\int_0^t\Big(\frac{\partial}{\partial s}+L\Big)v(s,\Pi_s)\,ds \tag{7.4.2}$$

is a $\mathbb{P}^\mu$-martingale and thus also a $\mathbb{Q}^\mu$-martingale for any function $v\in C^{1,\vartheta}([0,T];\mathcal{C}_0^s(\mathbb{R}^d))\cap C^\vartheta([0,T];\mathcal{C}_0^{s+\alpha}(\mathbb{R}^d))$ with $s,\vartheta\in(0,1)$. This is proved exactly as in Theorem 4.2.1, part "(i)⇒(ii)" of [49]. Note that L is a bounded operator from $\mathcal{C}_0^{s+\alpha}(\mathbb{R}^d)$ to $\mathcal{C}_0^s(\mathbb{R}^d)$ which is what we need. The conclusion "(i)⇒(ii)" does not depend on the local structure of the differential operator or another specific property.

The main result follows once the following equality

$$\int_0^T\varphi(s)\mathbb{E}_{\mathbb{P}^\mu}(\psi(\Pi_s))\,ds=\int_0^T\varphi(s)\mathbb{E}_{\mathbb{Q}^\mu}(\psi(\Pi_s))\,ds \tag{7.4.3}$$

is established for any $T>0$ and any choice of $\varphi\in C_0^\infty((0,T))$, $\psi\in C_0^\infty(\mathbb{R}^d)$. Here, $\mathbb{E}_{\mathbb{P}^\mu}$ and $\mathbb{E}_{\mathbb{Q}^\mu}$ denote the expectation with respect to $\mathbb{P}^\mu$ and $\mathbb{Q}^\mu$ respectively. Assertion (7.4.3) proves the equality of one-dimensional distributions, i.e. $\mathbb{P}^\mu_{\Pi_t}=\mathbb{Q}^\mu_{\Pi_t}$ for all $t>0$, which in light of 7.4.4 proves the desired uniqueness result. Equality (7.4.3) is proved as follows.

Setting $f(t,x)=\varphi(t)\psi(x)$. Condition (ii) ensures that there is a function v belonging to $C^{1,\vartheta}([0,T];\mathcal{C}_0^s(\mathbb{R}^d))\cap C^\vartheta([0,T];\mathcal{C}_0^{s+\alpha}(\mathbb{R}^d))$ and solving

$$\begin{aligned}\partial_t v+Lv&=f &&\text{in } (0,T)\times\mathbb{R}^n,\\ v(T,\cdot)&=0 &&\text{in } \mathbb{R}^n.\end{aligned}$$

Thus

$$-\int_0^T\varphi(s)\mathbb{E}_{\mathbb{P}^\mu}(\psi(\Pi_s))\,ds=-\mathbb{E}_{\mathbb{P}^\mu}\int_0^T f(s,\Pi_s)\,ds=\mathbb{E}_{\mathbb{P}^\mu}(M_T)=\mathbb{E}_{\mathbb{P}^\mu}(M_0)$$

$$= \mathbb{E}_{\mathbb{P}^\mu}(v(0,\Pi_0)) = \int\limits_{\mathcal{D}([0,\infty);\mathbb{R}^d)} v(0,\Pi_0(\omega))\mathbb{P}^\mu(d\omega) = \int\limits_{\mathbb{R}^d} v(0,x)\mu(dx)\,.$$

Since the same line with the same right-hand side holds true when $\mathbb{P}^\mu$ is replaced by $\mathbb{Q}^\mu$ equality (7.4.3) is established. The theorem is proved. □

7.5 Regularity Estimates in Hölder Spaces

In this final section, we explain how to prove Hölder estimates for solutions $u : \mathbb{R}^d \to \mathbb{R}$ to the equation $Lu = f$ in Ω, where $\Omega \subset \mathbb{R}^d$ is open and L is an operator of the form (7.4.1). We offer two approaches, one direct approach using the integrodifferential representation of L and one approach using the Markov process that is associated to L via the martingale problem. The material of the section in based on [8], [47] and [26]. We refer to [27, Section 2] and [46, Section 1A] for a detailed discussion of the literature and further results.

Before we can formulate the main result, we need to impose some condition on $n(x,h)$. The main assumption for this section is

$$\Lambda^{-1}|h|^{-d-\alpha} \leqslant n(x,h) \leqslant \Lambda|h|^{-d-\alpha} \qquad (h,x\in\mathbb{R}^d, 0<|h|\leqslant 1)\,, \tag{7.5.1}$$

for some $\Lambda \geqslant 1$, $\alpha \in (0,2)$. We will prove the main two results under this condition. The results of this section go back to [8] in the framework given by (7.5.1). Recently, the result of [8] was extended to a much wider class of problems. As explained in [26], Hölder-type regularity estimates can be proved if (7.5.1) is replaced by

$$\Lambda^{-1}\frac{\ell(|h|)}{|h|^d} \leqslant n(x,h) \leqslant \Lambda\frac{\ell(|h|)}{|h|^d} \qquad (h,x\in\mathbb{R}^d, 0<|h|\leqslant 1)\,, \tag{7.5.2}$$

where $\ell : (0,1) \to (0,\infty)$ is locally bounded and varies regularly at zero with index $-\alpha \in (-2,0]$. Possible choices for ℓ include $\ell(s) = s^{-\alpha}$ for some $\alpha \in (0,2)$, $\ell(s) = 1$ and $\ell(s) = \ln(2/s)^{\pm 1}$. Note that condition (7.5.2) equals (7.3.9) if $n(x,h)$ is independent of x.

The main regularity result of this section is the following.

Theorem 7.5.1 ([26]). *Assume (7.5.2) is satisfied. There exist constants $c \geqslant 1$, $\beta \in (0,1)$ such that for $f \in L^\infty(B_r)$, $0 < r < 1/2$, and $u \in C_b^2(\mathbb{R}^d)$ satisfying $Lu = f$ in B_r, the following holds true:*

$$\sup_{x,y\in B_{r/4}} \frac{|u(x)-u(y)|}{\mathcal{L}(|x-y|)^{-\beta}} \leqslant c\mathcal{L}(r)^\beta\|u\|_\infty + c\mathcal{L}(r)^{\beta-1}\|f\|_{L^\infty(B_r)}. \tag{7.5.3}$$

Here $\mathcal{L}(r) = \int_r^1 \frac{\ell(s)}{s}\, ds$.

Corollary 7.5.2 ([8], [47]). *Assume* (7.5.1) *holds true. Then* (7.5.3) *becomes*

$$\sup_{x,y\in B_{r/4}} \frac{|u(x)-u(y)|}{|x-y|^{\gamma}} \leqslant c r^{-\gamma}\|u\|_{\infty} + c r^{\alpha-\gamma}\|f\|_{L^{\infty}(B_r)}, \tag{7.5.4}$$

where $\gamma = \alpha\beta \in (0,1)$.

Remark 7.5. *Note that the constant c in the above estimates depends on the value of* α. *In fact, from the proofs one can see that c is unbounded for* $\alpha \to 2-$. *It is possible to prove* (7.5.4) *with a constant that is independent of* α *for* α *away from zero, cf. [14].*

The significance of these two results lies in the fact that they require almost no regularity assumption on the dependence of $n(x,h)$ on $x \in \mathbb{R}^d$. Note that 7.5.1 generalizes 7.5.2 to a large extent. The operators resp. stochastic processes that are covered by 7.5.1 allow for a rich structure with respect to scaling, see the examples mentioned above.

For the sequel of this section, we concentrate on the case that corresponds to (7.5.1) and $f = 0$. Let us begin with the more probabilistic approach.

7.5.1 Probabilistic approach

We assume that the martingale problem for $(L, C_c^{\infty}(\mathbb{R}^d))$ is well-posed, i.e., there exists a strong Markov process X associated to L. Implicitly, the assumption that the martingale problem for $(L, C_c^{\infty}(\mathbb{R}^d))$ is well-posed, imposes some (weak) restriction on $n(x,h)$. Let us first clarify how we are going to understand the equality $Lu = 0$ in Ω.

Definition 7.5.3. *A bounded function* $u : \mathbb{R}^d \to \mathbb{R}$ *is said to be harmonic with respect to L in an open subset* $\Omega \subset \mathbb{R}^d$, *if for every bounded open set* $B \Subset \Omega$ *and every* $x \in \mathbb{R}^d$ *the process* $(u(X_{\tau_B \wedge t}))_{t\geqslant 0}$ *is a* $\mathbb{P}_x$*-martingale. In this case we say* $Lu = 0$ *in* Ω.

Here τ_B denotes the random exit time for X leaving B. X denotes the standard coordinate process, which is denoted by Π in Section 7.4.

The main auxiliary result is the following.

Proposition 7.5.4. *Assume* (7.5.1). *Let* $\mathbb{P}_x$ *be the solution to the martingale problem for* $(L, C_c^{\infty}(\mathbb{R}^d))$ *with* $\mathbb{P}_x(X_0 = x) = 1$. *There is a constant* $c > 0$ *such that for every* $R > 0$, *every measurable set* $A \subset B_{2R}\setminus B_R$ *with* $|(B_{2R}\setminus B_R)\cap A| \geqslant \frac{1}{2}|B_{2R}\setminus B_R|$, *and every* $x \in B_{R/2}$

$$\mathbb{P}_x(T_A < \tau_{B_{2R}}) \geqslant \mathbb{P}_x(X_{\tau_{B_R}} \in A) \geqslant c.$$

Here T_A resp. τ_A denote the hitting resp. exit time for a mb. set $A \subset \mathbb{R}^d$. 7.5.4 says that, independent of the scale $R > 0$ and the starting point x from the inner ball B_R, there is

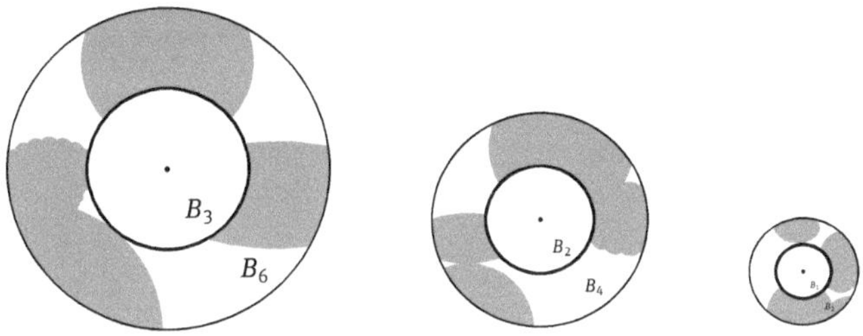

Fig. 7.1: Sets $A \subset B_{2R} \backslash B_R$ for different radii R, cf. 7.5.4

at least a fixed portion of paths that hit the set $A \subset B_{2R} \backslash B_R$ before they exit the larger ball B_{2R}.

Let us explain how to prove 7.5.4. First of all, one uses the martingale problem in order to establish some fundamental properties of the corresponding stochastic process.

Proposition 7.5.5. *(i) There exists a constant $C_1 \geqslant 1$ such that for $x_0 \in \mathbb{R}^d$, $r \in (0,1)$ and $t > 0$*

$$\mathbb{P}_{x_0}(\tau_{B_r(x_0)} \leqslant t) \leqslant C_1 t r^{-\alpha} .$$

(ii) There is a constant $C_2 \geqslant 1$ such that for $x_0 \in \mathbb{R}^d$

$$\sup_{x \in \mathbb{R}^d} \mathbb{E}_x \tau_{B_r(x_0)} \leqslant C_2 r^{\alpha} , \quad r \in (0, 1/2) .$$

(iii) There is a constant $C_3 \geqslant 1$ such that for $x_0 \in \mathbb{R}^d$ and

$$\inf_{x \in B_{r/2}(x_0)} \mathbb{E}_x \tau_{B_r(x_0)} \geqslant C_3 r^{\alpha} , \quad r \in (0, 1) .$$

This result allows to establish an estimate on the probability that paths perform a very large jump given that they perform a jump of fixed size.

Proposition 7.5.6. *There is a constant $C_4 \geqslant 1$ such that for all $x_0 \in \mathbb{R}^d$ and $r, s \in (0,1)$ satisfying $2r < s$*

$$\sup_{x \in B_r(x_0)} \mathbb{P}_x(X_{\tau_{B_r(x_0)}} \notin B_s(x_0)) \leqslant C_4 \left(\frac{r}{s}\right)^{\alpha} .$$

Proof. Let $x_0 \in \mathbb{R}^d$, $r, s \in (0,1)$ and $x \in B_r(x_0)$. Set $B_r := B_r(x_0)$. By the Lévy system formula, for $t > 0$

$$\begin{aligned} \mathbb{P}_x(X_{\tau_{B_r} \wedge t} \notin B_s) &= \mathbb{E}_x \sum_{v \leqslant \tau_{B_r} \wedge t} \mathbb{1}_{\{X_{v-} \in B_r, X_v \in B_s^c\}} \\ &= \mathbb{E}_x \int_0^{\tau_{B_r} \wedge t} \int_{B_s^c} n(X_v, z - X_v) \, \mathrm{d}z \, \mathrm{d}v . \end{aligned}$$

Let $y \in B_r$. Since $s \geqslant 2r$, it follows that $B_{s/2}(y) \subset B_s$ and hence

$$\int\limits_{B_s^c} n(y, z-y)\,\mathrm{d}z \leqslant \int\limits_{B_{s/2}(y)^c} n(y, z-y)\,\mathrm{d}z \leqslant c_1 s^{-\alpha}\,.$$

These considerations together with 7.5.5 imply

$$\mathbb{P}_x(X_{\tau_{B_r}\wedge t} \notin B_s) \leqslant c_2 s^{-\alpha}\mathbb{E}_x \tau_{B_r} \leqslant c_3\left(\frac{r}{s}\right)^{\alpha}.$$

Letting $t \to \infty$ we obtain the desired estimate. □

Finally, we can establish 7.5.4.

Proof. Assume $R > 0$. Let $A \subset B_{2R}\backslash B_R$ satisfy $|(B_{2R}\backslash B_R) \cap A| \geqslant \frac{1}{2}|B_{2R}\backslash B_R|$. Set $B_s := B_s(0)$ for $s > 0$ and let $x_0 \in B_{R/2}$. The first inequality follows from $\{X_{\tau_{B_R}} \in A\} \subset \{T_A < \tau_{B_{2R}}\}$ since $A \subset B_{2R}\backslash B_R$.

By the Lévy system formula, for $t > 0$,

$$\mathbb{P}_{x_0}(X_{\tau_{B_R}\wedge t} \in A) = \mathbb{E}_{x_0} \sum_{s \leqslant \tau_{B_R}\wedge t} \mathbb{1}_{\{X_{s-}\in B_R, X_s \in A\}} \tag{7.5.5}$$

$$= \mathbb{E}_{x_0} \int\limits_0^{\tau_{B_R}\wedge t} \int\limits_A n(X_s, z - X_s)\,\mathrm{d}z\,\mathrm{d}s\,. \tag{7.5.6}$$

Since $|z - x| \leqslant |z| + |x| \leqslant |z| + R \leqslant 2|z|$ for $x \in B_R$ and $z \in B_R^c$,

$$\mathbb{E}_{x_0} \int\limits_0^{\tau_{B_R}\wedge t} \int\limits_A n(X_s, z - X_s)\,\mathrm{d}z\,\mathrm{d}s \geqslant c_1 \mathbb{E}_{x_0}[\tau_{B_R} \wedge t] \int_A |z|^{-d-\alpha}\,\mathrm{d}z \tag{7.5.7}$$

$$\geqslant c_1 \mathbb{E}_{x_0}[\tau_{B_R} \wedge t] R^{-\alpha}|A| \geqslant \frac{c_1}{2}\mathbb{E}_{x_0}[\tau_{B_R} \wedge t] R^{-\alpha}|B_{2R}\backslash B_R|\,. \tag{7.5.8}$$

We conclude

$$\mathbb{P}_{x_0}(T_A < \tau_{B_{2R}}) \geqslant c_3 R^{-\alpha}\mathbb{E}_{x_0}[\tau_{B_R} \wedge t]\,.$$

Letting $t \to \infty$ and using the lower bound in 7.5.5 we get

$$\mathbb{P}_{x_0}(T_A < \tau_{B_{2R}}) \geqslant c_3 R^{-\alpha}\, \mathbb{E}_{x_0}\tau_{B_R} \geqslant c_3 C_3 R^{-\alpha}\, R^{\alpha} = c_4\,.$$

□

We omit the details of the proof that derives 7.5.2 from 7.5.4. This step is very similar to the proof of 7.5.2, which we explain below. In the probabilistic framework, one uses optimal stopping and the following decomposition

$$\begin{aligned} u(z_2) - u(z_1) &= \mathbb{E}_{z_2}[u(X_{\tau_n}) - u(z_1)] \\ &= \mathbb{E}_{z_2}[u(X_{\tau_n}) - u(z_1); X_{\tau_n} \in B_{n-1}] \end{aligned}$$

$$+\sum_{i=1}^{n-2} \mathbb{E}_{z_2}[u(X_{\tau_n}) - u(z_1); X_{\tau_n} \in B_{n-i-1}\backslash B_{n-i}]$$
$$+\mathbb{E}_{z_2}[u(X_{\tau_n}) - u(z_1); X_{\tau_n} \in B_1^c] = I_1 + I_2 + I_3 .$$

for $B_n = B_{r_n}(x_0)$, where r_n is a sequence of radii with $r_n \to 0$ and z_1, z_2 are points in B_{n+1}. The previous quantitative assertions on the stochastic process, in particular 7.5.2 are then uses in order to control the different terms.

7.5.2 Analytic approach

The main tool in the analytic approach is provided by the following result.

Lemma 7.5.7 ([47]). *Assume* (7.5.1) *holds true. There is* $\eta > 0$, $\vartheta \in (0,1)$ *such that, if* $u : \mathbb{R}^d \to \mathbb{R}$ *satisfies*

$$\begin{aligned} -Lu(x) &\leqslant 0 && \text{for } x \in B_1 , \\ u(x) &\leqslant 1 && \text{for } x \in B_1 , \\ u(x) &\leqslant 2|2x|^\eta - 1 && \text{for } x \in \mathbb{R}^d\backslash B_1 , \end{aligned}$$
$$|B_1 \cap \{u \leqslant 0\}| \geqslant \frac{1}{2}|B_1| ,$$

then $u \leqslant 1 - \vartheta$ *in* $B_{\frac{1}{2}}$.

From a probabilistic point of view, the lemma can easily be motivated: The condition $-Lu(x) \leqslant 0$ for $x \in B_1$ means that the function u is subharmonic in B_1. If $\{u \leqslant 0\}$ is denoted by M, then one can use optional stopping to obtain an estimate of the form

$$u(x) \leqslant 1 \cdot \mathbb{P}^x(\tau_{B_1} > T_M) \quad (x \in B_{\frac{1}{2}}),$$

where X denotes the corresponding strong Markov process. Since we know by 7.5.4 that a positive portion of all paths hits M before leaving B_1, the expression $\mathbb{P}^x(\tau_{B_1} > T_M)$ turns out to be strictly less than 1.

We omit the analytic proof of 7.5.7. Instead, we show how it implies the estimate that is asserted in 7.5.2.

Proof 7.5.2. We prove the result for $r = 1$, the general case follows from scaling. Assume $u \in C_b^2(\mathbb{R}^d)$ satisfies $Lu = 0$ in B_1. Without loss of generality we can assume $u \neq 0$ and $\|u\|_\infty \leqslant 1/2$. Let $x_0 \in B_{1/4}$. We may and we do assume that $u(x_0) > 0$. Our aim is to show

$$|u(x) - u(x_0)| \leqslant c|x - x_0|^\beta \qquad (x \in B_1), \tag{7.5.9}$$

for some $\beta \in (0,1)$ and $c \geqslant 1$.

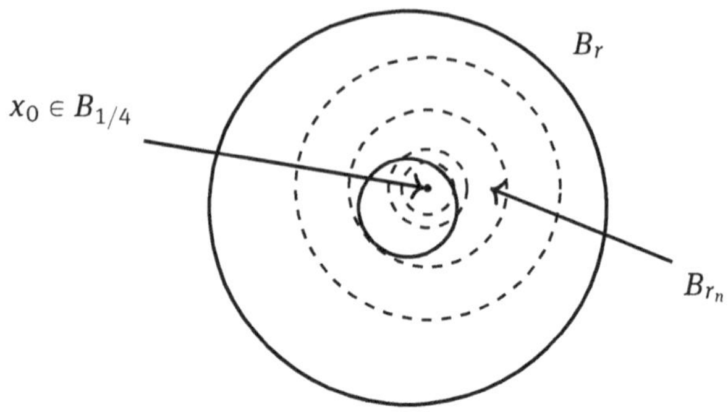

Fig. 7.2: Choice of $x_0 \in B_{1/4}$

Define $r_n = 2^{-n}$ for $n \in \mathbb{N}$. Set $\beta = \min\{\eta, \ln(\frac{2}{2-\vartheta})/\ln 2\}$, where η and ϑ are as in 7.5.7. We construct a nondecreasing sequence (c_n) and a nonincreasing sequence (d_n) of positive numbers such that

$$\begin{aligned} &c_n \leqslant u(y) \leqslant d_n \qquad \text{for } y \in B_{r_n} := B_{r_n}(x_0), \\ &d_n - c_n = 2^{-n\beta} \end{aligned} \tag{7.5.10}$$

This would complete the proof. Let us show how to construct the sequences (c_n), (d_n) inductively. Set

$$c_1 := \inf_{\mathbb{R}^d} u, \quad d_1 := c_1 + 1.$$

Let $n \in \mathbb{N}, n \geqslant 2$. Assume that c_k, d_k have been constructed for $k \leqslant n$, such that (7.5.10) holds for $k \leqslant n$. We will now construct c_{n+1} and d_{n+1}. Set $m = \frac{c_n+d_n}{2}$. By (7.5.10) it follows for $y \in B_{r_n}$

$$u(y) - m \leqslant \tfrac{1}{2}(d_n - c_n) = \tfrac{1}{2}2^{-n\beta}.$$

Define a function $v\colon \mathbb{R}^d \to \mathbb{R}$ by $v(x) := 2^{1+n\beta}(u(x_0 + 2^{-n}x) - m)$. Then $v(x) \leqslant 1$ for $x \in B_1$ and $Lv = 0$ in B_1.

Assume that $\big|\{x \in B_1 : v(x) \leqslant 0\}\big| \geqslant \frac{1}{2}|B_1|$. We recall that the ball $B_{1/2}$ has center 0 and the balls B_{r_n} have center x_0. Given $|x| > 1$, choose $k \in \mathbb{N}_0$ so that $2^k \leqslant |x| < 2^{k+1}$. Then by (7.5.10) we have

$$\begin{aligned} v(x) &= 2^{1+n\beta}(u(x_0 + 2^{-n}x) - m) \leqslant 2^{1+n\beta}(d_{n-k-1} - m) \\ &\leqslant 2^{1+n\beta}(d_{n-k-1} - c_{n-k-1} + c_n - m) \\ &= 2^{1+n\beta}(2^{-(n-k-1)\beta} - \tfrac{1}{2}2^{-n\beta}) \\ &\leqslant 2^{1+(k+1)\beta} - 1 \leqslant 2|2x|^\beta - 1. \end{aligned}$$

Next, we apply 7.5.7. Then we conclude $v \leqslant 1 - \vartheta$ in $B_{\frac{1}{2}}$. This is equivalent to

$$u \leqslant c_n + \frac{2-\vartheta}{2}2^{-n\beta} \text{ in } B_{r_{n+1}}.$$

In this case, we define $c_{n+1} = c_n$ and $d_{n+1} = c_n + 2^{(-n-1)\beta}$. Note that $u \leqslant d_{n+1}$ in $B_{2^{-n-1}}$ because of our choice of β. In the case $|\{x \in B_1 \colon v(x) \leqslant 0\}| < \frac{1}{2}|B_1|$, we perform analogous steps for $-v$ and set $d_{n+1} = d_n$ and $c_{n+1} = d_n - 2^{(-n-1)\beta}$.

□

Acknowledgment: The author thanks Mouhamed M. Fall and Tobias Weth for having organized the AIMS Fall-School in Mbour/Senegal, November 2016. Parts of this material were presented there.

Bibliography

[1] Helmut Abels and Moritz Kassmann. The Cauchy problem and the martingale problem for integro-differential operators with non-smooth kernels, 2006. https://arxiv.org/pdf/math/0610445.

[2] Helmut Abels and Moritz Kassmann. The Cauchy problem and the martingale problem for integro-differential operators with non-smooth kernels. *Osaka J. Math.*, 46(3):661–683, 2009.

[3] Richard F. Bass. Uniqueness in law for pure jump Markov processes. *Probab. Theory Related Fields*, 79(2):271–287, 1988.

[4] Richard F. Bass. Stochastic differential equations driven by symmetric stable processes. In *Séminaire de Probabilités, XXXVI*, volume 1801 of *Lecture Notes in Math.*, pages 302–313. Springer, Berlin, 2003.

[5] Richard F. Bass. Stochastic differential equations with jumps. *Probab. Surv.*, 1:1–19, 2004.

[6] Richard F. Bass, Krzysztof Burdzy, and Zhen-Qing Chen. Stochastic differential equations driven by stable processes for which pathwise uniqueness fails. *Stochastic Process. Appl.*, 111(1):1–15, 2004.

[7] Richard F. Bass and Zhen-Qing Chen. Systems of equations driven by stable processes. *Probab. Theory Related Fields*, 134(2):175–214, 2006.

[8] Richard F. Bass and David A. Levin. Harnack inequalities for jump processes. *Potential Analysis*, 17(4):375–388, Dec 2002.

[9] Jean Bertoin. *Lévy processes*, volume 121 of *Cambridge Tracts in Mathematics*. Cambridge University Press, Cambridge, 1996.

[10] Patrick Billingsley. *Convergence of probability measures*. John Wiley & Sons Inc., New York, 1968.

[11] Patrick Billingsley. *Convergence of probability measures*. Wiley Series in Probability and Statistics: Probability and Statistics. John Wiley & Sons Inc., New York, second edition, 1999. A Wiley-Interscience Publication.

[12] Krzysztof Bogdan, Pawel Sztonyk, and Victoria Knopova. Heat

kernel of anisotropic nonlocal operators. *ArXiv e-prints*, 2017. https://arxiv.org/pdf/1704.03705.

[13] Björn Böttcher. A parametrix construction for the fundamental solution of the evolution equation associated with a pseudo-differential operator generating a Markov process. *Math. Nachr.*, 278(11):1235–1241, 2005.

[14] Luis Caffarelli and Luis Silvestre. Regularity theory for fully nonlinear integro-differential equations. *Comm. Pure Appl. Math.*, 62(5):597–638, 2009.

[15] Zhen-Qing Chen, Renming Song, and Xicheng Zhang. Stochastic flows for Lévy processes with Hölder drifts, 2015. https://arxiv.org/pdf/1501.04758.

[16] Zhen-Qing Chen and Xicheng Zhang. Uniqueness of stable-like processes, 2016. https://arxiv.org/pdf/1604.02681.

[17] Stewart N. Ethier and Thomas G. Kurtz. *Markov processes*. Wiley Series in Probability and Mathematical Statistics: Probability and Mathematical Statistics. John Wiley & Sons Inc., New York, 1986.

[18] Nicolas Fournier. On pathwise uniqueness for stochastic differential equations driven by stable Lévy processes. *Ann. Inst. Henri Poincaré Probab. Stat.*, 49(1):138–159, 2013.

[19] Walter Hoh. The martingale problem for a class of pseudo-differential operators. *Math. Ann.*, 300(1):121–147, 1994.

[20] Walter Hoh. Pseudo differential operators with negative definite symbols of variable order. *Rev. Mat. Iberoamericana*, 16(2):219–241, 2000.

[21] Peter Imkeller and Niklas Willrich. Solutions of martingale problems for Lévy-type operators with discontinuous coefficients and related SDEs. *Stochastic Process. Appl.*, 126(3):703–734, 2016.

[22] N. Jacob. *Pseudo differential operators and Markov processes. Vol. I.* Imperial College Press, London, 2001. Fourier analysis and semigroups.

[23] Niels Jacob and Hans-Gerd Leopold. Pseudo-differential operators with variable order of differentiation generating Feller semigroups. *Integral Equations Operator Theory*, 17(4):544–553, 1993.

[24] Jean Jacod and Albert N. Shiryaev. *Limit theorems for stochastic processes*, volume 288 of *Grundlehren der Mathematischen Wissenschaften [Fundamental Principles of Mathematical Sciences]*. Springer-Verlag, Berlin, 2003.

[25] Jan Kallsen and Paul Krühner. On uniqueness of solutions to martingale problems — counterexamples and sufficient criteria, 2016. https://arxiv.org/pdf/1607.02998.

[26] Moritz Kassmann and Ante Mimica. Intrinsic scaling properties for nonlocal operators. *J. Eur. Math. Soc. (JEMS)*, 19(4):983–1011, 2017.

[27] Moritz Kassmann and Russell W. Schwab. Regularity results for nonlocal parabolic equations. *Riv. Math. Univ. Parma (N.S.)*, 5(1):183–212, 2014.

[28] Koji Kikuchi and Akira Negoro. On Markov process generated by pseudodifferential operator of variable order. *Osaka J. Math.*, 34(2):319–335, 1997.

[29] Victoria Knopova and Alexei Kulik. Parametrix construction for certain Lévy-type processes. *Random Oper. Stoch. Equ.*, 23(2):111–136, 2015.

[30] Vassili N. Kolokoltsov. On Markov processes with decomposable pseudo-differential generators. *Stoch. Stoch. Rep.*, 76(1):1–44, 2004.

[31] Takashi Komatsu. Markov processes associated with certain integro-differential operators. *Osaka J. Math.*, 10:271–303, 1973.

[32] Takashi Komatsu. On the martingale problem for generators of stable processes with perturbations. *Osaka J. Math.*, 21(1):113–132, 1984.

[33] Takashi Komatsu. Pseudodifferential operators and Markov processes. *J. Math. Soc. Japan*, 36(3):387–418, 1984.

[34] Franziska Kühn. On Martingale problems and Feller processes. *ArXiv e-prints*, 2017. https://arxiv.org/pdf/1706.04132.

[35] Franziska Kühn. Transition probabilities of Lévy-type processes: Parametrix construction. *ArXiv e-prints*, 2017. https://arxiv.org/pdf/1702.00778.

[36] Alexei Kulik. On weak uniqueness and distributional properties of a solution to an sde with α-stable noise. *ArXiv e-prints*, 2015. https://arxiv.org/pdf/1511.00106.

[37] Jean-Pierre Lepeltier and Bernard Marchal. Problème des martingales et équations différentielles stochastiques associées à un opérateur intégro-différentiel. *Ann. Inst. H. Poincaré Sect. B (N.S.)*, 12(1):43–103, 1976.

[38] Zenghu Li and Fei Pu. Strong solutions of jump-type stochastic equations. *Electron. Commun. Probab.*, 17:no. 33, 13, 2012.

[39] Remigijus Mikulevičius and Henrikas Pragarauskas. The martingale problem related to nondegenerate Lévy operators. *Liet. Mat. Rink.*, 32(3):377–396, 1992.

[40] Remigijus Mikulevičius and Henrikas Pragarauskas. On the Cauchy problem for certain integro-differential operators in Sobolev and Hölder spaces. *Liet. Mat. Rink.*, 32(2):299–331, 1992.

[41] Remigijus Mikulevičius and Henrikas Pragarauskas. On the uniqueness of solutions of the martingale problem that is associated with degenerate Lévy operators. *Liet. Mat. Rink.*, 33(4):455–475, 1993.

[42] Akira Negoro and Masaaki Tsuchiya. Stochastic processes and semigroups associated with degenerate Lévy generating operators. *Stochastics Stochastics Rep.*, 26(1):29–61, 1989.

[43] Enrico Priola. Pathwise uniqueness for singular SDEs driven by stable processes. *Osaka J. Math.*, 49(2):421–447, 2012.

[44] Enrico Priola. Stochastic flow for SDEs with jumps and irregular drift term. In *Stochastic analysis*, volume 105 of *Banach Center Publ.*, pages 193–210. Polish Acad. Sci. Inst. Math., Warsaw, 2015.

[45] Ken-iti Sato. *Lévy processes and infinitely divisible distributions*, volume 68 of *Cambridge Studies in Advanced Mathematics*. Cambridge University Press, Cambridge, 1999. Translated from the 1990 Japanese original, Revised by the author.

[46] Russell W. Schwab and Luis Silvestre. Regularity for parabolic integro-differential equations with very irregular kernels. *Anal. PDE*, 9(3):727–772, 2016.

[47] Luis Silvestre. Hölder estimates for solutions of integro-differential equations like the fractional Laplace. *Indiana Univ. Math. J.*, 55(3):1155–1174, 2006.

[48] Daniel W. Stroock. Diffusion processes associated with Lévy generators. *Z. Wahrscheinlichkeitstheorie und Verw. Gebiete*, 32(3):209–244, 1975.

[49] Daniel W. Stroock and S. R. Srinivasa Varadhan. *Multidimensional diffusion processes*, volume 233 of *Grundlehren der Mathematischen Wissenschaften [Fundamental Principles of Mathematical Sciences]*. Springer-Verlag, Berlin, 1979.

[50] Masaaki Tsuchiya. Lévy measure with generalized polar decomposition and the associated SDE with jumps. *Stochastics Stochastics Rep.*, 38(2):95–117, 1992.

[51] Pio Andrea Zanzotto. On stochastic differential equations driven by a Cauchy process and other stable Lévy motions. *Ann. Probab.*, 30(2):802–825, 2002.

Tuomo Kuusi, Giuseppe Mingione, and Yannick Sire

Regularity Issues Involving the Fractional p-Laplacian

Abstract: We survey on existence and regularity for nonlinear integro-differential equations involving measure data, focusing on zero order potential estimates. The nonlocal elliptic operators considered are possibly degenerate or singular and cover the case of the fractional p-Laplacian operator with measurable coefficients. We report on the results in [37] providing different, more streamlined proofs.

8.1 Introduction

We survey on recent results in [37], where existence and regularity of solutions to nonlinear nonlocal equations with measure data are obtained. We consider nonlocal elliptic equations written as

$$-\mathcal{L}_\Phi u = \mu \qquad \text{in } \Omega \subset \mathbb{R}^n\,, \tag{8.1.1}$$

where Ω is a bounded open subset for $n \geqslant 2$, $-\mathcal{L}_\Phi$ is a nonlocal operator defined by

$$\langle -\mathcal{L}_\Phi u, \varphi \rangle := \int_{\mathbb{R}^n}\int_{\mathbb{R}^n} \Phi(u(x)-u(y))(\varphi(x)-\varphi(y))K(x,y)\,dx\,dy\,, \tag{8.1.2}$$

for every smooth function φ with compact support. In (8.1.1) it is assumed that μ belongs to $\mathcal{M}(\mathbb{R}^n)$, that is the space of Borel measures with finite total mass on $\mathbb{R}^n$. The function $\Phi : \mathbb{R} \mapsto \mathbb{R}$ is assumed to be continuous, satisfying $\Phi(0) = 0$ together with the monotonicity property

$$\Lambda^{-1}|t|^p \leqslant \Phi(t)t \leqslant \Lambda|t|^p\,, \qquad \forall\, t \in \mathbb{R}\,. \tag{8.1.3}$$

Finally, the kernel $K\colon \mathbb{R}^n \times \mathbb{R}^n \to \mathbb{R}$ is assumed to be measurable, symmetric, and satisfying the following ellipticity/coercivity properties:

$$\frac{1}{\Lambda|x-y|^{n+sp}} \leqslant K(x,y) \leqslant \frac{\Lambda}{|x-y|^{n+sp}} \qquad \text{for a.e. } (x,y) \in \mathbb{R}^n \times \mathbb{R}^n, \tag{8.1.4}$$

where $\Lambda \geqslant 1$ and

$$s \in (0,1)\,, \qquad p > 2 - \frac{s}{n} =: p_*\,. \tag{8.1.5}$$

Tuomo Kuusi, Aalto University Institute of Mathematics, P.O. Box 11100 FI-00076 Aalto, Finland, E-mail: tuomo.kuusi@aalto.fi
Giuseppe Mingione, Dipartimento di Matematica e Informatica, Università di Parma, Parco Area delle Scienze 53/a, Campus, 43100 Parma, Italy, E-mail: giuseppe.mingione@unipr.it.
Yannick Sire, Johns Hopkins University, Krieger Hall, Baltimore, USA, E-mail: sire@math.jhu.edu

https://doi.org/10.1515/9783110571561-010

The lower bound $p > p_*$ comes from the fact that we are considering elliptic problems involving a measure. The domain of defintion of the operator $-\mathcal{L}_\Phi$ is the fractional Sobolev space $W^{s,p}(\mathbb{R}^n)$ in the sense that this is the largest space to which φ has to belong to in order to make the duality in (8.1.2) finite when $u \in W^{s,p}(\mathbb{R}^n)$. In the special case $\Phi(t) = |t|^{p-2}t$, we recover the fractional p-Laplacian operator with measurable coefficients (see for instance [4, 14, 15]). On the other hand, in the case $\Phi(t) = t$ we cover the special case of linear fractional operators with measurable coefficients $\mathcal{L}$ defined by

$$\langle -\mathcal{L}u, \varphi\rangle := \int_{\mathbb{R}^n}\int_{\mathbb{R}^n} (u(x)-u(y))(\varphi(x) - \varphi(y))K(x,y)\,dx\,dy \tag{8.1.6}$$

(see also [2, 10]). In connection to the equation (8.1.1) we shall consider the related Dirichlet problems, that is those of the form

$$\begin{cases} -\mathcal{L}_\Phi u = \mu & \text{in } \Omega \\ u = g & \text{in } \mathbb{R}^n \setminus \Omega\,, \end{cases} \tag{8.1.7}$$

where in general the "boundary datum" $g \in W^{s,p}(\mathbb{R}^n)$ must be prescribed on the whole complement of Ω. In this case, and when $\Phi(t) = |t|^{p-2}t$ and $\mu = 0$, we are essentially considering the Euler-Lagrange equation of the functional

$$v \mapsto \int_{\mathbb{R}^n}\int_{\mathbb{R}^n} |v(x)-v(y)|^p K(x,y)\,dx\,dy$$

minimized in the class of functions such that $v = g$ outside Ω. This survey is two-fold:

1. We first sketch the proof of the solvability of the Dirichlet problem (8.1.7). After introducing a suitable notion of solutions (called SOLA for Solutions Obtained as Limits of Approximations), we briefly describe an existence theorem for SOLA solutions.
2. Secondly, we describe the pointwise behaviour of these solutions by means of nonlinear potentials, namely Wolff potentials. We also provide sufficient conditions for continuity properties of solutions by means of μ.

We note that a few interesting existence and regularity results for the specific equation obtained for the fractional laplacian for powers $s > 1/2$ have been obtained in [24] and, in a different setting in [30]. More recent work in [12] deals again with fractional equations involving measures, this time for any $s > 0$, while the operator is given by the fractional Laplacean, and the analysis is carried out by means of fundamental solutions. A notion of renormalised solution for semilinear equations is proposed in [1]. Another approach in [24] is via duality. In the present survey, for sake of exposition, we will always put ourselves in the case $\Phi(t) = |t|^{p-2}t$ and K being a symmetric kernel.

Before going into the results, let us give a brief outlook of the existing literature. The references we give are by no means exhaustive.

– The Hölder regularity and Harnack estimates for weak solutions of fractional p-Laplacian type equations with measurable coefficients, employed in this survey, were obtained in [14, 15]. There are of course many related results established later on. Let us mention that the boundary regularity was obtained in [23] using barriers, and then later on in [34] using a different method. The latter one also contains Hölder regularity for the obstacle problem. For the higher regularity we would like to mention [9] showing higher differentiability properties of weak solutions. It should be remarked, however, that in view of the local theory one would expect $C^{1,\alpha}$-regularity from weak solutions, at least for a certain range of s and p. To our knowledge, this is an open problem. In the case $p = 2$ there is a vast literature about “viscosity" solutions, and in many cases the obtained regularity is optimal, see for instance [10, 48]. For regularity of variational solutions in the case $p = 2$ we mention [2, 25] and references therein.

– The existence of solutions is widely studied issue in the case of local equations, see [5, 6, 7, 8, 13]. Typical classes are solutions obtained via limiting approximation (SOLA), renormalized solutions, and entropy solutions. In the case of nonnegative measures all this coincide with superharmonic solutions, see [27]. The uniqueness in the measure data problems is still a major open problem. In the case of the nonlocal equations, we deal here with SOLAs. Their existence is sketched also in this survey paper. The uniqueness of solutions is an open problem also in the fractional setting.

– Nonlinear potential theory. The first contributions to pointwise potential estimates were given by Kilpeläinen and Malý [28, 29] in the beginning of 90’s. They proved that any nonnegative *superharmonic* function allows a two-sided estimate via Wolff-potentials:

$$c^{-1}\mathbf{W}^{\mu}_{1,p}(x,r) \leqslant u(x) \leqslant c\left(\mathbf{W}^{\mu}_{1,p}(x,r) + \inf_{B_r(x)} u\right),$$

where the Wolff-potential ([19, 20]) is defined as

$$\mathbf{W}^{\mu}_{\beta,p}(x_0,r) := \int_0^r \left(\frac{|\mu|(B_\varrho(x_0))}{\varrho^{n-\beta p}}\right)^{1/(p-1)} \frac{d\varrho}{\varrho}, \qquad \beta > 0. \tag{8.1.8}$$

The nonnegative measure μ is identified via Riesz’ representation theorem, that is $-\mathrm{div}(a(x,Du)) = \mu$. An alternative approach to the proof was given by Trudinger and Wang in [50], and later also in [31]. For further discussion, also to theory developed for pointwise gradient estimates, we refer to [35, 36].

– Nonlocal potential theory. Pointwise potential estimates can be very efficiently used in the context of potential theory. For instance, they can be used to prove so-called Wiener criterion giving necessary and sufficient geometric conditions for boundary points to guarantee continuity up to the boundary whenever boundary values are continuous at that point, see [21, 29, 40, 41, 42, 50]. As of writing this, the nonlocal Wiener criterion is still open. Nonetheless, there are some partial developments to the corresponding potential theory in [33, 32, 39].

– Lane-Emden type equations. Finally, we would like to mention that the potential theoretic approach was very successfully used in the context of so-called Lane-Emden type equations by Phuc and Verbitsky [46, 47] and also Jaye and Verbitsky [22]. A very natural question is to ask to what extent such results would generalize to the nonlocal setting.

– A very interesting open problem is still that is it possible to extend the results of [37] to more general kernels, which can be, for example, unbounded away from the diagonal $x = y$. To our knowledge, even in the case $p = 2$, the shape of fundamental solutions is generally not well-understood.

8.2 The Basic Existence Theorem and SOLA

We investigate the Dirichlet problem (8.1.7). SOLA are therefore defined following the approximation scheme settled in the local case, with an additional approximation for the boundary values, which is here allowed to be different than zero. Since problems of the type (8.1.7) are defined on the whole space $\mathbb{R}^n$, the analysis of solutions necessarily involves a quantification of the long-range interactions of the function u. A suitable quantity to control the interaction is the following *Tail*, which is initially defined whenever $v \in L^{p-1}_{\mathrm{loc}}(\mathbb{R}^n)$:

$$\mathrm{Tail}(v; x_0, r) := \left[r^{sp} \int_{\mathbb{R}^n \setminus B_r(x_0)} \frac{|v(x)|^{p-1}}{|x-x_0|^{n+sp}}\, dx \right]^{1/(p-1)} . \tag{8.2.1}$$

See [14, 15] where this quantity is instrumental in the derivation of Harnack inequalities and Hölder regularity. We accordingly define

$$L^{p-1}_{sp}(\mathbb{R}^n) := \left\{ v \in L^{p-1}_{\mathrm{loc}}(\mathbb{R}^n) \ : \ \mathrm{Tail}(v; z, r) < \infty \ \ \forall\, z \in \mathbb{R}^n \,, \forall\, r \in (0, \infty) \right\} . \tag{8.2.2}$$

Definition 8.1. *Let $\mu \in (W^{s,p}(\Omega))'$ and $g \in W^{s,p}(\mathbb{R}^n)$. A weak (energy) solution to the problem*

$$\begin{cases} -\mathcal{L}_{\Phi} u = \mu & \text{in } \Omega \\ \qquad\ u = g & \text{in } \mathbb{R}^n \setminus \Omega \end{cases} \tag{8.2.3}$$

is a function $u \in W^{s,p}(\mathbb{R}^n)$ such that

$$\int_{\mathbb{R}^n} \int_{\mathbb{R}^n} \Phi(u(x)-u(y))(\varphi(x) - \varphi(y)) K(x, y)\, dx\, dy = \langle \mu, \varphi \rangle$$

holds for any $\varphi \in C^\infty_0(\Omega)$ and such that $u = g$ a.e. in $\mathbb{R}^n \setminus \Omega$. Accordingly, we say that u is a weak subsolution (supersolution) to (8.2.3) if and only if

$$\int_{\mathbb{R}^n} \int_{\mathbb{R}^n} \Phi(u(x)-u(y))(\varphi(x) - \varphi(y)) K(x, y)\, dx\, dy \leqslant (\geqslant) \langle \mu, \varphi \rangle \tag{8.2.4}$$

holds for every non-negative $\varphi \in C^\infty_0(\Omega)$.

In the following we shall very often refer to a weak solution (sub/supersolution) to (8.2.3) saying that is a weak solutions to $-\mathcal{L}_\Phi u = \mu$ in a domain Ω, thereby omitting to specify the boundary value g. Moreover we have

Definition 8.2 (SOLA for the Dirichlet problem). *Let $\mu \in \mathcal{M}(\mathbb{R}^n)$, $g \in W^{s,p}_{\mathrm{loc}}(\mathbb{R}^n) \cap L^{p-1}_{sp}(\mathbb{R}^n)$ and let $-\mathcal{L}_\Phi$ be defined in* (8.1.2) *under assumptions* (8.1.3)-(8.1.5). *We say that a function $u \in W^{h,q}(\Omega)$ for*

$$h \in (0,s)\,, \qquad \max\{1, p-1\} =: q_* \leqslant q < \overline{q} := \min\left\{\frac{n(p-1)}{n-s}, p\right\}, \tag{8.2.5}$$

is a SOLA to (8.1.7) *if it is a distributional solution to $-\mathcal{L}_\Phi u = \mu$ in Ω, that is*

$$\int_{\mathbb{R}^n}\int_{\mathbb{R}^n} \Phi(u(x)-u(y))(\varphi(x)-\varphi(y))K(x,y)\,dx\,dy = \int_{\mathbb{R}^n} \varphi\,d\mu \tag{8.2.6}$$

holds whenever $\varphi \in C^\infty_0(\Omega)$, if $u = g$ a.e. in $\mathbb{R}^n \backslash \Omega$. Moreover it has to satisfy the following approximation property: There exists a sequence of functions $\{u_j\} \subset W^{s,p}(\mathbb{R}^n)$ weakly solving the approximate Dirichlet problems

$$\begin{cases} -\mathcal{L}_\Phi u_j = \mu_j & \text{in } \Omega \\ \quad u_j = g_j & \text{on } \mathbb{R}^n \backslash \Omega\,, \end{cases} \tag{8.2.7}$$

in the sense of Definition 8.1, such that u_j converges to u a.e. in $\mathbb{R}^n$ and locally in $L^q(\mathbb{R}^n)$. Here the sequence $\{\mu_j\} \subset C^\infty_0(\mathbb{R}^n)$ converges to μ weakly in the sense of measures in Ω and moreover satisfies

$$\limsup_{j\to\infty} |\mu_j|(B) \leqslant |\mu|(\overline{B}) \tag{8.2.8}$$

whenever B is a ball. The sequence $\{g_j\} \subset C^\infty_0(\mathbb{R}^n)$ converges to g in the following sense: For all balls $B_r \equiv B_r(z)$ with center in z and radius $r > 0$, it holds that

$$g_j \to g \quad \text{in } W^{s,p}(B_r)\,, \qquad \text{and} \qquad \lim_j \mathrm{Tail}(g_j - g; z, r) = 0\,. \tag{8.2.9}$$

Condition (8.2.8) can be easily seen to be satisfied if, for example, the sequence $\{\mu_j\}$ is obtained via convolutions with a family of standard mollifiers; as a matter of fact, this is a canonical way to construct the approximating sequence $\{\mu_j\}$ when showing the existence of SOLA.

A SOLA to (8.1.7) always exists, as stated in the next theorem.

Theorem 8.3 (Solvability). *Let $\mu \in \mathcal{M}(\mathbb{R}^n)$, $g \in W^{s,p}_{\mathrm{loc}}(\mathbb{R}^n) \cap L^{p-1}_{sp}(\mathbb{R}^n)$ and let $-\mathcal{L}_\Phi$ be defined in* (8.1.2) *under assumptions* (8.1.3)-(8.1.5). *Then there exists a* SOLA *u to* (8.1.7) *in the sense of Definition 8.2, such that $u \in W^{h,q}(\Omega)$ for every h and q as described in* (8.2.5).

In the following we take the opportunity to correct a technical point in the proof of Theorem 8.3 in [37]. Indeed, we in some instances used a fractional Sobolev inequality in bounded domains without ensuring that it holds.

8.3 The De Giorgi-Nash-Moser Theory for the Fractional p-Laplacian

8.3.1 Some recent results on nonlocal fractional operators

In this section, we recall some recent results for fractional weak solutions (and sub- and supersolutions), which we adapted to our framework for the sake of the reader; see [14, 15, 34] for the related proofs. Notice that the proofs of Theorems 8.5 and 8.7 are valid if we merely assume $u \in W^{s,p}_{\mathrm{loc}}(\Omega) \cap L^{p-1}_{sp}(\mathbb{R}^n)$ instead of $u \in W^{s,p}(\mathbb{R}^n)$.

Definition 8.4. *We say that v is a weak subsolution to $\widetilde{\mathcal{L}}v = 0$ in Ω if $v \in W^{s,p}_{\mathrm{loc}}(\Omega) \cap L^{p-1}_{sp}(\mathbb{R}^n)$ and it satisfies*

$$\int_{\mathbb{R}^n}\int_{\mathbb{R}^n} |v(x)-v(y)|^{p-2}(v(x)-v(y))(\varphi(x)-\varphi(y))\widetilde{K}(x,y)\,dx\,dy \leqslant 0 \qquad (8.3.1)$$

for every nonnegative $\varphi \in C^\infty_0(\Omega)$. Similarly, v is a weak supersolution to $\widetilde{\mathcal{L}}v = 0$ in Ω if $-v$ is a weak subsolution to the same equation. Finally, $v \in W^{s,p}_{\mathrm{loc}}(\Omega) \cap L^{p-1}_{sp}(\mathbb{R}^n)$ is a weak solution if the integral above is zero for every $\varphi \in C^\infty_0(\Omega)$.

Here the assumptions on the kernel $\widetilde{K}$ are of the same type as for K, that is, $\widetilde{K}(\cdot,\cdot)$ is measurable and satisfies $\Lambda^{-2} \leqslant \widetilde{K}(x,y)|x-y|^{n+sp} \leqslant \Lambda^2$ for almost every $(x,y) \in \mathbb{R}^n \times \mathbb{R}^n$.

Firstly, we state a general inequality which shows that the natural extension of the Caccioppoli inequality to the nonlocal framework has to take into account a suitable tail. For other fractional Caccioppoli-type inequalities, though not taking into account the tail contribution, see [44, 45], and also [17].

Theorem 8.5 (Caccioppoli estimate with tail). ([14, Theorem 1.4]). *Let v be a weak subsolution to $\widetilde{\mathcal{L}}v = 0$ in $B_r \equiv B_r(z)$. Then, for any nonnegative $\varphi \in C^\infty_0(B_r)$, the following estimate holds true:*

$$\begin{aligned}\int_{B_r}\int_{B_r} &|w_+(x)\varphi(x)-w_+(y)\varphi(y)|^p\widetilde{K}(x,y)\,dx\,dy\\ &\leqslant c\int_{B_r}\int_{B_r}\big(\max\{w_+(x),w_+(y)\}\big)^p|\varphi(x)-\varphi(y)|^p\widetilde{K}(x,y)\,dx\,dy \qquad (8.3.2)\\ &\quad + c\int_{B_r} w_+\varphi^p\,dx\left(\sup_{y\in\operatorname{supp}\varphi}\int_{\mathbb{R}^n\setminus B_r} w_+^{p-1}(x)\widetilde{K}(x,y)\,dx\right),\end{aligned}$$

where $w_+ := (v-k)_+$ for any $k \in \mathbb{R}$, and c depends only on p.

Remark 8.6. *Observe that the estimate in (8.3.2) holds by replacing w with $w := k-u$ in the case when v is a weak supersolution.*

A first natural consequence is the local boundedness of fractional weak subsolutions, as stated in the following

Theorem 8.7. (Local boundedness, [14, Theorem 1.1 and Remark 4.2]). *Let v be a weak subsolution to $\widetilde{\mathcal{L}}v = 0$ in $B_r \equiv B_r(x_0)$. Then, for all $\delta \in (0,1]$, we have*

$$\operatorname*{ess\,sup}_{B_{r/2}(x_0)} (v-k)_+ \leqslant \delta\,\mathrm{Tail}((v-k)_+; x_0, r/2) + c\,\delta^{-\gamma}\left(\fint_{B_r(x_0)} (v-k)_+^p\,dx\right)^{\frac{1}{p}}, \quad (8.3.3)$$

where $\gamma = (p-1)n/(sp^2)$ and the constant c depends only on n, p, s, and Λ.

It is worth noticing that the parameter δ in (8.3.3) allows a precise interpolation between the local and nonlocal terms. A well-known consequence of reverse Hölder inequalities, as in Theorem 8.7, is that they improve themselves. The result is presented in the following corollary.

Corrolary 8.8. *Let v be a weak solution to $\widetilde{\mathcal{L}}v = 0$ in $B_r \equiv B_r(x_0)$. Then, for all $k \in \mathbb{R}$,*

$$\sup_{B_{\sigma r}(x_0)} |v-k| \leqslant \frac{c}{(1-\sigma)^{\frac{np}{p-1}}}\left[\fint_{B_r(x_0)} |v-k|\,dx + \mathrm{Tail}(v-k; x_0, r/2)\right]$$

holds whenever $\sigma \in (0,1)$, with $c \equiv c(n,s,p,\Lambda)$.

Proof. Assume that $\sigma \geqslant 1/2$, and without loss of generality that $k = 0$. Let us consider numbers $\sigma \leqslant t < \gamma \leqslant 1$ and point $z \in B_{tr}(x_0)$. Applying Theorem 8.7 with the choice $B_r \equiv B_{(\gamma-t)r/100}(z)$, we gain

$$|v(z)| \leqslant \frac{c}{(\gamma-t)^{n/p}}\left(\fint_{B_{sr}(x_0)} |v|^p\,dx\right)^{1/p} + \mathrm{Tail}(v; z, (\gamma-t)r/200)\,. \quad (8.3.4)$$

We have used the fact that $B_{(\gamma-t)r/100}(z) \subset B_{\gamma r}(x_0)$ whenever $z \in B_{tr}(x_0)$. The tail term in (8.3.4) can be estimated, by splitting the integration domain of the corresponding integral in the sets $B_{r/2}(x_0)\backslash B_{(\gamma-t)r/200}(z)$ and $\mathbb{R}^n\backslash(B_{r/2}(x_0) \cup B_{(\gamma-t)r/200}(z))$, as follows:

$$\begin{aligned}\mathrm{Tail}(v; z, (\gamma-t)r/200) &\leqslant \frac{c}{(\gamma-t)^{\frac{n}{p-1}}}\left(\fint_{B_{r/2}(x_0)} |v|^{p-1}\,dx\right)^{1/(p-1)} \\ &\quad + \left[\frac{r^{sp}}{(\gamma-t)^n}\int_{\mathbb{R}^n\backslash(B_{r/2}(x_0)\cup B_{(\gamma-t)r/200}(z))} \frac{|v(x)|^{p-1}}{|x-x_0|^{n+sp}}\,dx\right]^{1/(p-1)} \\ &\leqslant \frac{c}{(\gamma-t)^{\frac{n}{p-1}}}\left[\left(\fint_{B_{\gamma r}(x_0)} |v|^p\,dx\right)^{1/p} + \mathrm{Tail}(v; x_0, r/2)\right].\end{aligned} \quad (8.3.5)$$

Notice that we have used the elementary estimate

$$\frac{|x-x_0|^{n+sp}}{|x-z|^{n+sp}} \leqslant \frac{c(n,p)}{(\gamma-t)^{n+sp}}$$

for $x \notin B_{(\gamma-t)r/200}(z)$. Using (8.3.4) and (8.3.5) we get

$$\sup_{B_{tr}(x_0)} |v| \leqslant \frac{c}{(\gamma-t)^{\frac{n}{p-1}}} \left[\left(\fint_{B_{\gamma r}(x_0)} |v|^p \, dx \right)^{1/p} + \mathrm{Tail}(v; x_0, r/2) \right],$$

which is valid whenever $1/2 \leqslant t < \gamma \leqslant 1$. Extracting $\sup v$ from the last integral on the right hand side and appealing to Young's inequality, we arrive at

$$\sup_{B_{tr}(x_0)} |v| \leqslant \frac{1}{2} \sup_{B_{\gamma r}(x_0)} |v| + \frac{c}{(\gamma-t)^{\frac{np}{p-1}}} \left[\fint_{B_r(x_0)} |v| \, dx + \mathrm{Tail}(v; x_0, r/2) \right].$$

A standard iteration argument, see e.g. [18], then finishes the proof. □

Lemma 8.3.1. (Logarithmic lemma, [14, Lemma 1.3]). *Let $p \in (1,\infty)$. Let $u \in W^{s,p}(\mathbb{R}^n)$ be a weak supersolution to $\widetilde{\mathcal{L}}v = 0$ such that $u \geqslant 0$ in $B_R \equiv B_R(x_0) \subset \Omega$. Then the following estimate holds for any $B_r \equiv B_r(x_0) \subset B_{R/2}(x_0)$ and any $d > 0$,*

$$\int_{B_r}\int_{B_r} \widetilde{K}(x,y) \left| \log\left(\frac{d+u(x)}{d+u(y)} \right) \right|^p dx\, dy \leqslant c\, r^{n-sp} \left\{ d^{1-p} \left(\frac{r}{R} \right)^{sp} [\mathrm{Tail}(u_-; x_0, R)]^{p-1} + 1 \right\}, \tag{8.3.6}$$

where $u_- = \max\{-u, 0\}$ is the negative part of the function u, and c depends only on n, p, s and Λ.

Combining Theorem 8.5 together with a nonlocal *Logarithmic-Lemma*, one can prove that both the p-minimizers and weak solutions enjoy oscillation estimates, which naturally yield Hölder continuity (see Theorem 8.9) and some natural Harnack estimates with tail, as the nonlocal weak Harnack estimate presented in Theorem 8.10 below.

Theorem 8.9. (Hölder continuity, [14, Theorem 1.2]) *Let u be a weak solution to $\widetilde{\mathcal{L}}v = 0$ in $B_r(x_0)$. Then u is locally Hölder continuous in $B_r(x_0)$. In particular, there are positive constants α, $\alpha < sp/(p-1)$, and c, both depending only on n, p, s, Λ, such that*

$$\operatorname*{osc}_{B_\varrho(x_0)} u \leqslant c \left(\frac{\varrho}{r} \right)^{\alpha} \left[\mathrm{Tail}(u; x_0, r) + \left(\fint_{B_r(x_0)} |u|^p \, dx \right)^{\frac{1}{p}} \right] \tag{8.3.7}$$

holds whenever $\varrho \in (0, r/2]$.

Theorem 8.10. (Nonlocal weak Harnack inequality, [15, Theorem 1.2]). *be a weak supersolution to $\widetilde{\mathcal{L}}v = 0$ such that $u \geqslant 0$ in $B_R \equiv B_R(x_0) \subset \Omega$. Let*

$$\bar{t} := \begin{cases} \frac{(p-1)n}{n-sp}, & 1 < p < \frac{n}{s}, \\ +\infty, & p \geqslant \frac{n}{s}. \end{cases} \tag{8.3.8}$$

Then, for any $B_r \equiv B_r(x_0) \subset B_{R/2}(x_0)$ and for any $t < \bar{t}$, we have

$$\left(\fint_{B_r} u^t \, dx \right)^{\frac{1}{t}} \leqslant c \operatorname*{ess\,inf}_{B_{2r}} u + c \left(\frac{r}{R} \right)^{\frac{sp}{p-1}} \operatorname{Tail}(u_-; x_0, R),$$

where the constant c depends only on n, p, s, and Λ.

To be precise, the case $p \geqslant \frac{n}{s}$ was not treated in the proof of the weak Harnack with tail in [15], but one may deduce the result in this case by straightforward modifications. As expected, the contribution given by the nonlocal tail has again to be considered and the result is analogous to the local case if u is nonnegative in the whole $\mathbb{R}^n$.

In the proof of our main estimates we need to transfer L^∞-oscillation estimates into L^q-estimates, reminiscent of the Campanato theory [11, 18]. The excess functional needs to track also the nonlocal contributions. For this we find it convenient to consider the following functional:

$$E(v; z, r) := \left(\fint_{B_r(z)} \left| v - (v)_{B_r(z)} \right|^{q^*} \right)^{1/q^*} + \operatorname{Tail}(v - (v)_{B_r(z)}; z, r), \tag{8.3.9}$$

where $q^* := \max\{1, p - 1\}$. Observe that E satisfies, for any $\eta, \zeta \in L^{q^*}(B_r(z)) \cap L^{p-1}_{sp}(\mathbb{R}^n)$ the trivial decay

$$E(\eta; z, \sigma r) \leqslant c(\sigma, n, s, p) E(\eta; z, r), \tag{8.3.10}$$

and the quasi-triangle inequality

$$E(\eta + \zeta; z, r) \leqslant c(p) \left(E(\eta; z, r) + E(\zeta; z, r) \right). \tag{8.3.11}$$

If $p \geqslant 2$, then $c(p) = 1$ in (8.3.11). Both properties are straightforward to check.

One of the key estimates is the following excess decay estimate.

Theorem 8.11 (Global excess decay). *Let v be a weak solution to $\widetilde{\mathcal{L}}v = 0$ in $B_r(x_0)$. Then there exist positive constants $\alpha \in (0, sp/(p-1))$, and c, both depending only on n, s, p, Λ, such that the following inequality holds whenever $0 < \varrho \leqslant r$:*

$$E(v; x_0, \varrho) \leqslant c \left(\frac{\varrho}{r} \right)^{\alpha} E(v; x_0, r). \tag{8.3.12}$$

Proof. In view of (8.3.10) we may assume that $\varrho \leqslant r/4$. The basic tool is provided by Theorem 8.9 and Corollary 8.8. Indeed, together with Hölder's inequality, they imply

$$\operatorname*{osc}_{B_t} v \leqslant c \left(\frac{t}{r} \right)^{\alpha} E(r) \tag{8.3.13}$$

for all $t \in [\varrho, r/4]$, where we have suppressed v and x_0 from the notation. In particular,

$$\left(\fint_{B_\varrho} \left| v - (v)_{B_\varrho} \right|^{q^*} \right)^{1/q^*} \leqslant \operatorname*{osc}_{B_\varrho} v \leqslant c \left(\frac{\varrho}{r} \right)^{\alpha} E(r) . \tag{8.3.14}$$

Let us then estimate the tail term appearing in the definition of $E(\varrho)$. We rewrite it as

$$\begin{aligned} \operatorname{Tail}(v - (v)_{B_\varrho}; \varrho)^{p-1} &\equiv \varrho^{sp} \int_{\mathbb{R}^n \setminus B_\varrho} \frac{|v(x) - (v)_{B_\varrho}|^{p-1}}{|x - x_0|^{n+sp}} \, dx \\ &= \varrho^{sp} \int_{B_{r/4} \setminus B_\varrho} \frac{|v(x) - (v)_{B_\varrho}|^{p-1}}{|x - x_0|^{n+sp}} \, dx + \varrho^{sp} \int_{\mathbb{R}^n \setminus B_{r/4}} \frac{|v(x) - (v)_{B_\varrho}|^{p-1}}{|x - x_0|^{n+sp}} \, dx . \end{aligned}$$

Now, by Corollary 8.8 and Hölder's inequality we have that

$$|(v)_{B_\varrho} - (v)_{B_r}| \leqslant \sup_{B_{r/2}} |v - (v)_{B_r}| \leqslant cE(r) .$$

Therefore we obtain, again by Hölder's inequality,

$$\begin{aligned} &\varrho^{sp} \int_{\mathbb{R}^n \setminus B_{r/4}} \frac{|v(x) - (v)_{B_\varrho}|^{p-1}}{|x - x_0|^{n+sp}} \, dx \\ &\quad \leqslant c\varrho^{sp} \int_{\mathbb{R}^n \setminus B_{r/4}} \frac{|v(x) - (v)_{B_r}|^{p-1}}{|x - x_0|^{n+sp}} \, dx + c \left(\frac{\varrho}{r} \right)^{sp} |(v)_{B_\varrho} - (v)_{B_r}| \\ &\quad \leqslant c\varrho^{sp} \int_{\mathbb{R}^n \setminus B_r} \frac{|v(x) - (v)_{B_r}|^{p-1}}{|x - x_0|^{n+sp}} \, dx + c \left(\frac{\varrho}{r} \right)^{sp} \fint_{B_r} |v - (v)_{B_r}|^{p-1} \, dx \\ &\quad\quad + c \left(\frac{\varrho}{r} \right)^{sp} E(r) \\ &\quad \leqslant c \left(\frac{\varrho}{r} \right)^{sp} E(r)^{p-1} . \end{aligned}$$

On the other hand, appealing to (8.3.13) we have

$$\begin{aligned} \varrho^{sp} \int_{B_{r/4} \setminus B_\varrho} \frac{|v(x) - (v)_{B_\varrho}|^{p-1}}{|x - x_0|^{n+sp}} \, dx &\leqslant c \int_\varrho^{r/4} \left(\frac{\varrho}{t} \right)^{sp} (\operatorname*{osc}_{B_t} v)^{p-1} \frac{dt}{t} \\ &\leqslant cE(r)^{p-1} \int_\varrho^{r/4} \left(\frac{\varrho}{t} \right)^{sp} \left(\frac{t}{r} \right)^{\alpha(p-1)} \frac{dt}{t} \\ &\leqslant \frac{c}{sp - \alpha(p-1)} \left(\frac{\varrho}{r} \right)^{\alpha(p-1)} E(r)^{p-1} , \end{aligned}$$

using also the fact that $\alpha(p-1) < sp$. Combining the estimates leads to

$$\operatorname{Tail}(v - (v)_{B_\varrho}; \varrho) \leqslant c \left(\frac{\varrho}{r} \right)^{\alpha} E(r) ,$$

again using $\alpha(p-1) < sp$. Together with (8.3.14) this concludes the proof. □

8.4 The "Harmonic Replacement", a Crucial Estimate and the Proof of Theorem 8.3

In this section we are going to consider operators of the type $\mathcal{L}_\Phi$ under the assumptions (8.1.3)-(8.1.5). With $B_{2r} \equiv B_{2r}(x_0) \subset \mathbb{R}^n$ being a fixed ball, we shall consider weak solutions $u \in W^{s,p}(\mathbb{R}^n)$ to the Dirichlet problem

$$\begin{cases} -\mathcal{L}_\Phi u = \mu \in C_0^\infty(\mathbb{R}^n) & \text{in} \quad B_{2r} \\ u = g \in W^{s,p}(\mathbb{R}^n) & \text{in} \quad \mathbb{R}^n \setminus B_{2r}\,. \end{cases} \tag{8.4.1}$$

We set

$$\widetilde{K}(x,y) := \begin{cases} \dfrac{|\Phi(u(x)-u(y))|}{|u(x)-u(y)|^{p-1}} K(x,y), & u(x) \neq u(y) \\ K(x,y), & u(x) = u(y)\,. \end{cases} \tag{8.4.2}$$

It is easy to verify that $\widetilde{K}(\cdot,\cdot)$ is measurable and satisfies $\Lambda^{-2} \leqslant \widetilde{K}(x,y)|x-y|^{n+sp} \leqslant \Lambda^2$ for almost every $(x,y) \in \mathbb{R}^n \times \mathbb{R}^n$. We then define the following weak comparison solution $v \in W^{s,p}(\mathbb{R}^n)$ solving the following Dirichlet problem:

$$\begin{cases} -\mathcal{L}_{\widetilde{K}} v = 0 & \text{in} \quad B_r \\ v = u & \text{in} \quad \mathbb{R}^n \setminus B_r\,, \end{cases} \tag{8.4.3}$$

where $\mathcal{L}_{\widetilde{K}}$ is the fractional p-Laplacian with the kernel $\widetilde{K}$. Throughout Lemmas 8.4.1-8.4.5, u and v will denote the solutions defined in (8.4.1) and (8.4.3), respectively, while we define $w := u - v$.

8.4.1 The crucial inequality

Now, the first crucial comparison lemma. The statement of the following lemma slightly differs from the one in [37]. We provide the proof of the necessary changes.

Lemma 8.4.1. *The following inequality holds for a constant $c \equiv c(n,p,\Lambda)$, whenever $\xi > 1$ and $d > 0$:*

$$\int_{\mathbb{R}^n}\int_{\mathbb{R}^n} \frac{(|u(x)-u(y)| + |v(x)-v(y)|)^{p-2}|w(x)-w(y)|^2}{(d+|w(x)|+|w(y)|)^\xi} \frac{dx\,dy}{|x-y|^{n+sp}} \leqslant \frac{cd^{1-\xi}|\mu|(B_r)}{(p-1)(\xi-1)}\,. \tag{8.4.4}$$

In particular, when $p \geqslant 2$ we have

$$\int_{\mathbb{R}^n}\int_{\mathbb{R}^n} \frac{|w(x)-w(y)|^p}{(d+|w(x)|+|w(y)|)^\xi} \frac{dx\,dy}{|x-y|^{n+sp}} \leqslant \frac{cd^{1-\xi}|\mu|(B_r)}{(p-1)(\xi-1)}\,. \tag{8.4.5}$$

Proof. With $w := u - v$ and $w_\pm := \max\{\pm w, 0\}$, we let

$$\varphi_\pm = \pm\left(d^{1-\xi} - (d + w_\pm)^{1-\xi}\right).$$

Note, in particular that $\varphi_\pm = 0$ on $\mathbb{R}^n \setminus B_r$; moreover this function is bounded and still belongs to $W^{s,p}(\mathbb{R}^n)$ since it is obtained via composition of w with a Lipschitz function. In the following we shall use the notation $A(t,s) := |t|^{p-2}t - |s|^{p-2}s$ for $t, s \in \mathbb{R}$. We choose $\varphi_\pm$ as test functions in the weak formulations of (8.4.1) and (8.4.3); subtracting them and using the fact that $u \equiv v$ outside B_r, we obtain

$$\int_{\mathbb{R}^n}\int_{\mathbb{R}^n} A(u(x)-u(y), v(x)-v(y))(\varphi_\pm(x)-\varphi_\pm(y))\widetilde{K}(x,y)\,dx\,dy \leqslant d^{1-\xi}|\mu|(B_r).$$

The integral on the left is treated as in [37] and we give a sketch of the proof. Since

$$\varphi_\pm(x)-\varphi_\pm(y) = \pm(w_\pm(x)-w_\pm(y))(\xi - 1)\int_0^1 (d + tw_\pm(y) + (1-t)w_\pm(x))^{-\xi}\,dt,$$

the quantity

$$\Psi_\pm(x,y) := \pm A(u(x)-u(y), v(x)-v(y))(w_\pm(x)-w_\pm(y)),$$

dictates the sign of the integrand. We have the identity

$$\begin{aligned} &A(u(x)-u(y), v(x)-v(y)) \\ &\quad = (p-1)\int_0^1 (t|u(x)-u(y)| + (1-t)|v(x)-v(y)|)^{p-2}\,dt\,(w(x)-w(y)), \end{aligned}$$

so that using the inequality

$$\begin{aligned} &\int_0^1 (t|u(x)-u(y)| + (1-t)|v(x)-v(y)|)^{p-2}\,dt \\ &\qquad\qquad \geqslant \frac{1}{c}(|u(x)-u(y)| + |v(x)-v(y)|)^{p-2} \end{aligned}$$

yields

$$\Psi_\pm(x,y) \geqslant \frac{(p-1)}{c}(|u(x)-u(y)| + |v(x)-v(y)|)^{p-2}(w_\pm(x)-w_\pm(y))^2.$$

Appealing now to

$$\int_0^1 (d + tw_\pm(y) + (1-t)w_\pm(x))^{-\xi}\,dt \geqslant \frac{1}{c}(d + w_\pm(y) + w_\pm(x))^{-\xi}$$

finishes the proof together with the properties of the kernel $\widetilde{K}$. □

8.4.2 Basic estimates in the case $p \geqslant 2$

We now proceed with the proof of the main a priori estimates for solutions, thereby introducing the following functions:

$$U_h(x,y) = \frac{|u(x)-u(y)|}{|x-y|^h}, \quad V_h(x,y) = \frac{|v(x)-v(y)|}{|x-y|^h}, \quad W_h(x,y) = \frac{|w(x)-w(y)|}{|x-y|^h} \tag{8.4.6}$$

for $x, y \in \mathbb{R}^n$, $x \neq y$, and for any $h \in (0, s]$. The first a priori estimate we are going to prove is concerned with the case $p \geqslant 2$ and is contained in the following:

Lemma 8.4.2. *Assume that $p \geqslant 2$, $h \in (0,s)$, and $q \in [1, \overline{q})$, where $\overline{q}$ has been defined in* (8.2.5). *Let $\delta := \min\{\overline{q} - q, s - h\}$. Then*

$$\left(\int_{B_{2r}} \fint_{B_{2r}} \frac{|w(x)-w(y)|^q}{|x-y|^{n+hq}}\, dx\, dy \right)^{1/q} \leqslant \frac{c}{r^h} \left[\frac{|\mu|(B_r)}{r^{n-sp}} \right]^{1/(p-1)} \tag{8.4.7}$$

holds for a constant $c \equiv c(n, s, p, \Lambda, \delta)$, which blows up as $\delta \to 0$.

Proof. With $d > 0$ and $\xi > 1$ to be chosen in a few lines, we start by rewriting

$$\begin{aligned} W_h^q(x,y) &= \left(\frac{W_s^p(x,y)}{(d+|w(x)|+|w(y)|)^{\xi}} \right)^{q/p} \\ &\quad \cdot \left[(d+|w(x)|+|w(y)|)^{\xi} |x-y|^{(s-h)p} \right]^{q/p}, \end{aligned}$$

and then apply Hölder's inequality in order to get

$$\begin{aligned} \frac{1}{|B_r|} \int_{\mathcal{B}_{2r}} W_h^q(x,y) \frac{dx\, dy}{|x-y|^n} &\leqslant \left[\frac{1}{|B_r|} \int_{\mathcal{B}_{2r}} \frac{W_s^p(x,y)}{(d+|w(x)|+|w(y)|)^{\xi}} \frac{dx\, dy}{|x-y|^n} \right]^{q/p} \\ &\quad \cdot \left[\frac{1}{|B_r|} \int_{\mathcal{B}_{2r}} \frac{(d+|w(x)|+|w(y)|)^{\xi q/(p-q)}}{|x-y|^{n-(s-h)qp/(p-q)}}\, dx\, dy \right]^{(p-q)/p}. \end{aligned}$$

Notice that we have used that $q < p$. Using (8.4.5) and Fubini's theorem then yields

$$\begin{aligned} \frac{1}{|B_r|} \int_{\mathcal{B}_{2r}} W_h^q(x,y) \frac{dx\, dy}{|x-y|^n} &\leqslant \left[\frac{c d^{1-\xi}}{\xi - 1} \frac{|\mu|(B_r)}{|B_r|} \right]^{q/p} \\ &\quad \cdot c r^{(s-h)q} \left[\fint_{B_{2r}} (d + |w(x)|)^{\xi q/(p-q)}\, dx \right]^{(p-q)/p}. \end{aligned}$$

We then choose

$$d := \left(\fint_{B_{2r}} |w|^{\xi q/(p-q)}\, dx \right)^{(p-q)/\xi q}. \tag{8.4.8}$$

Notice that we can assume that $d > 0$; otherwise (8.4.7) follows trivially. This gives

$$\frac{1}{|B_r|} \int_{\mathcal{B}_r} W_h^q(x,y) \frac{dx\, dy}{|x-y|^n} \leqslant c d^{q/p} \left(r^{(s-h)p} \frac{|\mu|(B_r)}{|B_r|} \right)^{q/p}. \tag{8.4.9}$$

By carefully choosing the parameter ξ (see [37] for details), we may apply the fractional Sobolev inequality (see e.g. [16]) to get

$$d \leqslant cr^h \left(\frac{1}{|B_r|} \int_{\mathcal{B}_{2r}} W_h^q(x,y) \, \frac{dx\,dy}{|x-y|^n} \right)^{1/q} . \tag{8.4.10}$$

Here is where we need a larger ball than B_r, and this was a missing technical point in [37]. Indeed, for small h the Sobolev inequality is not valid in B_r, but the fact that $w \equiv 0$ in $B_{2r} \setminus B_r$ allows us to apply it in the larger ball B_{2r}. Combining the last two displays yields the result. □

A straightforward application of the fractional Sobolev embedding theorem together with (8.4.7) gives the following:

Lemma 8.4.3. *Assume that $p \geqslant 2$, $\gamma \in [1, \gamma^*)$, where*

$$\gamma^* := \begin{cases} \dfrac{n(p-1)}{n-ps} & p < n/s \\ +\infty & p \geqslant n/s \, . \end{cases}$$

Then the inequality

$$\left(\fint_{B_r} |w|^\gamma \, dx \right)^{1/\gamma} \leqslant c \left[\frac{|\mu|(B_r)}{r^{n-sp}} \right]^{1/(p-1)}$$

holds for a constant $c \equiv c(n, s, p, \Lambda, \gamma^* - \gamma)$.

8.4.3 Basic estimates in the case $2 > p > 2 - s/n$

The counterpart of Lemma 8.4.2 in the case $p < 2$ turns out to be more involved. We just state the results, since the proofs are completely analogous to [37] taking into account the small changes presented above in the proof of Lemma 8.4.2.

Lemma 8.4.4. *Assume that $2 > p > p_* = 2 - s/n$; for every $q \in [1, \bar{q})$ (with $\bar{q} := n(p-1)/(n-s)$) there exists $h(q) \in (0, s)$, such that if $h(q) < h < s$ and $\delta := \min\{\bar{q} - q, p - p_*, s - h\}$ then*

$$\begin{aligned} \left(\int_{B_{2r}} \fint_{B_{2r}} \frac{|w(x)-w(y)|^q}{|x-y|^{n+hq}} \, dx\,dy \right)^{1/q} &\leqslant \frac{c}{r^h} \left[\frac{|\mu|(B_r)}{r^{n-sp}} \right]^{1/(p-1)} \\ &\quad + \frac{c}{r^{h(p-1)}} \left(\int_{B_{2r}} \fint_{B_{2r}} \frac{|u(x)-u(y)|^q}{|x-y|^{n+hq}} \, dx\,dy \right)^{(2-p)/q} \left[\frac{|\mu|(B_r)}{r^{n-sp}} \right] \end{aligned} \tag{8.4.11}$$

holds whenever $h(q) \leqslant h < s$, for a constant $c \equiv c(n, s, p, \Lambda, \delta)$, which blows up as $\delta \to 0$.

Lemma 8.4.5. *Assume that* $2 > p > p_* = 2 - n/s$ *and*

$$1 \leqslant \gamma < \gamma^* = \frac{n(p-1)}{n-ps}\,.$$

Then there exists a constant $c \equiv c(n, s, p, \Lambda, \gamma^* - \gamma, p - p_*)$ *such that the following inequality holds:*

$$\left(\fint_{B_r} |w|^\gamma\, dx\right)^{1/\gamma} \leqslant c\left[\frac{|\mu|(B_{2r})}{r^{n-sp}}\right]^{1/(p-1)} + c[E(u;x_0,2r)]^{2-p}\left[\frac{|\mu|(B_{2r})}{r^{n-sp}}\right]. \tag{8.4.12}$$

Moreover, for every $q \in [1, \overline{q})$ *there exists* $h(q) \in (0, s)$, *such that if* $h(q) < h < s$ *such that the inequality*

$$\left(\int_{B_{2r}} \fint_{B_{2r}} \frac{|w(x)-w(y)|^q}{|x-y|^{n+hq}}\, dx\, dy\right)^{1/q} \leqslant \frac{c}{r^h}\left[\frac{|\mu|(B_{2r})}{r^{n-sp}}\right]^{1/(p-1)} + \frac{c}{r^h}[E(u;x_0,2r)]^{2-p}\left[\frac{|\mu|(B_{2r})}{r^{n-sp}}\right] \tag{8.4.13}$$

holds for a constant c depending only on n, s, p, Λ *and* δ, *where the meaning of* $\overline{q}, \delta$ *is specified in Lemma 8.4.4.*

8.4.4 Proof of Theorem 8.3

We now come to the proof of Theorem 8.3. We just sketch it and refer to [37] for full details. We have first an approximation lemma.

Lemma 8.4.6 (Construction of the approximating boundary values g_j). *Let* $z \in \Omega$; *there exists a sequence* $\{g_j\} \subset C_0^\infty(\mathbb{R}^n)$ *such that for any* $R > 0$

$$g_j \to g \quad in\ W^{s,p}(B_R) \quad and \quad \int_{\mathbb{R}^n\setminus B_R(z)} \frac{|g_j(y) - g(y)|^{p-1}}{|y-z|^{n+sp}}\, dy \to 0 \tag{8.4.14}$$

as $j \to \infty$. *Moreover, for every* $\varepsilon > 0$ *there exist a radius* $\tilde{R}$ *and an index* $\tilde{j}$, *both depending on* ε, *such that*

$$\int_{\mathbb{R}^n\setminus B_R(z)} \frac{|g(y)|^{p-1} + |g_j(y)|^{p-1}}{|y-z|^{n+sp}}\, dy \leqslant \varepsilon \tag{8.4.15}$$

holds whenever $j \geqslant \tilde{j}$ *and* $R \geqslant \tilde{R}$. *Finally, for every* $R > 0$ *there exists a constant* c_R, *depending on* R *and* $g(\cdot)$, *such that*

$$\sup_j \|g_j\|_{W^{s,p}(B_R)} \leqslant c_R\,. \tag{8.4.16}$$

Recall the following estimates from the previous discussion:

Lemma 8.4.7. *Let $\Omega \subset B_R$ be an open and bounded domain. Let $u, v \in W^{s,p}(\mathbb{R}^n)$ be weak solutions to the problems*

$$\begin{cases} -\mathcal{L}_\Phi u = \mu \in C^\infty(\mathbb{R}^n) & \text{in} \quad \Omega \\ u = g & \text{in} \quad \mathbb{R}^n \setminus \Omega \end{cases} \quad \text{and} \quad \begin{cases} -\widetilde{\mathcal{L}}_{\widetilde{K}} v = 0 & \text{in} \quad \Omega \\ v = g & \text{in} \quad \mathbb{R}^n \setminus \Omega, \end{cases}$$

respectively, where $g \in W^{s,p}(\mathbb{R}^n)$ and $\widetilde{K}$ has been defined as in (8.4.2). Let $w := u - v$, $\overline{q} := n(p-1)/(n-s)$. Then:

- *When $p \geqslant 2$, for every $h \in (0,s)$ and $q \in [1, \overline{q})$, and with $\delta := \min\{\overline{q} - q, s - h\}$, the inequality*

$$\left(\int_{B_{2R}} \int_{B_{2R}} \frac{|w(x) - w(y)|^q}{|x-y|^{n+hq}} \, dx \, dy \right)^{1/q} \leqslant c \, [|\mu|(\Omega)]^{1/(p-1)}$$

holds for a constant $c \equiv c(n, s, p, \Lambda, \delta, \Omega)$.

- *When $2 > p > p_* = 2 - s/n$, for every $q \in [1, \overline{q})$ there exists $h(q) \in (0,s)$, such that if $h(q) < h < s$ and $\delta := \min\{\overline{q} - q, p - p_*, s - h\}$ then the inequality*

$$\begin{aligned} &\left(\int_{B_{2R}} \int_{B_{2R}} \frac{|w(x) - w(y)|^q}{|x-y|^{n+hq}} \, dx \, dy \right)^{1/q} \\ &\quad \leqslant c \, [|\mu|(\Omega)]^{1/(p-1)} + c \left(\int_{B_{2R}} \int_{B_{2R}} \frac{|u(x) - u(y)|^q}{|x-y|^{n+hq}} \, dx \, dy \right)^{(2-p)/q} [|\mu|(\Omega)] \end{aligned}$$

holds for a constant $c \equiv c(n, s, p, \Lambda, \delta, \Omega)$.

One gets the bounds

Lemma 8.4.8. *Let $p > 2$, $h \in (0,s)$ and $q \in [1, \overline{q})$, where $\overline{q}$ has been defined in (8.2.5). Then there exists a constant c depending only on $n, s, p, \Lambda, s - h, \overline{q} - q, g(\cdot), \Omega \subset B_R$ such that*

$$\left(\int_\Omega |u_j|^q \, dx \right)^{1/q} + \left(\int_{B_{2R} \times B_{2R}} (U_h^j)^q(x,y) \, \frac{dx \, dy}{|x-y|^n} \right)^{1/q} \leqslant c \tag{8.4.17}$$

holds for all $j \in \mathbb{N}$. In the case $2 > p > p_* = 2 - s/n$, for every $q \in [1, \overline{q})$ there exists $h(q) \in (0,s)$, such that if $h(q) < h < s$ then estimate (8.4.17) continues to hold and the constant c additionally depends on $p - p_*$.

From the results in the previous discussion we can conclude that, up to non-relabelled subsequences and using a diagonal argument, there exists $u \in W^{h,q}(\Omega)$ such that $u = g$ in $\mathbb{R}^n \setminus \Omega$ and such that

$$\begin{cases} u_j \rightharpoonup u & \text{in} \quad W^{h,q}(B_{2R}) \\ u_j \to u & \text{in} \quad L^q(\Omega) \\ u_j \to u & \text{a.e. in} \quad \mathbb{R}^n \end{cases} \tag{8.4.18}$$

hold for any given $h \in (0,s)$ and $q \in [p-1, \overline{q})$. This gives the desired result (see [37] for the rest of the proof).

8.5 Pointwise Behaviour of SOLA Solutions

With μ being a Borel measure, we define the (truncated) Wolff potential $\mathbf{W}^{\mu}_{\beta,p}$ of the measure μ as

$$\mathbf{W}^{\mu}_{\beta,p}(x_0,r) := \int_0^r \left(\frac{|\mu|(B_\varrho(x_0))}{\varrho^{n-\beta p}}\right)^{1/(p-1)} \frac{d\varrho}{\varrho}, \qquad \beta > 0 \tag{8.5.1}$$

whenever $x \in \mathbb{R}^n$ and $0 < r \leqslant \infty$. We have the following two theorems concerning the nonlinear potential estimates.

Theorem 8.12 (An upper bound via Wolff potential). *Let $\mu \in \mathcal{M}(\mathbb{R}^n)$, $g \in W^{s,p}_{\rm loc}(\mathbb{R}^n) \cap L^{p-1}_{sp}(\mathbb{R}^n)$ and let $-\mathcal{L}_\Phi$ be defined in (8.1.2) under assumptions (8.1.3)-(8.1.5). Let u be a SOLA to (8.1.7) and assume that for a ball $B_r(x_0) \subset \Omega$ the Wolff potential $\mathbf{W}^{\mu}_{s,p}(x_0,r)$ is finite. Then x_0 is a Lebesgue point of u in the sense that there exists the precise representative of u at x_0*

$$u(x_0) := \lim_{\varrho\to 0} (u)_{B_\varrho(x_0)} = \lim_{\varrho \to 0} \fint_{B_\varrho(x_0)} u\,dx \tag{8.5.2}$$

and the following estimate holds for a constant c depending only on n, s, p, Λ:

$$|u(x_0)| \leqslant c\mathbf{W}^{\mu}_{s,p}(x_0,r) + c\left(\fint_{B_r(x_0)} |u|^{q_*}\,dx\right)^{1/q_*} + c\,\mathrm{Tail}(u;x_0,r)\,, \tag{8.5.3}$$

where $q_* := \max\{1, p-1\}$.

Theorem 8.12 is actually itself a corollary of a more general result that we report below, and that in a sense quantifies the oscillations of the gradient averages around the considered point. For this we need to introduce another quantity that we shall extensively use throughout the paper. This is the following *global excess* functional, which is defined for functions $f \in L^{q_*}_{\rm loc}(\mathbb{R}^n) \cap L^{p-1}_{sp}(\mathbb{R}^n)$:

$$E(f;x_0,r) := \left(\fint_{B_r(x_0)} |f-(f)_{B_r(x_0)}|^{q_*}\,dx\right)^{1/q_*} + \mathrm{Tail}(f-(f)_{B_r(x_0)};x_0,r)\,, \tag{8.5.4}$$

where, as above, $q_* = \max\{1, p-1\}$. When the role of the point x_0 will be clear from the context we shall often denote $E(f;r) \equiv E(f;x_0,r)$. The global excess functional is the right tool to quantify, in an integral way, the oscillations of functions .Then Theorem 8.12 follows from a stronger regularity/decay property of the global excess:

Theorem 8.13 (Global excess decay). *Under the assumptions of Theorem 8.12 there exists a constant $c \equiv c(n,s,p,\Lambda)$ such that the following estimate:*

$$\int_0^r E(u;x_0,t)\frac{dt}{t} + |(u)_{B_r(x_0)} - u(x_0)| \leqslant c\mathbf{W}^{\mu}_{s,p}(x_0,r) + cE(u;x_0,r)\,, \tag{8.5.5}$$

holds whenever $\mathbf{W}^{\mu}_{s,p}(x_0,r)$ is finite.

Estimate (8.5.5) tells that finiteness of Wolff potentials at a point x_0 allows for a pointwise control on the oscillations of the solution averages and eventually implies that x_0 is a Lebesgue point for u. We will next provide the full proof of the previous estimates.

8.5.1 Proofs of Theorems 8.12 and 8.13

Observe first that were we able to show that the x_0 is a Lebesgue point of u and that (8.5.5) holds, then Theorem 8.12 follows immediately. Thus we proceed proving these.

Let $\{u_j\} \subset W^{s,p}(\mathbb{R}^n)$ be an approximating sequence for the SOLA u with measure μ_j and boundary values g_j, as described in Definition 8.2, and let $\widetilde{K}_j$ be defined as in (8.4.2) with u_j instead of u. For $\varrho \leqslant r$, we define the comparison solution $v_j \in W^{s,p}(\mathbb{R}^n)$ as

$$\begin{cases} -\mathcal{L}_{\widetilde{K}_j} v_{j,\varrho} = 0 & \text{in} \quad B_{\varrho/2}(x_0) \\ v_j = u_j & \text{in} \quad \mathbb{R}^n \setminus B_{\varrho/2}(x_0) \,, \end{cases}$$

eventually letting $w_j = u_j - v_j$; In the following we suppress the dependence on x_0 from the notation writing, for instance, $E(u;\varrho) \equiv E(u;x_0,\varrho)$, $\mathrm{Tail}(u;t) \equiv \mathrm{Tail}(u;x_0,t)$ and so on; we recall that $q_* = \max\{1, p-1\}$.

Our first goal is to get a oscillation decay estimate for u.

Lemma 8.5.1. *Let u be a SOLA-solution in $B_\varrho(x_0)$. Then there are constants $c \equiv c(n,s,p,\Lambda)$ and $\eta \equiv \eta(n,p)$ such that*

$$E(u;x_0,\sigma\varrho) \leqslant c\sigma^\alpha E(u;x_0,\varrho) + c\sigma^{-\eta}\left[\frac{|\mu|(\overline{B}_\varrho(x_0))}{\varrho^{n-sp}}\right]^{1/(p-1)} \tag{8.5.6}$$

holds for any $\sigma \in (0,1)$.

Proof. We have the quasi-triangle inequality (the normal triangle inequality when $p \geqslant 2$) whenever $\eta, \zeta \in L^{sp}_{p-1}(\mathbb{R}^n) \cap L^1(B_\varrho)$:

$$E(\eta,t) \leqslant c\,(E(\zeta,t) + E(\zeta-\eta,t)) \,.$$

The involved constant comes from the fact that in the definition of the $\mathrm{Tail}(\cdot)$ there is power $p-1$ present, which leads to concave functions when $p < 2$. Since $v_j = u_j$ outside of $B_{\varrho/2}$, we have, for $t < \varrho/2$, that

$$E(u_j - v_j, t) \leqslant c\left(\frac{\varrho}{t}\right)^{n/q*}\left(\fint_{B_\varrho} |u_j - v_j|^{q*}\,dx\right)^{1/q*} .$$

Therefore, Theorem 8.11 yields

$$E(u_j,\sigma\varrho) \quad \leqslant \quad cE(v_j,\sigma\varrho) + c\sigma^{-n/q*}\left(\fint_{B_\varrho} |u_j - v_j|^{q*}\,dx\right)^{1/q*}$$

$$\leqslant c\sigma^{\alpha}E(v_j,\varrho) + c\sigma^{-n/q^*}\left(\fint_{B_\varrho}|u_j - v_j|^{q^*}\,dx\right)^{1/q^*}$$
$$\leqslant c\sigma^{\alpha}E(u_j,\varrho) + c\sigma^{-n/q^*}\left(\fint_{B_\varrho}|u_j - v_j|^{q^*}\,dx\right)^{1/q^*}.$$

Now, by Lemma 8.4.3 for $p \geqslant 2$ we see that

$$\left(\fint_{B_\varrho}|u_j - v_j|^{q^*}\,dx\right)^{1/q^*} \leqslant c\left[\frac{|\mu_j|(B_\varrho)}{\varrho^{n-sp}}\right]^{1/(p-1)}.$$

Instead, for $2 - s/n < p < 2$, Lemma 8.4.5 and Young's inequality with conjugate exponents $(1/(2-p), 1/(p-1))$, yield

$$\left(\fint_{B_{\varrho/2}}|u_j - v_j|^{q^*}\,dx\right)^{1/q^*} \leqslant \delta E(u_j;\varrho) + c\delta^{(p-2)/(p-1)}\left[\frac{|\mu_j|(B_\varrho)}{\varrho^{n-sp}}\right]^{1/(p-1)}$$

for any $\delta \in (0,1)$. Choosing $\delta = \sigma^{\alpha+n/q^*}$ and combining the content of the last three displays then leads to (8.5.6), after letting $j \to \infty$. This finishes the proof. □

Next, integrating (8.5.6) against the Haar-measure gives

$$\int_\varrho^r E(u;\sigma t)\,\frac{dt}{t} \leqslant c\sigma^{\alpha}\int_\varrho^r E(u;t)\,\frac{dt}{t} + c\sigma^{-\eta}\int_\varrho^r\left[\frac{|\mu|(\overline{B}_t)}{t^{n-sp}}\right]^{1/(p-1)}\frac{dt}{t}. \tag{8.5.7}$$

Thus, choosing σ small enough so that $\sigma = (2c)^{-1/\alpha}$, which then depends only on n, s, p, Λ, we obtain after changing variables and reabsorption, also observing that $|\mu|(\partial B_t) = 0$ for almost every t, that

$$\int_{\sigma\varrho}^r E(u;t)\,\frac{dt}{t} \leqslant 2\int_{\sigma r}^r E(u;t)\,\frac{dt}{t} + c\int_\varrho^r\left[\frac{|\mu|(B_t)}{t^{n-sp}}\right]^{1/(p-1)}\frac{dt}{t}. \tag{8.5.8}$$

Furthermore, since

$$\int_{\sigma r}^r E(u;t)\,\frac{dt}{t} \leqslant cE(u,r),$$

we deduce an intermediate result

$$\int_\varrho^r E(u;t)\,\frac{dt}{t} \leqslant cE(u;r) + c\int_\varrho^r\left[\frac{|\mu|(B_t)}{t^{n-sp}}\right]^{1/(p-1)}\frac{dt}{t}. \tag{8.5.9}$$

This proves the part of estimate (8.5.5) concerning the first term on the left hand side.

We proceed to prove that the limit in (8.5.2) exists and to complete the proof of estimate (8.5.5). For this, let $0 < \tilde{\varrho} \leqslant \varrho/2 < r/8$ and find $k \in \mathbb{N}$ and $\vartheta \in (1/4, 1/2]$ such that $\tilde{\varrho} = \vartheta^k\varrho$. Then

$$|(u)_{B_\varrho} - (u)_{B_{\tilde{\varrho}}}| \leqslant \sum_{j=0}^{k-1}|(u)_{B_{\vartheta^j\varrho}} - (u)_{B_{\vartheta^{j+1}\varrho}}|$$

$$\leqslant \vartheta^{-n/q*} \sum_{j=0}^{k-1} E(u; \vartheta^j \varrho) .$$

Furthermore, using elementary properties of the excess decay functional, we have that

$$\begin{aligned} \sum_{j=0}^{k-1} E(u; \vartheta^j \varrho) &= \frac{1}{\log(1/\vartheta)} \sum_{j=0}^{k-1} \int_{\vartheta^j \varrho}^{\vartheta^{j-1}\varrho} E(u; \vartheta^j \varrho) \frac{dt}{t} \\ &\leqslant c \sum_{j=0}^{k-1} \int_{\vartheta^j \varrho}^{\vartheta^{j-1}\varrho} E(u; t) \frac{dt}{t} \leqslant c \int_{\tilde{\varrho}}^{\varrho/\vartheta} E(u; t) \frac{dt}{t} , \end{aligned}$$

so that , using the content of the last three displays and recalling that $\varrho \leqslant r/4 \leqslant r/\vartheta$, it follows

$$|(u)_{B_\varrho} - (u)_{B_{\tilde{\varrho}}}| \leqslant c \int_{\tilde{\varrho}}^{\varrho/\vartheta} E(u; t) \frac{dt}{t} . \tag{8.5.10}$$

In turn, recalling that $\varrho/\vartheta \leqslant r/(4\vartheta) \leqslant r$ and using (8.5.9) we have

$$|(u)_{B_\varrho} - (u)_{B_{\tilde{\varrho}}}| \leqslant cE(u; r) + c\mathbf{W}^{\mu}_{s,p}(x_0, r) . \tag{8.5.11}$$

On the other hand, by (8.5.9) the finiteness of $\mathbf{W}^{\mu}_{s,p}(x_0, r)$ implies the finiteness of the right hand side in (8.5.11) and therefore (8.5.10) readily implies that $\{(u)_{B_\varrho}\}$ is a Cauchy net. As a consequence, the limit in (8.5.2) exists and thereby defines the pointwise precise representative of u at x_0. Letting $\tilde{\varrho} \to 0$ in (8.5.11) and taking $\varrho = r/4$ then gives

$$|(u)_{B_{r/4}} - u(x_0)| \leqslant cE(u; r) + c\mathbf{W}^{\mu}_{s,p}(x_0, r) .$$

On the other hand, notice that we have

$$|(u)_{B_r} - (u)_{B_{r/4}}| \leqslant cE(u; r)$$

so that the last two displays and triangle inequality finally give (8.5.5) (recall also (8.5.9)). This completes the proof of Theorem 8.13. Finally, estimate (8.5.3) follows from (8.5.5) via elementary manipulations.

8.6 A Lower Bound via Wolff Potentials

Theorem 8.12 is sharp in describing the pointwise behaviour of the SOLA in the sense that the Wolff potential appearing on the right hand side of (8.5.3) cannot be replaced by any other potential. Indeed, if the measure μ is nonnegative, then we also have the following potential lower bound, that in turn implies an optimal description, in terms of Lebesgue points :

Theorem 8.14 (Potential lower bound and fine properties). *Let $\mu \in \mathcal{M}(\mathbb{R}^n)$ be a non-negative measure, $g \in W^{s,p}_{\mathrm{loc}}(\mathbb{R}^n) \cap L^{p-1}_{sp}(\mathbb{R}^n)$ and let $-\mathcal{L}_\Phi$ be defined in* (8.1.2) *under* *assumptions (8.1.3)-(8.1.5) with $p < n/s$. Let u be a SOLA to (8.1.7) which is non-negative in the ball $B_r(x_0) \subset \Omega$ and such that the approximating sequence $\{\mu_j\}$ for μ as described in Definition 8.2 is made of nonnegative functions. Then the estimate*

$$\mathbf{W}^{\mu}_{s,p}(x_0, r/8) \leqslant cu(x_0) + c\mathrm{Tail}(u_-; x_0, r/2) \tag{8.6.1}$$

holds for a constant $c \equiv c(n, s, p, \Lambda)$, as soon as $\mathbf{W}^{\mu}_{s,p}(x_0, r/8)$ is finite, where $u_- := \max\{-u, 0\}$. In this case, according to Theorem 8.12, $u(x_0)$ is defined as the precise representative of u at x_0 as in (8.5.2). Moreover, when $\mathbf{W}^{\mu}_{s,p}(x_0, r/8)$ is infinite, we have that

$$\lim_{t\to 0} (u)_{B_t(x_0)} = \infty\,. \tag{8.6.2}$$

Consequently, every point is a Lebesgue point for u.

We use the following results from [15].

Lemma 8.6.1 (Caccioppoli estimate [15]). *Let $p \in (1, \infty)$, $q \in (1, p)$, $d > 0$ and let u be a weak supersolution to (8.2.3) with $\mu = 0$, such that $u \geqslant 0$ in $B_R(x_0) \subset \Omega$. Then, for $w := (u + d)^{1-q/p}$ the following inequality:*

$$\begin{aligned}
&\int_{B_r}\int_{B_r} \frac{|w(x)\varphi(x)-w(y)\varphi(y)|^p}{|x-y|^{n+sp}}\,dx\,dy\\
&\quad\leqslant c\int_{B_r}\int_{B_r} (\max\{w(x), w(y)\})^p \frac{|\varphi(x)-\varphi(y)|^p}{|x-y|^{n+sp}}\,dx\,dy\\
&\qquad + c\left\{ \sup_{y\in\operatorname{supp}\varphi} \int_{\mathbb{R}^n\setminus B_r} |x-y|^{-n-sp}\,dx\right.\\
&\qquad\qquad \left. + d^{1-p}R^{-sp}[\mathrm{Tail}(u_-; x_0, R)]^{p-1}\right\} \left(\int_{B_r} (w\varphi)^p\,dx\right)
\end{aligned}$$

holds for $B_r \equiv B_r(x_0) \subset B_{3R/4}(x_0)$ and any nonnegative $\varphi \in C_0^\infty(B_r)$, where the constant c depends only on s, p, Λ, q and $u_- := \max\{-u, 0\}$.

An easy adaptation of Theorem 8.10 yields the following lemma.

Lemma 8.6.2. *For any $s \in (0, 1)$ and any $p \in (1, n/s)$, let u be a weak supersolution to (8.2.3) with $\mu = 0$ such that $u \geqslant 0$ in $B_R \equiv B_R(x_0) \subset \Omega$. Then the following estimate holds for any $B_r \equiv B_r(x_0) \subset B_{R/2}(x_0)$ and for any positive number $\gamma < n(p-1)/(n-sp)$:*

$$\left(\fint_{B_r} u^\gamma\right)^{1/\gamma} \leqslant c \inf_{B_{2r}} u + c\left(\frac{r}{R}\right)^{\frac{sp}{p-1}} \mathrm{Tail}(u_-; x_0, R),$$

where the constant c depends only on γ, n, s, p, Λ.

Lemma 8.6.3. *Let u be a weak solution to (8.2.3) with $\mu \in C_0^\infty(\mathbb{R}^n)$, such that $u \geqslant 0$ in $B_{4r} \equiv B_{4r}(x_0) \subset \Omega$ and $\mu \geqslant 0$. Then there exists a constant $c \equiv c(n, s, p, \Lambda)$ such that* *the following inequality holds:*

$$\frac{\mu(B_r)}{r^{n-sp}} \leqslant \frac{c}{r^{1-sp}} \int_{B_{3r/2}} \fint_{B_{3r/2}} \frac{|u(x)-u(y)|^{p-1}}{|x-y|^{n+sp-1}}\,dx\,dy + c\left[\inf_{B_r} u + \mathrm{Tail}(u_-; x_0, 4r)\right]^{p-1}. \tag{8.6.3}$$

Proof. Let $\varphi \in C_0^\infty(B_{5r/4})$ be such that $0 \leqslant \varphi \leqslant 1$, $\varphi = 1$ in B_r and $\|D\varphi\|_{L^\infty} \leqslant 16/r$. Taking such φ in (8.2.6) and using (8.1.4), we get

$$\begin{aligned}
\frac{\mu(B_r)}{r^{n-sp}} &\leqslant \frac{1}{r^{n-sp}} \int_{\mathbb{R}^n}\int_{\mathbb{R}^n} \Phi(u(x)-u(y))(\varphi(x)-\varphi(y))K(x,y)\,dx\,dy \\
&\leqslant |B_{3/2}|r^{sp} \int_{B_{3r/2}} \fint_{B_{3r/2}} |u(x)-u(y)|^{p-1}|\varphi(x)-\varphi(y)|K(x,y)\,dx\,dy \\
&\quad + |B_{3/2}|r^{sp} \int_{\mathbb{R}^n\setminus B_{3r/2}} \fint_{B_{3r/2}} (u(x)-u(y))_+^{p-1}\varphi(x)K(x,y)\,dx\,dy \\
&\quad + |B_{3/2}|r^{sp} \fint_{B_{3r/2}} \int_{\mathbb{R}^n\setminus B_{3r/2}} (u(x)-u(y))_+^{p-1}\varphi(y)K(y,x)\,dx\,dy \\
&\leqslant c r^{sp-1} \int_{B_{3r/2}} \fint_{B_{3r/2}} \frac{|u(x)-u(y)|^{p-1}}{|x-y|^{n+sp-1}}\,dx\,dy \\
&\quad + c r^{sp} \int_{\mathbb{R}^n\setminus B_{3r/2}} \fint_{B_{3r/2}} (u(x)-u(y))_+^{p-1}\varphi(x)K(x,y)\,dx\,dy.
\end{aligned} \tag{8.6.4}$$

We turn to the estimate of the last integral; we shall also use the fact that $|y-x_0| \leqslant 16|x-y|$ whenever $y \in \mathbb{R}^n\setminus B_{3r/2}$ and $x \in \operatorname{supp}\varphi$ (since $\operatorname{supp}\varphi \subset B_{5r/4}$). Recalling that $u \geqslant 0$ in B_{4r}, we have

$$\begin{aligned}
& r^{sp} \int_{\mathbb{R}^n\setminus B_{3r/2}} \fint_{B_{3r/2}} (u(x)-u(y))_+^{p-1}\varphi(x)K(x,y)\,dx\,dy \\
&\leqslant c r^{sp} \int_{\mathbb{R}^n\setminus B_{3r/2}} \fint_{B_{3r/2}} \left([u(x)]^{p-1} + [u_-(y)]^{p-1}\right)\varphi(x)K(x,y)\,dx\,dy \\
&\leqslant c \fint_{B_{3r/2}} u^{p-1}\,dx + c\,[\mathrm{Tail}(u_-; x_0, 4r)]^{p-1},
\end{aligned}$$

using also (8.1.4). Since μ is non-negative, then u is a non-negative weak supersolution to $-\mathcal{L}_\Phi u = 0$ in B_{4r}; at this point an application of Lemma 8.6.2 finishes the proof. □

We then estimate the integral appearing in (8.6.3).

Lemma 8.6.4. *Let u be a weak solution to* (8.2.3) *with $\mu \in C_0^\infty(\mathbb{R}^n)$, such that $u \geqslant 0$ in $B_{4r} \equiv B_{4r}(x_0) \subset \Omega$ and $\mu \geqslant 0$. Let $h \in (0, s)$, $q \in (0, \overline{q})$, where $\overline{q}$ has been defined in* (8.2.5). *Then there exists a constant $c \equiv c(n, s, p, \Lambda, s-h, \overline{q}-q)$ such that*

$$\left(\int_{B_{2r}} \fint_{B_{2r}} \frac{|u(x)-u(y)|^q}{|x-y|^{n+hq}}\,dx\,dy\right)^{1/q} \leqslant \frac{c}{r^h}\left[\inf_{B_r} u + \mathrm{Tail}(u_-; x_0, 4r)\right] \tag{8.6.5}$$

holds.

Proof. Let

$$d \equiv d_\delta := \inf_{B_r} u + \mathrm{Tail}(u_-; x_0, 4r) + \delta\,, \qquad \text{for } \delta > 0\,, \tag{8.6.6}$$

and set

$$\overline{u} = u + d\,, \qquad w := \overline{u}^{1-m/p}\,, \qquad \text{where } m \in (1,p)\,.$$

Applying Lemma 8.6.1 (with $2r$ instead of r and $R \equiv 4r$) with a cut-off function $\varphi \in C_0^\infty(B_{7r/4})$ such that $\varphi = 1$ in $B_{3r/2}$, $0 \leqslant \varphi \leqslant 1$ and $|D\varphi| \leqslant 16/r$, we obtain

$$\int_{B_{3r/2}} \int_{B_{3r/2}} \frac{|w(x)-w(y)|^p}{|x-y|^{n+sp}}\,dx\,dy \leqslant \frac{c}{r^{sp}} \int_{B_{2r}} w^p\,dx\,.$$

To evaluate the integral in the left-hand side, we start assuming, without loss of generality, $u(x) > u(y)$; we have

$$\begin{aligned} |w(x)-w(y)| &= |[\overline{u}(x)]^{1-m/p} - [\overline{u}(y)]^{1-m/p}| \\ &= [\overline{u}(x)]^{1-m/p}\left[1 - \left(\frac{\overline{u}(y)}{\overline{u}(x)}\right)^{1-m/p}\right] \geqslant \frac{p-m}{p}\,\frac{\overline{u}(x) - \overline{u}(y)}{[\overline{u}(x)]^{m/p}}\,. \end{aligned}$$

Notice that in the last lines we have used the elementary inequality $(1-t^\beta) \geqslant \beta(1-t)$, that holds in the case $t, \beta \in (0,1]$ and that follows by mean value theorem. Considering in a similar way also the case $u(y) > u(x)$ and recalling the definition of $\overline{u}$, we conclude that

$$\frac{|u(x)-u(y)|}{[\overline{u}(x)]^{m/p} + [\overline{u}(y)]^{m/p}} \leqslant \frac{cp}{p-m}|w(x)-w(y)|\,.$$

On the other hand, recalling again the definitions of w and d, Lemma 8.6.2 then implies

$$⨍_{B_{2r}} w^p\,dx \leqslant c d^{p-m}$$

since $m \in (1,p)$, where $c \equiv (n, s, p, \Lambda)$. Putting the last three displays above together yields, and using that $u \geqslant 0$ in B_{4r}, we have

$$\int_{B_{3r/2}} ⨍_{B_{3r/2}} \frac{|u(x)-u(y)|^p}{[\overline{u}(y)+\overline{u}(x)]^m}\,\frac{dx\,dy}{|x-y|^{n+sp}} \leqslant \frac{cd^{p-m}}{r^{sp}}\,.$$

Next, we get by Hölder's inequality that

$$\begin{aligned} &\int_{B_{3r/2}} ⨍_{B_{3r/2}} \frac{|u(x)-u(y)|^q}{|x-y|^{n+hq}}\,dx\,dy \\ &\quad\leqslant \left(\int_{B_{3r/2}} ⨍_{B_{3r/2}} \frac{|u(x)-u(y)|^p}{[\overline{u}(y)+\overline{u}(x)]^m}\,\frac{dx\,dy}{|x-y|^{n+sp}}\right)^{q/p} \\ &\qquad\cdot \left(\int_{B_{3r/2}} ⨍_{B_{3r/2}} \frac{[\overline{u}(y)+\overline{u}(x)]^{mq/(p-q)}\,dx\,dy}{|x-y|^{n+(h-s)qp/(p-q)}}\right)^{(p-q)/p}\,. \end{aligned}$$

Using the definition of d in (8.6.6), Lemma 8.6.2 then gives

$$\left(\int_{B_{3r/2}} \fint_{B_{3r/2}} \frac{[\overline{u}(y)+\overline{u}(x)]^{mq/(p-q)}}{|x-y|^{n+(h-s)qp/(p-q)}}\, dx\, dy\right)^{(p-q)/p}$$
$$\leqslant \frac{c}{r^{(h-s)q}} \left(\fint_{B_{3r/2}} \overline{u}^{mq/(p-q)}\, dx\right)^{(p-q)/p} \leqslant c\left(r^{(s-h)p} d^m\right)^{q/p}$$

with the last inequalities that hold provided

$$\frac{mq}{p-q} < \frac{n(p-1)}{n-sp} \qquad \text{and} \qquad h < s\,.$$

The first inequality in the above display is required in order to apply Lemma 8.6.2. Notice that we can find $m > 1$ satisfying the previous condition observing that

$$\frac{q}{p-q} < \frac{n(p-1)}{n-sp} \quad \Longleftrightarrow \quad q < \frac{n(p-1)}{n-s}\,.$$

We notice that the constant c depends where the constant depends on $n, s, p, \Lambda, s-h, \overline{q}-q$. Hence we arrive at

$$\fint_{B_{3r/2}} \frac{|u(x)-u(y)|^q}{|x-y|^{n+hq}}\, dx\, dy \leqslant \frac{cd^q}{r^{hq}}\,.$$

This finishes the proof letting $\delta \to 0$ and recalling the definition of d in (8.6.6). □

Lemma 8.6.5. *Let u be a weak solution to* (8.2.3) *with $\mu \in C_0^\infty(\mathbb{R}^n)$, such that $u \geqslant 0$ in $B_{4r} \equiv B_{4r}(x_0) \subset \Omega$ and $\mu \geqslant 0$. Then there exists a constant $c \equiv c(n, s, p, \Lambda)$ such that*

$$\left[\frac{\mu(B_r)}{r^{n-sp}}\right]^{1/(p-1)} \leqslant c\left[\inf_{B_r} u + \mathrm{Tail}(u_-; x_0, 4r)\right] \tag{8.6.7}$$

holds.

Proof. By (8.6.3) we obtain

$$\frac{\mu(B_r)}{r^{n-sp}} \leqslant \frac{c}{r^{1-sp}} \int_{B_{3r/2}} \fint_{B_{3r/2}} \frac{|u(x)-u(y)|^{p-1}}{|x-y|^{n+sp-1}}\, dx\, dy$$
$$+c\left[\inf_{B_r} u + \mathrm{Tail}(u_-; x_0, 4r)\right]^{p-1}.$$

Using the fact that in $B_{3r/2} \times B_{3r/2}$ we have

$$\left(\frac{|x-y|}{r}\right)^{1-sp} \leqslant c\left(\frac{|x-y|}{r}\right)^{-h(p-1)}$$

for some $h \in (0, s)$ provided that

$$1 - sp \geqslant -h(p-1) \quad \Longrightarrow \quad h \geqslant \frac{sp-1}{p-1}\,.$$

Looking h in the form $h = s - \varepsilon$, we get the condition $\varepsilon \leqslant \frac{1-s}{p-1}$, and thus we always find $h \in (0, s)$ satisfying conditions above. We then get from Lemma 8.6.4 that

$$\frac{1}{r^{1-sp}} \int_{B_{3r/2}} \fint_{B_{3r/2}} \frac{|u(x)-u(y)|^{p-1}}{|x-y|^{n+sp-1}}\, dx\, dy \leqslant c \left[\inf_{B_r} u + \mathrm{Tail}(u_-; x_0, 4r)\right]^{p-1} .$$

Combining estimates finishes the proof. □

Proof of Theorem 8.14. In the following all the balls will be centred at x_0. Let $\{u_j\}$ be an approximating sequence for the SOLA u as described in Definition 8.2, with the source terms μ_j being nonnegative, as prescribed in the assumption of Theorem 8.14. Since $-u_j$ is in particular a weak subsolution to (8.2.7) implies

$$\sup_{B_{r/2}}(u_j)_- \leqslant c \left[\fint_{B_r} (u_j)_-\, dx + \mathrm{Tail}((u_j)_-; x_0, r/2)\right] .$$

By the convergence properties of $\{u_j\}$ and the nonnegativity of u in B_r, and recalling that $(-u_j)_+ = (u_j)_-$, we can pass to the limit under the sign of integral thereby getting

$$\limsup_{j\to\infty} \sup_{B_{r/2}}(u_j)_- \leqslant c\mathrm{Tail}(u_-; x_0, r/2)\,, \tag{8.6.8}$$

for a constant c depending only on n, s, p, Λ. Now, the function

$$\tilde{u}_j := u_j - \inf_{B_{r/2}} u_j \tag{8.6.9}$$

is nonnegative in $B_{r/2}$. Denote

$$m_{\varrho,j} := \inf_{B_\varrho} \tilde{u}_j \qquad \text{and} \qquad T_{\varrho,j} := \mathrm{Tail}((\tilde{u}_j - m_{\varrho,j})_-; x_0, \varrho)$$

for $\varrho \in (0, r/2]$. Lemma 8.6.5 gives

$$\left[\frac{\mu_j(B_\varrho)}{\varrho^{n-sp}}\right]^{1/(p-1)} \leqslant c\left(m_{\varrho,j} - m_{4\varrho,j} + T_{4\varrho,j}\right) \tag{8.6.10}$$

for any $\varrho \in (0, r/8]$ for a constant depending only on n, s, p, Λ. Now, with $M \geqslant 1$ to be chosen in a few lines, we have that

$$\begin{aligned}
T_{\varrho,j} &= c\varrho^{sp/(p-1)} \left[\int_{\mathbb{R}^n\setminus B_\varrho} \frac{(\tilde{u}_j(x) - m_{\varrho,j})_-^{p-1}}{|x-x_0|^{n+sp}}\, dx\right]^{1/(p-1)} \\
&\leqslant c\varrho^{sp/(p-1)} \left[\int_{\mathbb{R}^n\setminus B_{M\varrho}} \frac{(\tilde{u}_j(x) - m_{\varrho,j})_-^{p-1}}{|x-x_0|^{n+sp}}\, dx\right]^{1/(p-1)} \\
&\quad + c\varrho^{sp/(p-1)} \left[\int_{B_{M\varrho}\setminus B_\varrho} \frac{(\tilde{u}_j(x) - m_{\varrho,j})_-^{p-1}}{|x-x_0|^{n+sp}}\, dx\right]^{1/(p-1)} \\
&\leqslant cM^{-sp/(p-1)}\left(m_{\varrho,j} - m_{M\varrho,j} + T_{M\varrho,j}\right) + c\left(m_{\varrho,j} - m_{M\varrho,j}\right)
\end{aligned}$$

$$\leqslant \quad cM^{-sp/(p-1)}T_{M\varrho,j} + c\left(m_{\varrho,j} - m_{M\varrho,j}\right) \tag{8.6.11}$$

holds for $\varrho \in (0, r/2)$. Notice that we have used the elementary inequality

$$(\tilde{u}_j(x) - m_{\varrho,j})_- \leqslant (\tilde{u}_j(x) - m_{M\varrho,j})_- + m_{\varrho,j} - m_{M\varrho,j} \,.$$

Now, with $t \in (0, r/(8M))$, by integrating (8.6.11) and changing variables we obtain

$$\begin{aligned}
\int_t^{r/(8M)} T_{4\varrho,j}\frac{d\varrho}{\varrho} &= \int_{4t}^{r/(2M)} T_{\varrho,j}\frac{d\varrho}{\varrho}\\
&\leqslant cM^{-sp/(p-1)}\int_t^{r/(2M)} T_{M\varrho,j}\frac{d\varrho}{\varrho} + c\int_t^{r/(2M)}(m_{\varrho,j} - m_{M\varrho,j})\frac{d\varrho}{\varrho}\\
&= cM^{-sp/(p-1)}\int_{tM}^{r/2} T_{\varrho,j}\frac{d\varrho}{\varrho} + c\left(\int_t^{Mt} m_{\varrho,j}\frac{d\varrho}{\varrho} - \int_{r/(2M)}^{r/2} m_{\varrho,j}\frac{d\varrho}{\varrho}\right).
\end{aligned}$$

Choosing $M \equiv M(s,p)$ so large that

$$cM^{-sp/(p-1)} = \frac{1}{2}\,,$$

we obtain, after easy manipulations,

$$\int_t^{r/(8M)} T_{4\varrho,j}\frac{d\varrho}{\varrho} \leqslant \int_{r/(2M)}^{r/2} T_{\varrho,j}\frac{d\varrho}{\varrho} + c\left(\int_t^{Mt} m_{\varrho,j}\frac{d\varrho}{\varrho} - \int_{r/(2M)}^{r/2} m_{\varrho,j}\frac{d\varrho}{\varrho}\right).$$

Using (8.6.11) again (this time with $M\varrho = r/2$) we obtain

$$\int_{r/(2M)}^{r/2} T_{\varrho,j}\frac{d\varrho}{\varrho} \leqslant cT_{r/2,j} + c\int_{r/(2M)}^{r/2}(m_{\varrho,j} - m_{r/2,j})\frac{d\varrho}{\varrho}\,,$$

and hence also, by changing variables, that

$$\begin{aligned}
\int_t^{r/8} T_{4\varrho,j}\frac{d\varrho}{\varrho} &= \int_t^{r/(8M)} T_{4\varrho,j}\frac{d\varrho}{\varrho} + \int_{r/(2M)}^{r/2} T_{\varrho,j}\frac{d\varrho}{\varrho}\\
&\leqslant c(m_{t,j} - m_{r/2,j}) + c(m_{r/(2M),j} - m_{r/2,j}) + cT_{r/2,j}\\
&\leqslant cm_{t,j} + cT_{r/2,j}
\end{aligned}$$

holds whenever $t < r/(8M)$, for a constant $c \equiv c(n,s,p,\Lambda)$; we have used that the function $\varrho \to m_{\varrho,j}$ is clearly non-increasing and that $m_{r/2,j} = 0$. Again integrating (8.6.10), and using the last inequality, we conclude with

$$\begin{aligned}
\int_t^{r/8}\left[\frac{\mu_j(B_\varrho)}{\varrho^{n-sp}}\right]^{1/(p-1)}\frac{d\varrho}{\varrho} &\leqslant c\int_t^{r/8}(m_{\varrho,j} - m_{4\varrho,j})\frac{d\varrho}{\varrho} + c\int_t^{r/8} T_{4\varrho,j}\frac{d\varrho}{\varrho}\\
&\leqslant c\left(\int_t^{r/8} m_{\varrho,j}\frac{d\varrho}{\varrho} - \int_{4t}^{r/2} m_{\varrho,j}\frac{d\varrho}{\varrho}\right) + c\int_t^{r/8} T_{4\varrho,j}\frac{d\varrho}{\varrho}\\
&\leqslant c\int_t^{4t} m_{\varrho,j}\frac{d\varrho}{\varrho} + c\int_t^{r/8} T_{4\varrho,j}\frac{d\varrho}{\varrho}
\end{aligned}$$

$$\leqslant \quad cm_{t,j} + cT_{r/2,j} \,,$$

for a constant c depending only on n, s, p, Λ. We now want to let $j \to \infty$ in the above estimate, using the properties of SOLA described in Definition 8.2. As for the first term $T_{r/8,j}$, by the definition in (8.6.9) and recalling that in particular we have that $t \leqslant r/2$ so that $\inf_{B_{r/2}} u_j \leqslant (u_j)_t$, we have

$$\begin{aligned}
\limsup_{j\to\infty} T_{r/2,j} &= \limsup_{j\to\infty} \mathrm{Tail}((\tilde{u}_j - m_{r/2,j})_-; x_0, r/2) \\
&= \limsup_{j\to\infty} \mathrm{Tail}\left(\left(u_j - \inf_{B_{r/2}} u_j\right)_- ; x_0, r/2\right) \\
&\leqslant c \limsup_{j\to\infty} (u_j)_t + c \limsup_{j\to\infty} \mathrm{Tail}((u_j)_-; x_0, r/2) \\
&\leqslant c(u)_t + c\mathrm{Tail}(u_-; x_0, r/2) \,.
\end{aligned}$$

Using also (8.6.8), we obtain

$$\begin{aligned}
\limsup_{j\to\infty} m_{t,j} &= \limsup_{j\to\infty} \inf_{B_t} \left(u_j - \inf_{B_{r/2}} u_j\right) \\
&\leqslant \limsup_{j\to\infty} (u_j)_{B_t} + \limsup_{j\to\infty} \sup_{B_{r/2}} (u_j)_- \\
&\leqslant c \limsup_{j\to\infty} (u_j)_{B_t} + c\mathrm{Tail}(u_-; x_0, r/2) \\
&\leqslant c(u)_{B_t} + c\mathrm{Tail}(u_-; x_0, r/2) \,.
\end{aligned}$$

Moreover, by the weak convergence of μ_j to μ and Lebesgue dominated convergence theorem, we obtain that

$$\int_t^{r/8} \left[\frac{\mu_j(B_\varrho)}{\varrho^{n-sp}}\right]^{1/(p-1)} \frac{d\varrho}{\varrho} \to \int_t^{r/8} \left[\frac{\mu(B_\varrho)}{\varrho^{n-sp}}\right]^{1/(p-1)} \frac{d\varrho}{\varrho}$$

as $j \to \infty$. Thus we arrive at

$$\int_t^{r/8} \left[\frac{\mu(B_\varrho)}{\varrho^{n-sp}}\right]^{1/(p-1)} \frac{d\varrho}{\varrho} \leqslant c(u)_{B_t} + c\mathrm{Tail}(u_-; x_0, r/2)$$

for any $t \in (0, r/(8M))$ and where $c \equiv c(n, s, p, \Lambda)$. Now, if $\mathbf{W}^{\mu}_{s,p}(x_0, r/8)$ is finite, then by Theorem 8.12 there exists a the precise representative of u at x_0, as described in (8.5.2). This finishes the proof of (8.6.1) after letting $t \to 0$ in the above display, when $\mathbf{W}^{\mu}_{s,p}(x_0, r/8) < \infty$. In a similar way, if $\mathbf{W}^{\mu}_{s,p}(x_0, r/8) = \infty$, then (8.6.2) follows, and in any case x_0 is a Lebesgue point for u. □

8.7 Continuity Conditions for SOLA

Theorem 8.13 gives us an easy way to characterize when SOLA solutions are continuous. The first qualitative result is the following.

Theorem 8.15 (Continuity criterion). *Let $\mu \in \mathcal{M}(\mathbb{R}^n)$, $g \in W^{s,p}_{\mathrm{loc}}(\mathbb{R}^n) \cap L^{p-1}_{sp}(\mathbb{R}^n)$ and let $-\mathcal{L}_\Phi$ be defined in* (8.1.2) *under assumptions* (8.1.3)-(8.1.5)*. Let u be a* SOLA *to* (8.1.7) *and $\Omega' \Subset \Omega$ be an open subset. If*

$$\lim_{t\to 0} \sup_{x\in\Omega'} \mathbf{W}^{\mu}_{s,p}(x,t) = 0\,, \tag{8.7.1}$$

then u is continuous in Ω'.

Proof. We are going to prove that for any $\varepsilon > 0$ and $z \in \Omega'$ there is small enough δ such that

$$\operatorname*{osc}_{B_\delta(z)\cap\Omega'} u \leqslant \varepsilon\,.$$

Let us fix $\varepsilon > 0$ and $z \in \Omega'$. Recording the estimate (8.5.5) in Theorem 8.13 we have

$$\int_0^r E(u;x,t)\,\frac{dt}{t} + |(u)_{B_r(x)} - u(x)| \leqslant c\mathbf{W}^{\mu}_{s,p}(x,r) + cE(u;x,r) \tag{8.7.2}$$

for all $x \in \Omega'$ and $r \in (0, \operatorname{dist}(\Omega', \partial\Omega))$. By (8.7.3) we find $r_0 \in (0, \operatorname{dist}(\Omega', \partial\Omega))$ such that

$$M_0 := \sup_{x\in\Omega'} \left(E(u;x,r_0) + \mathbf{W}^{\mu}_{s,p}(x,r)\right) < +\infty\,.$$

From these two estimates we deduce that

$$E(u;z,t) \to 0 \quad \text{as} \quad t \to 0\,.$$

Moreover, by the triangle inequality we get that there is a constant $c \equiv c(n,p,s)$ such that

$$E(u;\tilde{z},t) \leqslant cE(u;z,2t)$$

whenever $\tilde{z} \in B_t(z)$. Using above estimates we deduce that

$$\begin{aligned} |u(z) - u(\tilde{z})| &\leqslant |u(z) - (u)_{B_t(z)}| + |u(\tilde{z}) - (u)_{B_t(\tilde{z})}| + |(u)_{B_t(z)} - (u)_{B_t(\tilde{z})}| \\ &\leqslant c\left(E(u;z,2t) + \mathbf{W}^{\mu}_{s,p}(z,2t) + \mathbf{W}^{\mu}_{s,p}(\tilde{z},2t)\right) \end{aligned}$$

whenever $\tilde{z} \in B_t(z)$. Thus, using again (8.7.2) and the convergence $E(u;z,t) \to 0$ as $t \to 0$, we deduce that we find small enough t so that the term on the right is less than ε. We can hence take δ to be this t, and the proof is complete. □

If there is more information on the measure, we may improve the regularity. For this purpose, we define a fractional (restricted and centred) maximal function, for $\beta \in (0,n)$, as

$$M^{\mu}_{\beta}(x_0,r) := \sup_{0<\varrho<r} \varrho^{\beta-n}|\mu|(B_\varrho(x_0))\,.$$

Notice that, for all $\delta > 0$, we have that

$$M^{\mu}_{sp}(x_0,r)^{1/(p-1)} \leqslant c\mathbf{W}^{\mu}_{s,p}(x,2r) \leqslant c(\delta)M^{\mu}_{sp-\delta}(x_0,2r)^{1/(p-1)}.$$

The theorem tells that if the fractional maximal function is pointwise bounded, the regularity is dictated either by the maximal Hölder-regularity for solutions to homogeneous equations, or by the presence of concentrations in the measure.

Theorem 8.16 (Hölder continuity criterion). *Let* $\mu \in \mathcal{M}(\mathbb{R}^n)$, $g \in W^{s,p}_{\mathrm{loc}}(\mathbb{R}^n) \cap L^{p-1}_{sp}(\mathbb{R}^n)$ *and let* $-\mathcal{L}_\Phi$ *be defined in* (8.1.2) *under assumptions* (8.1.3)-(8.1.5). *Let u be a* SOLA *to* (8.1.7) *and* $\Omega' \Subset \Omega$ *be an open subset. If*

$$\sup_{x\in\Omega'} \left(E(u;x,r) + M^{\mu}_{sp-\delta}(x,r)^{1/(p-1)} \right) < \infty \tag{8.7.3}$$

for some $\delta \in (0, sp]$ *and* $r < \mathrm{dist}(\Omega', \partial\Omega)$, *then* $u \in C^{0,\beta}(\Omega')$ *for* $\beta = \delta/(p-1)$ *if* $\delta < \alpha(p-1)$, *and for* $\beta \in (0,\alpha)$ *otherwise.*

Proof. By Lemma 8.5.1 we deduce that

$$E(u;x,\sigma\varrho) \leqslant c\sigma^{\alpha} E(u;x,\varrho) + c\sigma^{-\eta} \left[\frac{|\mu|(\overline{B}_\varrho(x))}{\varrho^{n-sp}} \right]^{1/(p-1)}$$

provided that $B_\varrho(x) \subset \Omega$. Multiply the result with $(\sigma\varrho)^{-\beta}$, and notice that

$$(\sigma\varrho)^{-\beta} E(u;x,\sigma\varrho) \leqslant c\sigma^{\alpha-\beta}\varrho^{-\beta} E(u;x,\varrho) + c\sigma^{-\eta-\beta} \left[\frac{|\mu|(\overline{B}_\varrho(x))}{\varrho^{n-sp+\beta(p-1)}} \right]^{1/(p-1)} .$$

After taking supremum with respect to ϱ and choosing σ so that $2c\sigma^{\alpha-\beta} = 1$, we get after reabsorption

$$\sup_{0<\varrho<r} \varrho^{-\beta} E(u;x,\varrho) \leqslant cr^{-\beta} E(u;x,r) + cM^{\mu}_{sp-\beta(p-1)}(x,r)^{1/(p-1)} .$$

The desired Hölder regularity now follows by the standard Campanato theory. □

Bibliography

[1] Alibaud N. & Andreianov B. & Bendahmane M.: Renormalized solutions of the fractional Laplace equation. *C. R. Math. Acad. Sci. Paris* 348 (2010), 759–762.

[2] Barles G. & Chasseigne E. & Imbert C.: Hölder continuity of solutions of second-order non-linear elliptic integro-differential equations. *J. Eur. Math. Soc. (JEMS)* 13 (2011), 1–26.

[3] Bénilan P. & Boccardo L. & Gallouët T. & Gariepy R. & Pierre M. & Vázquez J. L.: An L^1-theory of existence and uniqueness of solutions of nonlinear elliptic equations. *Ann. Scuola Norm. Sup. Pisa Cl. Sci. (IV)* 22 (1995), 241–273.

[4] Bjorland C. & Caffarelli L. & Figalli A.: Non-local tug-of-war and the infinity fractional Laplacian. *Adv. Math.* 230 (2012), 1859–1894.

[5] Boccardo L. & Gallouët T.: Nonlinear elliptic and parabolic equations involving measure data. *J. Funct. Anal.* 87 (1989), 149–169.
[6] Boccardo L. & Gallouët T.: Nonlinear elliptic equations with right-hand side measures. *Comm. Partial Differential Equations* 17 (1992), 641–655.
[7] Boccardo L. & Gallouët T.: $W^{1,1}$-solutions in some borderline cases of Calderón-Zygmund theory. *J. Differential Equations* 253 (2012), 2698–2714.
[8] Boccardo L. & Gallouët T. & Orsina L.: Existence and uniqueness of entropy solutions for nonlinear elliptic equations with measure data. *Ann. Inst. H. Poincarè Anal. Non Linéaire* 13 (1996), 539–551.
[9] Brasco L. & Lindgren E.: Higher Sobolev regularity for the fractional p-Laplace equation in the superquadratic case. *Adv. Math.* 304(2) (2017), 300–354.
[10] Caffarelli L. & Silvestre L.: Regularity results for nonlocal equations by approximation. *Arch. Ration. Mech. Anal.* 200 (2011), no. 1, 59–88.
[11] Campanato S.: Equazioni ellittiche del II ordine e spazi $\mathfrak{L}^{(2,\lambda)}$. *Ann. Mat. Pura Appl. (IV)* 69 (1965), 321–381.
[12] Chen H. & Veron L.: Semilinear fractional elliptic equations involving measures. *J. Differential Equations* 25 (1973), 565–590.
[13] Dal Maso G. & Murat F. & Orsina L. & Prignet A.: Renormalized solutions of elliptic equations with general measure data. *Ann. Scuola Norm. Sup. Pisa Cl. Sci. (IV)* 28 (1999), 741–808.
[14] Di Castro A. & Kuusi T. & Palatucci G.: Local behavior of fractional p-minimizers. *Ann. Inst. H. Poincaré Anal. Non Linéaire* 33(5) (2016), 1279–1299.
[15] Di Castro A. & Kuusi T. & Palatucci G.: Nonlocal Harnack inequalities. *J. Funct. Anal.* 267 (2014), 1807–1836.
[16] Di Nezza E. & Palatucci G. & Valdinoci E.: Hitchhiker's guide to the fractional Sobolev spaces. *Bull. Sci. Math.* 136 (2012), 521–573.
[17] Franzina G. & Palatucci G: Fractional p-eigenvalues. *Riv. Mat. Univ. Parma* 5(2) (2014), 315–328.
[18] Giusti E.: *Direct methods in the calculus of variations.* World Scientific Publishing Co., Inc., River Edge, NJ, 2003.
[19] Havin M. & Maz'ja V. G.: Non-linear potential theory. *Russ. Math. Surveys* 27 (1972), 71–148.
[20] Hedberg L. & Wolff T.: Thin sets in nonlinear potential theory. *Ann. Inst. Fourier (Grenoble)* 33 (1983), 161–187.
[21] Heinonen J. & Kilpeläinen T. & Martio O.: *Nonlinear potential theory of degenerate elliptic equations.* Oxford Mathematical Monographs., New York, 1993.
[22] Jaye B. & Verbitsky I.: Local and global behaviour of solutions to nonlinear equations with natural growth terms. *Arch. Rat. Mech. Anal.* 204 (2012), 627–681.
[23] Iannizzotto A. & Mosconi S. & Squassina M: Global Hölder regularity for the fractional p-Laplacian. *Rev. Mat. Iberoam.* 32(4), (2016), 1355–1394.
[24] Karlsen K. H. & Petitta F. & Ulusoy S.: A duality approach to the fractional Laplacian with measure data. *Publ. Mat.* 55 (2011), 151–161.

[25] Kassmann M.: A priori estimates for integro-differential operators with measurable kernels. *Calc. Var. Partial Differential Equations* 34 (2009), 1–21.
[26] Kilpeläinen T.: Hölder continuity of solutions to quasilinear elliptic equations involving measures. *Potential Anal.* 3 (1994), 265–272.
[27] Kilpeläinen, T. & Kuusi, T. & Tuhola-Kujanpää, A.: Superharmonic functions are locally renormalized solutions. *Ann. Inst. H. Poincaré, Anal. Non Linèaire* 28 (2011), 775–795.
[28] Kilpeläinen T. & Malý J.: Degenerate elliptic equations with measure data and nonlinear potentials. *Ann. Scuola Norm. Sup. Pisa Cl. Sci. (IV)* 19 (1992), 591–613.
[29] Kilpeläinen T. & Malý J.: The Wiener test and potential estimates for quasilinear elliptic equations. *Acta Math.* 172 (1994), 137–161.
[30] Klimsiak T. & Rozkosz A.: Dirichlet forms and semilinear elliptic equations with measure data. *J. Funct. Anal.* 265 (2013), 890–925.
[31] Korte R. & Kuusi T.: A note on the Wolff potential estimate for solutions to elliptic equations involving measures. *Adv. Calc. Var.* 3 (2010), 99–113.
[32] Korvenpää J. & Kuusi T. & Lindgren E.: Equivalence of solutions to fractional p-Laplace type equations. *J. Math. Pures Appl.*, to appear.
[33] Korvenpää J. & Kuusi T. & Palatucci G.: Fractional superharmonic functions and the Perron method for nonlinear integro-differential equations. *Math. Ann.*, DOI:10.1007/s00208-016-1495-x
[34] Korvenpää J. & Kuusi T. & Palatucci G.: The obstacle problem for nonlinear integro-differential operators. *Calc. Var. Partial Differential Equations* 55 (2016), Art. 63, 30 pp.
[35] Kuusi T. & Mingione G.: Linear potentials in nonlinear potential theory. *Arch. Rat. Mech. Anal.* 207 (2013), 215–246.
[36] Kuusi T. & Mingione G.: Guide to nonlinear potential estimates. *Bull. Math. Sci.* 4 (2014), 1–82.
[37] Kuusi T. & Mingione G. & Sire Y.: Nonlocal problems with measure data *Comm. Math. Phys.* 337 (2015), 3, 1317–1368.
[38] Lieberman G. M.: Sharp forms of estimates for subsolutions and supersolutions of quasilinear elliptic equations involving measures. *Comm. Partial Differential Equations* 18 (1993), 1191–1212.
[39] Lindgren E. & Lindqvist P.: Perron's Method and Wiener's Theorem for a nonlocal equation. *Pot. Anal.* 46(4) (2017), 705–737.
[40] Lindqvist P.: On the definition and properties of p-superharmonic functions. *J. reine angew. Math. (Crelles J.)* 365 (1986), 67–79.
[41] Lindqvist P. & Martio O.: Two theorems of N. Wiener for solutions of quasilinear elliptic equations. *Acta Math.* 155 (1985), 153–171.
[42] Maz'ya, V.: The continuity at a boundary point of the solutions of quasi-linear elliptic equations. (Russian) *Vestnik Leningrad. Univ.* 25 (1970), 42–55.
[43] Mingione G.: Nonlinear measure data problems. *Milan J. Math.* 79 (2011), 429–496.

[44] Mingione G.: The Calderón-Zygmund theory for elliptic problems with measure data. *Ann. Sc. Norm. Super. Pisa Cl. Sci. (5)* **6** (2007), 195–261.
[45] Mingione G.: Gradient potential estimates. *J. Eur. Math. Soc.* **13** (2011), 459–486.
[46] Phuc N. C. & Verbitsky I. E.: Quasilinear and Hessian equations of Lane-Emden type. *Ann. of Math. (II)* 168 (2008), 859–914.
[47] Phuc N. C. & Verbitsky I. E.: Singular quasilinear and Hessian equations and inequalities. *J. Funct. Anal.* 256 (2009), 1875–1906.
[48] Ros-Oton X. & Serra J.: Regularity theory for general stable operators. *J. Differential Equations* 260 (2016), 8675–8715.
[49] L. Silvestre: Hölder estimates for solutions of integro-differential equations like the fractional Laplace. *Indiana Univ. Math. J.* 55(3) (2006), 1155–1174.
[50] Trudinger N.S. & Wang X.J.: On the weak continuity of elliptic operators and applications to potential theory. *Amer. J. Math.* 124 (2002), 369–410.

Xavier Ros-Oton

Boundary Regularity, Pohozaev Identities and Nonexistence Results

Abstract: In this expository paper we survey some recent results on Dirichlet problems of the form $Lu = f(x, u)$ in Ω, $u \equiv 0$ in $\mathbb{R}^n \setminus \Omega$. We first discuss in detail the boundary regularity of solutions, stating the main known results of Grubb and of the author and Serra. We also give a simplified proof of one of such results, focusing on the main ideas and on the blow-up techniques that we developed in [26, 27]. After this, we present the Pohozaev identities established in [24, 29, 16] and give a sketch of their proofs, which use strongly the fine boundary regularity results discussed previously. Finally, we show how these Pohozaev identities can be used to deduce nonexistence of solutions or unique continuation properties.
The operators L under consideration are integro-differential operator of order $2s$, $s \in (0, 1)$, the model case being the fractional Laplacian $L = (-\Delta)^s$.

Keywords: Integro-differential equations; bounded domains; boundary regularity; Pohozaev identities; nonexistence.

MSc: 47G20; 35B33; 35J61.

9.1 Introduction

This expository paper is concerned with the study of solutions to

$$\begin{cases} Lu = f(u) & \text{in } \Omega \\ u = 0 & \text{in } \mathbb{R}^n \setminus \Omega, \end{cases} \tag{9.1.1}$$

where $\Omega \subset \mathbb{R}^n$ is a bounded domain, and L is an elliptic integro-differential operator of the form

$$Lu(x) = \text{P.V.} \int_{\mathbb{R}^n} \big(u(x) - u(x+y)\big) K(y) dy,$$

$$K \geqslant 0, \qquad K(y) = K(-y), \qquad \text{and} \qquad \int_{\mathbb{R}^n} \min\{|y|^2, 1\} K(y) dy < \infty. \tag{9.1.2}$$

Such operators appear in the study of stochastic process with jumps: Lévy processes. In the context of integro-differential equations, Lévy processes play the same role that Brownian motion plays in the theory of second order PDEs. In particular, the study of such processes leads naturally to problems posed in bounded domains like (9.1.1).

Xavier Ros-Oton, The University of Texas at Austin, Department of Mathematics, 2515 Speedway, Austin, TX 78751, USA, E-mail: ros.oton@math.utexas.edu

https://doi.org/10.1515/9783110571561-011

Solutions to (9.1.1) are critical points of the nonlocal energy functional

$$\mathcal{E}(u)=\frac{1}{2}\int_{\mathbb{R}^n}\int_{\mathbb{R}^n}\big(u(x)-u(x+y)\big)^2K(y)dy\,dx-\int_\Omega F(u)dx$$

among functions $u\equiv 0$ in $\mathbb{R}^n\backslash\Omega$. Here, $F'=f$.

Here, we will work with operators L of order $2s$, with $s\in(0,1)$. In the simplest case we will have $K(y)=c_{n,s}|y|^{-n-2s}$, which corresponds to $L=(-\Delta)^s$, the fractional Laplacian. More generally, a typical assumption would be

$$0<\frac{\lambda}{|y|^{n+2s}}\leqslant K(y)\leqslant\frac{\Lambda}{|y|^{n+2s}}.$$

Under such assumption, operators (9.1.2) can be seen as uniformly elliptic operators of order $2s$, for which Harnack inequality and other regularity properties are well understood; see for example [22].

For the Laplace operator, (9.1.1) becomes

$$\begin{cases}-\Delta u &= f(u) \quad \text{in } \Omega\\ \quad u &= 0 \qquad \text{on } \partial\Omega.\end{cases}\tag{9.1.3}$$

A model case for (9.1.3) is the power-type nonlinearity $f(u)=|u|^{p-1}u$, with $p>1$. In this case, it is well known that the mountain pass theorem yields the existence of (nonzero) solutions for $p<\frac{n+2}{n-2}$, while for powers $p\geqslant\frac{n+2}{n-2}$ the only bounded solution in star-shaped domains is $u\equiv 0$. In other words, one has existence of solutions in the subcritical regime, and non-existence of solutions in star-shaped domains in the critical or supercritical regimes.

An important tool in the study of solutions to (9.1.3) is the *Pohozaev identity* [21]. This celebrated result states that any bounded solution to this problem satisfies the identity

$$\int_\Omega\{2nF(u)-(n-2)uf(u)\}dx=\int_{\partial\Omega}\left(\frac{\partial u}{\partial\nu}\right)^2(x\cdot\nu)d\sigma(x),\tag{9.1.4}$$

where

$$F(u)=\int_0^u f(t)dt.$$

When $f(u)=|u|^{p-1}u$ then the identity becomes

$$\left(\frac{2n}{p+1}-(n-2)\right)\int_\Omega|u|^{p+1}dx=\int_{\partial\Omega}\left(\frac{\partial u}{\partial\nu}\right)^2(x\cdot\nu)d\sigma(x).$$

When $p\geqslant\frac{n+2}{n-2}$, the left hand side of this identity is negative or zero, while the right hand side is strictly positive for nonzero solutions in star-shaped domains. Thus, the nonexistence of solutions follows.

The proof of the identity (9.1.4) is based on the following integration-by-parts type formula

$$2\int_\Omega(x\cdot\nabla u)\Delta u\,dx=(2-n)\int_\Omega u\,\Delta u\,dx+\int_{\partial\Omega}\left(\frac{\partial u}{\partial\nu}\right)^2(x\cdot\nu)d\sigma(x),\tag{9.1.5}$$

which holds for any C^2 function with $u = 0$ on $\partial\Omega$. This identity is an easy consequence of the *divergence theorem*. Indeed, using that

$$\Delta(x \cdot \nabla u) = x \cdot \nabla \Delta u + 2\Delta u$$

and that

$$x \cdot \nabla u = (x \cdot \nu)\frac{\partial u}{\partial \nu} \qquad \text{on} \quad \partial\Omega,$$

then integrating by parts (three times) we find

$$\begin{aligned}
\int_\Omega (x \cdot \nabla u)\Delta u \, dx &= -\int_\Omega \nabla(x \cdot \nabla u) \cdot \nabla u \, dx + \int_{\partial\Omega} (x \cdot \nabla u)\frac{\partial u}{\partial \nu} \, d\sigma \\
&= \int_\Omega \Delta(x \cdot \nabla u) \, u \, dx + \int_{\partial\Omega} (x \cdot \nabla u)\frac{\partial u}{\partial \nu} \, d\sigma \\
&= \int_\Omega \{x \cdot \nabla \Delta u + 2\Delta u\} u \, dx + \int_{\partial\Omega} (x \cdot \nu) \left(\frac{\partial u}{\partial \nu}\right)^2 d\sigma \\
&= \int_\Omega \{-\mathrm{div}(xu)\Delta u + 2u\Delta u\} dx + \int_{\partial\Omega} (x \cdot \nu) \left(\frac{\partial u}{\partial \nu}\right)^2 d\sigma \\
&= \int_\Omega \{-(x \cdot \nabla u)\Delta u - (n-2)u\Delta u\} dx + \int_{\partial\Omega} (x \cdot \nu) \left(\frac{\partial u}{\partial \nu}\right)^2 d\sigma,
\end{aligned}$$

and hence (9.1.5) follows.

Identities of Pohozaev-type like (9.1.4) and (9.1.5) have been used widely in the analysis of elliptic PDEs: they yield to monotonicity formulas, unique continuation properties, radial symmetry of solutions, and uniqueness results. Moreover, they are also used in other contexts such as hyperbolic equations, harmonic maps, control theory, and geometry.

The aim of this paper is to show what are the *nonlocal* analogues of these identities, explain the main ideas appearing in their proofs, and give some immediate consequences concerning the nonexistence of solutions. Furthermore, we will also discuss a very related issue: the *boundary regularity* of solutions.

• **A simple case.** In order to have a first hint of what should be the analogue of (9.1.5) for integro-differential operators (9.1.2), let us look at the simplest case $L = (-\Delta)^s$, and let us assume that $u \in C_c^\infty(\Omega)$. In this case, a standard computation shows that

$$(-\Delta)^s(x \cdot \nabla u) = x \cdot \nabla(-\Delta)^s u + 2s\,(-\Delta)^s u.$$

This is a pointwise equality that holds at every point $x \in \mathbb{R}^n$. This, combined with the global integration by parts identity in all of $\mathbb{R}^n$

$$\int_{\mathbb{R}^n} u\,(-\Delta)^s v \, dx = \int_{\mathbb{R}^n} (-\Delta)^s u\, v \, dx, \tag{9.1.6}$$

leads to

$$2\int_\Omega (x \cdot \nabla u)(-\Delta)^s u \, dx = (2s - n)\int_\Omega u(-\Delta)^s u \, dx, \qquad \text{for} \quad u \in C_c^\infty(\Omega). \tag{9.1.7}$$

Indeed, taking $v = x \cdot \nabla u$ one finds

$$\begin{aligned}\int_{\mathbb{R}^n} (x \cdot \nabla u)(-\Delta)^s u\, dx &= \int_{\mathbb{R}^n} u\,(-\Delta)^s (x \cdot \nabla u)\, dx \\ &= \int_{\mathbb{R}^n} u\,\{x \cdot \nabla(-\Delta)^s u + 2s\,(-\Delta)^s u\}\, dx \\ &= \int_{\mathbb{R}^n} \{-\mathrm{div}(xu)(-\Delta)^s u + 2s\, u(-\Delta)^s u\}\, dx \\ &= \int_{\mathbb{R}^n} \{-(x \cdot \nabla u)(-\Delta)^s u + (2s - n)\, u(-\Delta)^s u\}\, dx,\end{aligned}$$

and thus (9.1.7) follows.

This identity has no boundary term (recall that we assumed that u and all its derivatives are zero on $\partial\Omega$), but it is a first approximation towards a nonlocal version of (9.1.5). The only term that is missing is the boundary term.

As we showed above, when $s = 1$ and $u \in C_0^2(\overline{\Omega})$, the use of the divergence theorem in Ω (instead of the global identity (9.1.6)) leads to the Pohozaev-type identity (9.1.5), with the boundary term. However, in case of nonlocal equations there is no divergence theorem in bounded domains, and this is why at first glance there is no clear candidate for a nonlocal analogue of the boundary term in (9.1.5).

In order to get such a Pohozaev-type identity for solutions to (9.1.1), with the right boundary term, we first need to answer the following:

What is the *boundary regularity* of solutions to (9.1.1)?

Once this is well understood, we will come back to the study of Pohozaev identities and we will present the nonlocal analogues of (9.1.4)-(9.1.5) established in [24, 29].

The paper is organized as follows:

In Section 9.2 we discuss the boundary regularity of solutions to (9.1.1). We will state the main known results, and give a sketch of the proofs and their main ingredients. Then, in Section 9.3 we present the Pohozaev identities of [24, 29] and give some ideas of their proofs. Finally, in Section 9.4 we give some consequences of such Pohozaev identities.

9.2 Boundary Regularity

The study of integro-differential equations started already in the fifties with the works of Getoor, Blumenthal, and Kac, among others [4, 13]. Due to the relation with Lévy processes, they studied Dirichlet problems

$$\begin{cases} Lu = g(x) & \text{in } \Omega \\ u = 0 & \text{in } \mathbb{R}^n \setminus \Omega, \end{cases} \tag{9.2.1}$$

and proved some basic properties of solutions, estimates for the Green function, and the asymptotic distribution of eigenvalues. Moreover, in the simplest case of the fractional Laplacian $(-\Delta)^s$, the following explicit solutions were found:

$$u_0(x) = (x_+)^s \qquad \text{solves} \qquad \begin{cases} (-\Delta)^s u_0 = 0 & \text{in } (0, \infty) \\ u_0 = 0 & \text{in } (-\infty, 0) \end{cases}$$

and

$$u_1(x) = \kappa_{n,s}\left(1 - |x|^2\right)_+^s \qquad \text{solves} \qquad \begin{cases} (-\Delta)^s u_0 = 1 & \text{in } B_1 \\ u_0 = 0 & \text{in } \mathbb{R}^n \backslash B_1, \end{cases} \tag{9.2.2}$$

for certain constant $\kappa_{n,s}$; see [6].

The interior regularity of solutions for $L = (-\Delta)^s$ is by now well understood. Indeed, potential theory for this operator enjoys an explicit formulation in terms of the Riesz potential, and thus it is similar to that of the Laplacian; see the classical book of Landkov [20].

For more general linear operators (9.1.2), the interior regularity theory has been developed in the last years, and it is now quite well understood for operators satisfying

$$0 < \frac{\lambda}{|y|^{n+2s}} \leqslant K(y) \leqslant \frac{\Lambda}{|y|^{n+2s}}; \tag{9.2.3}$$

see for example the results of Bass [1], Serra [31], and also the survey [22] for regularity results in Hölder spaces.

Concerning the boundary regularity theory for the fractional Laplacian, fine estimates for the Green's function near $\partial\Omega$ were established by Kulczycki [19] and Chen-Song [9]; see also [4]. These results imply that, in $C^{1,1}$ domains, all solutions u to (9.2.1) are comparable to d^s, where $d(x) = \operatorname{dist}(x, \mathbb{R}^n \backslash \Omega)$. More precisely,

$$|u| \leqslant Cd^s \tag{9.2.4}$$

for some constant C, and from this bound one can deduce an estimate of the form

$$\|u\|_{C^s(\overline{\Omega})} \leqslant C\|g\|_{L^\infty(\Omega)}$$

for (9.2.1). Moreover, when $g > 0$, then $u \geqslant c\, d^s$ for some $c > 0$ —recall the example (9.2.2). In particular, solutions u are C^s up to the boundary, and this is the optimal Hölder exponent for the regularity of u, in the sense that in general $u \notin C^{s+\varepsilon}(\overline{\Omega})$ for any $\varepsilon > 0$.

More generally, when the equation and the boundary data are given only in a subregion of $\mathbb{R}^n$, one has the following estimate, whose proof we sketch below. Notice that the following estimate is for a general class of nonlocal operators L, which includes the fractional Laplacian.

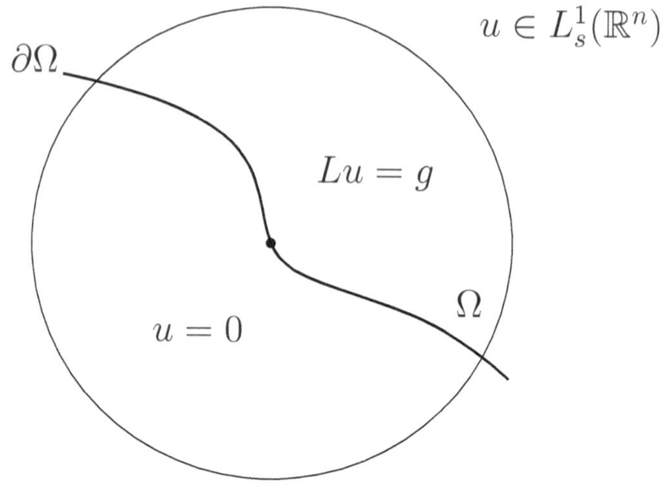

Proposition 9.1 ([27]). *Let $\Omega \subset \mathbb{R}^n$ be any bounded $C^{1,1}$ domain with $0 \in \partial\Omega$, and L be any operator* (9.1.2)-(9.2.3), *with $K(y)$ homogeneous. Let u be any bounded solution to*

$$\begin{cases} Lu = g & \text{in } \Omega \cap B_1 \\ u = 0 & \text{in } B_1 \setminus \Omega, \end{cases}$$

with $g \in L^\infty(\Omega \cap B_1)$. Then,[9.1]

$$\|u\|_{C^s(B_{1/2})} \leqslant C\big(\|g\|_{L^\infty(\Omega\cap B_1)} + \|u\|_{L^\infty(B_1)} + \|u\|_{L^1_s(\mathbb{R}^n)}\big),$$

with C depending only on n, s, Ω, and ellipticity constants.

Proof. We give a short sketch of this proof. For more details, see [23] or [27].

First of all, truncating u and dividing it by a constant if necessary, we may assume that $\|g\|_{L^\infty(\Omega\cap B_1)} + \|u\|_{L^\infty(\mathbb{R}^n)} \leqslant 1$. Second, by constructing a supersolution (using for example Lemma 9.3 below) one can show that

$$|u| \leqslant Cd^s \qquad \text{in} \quad \Omega. \tag{9.2.5}$$

Then, once we have this we need to show that

$$|u(x) - u(y)| \leqslant C|x-y|^s \qquad \forall x, y \in \overline{\Omega}. \tag{9.2.6}$$

We separate two cases, depending on whether $r = |x-y|$ is bigger or smaller than $\rho = \min\{d(x), d(y)\}$.

9.1 Here, we denote $\|w\|_{L^1_s(\mathbb{R}^n)} := \int_{\mathbb{R}^n} \frac{w(x)}{1+|x|^{n+2s}}\,dx$.

More precisely, if $2r \leqslant \rho$, then $y \in B_{\rho/2}(x) \subset B_\rho(x) \subset \Omega$ (we assume without loss of generality that $\rho = d(x) \leqslant d(y)$ here). Therefore, one can use known interior estimates (rescaled) and (9.2.5) to get

$$[u]_{C^s(B_{\rho/2}(x))} \leqslant C. \tag{9.2.7}$$

Indeed, by (9.2.5) we have that the rescaled function $u_\rho(z) := u(x + \rho z)$ satisfies

$$\|u_\rho\|_{L^\infty(B_1)} \leqslant C\rho^s, \qquad \|u_\rho\|_{L^1_s(\mathbb{R}^n)} \leqslant C\rho^s, \qquad \text{and} \qquad \|Lu_\rho\|_{L^\infty(B_1)} \leqslant C\rho^{2s},$$

and therefore by interior estimates

$$\begin{aligned}\rho^s[u]_{C^s(B_{\rho/2}(x))} = [u_\rho]_{C^s(B_{1/2})} &\leqslant C\big(\|u_\rho\|_{L^\infty(B_1)} + \|u_\rho\|_{L^1_s(\mathbb{R}^n)} + \|Lu_\rho\|_{L^\infty(B_1)}\big)\\ &\leqslant C(\rho^s + \rho^s + \rho^{2s}) \leqslant C\rho^s.\end{aligned}$$

In particular, it follows from (9.2.7) that $|u(x) - u(y)| \leqslant C|x - y|^s$.

On the other hand, in case $2r > \rho$ then we just use (9.2.5) to get

$$\begin{aligned}|u(x) - u(y)| &\leqslant |u(x)| + |u(y)| \leqslant Cd^s(x) + Cd^s(y)\\ &\leqslant Cd^s(x) + C\big(d^s(x) + |x - y|^s\big) \leqslant C\rho^s + C(r + \rho)^s\\ &\leqslant Cr^s = C|x - y|^s.\end{aligned}$$

In any case, we get (9.2.6), as desired. □

9.2.1 Higher order boundary regularity estimates

Unfortunately, in the study of Pohozaev identities the bound (9.2.4) is not enough, and finer regularity results are needed. A more precise description of solutions near $\partial\Omega$ is needed.

For second order (local) equations, solutions to the Dirichlet problem are known to be $C^\infty(\overline{\Omega})$ whenever Ω and the right hand side g are C^∞. In case $g \in L^\infty(\Omega)$, then $u \in C^{2-\varepsilon}(\overline{\Omega})$ for all $\varepsilon > 0$. This, in particular, yields a fine description of solutions u near $\partial\Omega$: for any $z \in \partial\Omega$ one has

$$\big|u(x) - c_z d(x)\big| \leqslant C|x - z|^{2-\varepsilon},$$

where $d(x) = \operatorname{dist}(x, \Omega^c)$ and $c_z \in \mathbb{R}$. This is an expansion of order $2 - \varepsilon$, which holds whenever $g \in L^\infty$ and Ω is $C^{1,1}$. When g and Ω are C^∞, then one has analogue higher order expansions that essentially say that $u/d \in C^\infty(\overline{\Omega})$.

The question for nonlocal operators was: are there any nonlocal analogues of such higher order boundary regularity estimates?

The first result in this direction was obtained by the author and Serra in [23] for the fractional Laplacian $L = (-\Delta)^s$; we showed that $u/d^s \in C^\alpha(\overline{\Omega})$ for some small

$\alpha > 0$. Such result was later improved and extended to more general operators by Grubb [14, 15] and by the author and Serra [26, 27]. These results may be summarized as follows.

Theorem 9.2 ([14, 15, 26, 27]). *Let $\Omega \subset \mathbb{R}^n$ be any bounded domain, and L be any operator* (9.1.2)-(9.2.3). *Assume in addition that $K(y)$ homogeneous, that is,*

$$K(y) = \frac{a\left(y/|y|\right)}{|y|^{n+2s}}. \tag{9.2.8}$$

Let u be any bounded solution to (9.2.1), *and*[9.2] $d(x) = \mathrm{dist}(x, \mathbb{R}^n \backslash \Omega)$. *Then,*

(a) If Ω is $C^{1,1}$, then

$$g \in L^\infty(\Omega) \quad \Longrightarrow \quad u/d^s \in C^{s-\varepsilon}(\overline{\Omega}) \qquad \text{for all } \varepsilon > 0,$$

(b) If Ω is $C^{2,\alpha}$ and $a \in C^{1,\alpha}(S^{n-1})$, then

$$g \in C^\alpha(\overline{\Omega}) \quad \Longrightarrow \quad u/d^s \in C^{\alpha+s}(\overline{\Omega}) \qquad \text{for small } \alpha > 0,$$

whenever $\alpha + s$ is not an integer.

(c) If Ω is C^∞ and $a \in C^\infty(S^{n-1})$, then

$$g \in C^\alpha(\overline{\Omega}) \quad \Longrightarrow \quad u/d^s \in C^{\alpha+s}(\overline{\Omega}) \qquad \text{for all } \alpha > 0,$$

whenever $\alpha + s \notin \mathbb{Z}$. In particular, $u/d^s \in C^\infty(\overline{\Omega})$ whenever $g \in C^\infty(\overline{\Omega})$.

It is important to remark that the above theorem is just a particular case of the results of [14, 15] and [26, 27]. Indeed, part (a) was proved in [27] for any $a \in L^1(S^{n-1})$ (without the assumption (9.2.3)); (b) was established in [26] in the more general context of fully nonlinear equations; and (c) was established in [14, 15] for all pseudodifferential operators satisfying the s-transmission property. Furthermore, when $s + \alpha$ is an integer in (c), more information is given in [15] in terms of Hölder-Zygmund spaces C^k_*.

When $g \in L^\infty(\Omega)$ and Ω is $C^{1,1}$, the above result yields a fine description of solutions u near $\partial\Omega$: for any $z \in \partial\Omega$ one has

$$\left|u(x) - c_z d^s(x)\right| \leqslant C|x - z|^{2s-\varepsilon},$$

where $d(x) = \mathrm{dist}(x, \Omega^c)$. This is an expansion of order $2s - \varepsilon$, which is analogue to the one described above for the Laplacian.

In case of second order (local) equations, only the regularity of g and $\partial\Omega$ play a role in the result. In the nonlocal setting of operators of the form (9.1.2)-(9.2.3), a third ingredient comes into play: the regularity of the kernels $K(y)$ in the y-variable. This

9.2 In fact, to avoid singularities inside Ω, we define $d(x)$ as a positive function that coincides with $\mathrm{dist}(x, \mathbb{R}^n \backslash \Omega)$ in a neighborhood of $\partial\Omega$ and is as regular as $\partial\Omega$ inside Ω.

is why in parts (b) and (c) of Theorem 9.2 one has to assume some regularity of K. This is a purely nonlocal feature, and cannot be avoided. In fact, when the kernels are not regular then counterexamples to higher order regularity can be constructed, both to interior and boundary regularity; see [31, 27]. Essentially, when the kernels are not regular, one can expect to get regularity results up to order $2s$, but not higher. We refer the reader to [22], where this is discussed in more detail.

Let us now sketch some ideas of the proof of Theorem 9.2. We will focus on the simplest case and try to show the main ideas appearing in its proof.

9.2.2 Sketch of the proof of Theorem 9.2(a)

A first important ingredient in the proof of Theorem 9.2(a) is the following computation.

Lemma 9.3. *Let Ω be any $C^{1,1}$ domain, $s \in (0,1)$, and L be any operator of the form (9.1.2)-(9.2.3) with $K(y)$ homogeneous, i.e., of the form (9.2.8). Let $d(x)$ be any positive function that coincides with $\mathrm{dist}(x, \mathbb{R}^n \backslash \Omega)$ in a neighborhood of $\partial\Omega$ and is $C^{1,1}$ inside Ω. Then,*

$$|L(d^s)| \leqslant C_\Omega \qquad \text{in} \quad \Omega, \tag{9.2.9}$$

where C_Ω depends only on n, s, Ω, and ellipticity constants.

Proof. Let $x_0 \in \Omega$ and $\rho = d(x)$. Notice that when $\rho \geqslant \rho_0 > 0$ then d^s is $C^{1,1}$ in a neighborhood of x_0, and thus $L(d^s)(x_0)$ is bounded by a constant depending only on ρ_0. Thus, we may assume that $\rho \in (0, \rho_0)$, for some small $\rho_0 > 0$.

Let us denote

$$\ell(x) := \left(d(x_0) + \nabla d^s(x_0) \cdot (x - x_0)\right)_+,$$

and notice that ℓ^s is a translated and rescaled version of the 1-D solution $(x_n)_+^s$. Thus, we have

$$L(\ell^s) = 0 \qquad \text{in} \quad \{\ell > 0\};$$

see [26, Section 2]. Moreover, notice that by construction of ℓ we have

$$d(x_0) = \ell(x_0) \qquad \text{and} \qquad \nabla d(x_0) = \nabla \ell(x_0).$$

Using this, it is not difficult to see that

$$\left|d(x_0 + y) - \ell(x_0 + y)\right| \leqslant C|y|^2,$$

and this yields

$$\left|d^s(x_0 + y) - \ell^s(x_0 + y)\right| \leqslant C|y|^2 \left(d^{s-1}(x_0 + y) + \ell^{s-1}(x_0 + y)\right).$$

On the other hand, for $|y| > 1$ we clearly have

$$|d^s(x_0+y) - \ell^s(x_0+y)| \leqslant C|y|^s \qquad \text{in} \quad \mathbb{R}^n \backslash B_1.$$

Using the last two inequalities, and recalling that $L(\ell^s)(x_0) = 0$ and that $d(x_0) = \ell(x_0)$, we find

$$\begin{aligned} |L(d^s)(x_0)| &= |L(d^s - \ell^s)(x_0)| = \left| \int_{\mathbb{R}^n} (d^s - \ell^s)(x_0+y) K(y) dy \right| \\ &\leqslant C \int_{B_1} |y|^2 \left(d^{s-1}(x_0+y) + \ell^{s-1}(x_0+y)\right) \frac{dy}{|y|^{n+2s}} + C \int_{B_1^c} |y|^s \frac{dy}{|y|^{n+2s}} \\ &\leqslant C \int_{B_1} \left(d^{s-1}(x_0+y) + \ell^{s-1}(x_0+y)\right) \frac{dy}{|y|^{n+2s-2}} + C. \end{aligned}$$

Such last integral can be bounded by a constant C depending only on s and Ω, exactly as in [28, Lemma 2.5], and thus it follows that

$$|L(d^s)(x_0)| \leqslant C,$$

as desired. □

Another important ingredient in the proof of Theorem 9.2(a) is the following classification result for solutions in a half-space.

Proposition 9.4 ([27]). *Let $s \in (0,1)$, and L be any operator of the form* (9.1.2)-(9.2.3) *with $K(y)$ homogeneous. Let u be any solution of*

$$\begin{cases} Lv &= 0 \quad \text{in } \{x \cdot e > 0\} \\ v &= 0 \quad \text{in } \{x \cdot e \leqslant 0\}. \end{cases} \tag{9.2.10}$$

Assume that, for some $\beta < 2s$, u satisfies the growth control

$$|v(x)| \leqslant C(1 + |x|^\beta) \qquad \text{in} \quad \mathbb{R}^n. \tag{9.2.11}$$

Then,

$$v(x) = K(x \cdot e)_+^s$$

for some constant $K \in \mathbb{R}$.

Proof. The idea is to differentiate v in the directions that are orthogonal to e, to find that v is a 1D function $v(x) = \overline{v}(x \cdot e)$. Then, for 1D functions any operator L with kernel (9.2.8) is just a multiple of the 1D fractional Laplacian, and thus one only has to show the result in dimension 1. Let us next explain the whole argument in more detail.

Given $R \geqslant 1$ we define

$$v_R(x) := R^{-\beta} v(Rx).$$

It follows from the growth condition on v that

$$|v_R(x)| \leqslant C(1 + |x|^\beta) \qquad \text{in} \quad \mathbb{R}^n,$$

and moreover v_R satisfies (9.2.10), too.

Since $\beta < 2s$, then $\|v_R\|_{L^1_s(\mathbb{R}^n)} \leqslant C$, and thus by Proposition 9.1 we get

$$\|v_R\|_{C^s(B_{1/2})} \leqslant C,$$

with C independent of R. Therefore, using $[v]_{C^s(B_{R/2})} = R^{\beta-s}[v_R]_{C^s(B_{1/2})}$, we find

$$[v]_{C^s(B_{R/2})} \leqslant CR^{\beta-s} \quad \text{for all} \quad R \geqslant 1. \tag{9.2.12}$$

Now, given $\tau \in S^{n-1}$ such that $\tau \cdot e = 0$, and given $h > 0$, consider

$$w(x) := \frac{v(x+h\tau) - v(x)}{h^s}.$$

By (9.2.12) we have that w satisfies the growth condition

$$\|w\|_{L^\infty(B_R)} \leqslant CR^{\beta-s} \quad \text{for all} \quad R \geqslant 1.$$

Moreover, since τ is a direction which is parallel to $\{x \cdot e = 0\}$, then w satisfies the same equation as v, namely $Lw = 0$ in $\{x \cdot e > 0\}$, and $w = 0$ in $\{x \cdot e \leqslant 0\}$. Thus, we can repeat the same argument above with v replaced by w (and β replaced by $\beta - s$), to find

$$[w]_{C^s(B_{R/2})} \leqslant CR^{\beta-2s} \quad \text{for all} \quad R \geqslant 1.$$

Since $\beta < 2s$, letting $R \to \infty$ we find that

$$w \equiv 0 \quad \text{in} \quad \mathbb{R}^n.$$

This means that v is a 1D function, $v(x) = \overline{v}(x \cdot e)$. But then (9.2.10) yields that such function $\overline{v} : \mathbb{R} \to \mathbb{R}$ satisfies

$$\begin{cases} (-\Delta)^s \overline{v} &= 0 \quad \text{in } (0, \infty) \\ \overline{v} &= 0 \quad \text{in } (-\infty, 0], \end{cases}$$

with the same growth condition (9.2.11). Using [26, Lemma 5.2], we find that $\overline{v}(x) = K(x_+)^s$, and thus

$$v(x) = K(x \cdot e)_+^s,$$

as desired. □

Using the previous results, let us now give the:

• **Sketch of the proof of Theorem 9.2(a).** In Proposition 9.2.6 we saw how combining (9.2.5) with interior estimates (rescaled) one can show that $u \in C^s(\overline{\Omega})$. In other words, in order to prove the C^s regularity up to the boundary, one only needs the bound $|u| \leqslant Cd^s$ and interior estimates.

Similarly, it turns out that in order to show that $u/d^s \in C^\gamma(\overline{\Omega})$, $\gamma = s - \varepsilon$, we just need an expansion of the form

$$\left|u(x) - Q_z d^s(x)\right| \leqslant C|x-z|^{s+\gamma}, \qquad z \in \partial\Omega, \qquad Q_z \in \mathbb{R}. \tag{9.2.13}$$

Once this is done, one can combine (9.2.13) with interior estimates and get $u/d^s \in C^\gamma(\overline{\Omega})$; see [27, Proof of Theorem 1.2] for more details.

Thus, we need to show (9.2.13).

The proof of (9.2.13) is by contradiction, using a blow-up argument. Indeed, assume that for some $z \in \partial\Omega$ the expansion (9.2.13) does not hold for any $Q \in \mathbb{R}$. Then, we clearly have

$$\sup_{r>0} r^{-s-\gamma} \|u - Qd^s\|_{L^\infty(B_r(z))} = \infty \qquad \text{for all} \quad Q \in \mathbb{R}.$$

Then, one can show (see [27, Lemma 5.3]) that this yields

$$\sup_{r>0} r^{-s-\gamma} \|u - Q(r)d^s\|_{L^\infty(B_r(z))} = \infty, \qquad \text{with} \qquad Q(r) = \frac{\int_{B_r(z)} u\, d^s}{\int_{B_r(z)} d^{2s}}.$$

Notice that this choice of $Q(r)$ is the one which minimizes the L^2 distance between u and Qd^s in $B_r(z)$.

We define the monotone quantity

$$\vartheta(r) := \max_{r' \geqslant r} (r')^{-s-\gamma} \|u - Q(r')d^s\|_{L^\infty(B_{r'}(z))}.$$

Since $\vartheta(r) \to \infty$ as $r \to 0$, then there exists a sequence $r_m \to 0$ such that

$$(r_m)^{-s-\gamma} \|u - Q(r_m)d^s\|_{L^\infty(B_{r_m})} = \vartheta(r_m).$$

We now define the blow-up sequence

$$v_m(x) := \frac{u(z + r_m x) - Q(r_m)d^s(z + r_m x)}{(r_m)^{s+\gamma}\vartheta(r_m)}.$$

By definition of $Q(r_m)$ we have

$$\int_{B_1} v_m(x)\, d^s(z + r_m x)dx = 0, \tag{9.2.14}$$

and by definition of r_m we have

$$\|v_m\|_{L^\infty(B_1)} = 1 \tag{9.2.15}$$

Moreover, it can be shown that we have the growth control

$$\|v_m\|_{L^\infty(B_R)} \leqslant CR^{s+\gamma} \qquad \text{for all } R \geqslant 1.$$

To prove this, one first shows that

$$|Q(Rr) - Q(r)| \leqslant C(rR)^\gamma \vartheta(r),$$

and then use the definitions of v_m and ϑ to get

$$\begin{aligned}
\|v_m\|_{L^\infty(B_R)} &= \frac{1}{(r_m)^{s+\gamma}\vartheta(r_m)}\|u - Q(r_m)d^s\|_{L^\infty(B_{r_mR})} \\
&\leqslant \frac{1}{(r_m)^{s+\gamma}\vartheta(r_m)}\left\{\|u - Q(r_mR)d^s\|_{L^\infty(B_{r_mR})} + |Q(r_mR) - Q(r_m)|(r_mR)^s\right\} \\
&\leqslant \frac{1}{(r_m)^{s+\gamma}\vartheta(r_m)}\vartheta(r_mR)(r_mR)^{s+\gamma} + \frac{C}{(r_m)^{s+\gamma}\vartheta(r_m)}(r_mR)^\gamma\vartheta(r_m)(r_mR)^s \\
&\leqslant R^{s+\gamma} + CR^{s+\gamma}.
\end{aligned}$$

In the last inequality we used $\vartheta(r_mR) \leqslant \vartheta(r_m)$, which follows from the monotonicity of ϑ and the fact that $R \geqslant 1$.

On the other hand, the functions v_m satisfy

$$|Lv_m(x)| = \frac{(r_m)^{2s}}{(r_m)^{s+\gamma}\vartheta(r_m)}|Lu(z + r_mx) - L(d^s)(z + r_mx)| \quad \text{in} \quad \Omega_m,$$

where the domain $\Omega_m = (r_m)^{-1}(\Omega - z)$ converges to a half-space $\{x \cdot e > 0\}$ as $m \to \infty$. Here $e \in S^{n-1}$ is the inward normal vector to $\partial\Omega$ at z.

Since Lu and $L(d^s)$ are bounded, and $\gamma < s$, then it follows that

$$Lv_m \to 0 \quad \text{uniformly in compact sets in } \{x \cdot e > 0\}.$$

Moreover, $v_m \to 0$ uniformly in compact sets in $\{x \cdot e < 0\}$, since $u = 0$ in Ω^c.

Now, by C^s regularity estimates up to the boundary and the Arzelà-Ascoli theorem the functions v_m converge (up to a subsequence) to a function $v \in C(\mathbb{R}^n)$. The convergence is uniform in compact sets of $\mathbb{R}^n$. Therefore, passing to the limit the properties of v_m, we find

$$\|v\|_{L^\infty(B_R)} \leqslant CR^{s+\gamma} \qquad \text{for all } R \geqslant 1, \tag{9.2.16}$$

and

$$\begin{cases} Lv = 0 & \text{in } \{x \cdot e > 0\} \\ v = 0 & \text{in } \{x \cdot e < 0\}. \end{cases} \tag{9.2.17}$$

Now, thanks to Proposition 9.4, we find

$$v(x) = K(x \cdot e)_+^s \qquad \text{for some} \quad K \in \mathbb{R}. \tag{9.2.18}$$

Finally, passing to the limit (9.2.14) —using that $d^s(z + r_mx)/(r_m)^s \to (x \cdot e)_+^s$ — we find

$$\int_{B_1} v(x)\,(x \cdot e)_+^s\,dx = 0, \tag{9.2.19}$$

so that $K \equiv 0$ and $v \equiv 0$. But then passing to the limit (9.2.15) we get a contradiction, and hence (9.2.13) is proved. □

It is important to remark that in [27] we show (9.2.13) with a constant C depending only on n, s, $\|g\|_{L^\infty}$, the $C^{1,1}$ norm of Ω, and ellipticity constants. To do that, the idea of the proof is exactly the same, but one needs to consider sequences of functions u_m, domains Ω_m, points $z_m \in \partial\Omega_m$, and operators L_m.

9.2.3 Comments, remarks, and open problems

Let us next give some final comments and remarks about Theorem 9.2, as well as some related open problems.

• **Singular kernels.** Theorem 9.2 (a) was proved in [27] for operators L with general homogeneous kernels of the form (9.2.8) with $a \in L^1(S^{n-1})$, not necessarily satisfying (9.2.3). In fact, a could even be a singular measure. In such setting, it turns out that Lemma 9.3 is in general false, even in C^∞ domains. Because of this difficulty, the proof of Theorem 9.2(a) given in [27] is in fact somewhat more involved than the one we sketched above.

• **Counterexamples for non-homogeneous kernels.** All the results above are for kernels K satisfying (9.2.3) and such that $K(y)$ is *homogeneous*. As said above, for the interior regularity theory one does not need the homogeneity assumption: the interior regularity estimates are the same for homogeneous or non-homogeneous kernels. However, it turns out that something different happens in the boundary regularity theory. Indeed, for operators with x-dependence

$$Lu(x) = \text{P.V.} \int_{\mathbb{R}^n} (u(x) - u(x+y))K(x,y)dy,$$

$$0 < \frac{\lambda}{|y|^{n+2s}} \leqslant K(x,y) \leqslant \frac{\Lambda}{|y|^{n+2s}}, \qquad K(x,y) = K(x,-y),$$

we constructed in [26] solutions to $Lu = 0$ in Ω, $u = 0$ in $\mathbb{R}^n \setminus \Omega$, that are *not* comparable to d^s. More precisely, we showed that in dimension $n = 1$ there are $\beta_1 < s < \beta_2$ for which the functions $(x_+)^{\beta_i}$, solve an equation of the form $Lu = 0$ in $(0, \infty)$, $u = 0$ in $(-\infty, 0]$. Thus, no fine boundary regularity like Theorem 9.2 can be expected for non-homogeneous kernels; see [26, Section 2] for more details.

• **On the proof of Theorem 9.2 (b).** The proof of Theorem 9.2(b) in [26] has a similar structure as the one sketched above, in the sense that we show first $L(d^s) \in C^\alpha(\overline{\Omega})$ and then prove an expansion of order $2s + \alpha$ similar to (9.2.13). However, there are extra difficulties coming from the fact that we would get exponent $2s + \alpha$ in (9.2.16), and thus the operator L is not defined on functions that grow that much. Thus, the blow-up procedure needs to be done with incremental quotients, and the global equation (9.2.17) is replaced by [26, Theorem 1.4].

• **On the proof of Theorem 9.2 (c).** Theorem 9.2(c) was proved in [14, 15] by Fourier transform methods, completely different from the techniques presented above. Namely, the results in [14, 15] are for general pseudodifferential operators satisfying the so-called s-transmission property. A key ingredient in those proofs is the existence of a factorization of the principal symbol, which leads to the boundary regularity properties for such operators.

• **Open problem: Regularity in $C^{k,\alpha}$ domains.** After the results of [14, 15, 26, 27], a natural question remains open: what happens in $C^{k,\alpha}$ domains?

Our results in [26, 27] give sharp regularity estimates in $C^{1,1}$ and $C^{2,\alpha}$ domains –Theorem 9.2 (a) and (b)–, while the results of Grubb [14, 15] give higher order estimates in C^{∞} domains –Theorem 9.2 (c). It is an open problem to establish sharp boundary regularity results in $C^{k,\alpha}$ domains, with $k \geqslant 3$, for operators (9.1.2)-(9.2.3) with homogeneous kernels.

For the fractional Laplacian, sharp estimates in $C^{k,\alpha}$ domains have been recently established in [17], by using the extension problem for the fractional Laplacian. For more general operators, this is only known for $k = 1$ [28] and $k = 2$ [26].

The development of sharp boundary regularity results in $C^{k,\alpha}$ domains for integro-differential operators L would lead to the higher regularity of the free boundary in obstacle problems such operators; see [10], [17], [8].

• **Open problem: Parabolic equations.** Part (a) of Theorem 9.2 was recently extended to parabolic equations in [12]. A natural open question is to understand the higher order boundary regularity of solutions for parabolic equations of the form

$$\partial_t u + Lu = f(t, x).$$

Are there analogous estimates to those in Theorem 9.2 (b) and (c) in the parabolic setting?

This could lead to the higher regularity of the free boundary in parabolic obstacle problems for integro-differential operators; see [7, 2].

• **Open problem: Operators with different scaling properties.** An interesting open problem concerning the boundary regularity of solutions is the following: What happens with operators (9.1.2) with kernels having a different type of singularity near $y = 0$? For example, what happens with operators with kernels $K(y) \approx |y|^{-n}$ for $y \approx 0$? This type of kernels appear when considering geometric stable processes; see [33]. The interior regularity theory has been developed by Kassmann-Mimica in [18] for very general classes of kernels, but much less is known about the boundary regularity; see [5] for some results in that direction.

9.3 Pohozaev Identities

Once the boundary regularity is known, we can now come back to the Pohozaev identities. We saw in the previous section that solutions u to

$$\begin{cases} Lu = f(x,u) & \text{in } \Omega \\ u = 0 & \text{in } \mathbb{R}^n \backslash \Omega. \end{cases} \tag{9.3.1}$$

are not C^1 up to the boundary, but the quotient u/d^s is Hölder continuous up to the boundary. In particular, for any $z \in \partial\Omega$ there exists the limit

$$\frac{u}{d^s}(z) := \lim_{\Omega \ni x \to z} \frac{u(x)}{d^s(x)}.$$

As we will see next, this function $u/d^s|_{\partial\Omega}$ plays the role of the normal derivative $\partial u/\partial\nu$ in the nonlocal analogues of (9.1.5)-(9.1.4).

Theorem 9.5 ([24, 29]). *Let Ω be any bounded $C^{1,1}$ domain, and L be any operator of the form (9.1.2), with*

$$K(y) = \frac{a(y/|y|)}{|y|^{n+2s}}.$$

and $a \in L^\infty(S^{n-1})$. Let f be any locally Lipschitz function, u be any bounded solution to (9.3.1). Then, the following identity holds

$$-2\int_\Omega (x\cdot\nabla u)Lu\,dx = (n-2s)\int_\Omega u\,Lu\,dx + \Gamma(1+s)^2\int_{\partial\Omega} \mathcal{A}(\nu)\left(\frac{u}{d^s}\right)^2(x\cdot\nu)d\sigma. \quad (9.3.2)$$

Moreover, for all $e \in \mathbb{R}^n$, we have[9.3]

$$-\int_\Omega \partial_e u\,Lu\,dx = \frac{\Gamma(1+s)^2}{2}\int_{\partial\Omega} \mathcal{A}(\nu)\left(\frac{u}{d^s}\right)^2(\nu\cdot e)\,d\sigma. \quad (9.3.3)$$

Here

$$\mathcal{A}(\nu) = c_s\int_{S^{n-1}} |\nu\cdot\vartheta|^{2s}a(\vartheta)d\vartheta, \quad (9.3.4)$$

$a(\vartheta)$ is the function in (9.2.8), and c_s is a constant that depends only on s. For $L = (-\Delta)^s$, we have $\mathcal{A}(\nu) \equiv 1$.

When the nonlinearity $f(x,u)$ does not depend on x, the previous theorem yields the following analogue of (9.1.4)

$$\int_\Omega \{2n\,F(u) - (n-2s)uf(u)\}dx = \Gamma(1+s)^2\int_{\partial\Omega} \mathcal{A}(\nu)\left(\frac{u}{d^s}\right)^2(x\cdot\nu)d\sigma(x).$$

Before our work [24], no Pohozaev identity for the fractional Laplacian was known (not even in dimension $n = 1$). Theorem 9.5 was first found and established for $L = (-\Delta)^s$ in [24], and later the result was extended to more general operators in [29]. A surprising feature of this result is that, even if the operators (9.1.2) are nonlocal, the identities (9.3.2)-(9.3.3) have completely local boundary terms.

Let us give now a sketch of the proof of the Pohozaev identity (9.3.2). In order to focus on the main ideas, no technical details will be discussed.

9.3 In (9.3.3), we have corrected the sign on the boundary contribution, which was incorrectly stated in [24, Theorem 1.9].

9.3.1 Sketch of the proof

For simplicity, let us assume that Ω is C^∞ and that $u/d^s \in C^\infty(\overline{\Omega})$.

Step 1. We first assume that Ω is strictly star-shaped; later we will deduce the general case from this one. Translating Ω if necessary, we may assume it is strictly star-shaped with respect to the origin.

Let us define

$$u_\lambda(x) = u(\lambda x), \qquad \lambda > 1,$$

and let us write the right hand side of (9.3.2) as

$$2\int_\Omega (x\cdot\nabla u)Lu = 2\left.\frac{d}{d\lambda}\right|_{\lambda=1^+}\int_\Omega u_\lambda Lu.$$

This follows from the fact that $\left.\frac{d}{d\lambda}\right|_{\lambda=1^+} u_\lambda(x) = (x\cdot\nabla u)$ and the dominated convergence theorem. Then, since u_λ vanishes outside Ω, we will have

$$\int_\Omega u_\lambda Lu = \int_{\mathbb{R}^n} u_\lambda Lu = \int_{\mathbb{R}^n} L^{\frac{1}{2}}u_\lambda L^{\frac{1}{2}}u,$$

and therefore

$$\begin{aligned}\int_\Omega u_\lambda Lu = \int_{\mathbb{R}^n} L^{\frac{1}{2}}u_\lambda L^{\frac{1}{2}}u &= \lambda^s\int_{\mathbb{R}^n}\left(L^{\frac{1}{2}}u\right)(\lambda x)L^{\frac{1}{2}}u(x)\,dx\\ &= \lambda^s\int_{\mathbb{R}^n} w(\lambda x)w(x)\,dx\\ &= \lambda^{\frac{2s-n}{2}}\int_{\mathbb{R}^n} w(\lambda^{\frac{1}{2}}y)w(\lambda^{-\frac{1}{2}}y)\,dy\end{aligned}$$

where $w(x) = L^{\frac{1}{2}}u(x)$.

Now, since $2\left.\frac{d}{d\lambda}\right|_{\lambda=1^+}\lambda^{\frac{2s-n}{2}} = 2s-n$, the previous identities (and the change $\sqrt{\lambda}\mapsto\lambda$) yield

$$2\int_\Omega (x\cdot\nabla u)Lu = (2s-n)\int_{\mathbb{R}^n} w^2 + \left.\frac{d}{d\lambda}\right|_{\lambda=1^+}\int_{\mathbb{R}^n} w_\lambda w_{1/\lambda}.$$

Moreover, since

$$\int_{\mathbb{R}^n} w^2 = \int_{\mathbb{R}^n} L^{1/2}u\,L^{1/2}u = \int_{\mathbb{R}^n} u\,Lu = \int_\Omega u\,Lu,$$

then we have

$$-2\int_\Omega (x\cdot\nabla u)Lu = (n-2s)\int_\Omega u\,Lu + \mathcal{I}(w), \tag{9.3.5}$$

where

$$\mathcal{I}(w) = -\left.\frac{d}{d\lambda}\right|_{\lambda=1^+}\int_{\mathbb{R}^n} w_\lambda w_{1/\lambda}, \tag{9.3.6}$$

$w_\lambda(x) = w(\lambda x)$, and $w(x) = L^{\frac{1}{2}}u(x)$.

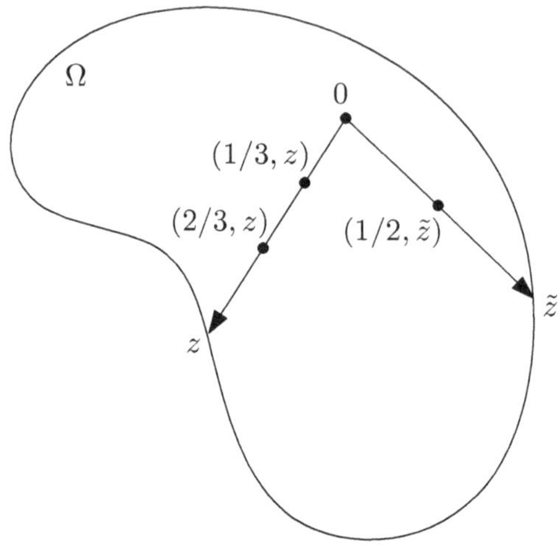

Fig. 9.1: Star-shaped coordinates $x = tz$, with $z \in \partial\Omega$.

At this point one should compare (9.3.2) and (9.3.5). In order to establish (9.3.2), we "just" need to show that $\mathfrak{I}(w)$ is exactly the boundary term we want.

Let us take a closer look at the operator defined by (9.3.6). The first thing one may observe by differentiating under the integral sign is that

$$\varphi \text{ is "nice enough"} \qquad \Longrightarrow \qquad \mathfrak{I}(\varphi) = 0.$$

In particular, one can also show that $\mathfrak{I}(\varphi + h) = \mathfrak{I}(\varphi)$ whenever h is "nice enough".

The function $w = L^{1/2}u$ is smooth inside Ω and also in $\mathbb{R}^n \setminus \overline{\Omega}$, but it has a singularity along $\partial\Omega$. In order to compute $\mathfrak{I}(w)$, we have to study carefully the behavior of $w = L^{1/2}u$ near $\partial\Omega$, and try to compute $\mathfrak{I}(w)$ by using (9.3.6). The idea is that, since u/d^s is smooth, then we will have

$$w = L^{1/2}u = L^{1/2}\left(d^s \frac{u}{d^s}\right) = L^{1/2}(d^s)\frac{u}{d^s} + \text{"nice terms"}, \tag{9.3.7}$$

and thus the behavior of w near $\partial\Omega$ will be that of $L^{1/2}(d^s)\frac{u}{d^s}$.

Using the previous observation, and writing the integral in (9.3.6) in the "star-shaped coordinates" $x = tz,\ z \in \partial\Omega,\ t \in (0, \infty)$, we find

$$\begin{aligned} -\mathfrak{I}(w) = \left.\frac{d}{d\lambda}\right|_{\lambda=1^+} \int_{\mathbb{R}^n} w_\lambda w_{1/\lambda} &= \left.\frac{d}{d\lambda}\right|_{\lambda=1^+} \int_{\partial\Omega} (z\cdot\nu)d\sigma(z) \int_0^\infty t^{n-1} w(\lambda tz) w\left(\frac{tz}{\lambda}\right) dt \\ &= \int_{\partial\Omega} (z\cdot\nu)d\sigma(z) \left.\frac{d}{d\lambda}\right|_{\lambda=1^+} \int_0^\infty t^{n-1} w(\lambda tz) w\left(\frac{tz}{\lambda}\right) dt. \end{aligned}$$

Now, a careful analysis of $L^{1/2}(d^s)$ leads to the formula

$$L^{1/2}(d^s)(tz) = \varphi_s(t)\sqrt{\mathcal{A}(\nu(z))} + \text{"nice terms"}, \tag{9.3.8}$$

where $\varphi_s(t) = c_1\{\log^-|t-1| + c_2\chi_{(0,1)}(t)\}$, and c_1, c_2 are explicit constants that depend only on s. Here, χ_A denotes the characteristic function of the set A.

This, combined with (9.3.7), gives

$$w(tz) = \varphi_s(t)\sqrt{\mathcal{A}(\nu(z))}\,\frac{u}{d^s}(z) + \text{“nice terms”}. \tag{9.3.9}$$

Using the previous two identities we find

$$\begin{aligned}
\mathcal{I}(w) &= -\int_{\partial\Omega}(z\cdot\nu)d\sigma(z)\,\frac{d}{d\lambda}\bigg|_{\lambda=1^+}\int_0^\infty t^{n-1}w(\lambda tz)w\left(\frac{tz}{\lambda}\right)dt\\
&= -\int_{\partial\Omega}(z\cdot\nu)d\sigma(z)\,\frac{d}{d\lambda}\bigg|_{\lambda=1^+}\int_0^\infty t^{n-1}\varphi_s(\lambda t)\varphi_s\left(\frac{t}{\lambda}\right)\mathcal{A}(\nu(z))\left(\frac{u}{d^s}(z)\right)^2 dt\\
&= \int_{\partial\Omega}\mathcal{A}(\nu)\left(\frac{u}{d^s}\right)^2(z\cdot\nu)d\sigma(z)\,C(s),
\end{aligned} \tag{9.3.10}$$

where

$$C(s) = -\frac{d}{d\lambda}\bigg|_{\lambda=1^+}\int_0^\infty t^{n-1}\varphi_s(\lambda t)\varphi_s\left(\frac{t}{\lambda}\right)dt$$

is a (positive) constant that can be computed explicitly. Thus, (9.3.2) follows from (9.3.5) and (9.3.10).

Step 2. Let now Ω be any $C^{1,1}$ domain. In that case, the above proof does not work, since the assumption that Ω was star-shaped was very important in such proof. Still, as shown next, once the identity (9.3.2) is established for star-shaped domains, then the identity for general $C^{1,1}$ domains follows from an argument involving a partition of unity and the fact that every $C^{1,1}$ domain is locally star-shaped.

Let B_i be a finite collection of small balls covering Ω. Then, we consider a family of functions $\psi_i \in C_c^\infty(B_i)$ such that $\sum_i \psi_i = 1$, and we let $u_i = u\psi_i$.

We claim that for every i, j we have the following bilinear identity

$$\begin{aligned}
-\int_\Omega(x\cdot\nabla u_i)Lu_j\,dx - \int_\Omega(x\cdot\nabla u_j)Lu_i\,dx &= \frac{n-2s}{2}\int_\Omega u_iLu_j\,dx +\\
+\frac{n-2s}{2}\int_\Omega u_jLu_i\,dx &+ \Gamma(1+s)^2\int_{\partial\Omega}\mathcal{A}(\nu)\frac{u_i}{\delta^s}\frac{u_j}{\delta^s}(x\cdot\nu)\,d\sigma.
\end{aligned} \tag{9.3.11}$$

To prove this, we separate two cases. In case $\overline{B}_i \cap \overline{B}_j \neq \varnothing$ then it turns out that u_i and u_j satisfy the hypotheses of Step 1, and thus they satisfy the identity (9.3.2) —here we are using that the intersection of the $C^{1,1}$ domain Ω with a small ball is always star-shaped. Then, applying (9.3.2) to the functions $(u_i + u_j)$ and $(u_i - u_j)$ and subtracting such two identities, one gets (9.3.11). On the other hand, in case $\overline{B}_i \cap \overline{B}_j = \varnothing$ then the identity (9.3.11) is a simple computation similar to (9.1.7), since in this case we have $u_iu_j = 0$ and thus there is no boundary term in (9.3.11). Hence, we got (9.3.11) for all i, j. Therefore, summing over all i and all j and using that $\sum_i u_i = u$, (9.3.2) follows.

Step 3. Let us finally show the second identity (9.3.3). For this, we just need to apply the identity that we already proved, (9.3.2), with a different origin $e \in \mathbb{R}^n$. We get

$$-2\int_\Omega ((x-e)\cdot\nabla u)Lu\,dx = (n-2s)\int_\Omega u\,Lu\,dx + \Gamma(1+s)^2\int_{\partial\Omega} \mathcal{A}(\nu)\left(\frac{u}{d^s}\right)^2((x-e)\cdot\nu)\,d\sigma. \quad (9.3.12)$$

Subtracting (9.3.2) and (9.3.12) we get (9.3.3), as desired. □

9.3.2 Comments and further results

Let us next give some final remarks about Theorem 9.5.

• **On the proof of Theorem 9.5.** First, notice that the smoothness of u/d^s and $\partial\Omega$ is hidden in (9.3.9). In fact, the proof of (9.3.8)-(9.3.9) requires a very fine analysis, even if one assumes that both u/d^s and $\partial\Omega$ are C^∞. Furthermore, even in this smooth case, the "nice terms" in (9.3.9) are not even C^1 near $\partial\Omega$, and a delicate result for $\mathcal{I}$ is needed in order to ensure that $\mathcal{I}$("nice terms") $= 0$; see Proposition 1.11 in [24].

Second, note that the kernel of the operator $L^{1/2}$ has an explicit expression in case $L = (-\Delta)^s$, but not for general operators with kernels (9.2.8). Because of this, the proofs of (9.3.8) and (9.3.9) are simpler for $L = (-\Delta)^s$, and some new ideas are required to treat the general case, in which we obtain the extra factor $\sqrt{\mathcal{A}(\nu(z))}$.

• **Extension to more general operators.** After the results of [24, 29], a last question remained to be answered: what happens with more general operators (9.1.2)? For example, is there any Pohozaev identity for the class of operators $(-\Delta + m^2)^s$, with $m > 0$? And for operators with x-dependence?

In a recent work [16], G. Grubb obtained integration-by-parts formulas as in Theorem 9.5 for pseudodifferential operators P of the form

$$Pu = \mathrm{Op}(p(x,\xi))u = \mathcal{F}^{-1}_{\xi\to x}(p(x,\xi)(\mathcal{F}u)(\xi)), \quad (9.3.13)$$

where $\mathcal{F}$ is the Fourier transform $(\mathcal{F}u)(\xi) = \int_{\mathbb{R}^n} e^{-ix\cdot\xi}u(x)\,dx$. The symbol $p(x,\xi)$ has an asymptotic expansion $p(x,\xi) \sim \sum_{j\in\mathbb{N}_0} p_j(x,\xi)$ in homogeneous terms: $p_j(x,t\xi) = t^{2s-j}p_j(x,\xi)$, and p is *even* in the sense that $p_j(x,-\xi) = (-1)^j p_j(x,\xi)$ for all j.

When a in (9.2.8) is $C^\infty(S^{n-1})$, then the operators (9.1.2)-(9.2.8) are pseudodifferential operators of the form (9.3.13). For these operators (9.1.2)-(9.2.8), the lower-order terms p_j ($j \geqslant 1$) vanish and p_0 is real and x-independent. Here $p_0(\xi) = \mathcal{F}_{y\to\xi}K(y)$, and $\mathcal{A}(\nu) = p_0(\nu)$. The fractional Laplacian $(-\Delta)^s$ corresponds to $a \equiv 1$ in (9.1.2)-(9.2.8), and to $p(x,\xi) = |\xi|^{2s}$ in (9.3.13).

In case of operators (9.3.13) with no x-dependence and with real symbols $p(\xi)$, the analogue of (9.3.2) proved in [16] is the following identity

$$-2\int_\Omega (x\cdot\nabla u)Pu\,dx = \Gamma(1+s)^2\int_{\partial\Omega} p_0(\nu)\left(\frac{u}{d^s}\right)^2(x\cdot\nu)d\sigma + \\ + n\int_\Omega u\,Pu\,dx - \int_\Omega u\,\mathrm{Op}(\xi\cdot\nabla p(\xi))u\,dx,$$

where $p_0(\nu)$ is the principal symbol of P at ν. Note that when the symbol $p(\xi)$ is homogeneous of degree $2s$ (hence equals $p_0(\xi)$), then $\xi\cdot\nabla p(\xi) = 2s\,p(\xi)$, and thus we recover the identity (9.3.2).

The previous identity can be applied to operators $(-\Delta+m^2)^s$. Furthermore, the results in [16] allow x-dependent operators P, which result in extra integrals over Ω. The methods in [16] are complex and quite different from the ones we use in [24, 29]. The domain Ω is assumed C^∞ in [16].

9.4 Nonexistence Results and Other Consequences

As in the case of the Laplacian Δ, the Pohozaev identity (9.3.2) gives as an immediate consequence the following nonexistence result for operators (9.1.2)-(9.2.8): If $f(u) = |u|^{p-1}u$ in (9.1.1), then

- If Ω is star-shaped and $p = \frac{n+2s}{n-2s}$, the only nonnegative weak solution is $u \equiv 0$.
- If Ω is star-shaped and $p > \frac{n+2s}{n-2s}$, the only bounded weak solution is $u \equiv 0$.

This nonexistence result was first established by Fall and Weth in [11] for $L = (-\Delta)^s$. They used the extension property of the fractional Laplacian, combined with the method of moving spheres.

On the other hand, the existence of solutions for subcritical powers $1 < p < \frac{n+2s}{n-2s}$ was proved by Servadei-Valdinoci [32] for the class of operators (9.1.2)-(9.2.3). Moreover, for the critical power $p = \frac{n+2s}{n-2s}$, the existence of solutions in an annular-type domains was obtained in [30].

The methods introduced in [24] to prove the Pohozaev identity (9.3.2) were used in [25] to show nonexistence results for much more general operators L, including for example the following.

Proposition 9.6 ([25]). *Let L be any operator of the form*

$$Lu(x) = -\sum_{i,j} a_{ij}\partial_{ij}u + \mathrm{PV}\int_{\mathbb{R}^n}(u(x)-u(x+y))K(y)dy, \tag{9.4.1}$$

where (a_{ij}) is a positive definite symmetric matrix and K satisfies the conditions in (9.1.2). *Assume in addition that*

$$K(y)|y|^{n+2} \text{ is nondecreasing along rays from the origin.} \tag{9.4.2}$$

and that

$$|\nabla K(y)| \leqslant C\,\frac{K(y)}{|y|} \quad \text{for all } y \neq 0,$$

Let Ω be any bounded star-shaped domain, and u be any bounded solution of (9.1.1) *with $f(u) = |u|^{p-1}u$. If $p \geqslant \frac{n+2}{n-2}$, then $u \equiv 0$.*

Similar nonexistence results were obtained in [25] for other types of nonlocal equations, including: kernels without homogeneity (such as sums of fractional Laplacians of different orders), nonlinear operators (such as fractional p-Laplacians), and operators of higher order ($s > 1$).

Finally, let us give another immediate consequence of the Pohozaev identity (9.3.2).

Proposition 9.7 ([29]). *Let L be any operator of the form* (9.1.2)-(9.2.3)-(9.2.8), *Ω be any bounded $C^{1,1}$ domain, and φ be any bounded solution to*

$$\begin{cases} L\varphi = \lambda\varphi & \text{in } \Omega \\ \varphi = 0 & \text{in } \mathbb{R}^n \setminus \Omega, \end{cases}$$

for some real λ. Then, φ/d^s is Hölder continuous up to the boundary, and the following unique continuation principle holds:

$$\left.\frac{\varphi}{d^s}\right|_{\partial\Omega} \equiv 0 \quad \text{on} \quad \partial\Omega \qquad \Longrightarrow \qquad \varphi \equiv 0 \quad \text{in} \quad \Omega.$$

The same unique continuation property holds for any *subcritical* nonlinearity $f(x, u)$; see Corollary 1.4 in [29].

Acknowledgment: The author was supported by NSF grant DMS-1565186 (US) and by MINECO grant MTM2014-52402-C3-1-P (Spain).

Bibliography

[1] R. Bass, *Regularity results for stable-like operators*, J. Funct. Anal. 257 (2009), 2693-2722.

[2] B. Barrios, A. Figalli, X. Ros-Oton, *Free boundary regularity in the parabolic fractional obstacle problem*, Comm. Pure Appl. Math., in press (2017).

[3] R. M. Blumenthal, R. K. Getoor, D. B. Ray, *On the distribution of first hits for the symmetric stable processes*, Trans. Amer. Math. Soc. 99 (1961), 540-554.

[4] K. Bogdan, T. Grzywny, M. Ryznar, *Heat kernel estimates for the fractional Laplacian with Dirichlet conditions*, Ann. of Prob. 38 (2010), 1901-1923.
[5] K. Bogdan, T. Grzywny, M. Ryznar, *Barriers, exit time and survival probability for unimodal Lévy processes*, Probab. Theory Relat. Fields 162 (2015), 155-198.
[6] C. Bucur, E. Valdinoci, *Nonlocal Diffusion and Applications*, Lecture Notes of the Unione Matematica Italiana, Vol. 20, 2016.
[7] L. Caffarelli, A. Figalli, *Regularity of solutions to the parabolic fractional obstacle problem*, J. Reine Angew. Math., 680 (2013), 191-233.
[8] L. Caffarelli, X. Ros-Oton, J. Serra, *Obstacle problems for integro-differential operators: regularity of solutions and free boundaries*, Invent. Math. 208 (2017), 1155-1211.
[9] Z.-Q. Chen, R. Song, *Estimates on Green functions and Poisson kernels for symmetric stable processes*, Math. Ann. 312 (1998), 465-501.
[10] D. De Silva, O. Savin, *A note on higher regularity boundary Harnack inequality*, Disc. Cont. Dyn. Syst. 35 (2015), 6155-6163.
[11] M. M. Fall, T. Weth, *Nonexistence results for a class of fractional elliptic boundary value problems*, J. Funct. Anal. 263 (2012), 2205-2227.
[12] X. Fernandez-Real, X. Ros-Oton. *Regularity theory for general stable operators: parabolic equations*, J. Funct. Anal. 272 (2017), 4165-4221.
[13] R. K. Getoor, *First passage times for symmetric stable processes in space*, Trans. Amer. Math. Soc. 101 (1961), 75-90.
[14] G. Grubb, *Fractional Laplacians on domains, a dvelopment of Hörmander's theory of μ-transmission pseudodifferential operators*, Adv. Math. 268 (2015), 478-528.
[15] G. Grubb, *Local and nonlocal boundary conditions for μ-transmission and fractional elliptic pseudodifferential operators*, Anal. PDE 7 (2014), 1649-1682.
[16] G. Grubb, *Integration by parts and Pohozaev identities for space-dependent fractional-order operators*, J. Differential Equations 261 (2016), 1835-1879.
[17] Y. Jhaveri, R. Neumayer, *Higher regularity of the free boundary in the obstacle problem for the fractional Laplacian*, Adv. Math. 311 (2017), 748-795.
[18] M. Kassmann, A. Mimica, *Intrinsic scaling properties for nonlocal operators*, J. Eur. Math. Soc. 19 (2017), 983-1011.
[19] T. Kulczycki, *Properties of Green function of symmetric stable processes*, Probab. Math. Statist. 17 (1997), 339-364.
[20] N. S. Landkof, *Foundations of Modern Potential Theory*, Springer, New York, 1972.
[21] S. I. Pohozaev, *On the eigenfunctions of the equation $\Delta u + \lambda f(u) = 0$*, Dokl. Akad. Nauk SSSR 165 (1965), 1408-1411.
[22] X. Ros-Oton, *Nonlocal elliptic equations in bounded domains: a survey*, Publ. Mat. 60 (2016), 3-26.
[23] X. Ros-Oton, J. Serra, *The Dirichlet problem for the fractional Laplacian: regularity up to the boundary*, J. Math. Pures Appl. 101 (2014), 275-302.
[24] X. Ros-Oton, J. Serra, *The Pohozaev identity for the fractional Laplacian*, Arch. Rat. Mech. Anal 213 (2014), 587-628.

[25] X. Ros-Oton, J. Serra. *Nonexistence results for nonlocal equations with critical and supercritical nonlinearities*, Comm. Partial Differential Equations 40 (2015), 115-133.

[26] X. Ros-Oton, J. Serra, *Boundary regularity for fully nonlinear integro-differential equations*, Duke Math. J. 165 (2016), 2079-2154.

[27] X. Ros-Oton, J. Serra, *Regularity theory for general stable operators*, J. Differential Equations 260 (2016), 8675-8715.

[28] X. Ros-Oton, J. Serra, *Boundary regularity estimates for nonlocal elliptic equations in C^1 and $C^{1,\alpha}$ domains*, Ann. Mat. Pura Appl. 196 (2017), 1637-1668.

[29] X. Ros-Oton, J. Serra, E. Valdinoci, *Pohozaev identities for anisotropic integro-differential operators*, Comm. Partial Differential Equations 42 (2017), 1290-1321

[30] S. Secchi, N. Shioji, M. Squassina, *Coron problem for fractional equations*, Differential Integral Equations 28 (2015), 103-118.

[31] J. Serra, *$C^{\sigma+\alpha}$ regularity for concave nonlocal fully nonlinear elliptic equations with rough kernels*, Calc. Var. Partial Differential Equations 54 (2015), 3571-3601.

[32] R. Servadei, E. Valdinoci, *Mountain pass solutions for non-local elliptic operators*, J. Math. Anal. Appl. 389 (2012), 887-898.

[33] H. Sikic, R. Song, Z. Vondracek, *Potential theory of geometric stable processes*, Probab. Theory Relat. Fields 135 (2006), 547-575.

Giovanni Molica Bisci

Variational and Topological Methods for Nonlocal Fractional Periodic Equations

Keywords: Variational techniques, Critical point theory, fractional Laplacian

MSC: 49J35, 35A15, 35S15; Secondary: 47G20, 45G05.

"I know nothing in the world
that has as much power as a word.
Sometimes I write one, and I look at it,
until it begins to shine"
Emily Dickinson

10.1 Introduction

A very interesting area of nonlinear analysis lies in the study of elliptic equations involving fractional operators. Recently, a great attention has been focused on these problems, both for the pure mathematical research and in view of concrete real–world applications. Indeed, these types of operators appear in a quite natural way in different contexts, such as the description of several physical phenomena.

Moreover, rich mathematical concepts allow in general several approaches, and this is the case for the fractional Laplacian, which can be defined using Fourier analysis, functional calculus, singular integrals or Lévy processes. Its inverse is closely related to the famous potentials introduced by Marcel Riesz in the late 1930s. In contrast to the Laplacian, which is a local operator, the fractional Laplacian is a paradigm of the family of nonlocal linear operators, and this has immediate consequences in the formulation of basic questions like the Dirichlet problem (see [91]).

Of course, it is impossible to cite all the papers involved in this branch of research, but we can not avoid mentioning the pioneering works [33, 34, 35, 36, 37, 38, 117] and the more recent papers [48, 49, 50, 65, 66]; see also the manuscripts [98, 99, 118, 119].

Here, we are interested in nonlocal fractional periodic equations. On the contrary of the classical literature dedicated to periodic boundary value problems involving the Laplace operator or some of its generalizations, up to now, to our knowledge, just a few

---

Giovanni Molica Bisci, Dipartimento P.A.U., Università degli Studi Mediterranea di Reggio Calabria, Salita Melissari - Feo di Vito, 89124 Reggio Calabria, Italy, E-mail: gmolica@unirc.it

https://doi.org/10.1515/9783110571561-012

numbers of papers consider nonlocal fractional periodic equations (see, for instance, [47, 64, 113, 114], as well as [5, 6, 7, 8, 9] and [10, 11, 12, 13, 14]).

More precisely, this paper is concerned with the existence of weak solutions of fractional nonlinear problems with periodic boundary conditions, whose simple prototype is

$$\begin{cases} (-\Delta+m^2)^s u = f(x,u) & \text{in } (0,T)^N \\ u(x+Te_i) = u(x) & \text{for all } x \in \mathbb{R}^N \text{ and } i=1,\dots,N \end{cases},$$

where $s \in (0,1)$, $N \geqslant 2$, $m > 0$, $(0,T)^N := (0,T) \times \ldots \times (0,T)$ (N–times), $\{e_i\}_{i=1}^N$ is the canonical basis in $\mathbb{R}^N$, and $f : \mathbb{R}^N \times \mathbb{R} \to \mathbb{R}$ is a suitable T–periodic Carathéodory function. It is well–known that these equations can be realized as local degenerate elliptic problems in a half–cylinder of $\mathbb{R}^{N+1}_+$ together with a nonlinear Neumann boundary condition, through the extension technique in periodic setting (see [5]). Thanks to this identification, the main tools employed here to get existence and multiplicity results are critical point theory and topological methods.

In order to define the nonlocal operator $(-\Delta+m^2)^s$ we proceed as follows: let $u \in C^\infty_T(\mathbb{R}^N)$, that is u is infinitely differentiable in $\mathbb{R}^N$ and T–periodic in each variable. Then $u \in C^\infty_T(\mathbb{R}^N)$ can be represented via Fourier series expansion:

$$u(x) = \sum_{k\in\mathbb{Z}^N} c_k \frac{e^{\imath\omega k\cdot x}}{\sqrt{T^N}} \quad (x \in \mathbb{R}^N)$$

where

$$\omega := \frac{2\pi}{T} \text{ and } c_k := \frac{1}{\sqrt{T^N}} \int_{(0,T)^N} u e^{-\imath\omega k\cdot x} dx \quad (k \in \mathbb{Z}^N)$$

are the Fourier coefficients of the function u.

With the above notations the nonlocal operator $(-\Delta+m^2)^s$ is given by

$$(-\Delta+m^2)^s u(x) := \sum_{k\in\mathbb{Z}^N} c_k(\omega^2|k|^2+m^2)^s \frac{e^{\imath\omega k\cdot x}}{\sqrt{T^N}}.$$

Furthermore, if $u(x) := \sum_{k\in\mathbb{Z}^N} c_k \frac{e^{\imath\omega k\cdot x}}{\sqrt{T^N}}$ and $v(x) := \sum_{k\in\mathbb{Z}^N} d_k \frac{e^{\imath\omega k\cdot x}}{\sqrt{T^N}}$, we have that the quadratic form

$$\mathcal{Q}(u,v) := \sum_{k\in\mathbb{Z}^N} (\omega^2|k|^2+m^2)^s c_k \bar{d}_k$$

can be extended by density on the Hilbert space

$$\mathbb{H}^s_T := \left\{ u(x) = \sum_{k\in\mathbb{Z}^N} c_k \frac{e^{\imath\omega k\cdot x}}{\sqrt{T^N}} \in L^2(0,T)^N : \sum_{k\in\mathbb{Z}^N} (\omega^2|k|^2+m^2)^s |c_k|^2 < +\infty \right\}$$

endowed with the norm

$$|u|_{\mathbb{H}^s_T} := \left(\sum_{k\in\mathbb{Z}^N} (\omega^2|k|^2 + m^2)^s |c_k|^2 \right)^{1/2}.$$

Let us recall that in $\mathbb{R}^N$ the operator $(-\Delta + m^2)^s$ is strictly related to the quantum mechanics; in fact, when $s = 1/2$, $\sqrt{-\Delta + m^2} - m$ corresponds to the Hamiltonian of a free relativistic particle of mass m; see, for instance, [71] and [62] where the authors studied the existence of solitary wave solutions of the pseudo-relativistic Hartree equation

$$\imath\partial_t\psi = (\sqrt{-\Delta + m^2} - m)\psi - (|x|^{-1} * |\psi|^2)\psi$$

on $\mathbb{R}^3$ (see also [45, 46] and references therein).

There is also a deep connection between $(-\Delta + m^2)^s - m^{2s}$ and the Stochastic Processes theory; the operator in question is an infinitesimal generator of a Lévy process $\{X_t^m\}_{t\geqslant 0}$ called the relativistic $2s$-stable process

$$\mathbb{E}(e^{i\xi\cdot X_t^m}) := e^{-t[(m^2+|\xi|^2)^s - m^{2s}]} \quad (\xi \in \mathbb{R}^N);$$

for details we refer to [42] and [116].

The present manuscript is divided into three parts. The first part deals with some basic facts about periodic fractional Sobolev spaces (see [6, 7, 8] for a detailed introduction of this concepts) and the second one is dedicated to the analysis of fractional elliptic problems involving subcritical nonlinearities, via classical variational methods and other novel approaches. At the end of the paper we present some recent results on (super)critical periodic fractional equations, studied in the recent literature, also in relation with the celebrated Brezis–Nirenberg problem (see [30]).

In particular, for subcritical problems mountain pass and linking type non-trivial solutions are obtained, as well as Ricceri's solutions for parametric problems, followed by equations at resonance, and the obtention of multiple solutions using pseudo-index theory. For critical equations, emphasis is put on extending to the new setting some results in connection to the famous critical Yamabe problem. While, for parametric supercritical equations by combining the Moser iteration scheme in the nonlocal framework with an abstract multiplicity result valid for differentiable functionals, we show that the problem under consideration admits at least three periodic solutions with the property that their Sobolev norms are bounded by a suitable constant. Finally, we provide a concrete estimate of the range of these parameters by using some properties of the fractional calculus on a specific family of test functions (defined in Section 10.2). This estimate turns out to be deeply related to the geometry of the domain.

The material presented in Section 10.2 comes from the results contained in [6, 7, 8, 13]. The other sections are based on very recent results obtained by ourselves or through direct cooperation with other mathematicians. Further, a complete and exhaustive description of the arguments introduced along this manuscript will be developed in a forthcoming book.

For a detailed introduction on fractional nonlocal problems we refer to the paper [50] and to the monograph [91] for variational methods on the nonlocal fractional framework; see also the classical book [79] in which the authors studied periodic problems by using fundamental techniques of critical point theory.

10.2 Nonlocal Periodic Setting

In this section we sketch the basic facts on fractional Sobolev spaces and fractional nonlocal operators. Our treatment is mostly self–contained and we tacitly assume that the reader has some knowledge with the basic objects discussed here. More precisely, the main purpose is to present some results on fractional Sobolev spaces and nonlocal operators in the form in which they will be exploited later on. Since this is an introductory part to convey the framework we work in, the rigorous proof will be kept minimum. Some extra reading of the references may be necessary to truly learn the material.

10.2.1 Fractional Sobolev spaces

In order to give the weak formulation of subcritical fractional periodic problems, we need to work in a special functional space. Indeed, one of the difficulties in treating these equations is related to encoding the periodic boundary condition in the variational formulation. In this respect the standard fractional Sobolev spaces are not sufficient in order to study the fractional periodic case. We overcome this difficulty by working in a new functional space, whose definition is recalled here. The reader familiar with this topic may skip this section and go directly to the next one.

From now on, we assume $s \in (0,1)$ and $N \geqslant 2$. Let

$$\mathbb{R}^{N+1}_{+} := \{(x,y) \in \mathbb{R}^{N+1} : x \in \mathbb{R}^N, y > 0\}$$

be the upper half–space in $\mathbb{R}^{N+1}$.

Let $\mathcal{S}_T := (0,T)^N \times (0,+\infty)$ be the half–cylinder in $\mathbb{R}^{N+1}_{+}$ with basis $\partial^0 \mathcal{S}_T := (0,T)^N \times \{0\}$ and we denote by $\partial_L \mathcal{S}_T := \partial(0,T)^N \times [0,+\infty)$ the lateral boundary of $\mathcal{S}_T$. With $\|v\|_{L^r(\mathcal{S}_T)}$ we will always denote the norm of $v \in L^r(\mathcal{S}_T)$ and with $|u|_{L^r(0,T)^N}$ the $L^r(0,T)^N$ norm of $u \in L^r(0,T)^N$.

Let $C^\infty_T(\mathbb{R}^N)$ be the space of functions $u \in C^\infty(\mathbb{R}^N)$ such that u is T–periodic in each variable, that is

$$u(x + Te_i) = u(x) \quad \text{for every } x \in \mathbb{R}^N,$$

and $i = 1, \dots, N$.

We define the fractional Sobolev space $\mathbb{H}^s_T$ as the closure of $C^\infty_T(\mathbb{R}^N)$ endowed by the norm

$$|u|_{\mathbb{H}^s_T} := \sqrt{\sum_{k \in \mathbb{Z}^N} (\omega^2 |k|^2 + m^2)^s \, |\beta_k|^2}.$$

Let us introduce the functional space $\mathbb{X}^s_T$ defined as the completion of

$$C^\infty_T(\overline{\mathbb{R}^{N+1}_+}) := \Big\{ v \in C^\infty(\overline{\mathbb{R}^{N+1}_+}) : v(x + Te_i, y) = v(x, y) \\ \text{for every } (x, y) \in \overline{\mathbb{R}^{N+1}_+}, i = 1, \dots, N \Big\}$$

under the $H^1(\mathcal{S}_T, y^{1-2s})$ norm given by

$$\|v\|_{\mathbb{X}^s_T} := \sqrt{\iint_{\mathcal{S}_T} y^{1-2s} (|\nabla v|^2 + m^2 v^2) \, dx dy}\,.$$

We notice that the Sobolev space $\mathbb{X}^s_T$ is separable. Indeed, the map T : $C^\infty_T(\overline{\mathbb{R}^{N+1}_+}) \to (L^2(\mathcal{S}_T))^N \times L^2(\mathcal{S}_T)$ defined by

$$T(v) := \left(y^{\frac{1-2s}{2}} \nabla v, y^{\frac{1-2s}{2}} v \right), \qquad \forall \, v \in C^\infty_T(\overline{\mathbb{R}^{N+1}_+})$$

is a linear isometry and the space

$$T(C^\infty_T(\overline{\mathbb{R}^{N+1}_+})) = (L^2(\mathcal{S}_T))^N \times L^2(\mathcal{S}_T)$$

is separable. Consequently, it follows that $C^\infty_T(\overline{\mathbb{R}^{N+1}_+})$ is separable and, by density arguments, $\mathbb{X}^s_T$ also does.

We recall that it is possible to define a *trace operator* between $\mathbb{X}^s_T$ and $\mathbb{H}^s_T$ (see [6, 7] for details):

Theorem 10.1. *There exists a surjective linear operator* $\mathrm{Tr} : \mathbb{X}^s_T \to \mathbb{H}^s_T$ *such that:*
(*i*) $\mathrm{Tr}(v) = v|_{\partial^0 \mathcal{S}_T}$ *for all* $v \in C^\infty_T(\overline{\mathbb{R}^{N+1}_+}) \cap \mathbb{X}^s_T$;
(*ii*) Tr *is bounded and*

$$\sqrt{\kappa_s} |\mathrm{Tr}(v)|_{\mathbb{H}^s_T} \leqslant \|v\|_{\mathbb{X}^s_T}, \tag{10.2.1}$$

for every $v \in \mathbb{X}^s_T$, *where*

$$\kappa_s := 2^{1-2s} \frac{\Gamma(1-s)}{\Gamma(s)}. \tag{10.2.2}$$

In particular, equality holds in (10.2.1) *for some* $v \in \mathbb{X}^s_T$ *if and only if* v *weakly solves the following equation*

$$-\operatorname{div}(y^{1-2s} \nabla v) + m^2 y^{1-2s} v = 0 \ \text{ in } \mathcal{S}_T.$$

We have the following embedding result:

Theorem 10.2. *Let $N \geqslant 2$ and $s \in (0,1)$. Then, the space $\mathrm{Tr}(\mathbb{X}^s_T)$ is continuously embedded in $L^q(0,T)^N$ for any $1 \leqslant q \leqslant 2^\sharp_s$, where*

$$2^\sharp_s := \frac{2N}{N-2s}$$

denotes the fractional critical Sobolev exponent.

Moreover, $\mathrm{Tr}(\mathbb{X}^s_T)$ is compactly embedded in $L^q(0,T)^N$ for any $1 \leqslant q < 2^\sharp_s$.

Let $g \in \mathbb{H}^{-s}_T$, where

$$\mathbb{H}^{-s}_T := \left\{ g = \sum_{k\in\mathbb{Z}^N} g_k \frac{e^{\imath\omega k\cdot x}}{\sqrt{T^N}} : \sum_{k\in\mathbb{Z}^N} \frac{|g_k|^2}{(\omega^2|k|^2+m^2)^s} < +\infty \right\}$$

is the dual of $\mathbb{H}^s_T$, and consider the following two problems:

$$\begin{cases} (-\Delta+m^2)^s u = g & \text{in } (0,T)^N \\ u(x+Te_i) = u(x) & \text{for all } x\in\mathbb{R}^N \text{ and } i=1,\dots,N \end{cases} \tag{10.2.3}$$

and

$$\begin{cases} -\operatorname{div}(y^{1-2s}\nabla v) + m^2 y^{1-2s} v = 0 & \text{in } \mathcal{S}_T \\ v_{|\{x_i=0\}} = v_{|\{x_i=T\}} & \text{on } \partial_L\mathcal{S}_T \\ \partial^{1-2s}_\nu v = \kappa_s g(x) & \text{on } \partial^0\mathcal{S}_T \end{cases}, \tag{10.2.4}$$

where the notation $v_{|\{x_i=0\}} = v_{|\{x_i=T\}}$ on $\partial_L\mathcal{S}_T$ means

$$v(x_1,\dots,x_{i-1},0,x_{i+1},\dots,x_N,y) = v(x_1,\dots,x_{i-1},T,x_{i+1},\dots,x_N,y),$$

for every $i \in \{1,\dots,N\}$ and $y \geqslant 0$. Further,

$$\partial^{1-2s}_\nu v(\cdot) := -\lim_{y\to 0^+} y^{1-2s}\frac{\partial v}{\partial y}(\cdot,y)$$

denotes the *conormal exterior derivative* of v.

We have the following definitions of weak solutions to (10.2.4) and (10.2.3) respectively:

Definition 10.3. *We say that $v \in \mathbb{X}^s_T$ is a weak solution to* (10.2.4) *if*

$$\iint_{\mathcal{S}_T} y^{1-2s}(\nabla v\nabla\varphi + m^2 v\varphi)\,dxdy = \kappa_s\langle g,\mathrm{Tr}(\varphi)\rangle,$$

for every $\varphi \in \mathbb{X}^s_T$, where $\langle\cdot,\cdot\rangle$ denotes the duality pairing between $\mathbb{H}^s_T$ and its dual $\mathbb{H}^{-s}_T$.

Definition 10.4. *A function $u \in \mathbb{H}^s_T$ is a weak solution to* (10.2.3) *if $u = \mathrm{Tr}(v)$ and $v \in \mathbb{X}^s_T$ is a weak solution to* (10.2.4).

Taking into account Theorem 10.1 and Theorem 10.2, it is possible to introduce the notion of *extension* for a function $u \in \mathbb{H}^s_T$.

More precisely, the next result holds:

Theorem 10.5. *Let $u \in \mathbb{H}^s_T$. Then, there exists a unique $v \in \mathbb{X}^s_T$ such that*

$$\begin{cases} -\operatorname{div}(y^{1-2s}\nabla v) + m^2 y^{1-2s} v = 0 & \text{in } \mathcal{S}_T \\ v_{|\{x_i=0\}} = v_{|\{x_i=T\}} & \text{on } \partial_L \mathcal{S}_T \\ v(\cdot, 0) = u & \text{on } \partial^0 \mathcal{S}_T \end{cases} \tag{10.2.5}$$

and

$$-\lim_{y\to 0^+} y^{1-2s} \frac{\partial v}{\partial y}(x,y) = \kappa_s(-\Delta + m^2)^s u(x) \text{ in } \mathbb{H}^{-s}_T.$$

We call $v \in \mathbb{X}^s_T$ the extension of $u \in \mathbb{H}^s_T$.

In particular, if $u = \sum_{k\in\mathbb{Z}^N} \beta_k \frac{e^{\imath\omega k\cdot x}}{\sqrt{T^N}}$, then v is given by

$$v(x,y) = \sum_{k\in\mathbb{Z}^N} \beta_k \Theta_k(y) \frac{e^{\imath\omega k\cdot x}}{\sqrt{T^N}},$$

where $\Theta_k(y) := \Theta(\sqrt{\omega^2|k|^2 + m^2}\, y)$ and $\Theta \in H^1(\mathbb{R}_+, y^{1-2s})$ solves the following ODE

$$\begin{cases} \Theta'' + \frac{1-2s}{y}\Theta' - \Theta = 0 \text{ in } \mathbb{R}_+ \\ \Theta(0) = 1 \text{ and } \Theta(+\infty) = 0 \end{cases}.$$

Moreover, v satisfies the following properties:

(i) v is smooth for $y > 0$ and T-periodic in x;

(ii) $\|v\|_{\mathbb{X}^s_T} \leqslant \|z\|_{\mathbb{X}^s_T}$ for any $z \in \mathbb{X}^s_T$ such that $\mathrm{Tr}(z) = u$;

(iii) $\|v\|_{\mathbb{X}^s_T} = \sqrt{\kappa_s}|u|_{\mathbb{H}^s_T}$.

We notice that, after the work of Caffarelli and Silvestre [36], several authors have considered a definition of the fractional Laplacian operator in a bounded domain with zero Dirichlet boundary data by means of an auxiliary variable (see the papers [25, 28, 31, 32, 41, 88] and references therein). For completeness, among other results appeared in the current literature, we recall that Stinga and Torrea in [127] showed that any fractional operator L^s, where $s \in (0,1)$ and L is a nonnegative self–adjoint linear second order partial differential operator, can be seen as a Dirichlet to Neumann operator associated to some degenerate boundary value problem in $\Omega \times (0, +\infty)$, where Ω is a smooth and bounded domain of $\mathbb{R}^N$. We use this strategy here in order to study nonlocal periodic equations.

We also observe that the function Θ in Theorem 10.5 has the explicit form

$$\Theta(y) = \frac{2}{\Gamma(s)}\left(\frac{y}{2}\right)^s K_s(y),$$

where Γ is the Gamma function defined by

$$\Gamma(t) := \int_0^{+\infty} z^{t-1} e^{-w}\, dw, \quad \forall t > 0$$

and K_s denotes the modified Bessel function of the second kind with order s, given by

$$K_s(y) := \pi \frac{I_{-s}(y) - I_s(y)}{2\sin(s\pi)},$$

where

$$I_s(y) := \sum_{k=0}^{+\infty} \frac{(y/2)^{2k+s}}{k!\Gamma(k+s+1)}, \quad \text{and} \quad I_{-s}(y) := \sum_{k=0}^{+\infty} \frac{(y/2)^{2k-s}}{k!\Gamma(k-s+1)}.$$

Further, it is known (see [53]) that

$$K_s'(y) = \frac{s}{y} K_s(y) - K_{s-1}(y), \tag{10.2.6}$$

for every $y > 0$.

The above results are useful in proving suitable estimates on the norm of some truncated test functions, that we will manage in Sections 10.4 and 10.5, when studying multiplicity results for periodic fractional problems.

More precisely, fix $\sigma \in \mathbb{R}$ and let us define $w_\sigma \in \mathbb{X}_T^s$ as follows

$$w_\sigma(x,y) := \frac{2}{\Gamma(s)} \left(\frac{my}{2}\right)^s K_s(my)\sigma, \quad \forall (x,y) \in \mathbb{R}_+^{N+1} \tag{10.2.7}$$

that is

$$w_\sigma(x,y) = \frac{\pi}{\Gamma(s)} \left(\frac{my}{2}\right)^s \left(\sum_{k=0}^{+\infty} \frac{(my/2)^{2k-s}}{k!\Gamma(k-s+1)} - \sum_{k=0}^{+\infty} \frac{(my/2)^{2k+s}}{k!\Gamma(k+s+1)} \right) \frac{\sigma}{\sin(s\pi)}.$$

Hence, it follows that

$$\lim_{y \to 0^+} w_\sigma(x,y) = \sigma, \tag{10.2.8}$$

that is $\mathrm{Tr}(w_\sigma) = \sigma$.

Moreover, by (10.2.6) one has

$$\kappa_s = \int_0^{+\infty} y^{1-2s} \left(|\Theta'(y)| + m^2 |\Theta(my)|^2 \right) dy,$$

and

$$\begin{aligned} \|w_\sigma\|_{\mathbb{X}_T^s}^2 &= \sigma^2 T^N \int_0^{+\infty} y^{1-2s} \left(|\Theta'(my)| + m^2 |\Theta(my)|^2 \right) dy \\ &= \sigma^2 m^{2s} T^N \int_0^{+\infty} y^{1-2s} \left(|\Theta'(y)| + m^2 |\Theta(my)|^2 \right) dy \\ &= \sigma^2 m^{2s} T^N \kappa_s. \end{aligned} \tag{10.2.9}$$

See [7, Theorem 17] for details.

10.2.2 Weak solutions

Let us consider the following problem

$$\begin{cases} (-\Delta + m^2)^s u = f(x,u) & \text{in } (0,T)^N \\ u(x+Te_i) = u(x) & \text{for all } x \in \mathbb{R}^N \text{ and } i = 1,\dots,N \end{cases}, \tag{10.2.10}$$

where f is a Carathéodory T–periodic (with respect to $x \in \mathbb{R}^N$) and (sub)critical function.

Instead of problem (10.2.10), we investigate the following problem

$$\begin{cases} -\operatorname{div}(y^{1-2s}\nabla v) + m^2 y^{1-2s} v = 0 & \text{in } \mathcal{S}_T \\ v_{|\{x_i=0\}} = v_{|\{x_i=T\}} & \text{on } \partial_L \mathcal{S}_T \\ \partial_\nu^{1-2s} v = \kappa_s f(x,v) & \text{on } \partial^0 \mathcal{S}_T \end{cases}. \tag{10.2.11}$$

According to the linear case we have the following

Definition 10.6. *A function $v \in \mathbb{X}_T^s$ is a weak solution of problem* (10.2.11) *if*

$$\iint_{\mathcal{S}_T} y^{1-2s}(\nabla v \nabla \varphi + m^2 v\varphi)\,dxdy = \kappa_s \int_{\partial^0 \mathcal{S}_T} f(x, \operatorname{Tr}(v))\operatorname{Tr}(\varphi)\,dx,$$

for every $\varphi \in \mathbb{X}_T^s$.

Thanks to the above definition, the notion of *weak solution* of problem (10.2.10) can be easily derived.

Definition 10.7. *A function $u \in \mathbb{H}_T^s$ is a weak solution of problem* (10.2.10) *if $u = \operatorname{Tr}(v)$ and $v \in \mathbb{X}_T^s$ is a weak solution of problem* (10.2.11).

In order to obtain the existence of weak solutions for problem (10.2.10) we study the existence of critical points for the energy functional $\mathcal{J} : \mathbb{X}_T^s \to \mathbb{R}$ defined by

$$\mathcal{J}(v) := \frac{1}{2}\|v\|_{\mathbb{X}_T^s}^2 - \kappa_s \int_{\partial^0 \mathcal{S}_T} F(x, \operatorname{Tr}(v))\,dx, \tag{10.2.12}$$

where

$$F(x,t) := \int_0^t f(x,w)\,dw.$$

Indeed, standard arguments ensure that $\mathcal{J} \in C^1(\mathbb{X}_T^s;\mathbb{R})$ and its Fréchet derivatives at $v \in \mathbb{X}_T^s$ is given by

$$\langle \mathcal{J}'(v), \varphi\rangle = \iint_{\mathcal{S}_T} y^{1-2s}(\nabla v \nabla \varphi + m^2 v\varphi)\,dxdy - \kappa_s \int_{\partial^0 \mathcal{S}_T} f(x, \operatorname{Tr}(v))\operatorname{Tr}(\varphi)\,dx,$$

for every $\varphi \in \mathbb{X}_T^s$.

Finally, we notice that the constant function $u(x) = \sigma \in \mathbb{R}$, for every $x \in (0, T)^N$, is a (trivial) solution of (10.2.10) if and only if

$$\sigma m^{2s} = f(x, \sigma) \text{ in } (0, T)^N.$$

Thus, constant solutions of problem (10.2.10) (if they exist) as fixed points of the real function

$$\sigma \mapsto \frac{f(x, \sigma)}{m^{2s}}.$$

10.2.3 Spectral properties of $(-\Delta + m^2)^s$

The study of the eigenvalues of a linear operator is a classical topic and many functional analytic tools of general flavor may be used to deal with it. The result that we give here is, indeed, more general and more precise than what we need, strictly speaking, for the proofs of our main results: nevertheless we believe it is good to have a result stated in detail also for further reference.

We focus on the following eigenvalue problem

$$\begin{cases} (-\Delta + m^2)^s u = \lambda u & \text{in } (0, T)^N \\ u(x + Te_i) = u(x) & \text{for all } x \in \mathbb{R}^N \text{ and } i = 1, \dots, N \end{cases}. \tag{10.2.13}$$

More precisely, we discuss the weak formulation of (10.2.13), which consists in the following eigenvalue problem:

Find $u \in \mathbb{H}_T^s$ such that $u := \operatorname{Tr}(v)$, where $v \in \mathbb{X}_T^s$ and

$$\iint_{\mathcal{S}_T} y^{1-2s}(\nabla v \nabla \varphi + m^2 v \varphi)\, dxdy = \lambda \kappa_s \int_{\partial^0 \mathcal{S}_T} \operatorname{Tr}(v)\operatorname{Tr}(\varphi) dx,$$

for every $\varphi \in \mathbb{X}_T^s$.

In other words, $v \in \mathbb{X}_T^s$ is a weak solution of the extended problem

$$\begin{cases} -\operatorname{div}(y^{1-2s}\nabla v) + m^2 y^{1-2s} v = 0 & \text{in } \mathcal{S}_T \\ v_{|\{x_i=0\}} = v_{|\{x_i=T\}} & \text{on } \partial_L \mathcal{S}_T \\ \partial_\nu^{1-2s} v = \lambda \kappa_s v & \text{on } \partial^0 \mathcal{S}_T \end{cases}. \tag{10.2.14}$$

We recall that $\lambda \in \mathbb{R}$ is an *eigenvalue* of $(-\Delta + m^2)^s$ provided there exists a non–zero weak solution of (10.2.13).

Following the classical spectral theory [112, 131], the powers of a positive operator in a bounded domain are defined through the spectral decomposition using the powers of the eigenvalues of the original operator.

Then, it is easy to derive the following result:

Lemma 10.8. *With the above notations the following facts hold:*

(i) *the operator* $(-\Delta + m^2)^s$ *has a countable family of eigenvalues* $\{\lambda_\ell\}_{\ell\in\mathbb{N}}$ *which can be written as an increasing sequence of positive numbers*

$$0 < \lambda_1 < \lambda_2 \leqslant \ldots \leqslant \lambda_\ell \leqslant \lambda_{\ell+1} \leqslant \ldots$$

Each eigenvalue is repeated according to its multiplicity (which is finite);

(ii) $\lambda_\ell = \mu_\ell^s$ *for all* $\ell \in \mathbb{N}$, *where* $\{\mu_\ell\}_{\ell\in\mathbb{N}}$ *is the increasing sequence of eigenvalues of* $-\Delta + m^2$;

(iii) λ_1 *is simple,* $\lambda_\ell = \mu_\ell^s \to +\infty$ *as* $\ell \to +\infty$;

(iv) *the sequence* $\{u_\ell\}_{\ell\in\mathbb{N}}$ *of eigenfunctions corresponding to* λ_ℓ *is an orthonormal basis of* $L^2(0,T)^N$ *and an orthogonal basis of the Sobolevspace* $\mathbb{H}_T^s$. *Let us note that* $\{u_\ell, \mu_\ell\}_{\ell\in\mathbb{N}}$ *are the eigenfunctions and eigenvalues of* $-\Delta + m^2$ *under periodic boundary conditions;*

(v) *for any* $h \in \mathbb{N}$, λ_ℓ *has finite multiplicity, and there holds*

$$\lambda_\ell = \min_{u\in\mathbb{P}_\ell\setminus\{0\}} \frac{|u|^2_{\mathbb{H}_T^s}}{|u|^2_{L^2(0,T)^N}} \qquad \textit{(Rayleigh's principle)}$$

where

$$\mathbb{P}_\ell := \{u \in \mathbb{H}_T^s : \langle u, u_j\rangle_{\mathbb{H}_T^s} = 0, \ \textit{for } j = 1, \ldots, \ell - 1\}.$$

Now, we aim to find some useful relation between the eigenvalues $\{\lambda_j\}_{j\in\mathbb{N}}$ of $(-\Delta+m^2)^s$ and the corresponding extended eigenvalue problem in the half–cylinder $\mathcal{S}_T$, given by problem (10.2.14).

Let us introduce the following notations. Set

$$\mathbb{V}_h := Span\{v_1, \ldots, v_h\}, \tag{10.2.15}$$

where v_j solves (10.2.14). Clearly $\mathrm{Tr}(v_j) = u_j$ for all $j \in \mathbb{N}$, where $\{u_j\}_{j\in\mathbb{N}}$ is the basis of eigenfunctions in $\mathbb{H}_T^s$, defined in Lemma 10.8. For any $h \in \mathbb{N}$ we define

$$\mathbb{V}_h^\perp := \{v \in \mathbb{X}_T^s : \langle v, v_j\rangle_{\mathbb{X}_T^s} = 0, \ \text{for } j = 1, \ldots, h\}. \tag{10.2.16}$$

Since v_j solves (10.2.14), then we deduce that

$$\mathbb{V}_h^\perp = \{v \in \mathbb{X}_T^s : \langle \mathrm{Tr}(v), \mathrm{Tr}(v_j)\rangle_{L^2(0,T)^N} = 0, \ \text{for } j = 1, \ldots, h\}.$$

Then $\mathbb{X}_T^s = \mathbb{V}_h \bigoplus \mathbb{V}_h^\perp$. Let us point out that the trace operator is bijective on

$$S := \{v \in \mathbb{X}_T^s : v \text{ weakly solves } (10.2.5)\}.$$

Indeed, if $\tilde{v}_1$ and $\tilde{v}_2$ are the extension of $\tilde{u}_1, \tilde{u}_2 \in \mathbb{H}_T^s$ respectively, then

$$\langle \tilde{v}_i, \varphi\rangle_{\mathbb{X}_T^s} = k_s\langle \tilde{u}_i, \mathrm{Tr}(\varphi)\rangle_{\mathbb{H}_T^s} \quad \forall \varphi \in \mathbb{X}_T^s, i = 1, 2. \tag{10.2.17}$$

If $\tilde{u}_1 = \mathrm{Tr}(\tilde{v}_1) = \mathrm{Tr}(\tilde{v}_2) = \tilde{u}_2$, by (10.2.17) it follows that

$$\langle \tilde{v}_1 - \tilde{v}_2, \varphi \rangle_{\mathbb{X}_T^s} = 0 \quad \forall \varphi \in \mathbb{X}_T^s,$$

so we deduce that $\tilde{v}_1 = \tilde{v}_2$, that is Tr is injective on S.

This fact and the linearity of the trace operator Tr yields

$$\dim \mathbb{V}_h = \dim Span\{\mathrm{Tr}(v_1), \cdots, \mathrm{Tr}(v_h)\} = h.$$

Now, we prove that $\|\cdot\|_{\mathbb{X}_T^s}$ and $|\cdot|_2$ are equivalent norms on the finite dimensional linear space $\mathbb{V}_h$.

More precisely, for any $v \in \mathbb{V}_h$, it results

$$\kappa_s m^{2s} |\mathrm{Tr}(v)|^2_{L^2(0,T)^N} \leqslant \|v\|^2_{\mathbb{X}_T^s} \leqslant \kappa_s \lambda_h |\mathrm{Tr}(v)|^2_{L^2(0,T)^N}. \tag{10.2.18}$$

Firstly, we notice that $\{v_j\}_{j\in\mathbb{N}}$ is an orthogonal system in $\mathbb{X}_T^s$, since $\{\mathrm{Tr}(v_j)\}_{j\in\mathbb{N}}$ is an orthonormal system in $L^2(0,T)^N$, and v_j satisfies

$$\langle z, v_j \rangle_{\mathbb{X}_T^s} = \kappa_s \lambda_j \langle \mathrm{Tr}(z), \mathrm{Tr}(v_j) \rangle_{L^2(0,T)^N},$$

for every $z \in \mathbb{X}^s$ and $j \in \mathbb{N}$.

Then, by using the fact that $\{\lambda_j\}_{j\in\mathbb{N}}$ is an increasing sequence (see (i) of Lemma 10.8), and thanks to the trace inequality (10.2.1), for $v = \sum_{j=1}^h \alpha_j v_j \in \mathbb{V}_h$ we have

$$\begin{aligned}
\kappa_s m^{2s} |\mathrm{Tr}(v)|^2_{L^2(0,T)^N} &\leqslant \|v\|^2_{\mathbb{X}_T^s} = \sum_{j=1}^h \alpha_j^2 \|v_j\|^2_{\mathbb{X}_T^s} \\
&= \kappa_s \sum_{j=1}^h \lambda_j \alpha_j^2 |\mathrm{Tr}(v_j)|^2_{L^2(0,T)^N} \leqslant \kappa_s \lambda_h \sum_{j=1}^h \alpha_j^2 |\mathrm{Tr}(v_j)|^2_{L^2(0,T)^N} \\
&= \kappa_s \lambda_h |\mathrm{Tr}(v)|^2_{L^2(0,T)^N} \leqslant \frac{\lambda_h}{m^{2s}} \|v\|^2_{\mathbb{X}_T^s}.
\end{aligned}$$

Finally, we prove that, for any $v \in \mathbb{V}_h^\perp$, the following inequality holds

$$\lambda_{h+1} |\mathrm{Tr}(v)|^2_{L^2(0,T)^N} \leqslant \frac{1}{\kappa_s} \|v\|^2_{\mathbb{X}_T^s}.$$

Fix $v \in \mathbb{V}_h^\perp$. Then $\mathrm{Tr}(v) \in \mathbb{P}_{h+1}$. Indeed $\mathrm{Tr}(v_j) = u_j$ weakly solves $(-\Delta + m^2)^s u = \lambda_j u$ and by using the fact that

$$\langle \mathrm{Tr}(v), \mathrm{Tr}(v_j) \rangle_{L^2(0,T)^N} = 0 \ \text{ for every } j = 1, \ldots, h$$

we can infer that $\langle \mathrm{Tr}(v), \mathrm{Tr}(v_j) \rangle_{\mathbb{H}_T^s} = 0$, for every $j = 1, \ldots, h$.

As a consequence, by using the variational characterization (v) of Lemma 10.8 and the trace inequality (10.2.1), we get

$$\lambda_{h+1} |\mathrm{Tr}(v)|^2_{L^2(0,T)^N} \leqslant |\mathrm{Tr}(v)|^2_{\mathbb{H}_T^s} \leqslant \frac{1}{\kappa_s} \|v\|^2_{\mathbb{X}_T^s}. \tag{10.2.19}$$

We will denote respectively by $\sigma((-\Delta+m^2)^s)$ and by

$$0<\lambda_1<\lambda_2\leqslant\ldots\leqslant\lambda_k\leqslant\ldots$$

the spectrum and the non–decreasing, diverging sequence of problem (10.2.13) repeated according to their multiplicity.

Lemma 10.8, as well as a careful analysis of the spectrum $\sigma((-\Delta+m^2)^s)$, will be crucial in Subsection 10.4.5. Finally, in the next proposition, an explicit computation of λ_1 is given.

Proposition 10.9. *Let λ_1 be the first eigenvalue of* (10.2.13). *Then*

$$\lambda_1=\min_{u\in\mathbb{H}^s_T\setminus\{0\}}\frac{|u|^2_{\mathbb{H}^s_T}}{|u|^2_{L^2(0,T)^N}}=m^{2s}. \tag{10.2.20}$$

Proof. The proof is elementary. Indeed, let us observe that

$$\begin{aligned}m^{2s}|u|^2_{L^2(0,T)^N}&=m^{2s}\sum_{k\in\mathbb{Z}^N}|c_k|^2\\&\leqslant\sum_{k\in\mathbb{Z}^N}(\omega^2|k|^2+m^2)^s|c_k|^2=|u|^2_{\mathbb{H}^s_T},\end{aligned}$$

for any $u=\sum_{k\in\mathbb{Z}^N}c_k\dfrac{e^{\imath\omega k\cdot x}}{\sqrt{T^N}}\in\mathbb{H}^s_T$. Moreover, if u_0 is a constant function, then $u_0\in\mathbb{H}^s_T$ and

$$|u_0|^2_{L^2(0,T)^N}=m^{2s}|u_0|^2_{\mathbb{H}^s_T}.$$

This complete the proof. □

10.3 Existence Results

Nonlinear elliptic equations have proven to be a fruitful area of application of variational and topological methods. Such approaches usually exploit the special form of the nonlinear terms, for instance their symmetries, in order to obtain not just the existence of a solution, but also allow to one to acquire knowledge about the behaviour of the solutions (for instance sign and regularity).

The purpose of this section is to study the existence of one (non–trivial) solution for elliptic problems driven by the fractional nonlocal operator $(-\Delta+m^2)^s$ with periodic boundary conditions, exploiting some classical and more recent abstract critical point methods.

10.3.1 A Mountain Pass solution

Starting from the well–known Mountain Pass Theorem (briefly, MPT) of Ambrosetti and Rabinowitz [4], many authors were interested in finding critical points of real–valued functionals defined on an infinite dimensional Banach space, obtaining several generalizations of this famous result, which allow to solve wide classes of ordinary or partial differential equations. Along this direction, in this subsection, we prove exhibit the existence of one weak solution for fraction nonlocal periodic equations.

Let $f : \mathbb{R}^N \times \mathbb{R} \to \mathbb{R}$ be a continuous T–periodic (with respect to $x \in \mathbb{R}^N$) function verifying the following conditions

$$\begin{gathered} \text{there exist } a_1, a_2 > 0 \text{ and } q \in (2, 2^\sharp_s),\ 2^\sharp_s := 2N/(N-2s) \text{ such that} \\ |f(x,t)| \leqslant a_1 + a_2|t|^{q-1} \text{ for every } x \in \mathbb{R}^N, t \in \mathbb{R}\,; \end{gathered} \tag{10.3.1}$$

$$\lim_{t\to 0} \frac{f(x,t)}{t} = 0 \text{ uniformly in } x \in \mathbb{R}^N\,; \tag{10.3.2}$$

$$\begin{gathered} \text{there exist } \mu > 2 \text{ and } r > 0 \text{ such that for every } x \in \mathbb{R}^N, t \in \mathbb{R},\ |t| \geqslant r \\ 0 < \mu F(x,t) \leqslant t f(x,t)\,, \end{gathered} \tag{10.3.3}$$

where the function F is the primitive of f with respect to the second variable, that is

$$F(x,t) := \int_0^t f(x,w)dw\,.$$

As a model for f we can take the function $f(x,t) = \alpha(x)|t|^{q-2}t$, where $\alpha : \mathbb{R}^N \to \mathbb{R}$ is a T–periodic continuous function, and $q \in (2, 2^\sharp_s)$.

We study here the following problem

$$\begin{cases} \displaystyle\iint_{\mathcal{S}_T} y^{1-2s}(\nabla v \nabla \varphi + m^2 v\varphi)\, dxdy \\ \qquad = \kappa_s \gamma \displaystyle\int_{\partial^0 \mathcal{S}_T} \mathrm{Tr}(v)\mathrm{Tr}(\varphi)dx + \kappa_s \int_{\partial^0 \mathcal{S}_T} f(x, \mathrm{Tr}(v))\mathrm{Tr}(\varphi)dx, \quad \forall\, \varphi \in \mathbb{X}^s_T \\ v \in \mathbb{X}^s_T\,, \end{cases} \tag{10.3.4}$$

where $\gamma \in \mathbb{R}$, with $\gamma < m^{2s}$.

In our setting (10.3.4) represents the weak formulation of the following problem

$$\begin{cases} -\operatorname{div}(y^{1-2s}\nabla v) + m^2 y^{1-2s} v = 0 & \text{in } \mathcal{S}_T \\ v_{|\{x_i=0\}} = v_{|\{x_i=T\}} & \text{on } \partial_L \mathcal{S}_T \\ \partial^{1-2s}_\nu v = \kappa_s[\gamma v + f(x,v)] & \text{on } \partial^0 \mathcal{S}_T \end{cases}. \tag{10.3.5}$$

In order to find solutions for problem (10.3.5), we make use of the Mountain Pass Theorem (see [4, 101, 129]). For this, we have to check that the energy functional $\mathcal{J} : \mathbb{X}^s_T \to \mathbb{R}$ given by

$$\mathcal{J}(v) := \frac{1}{2}\left(\|v\|^2_{\mathbb{X}^s_T} - \gamma\kappa_s |\mathrm{Tr}(v)|^2_{L^2(0,T)^N}\right) - \kappa_s \int_{\partial^0 \mathcal{S}_T} F(x, \mathrm{Tr}(v))dx, \quad \forall\, v \in \mathbb{X}^s_T$$

has a particular geometric structure (as stated, e.g., in conditions (1°)–(3°) of [129, Theorem 6.1]) and satisfies the classical Palais–Smale compactness condition (see, for instance, [129, page 70]).

The main existence result for problem (10.3.4) reads as follows:

Theorem 10.10. *Let f be a continuous T–periodic (with respect to $x \in \mathbb{R}^N$) function verifying* (10.3.1) *and* (10.3.2) *in addition to* (10.3.3). *Then, problem* (10.3.4) *admits a Mountain Pass type solution $v \in \mathbb{X}^s_T$ which is not identically zero.*

We notice that Theorem 10.10 is an existence theorem for equations driven by the nonlocal fractional operator $(-\Delta+m^2)^s$. More precisely, Theorem 10.10 gives the existence of one non–zero weak solution $u \in \mathbb{H}^s_T$ for the following nonlocal problem

$$\begin{cases} (-\Delta + m^2)^s u = \gamma u + f(x,u) & \text{in } (0,T)^N \\ u(x+Te_i) = u(x) & \text{for all } x \in \mathbb{R}^N \text{ and } i = 1,\dots,N \end{cases}.$$

The nonlocal analysis that we perform in order to use the Mountain Pass Theorem for the proof of Theorem 10.10 is quite standard and it has been used in the literature obtaining the existence of one non–zero solution for different classes of elliptic problems; see, for instance, the classical result in [129, Theorem 6.2] (see also [4, 101]) and the recent papers [123, 125], where the authors studied some fractional nonlocal problems under the Dirichlet boundary condition. When applying the MPT, both in the nonlocal framework and in the classical Laplacian setting, a crucial role, is played by the Ambrosetti–Rabinowitz condition (10.3.3). See [97] and references therein for related topics.

We emphasize that, for our goal, it is fundamental to use the analytical setting recalled in Section 10.2 in order to correctly encode the periodic boundary datum in the variational formulation of problem (10.3.5). Finally, we point out that, in the same spirit adopted here, in [7] the author, by using a linking approach (see [101, Theorem 5.3]), studied the interesting case $\gamma = m^{2s}$ obtaining the existence of one non–trivial solution for the problem

$$\begin{cases} (-\Delta + m^2)^s u = m^{2s} u + f(x,u) & \text{in } (0,T)^N \\ u(x+Te_i) = u(x) & \text{for all } x \in \mathbb{R}^N \text{ and } i = 1,\dots,N \end{cases}.$$

Remark 10.11. The statement of Theorem 10.10 remains valid if, instead of condition (10.3.2), we require that

$$\limsup_{t\to 0} \frac{F(x,t)}{t^2} < \frac{1}{2} m^{2s},$$

uniformly in $x \in (0,T)^N$. See [51, Theorem 3.6] and [52, Theorem 5.2] for related topics in the classical elliptic Dirichlet case.

Remark 10.12. If, in addition to (10.3.1) and (10.3.3) we also assume that f is odd in the second variable, applying the $\mathbb{Z}_2$–symmetric version of the Mountain Pass Theorem for even functionals (see [101, Theorem 9.12]), problem (10.3.4) admits a sequence of infinitely many weak solutions $\{v_j\}_{j\in\mathbb{N}} \subset \mathbb{X}^s_T$. Note that the symmetry hypothesis on f allows to remove any asymptotic condition at the origin (see [84] for related topics). For completeness we recall that, by adapting the classical variational techniques used in order to study the standard Laplace equation with subcritical growth nonlinearities to the nonlocal periodic framework, in [10] the author proved that the following problem

$$\begin{cases} (-\Delta + m^2)^s u = \gamma u + f(x,u) + h(x) & \text{in } (0,T)^N \\ u(x + Te_i) = u(x) & \text{for all } x \in \mathbb{R}^N \text{ and } i = 1, \dots, N \end{cases},$$

where h is a T–periodic L^∞ function, admits infinitely many weak solutions $\{u_j\}_{j\in\mathbb{N}} \subset \mathbb{H}^s_T$, with the property that their Sobolev norm goes to infinity as $j \to +\infty$, provided the exponent

$$q < 2^\sharp_s - \frac{2s}{N - 2s}.$$

This result represents the periodic nonlocal counterpart of the classical Laplacian case studied in [17, 18, 128] where the existence of infinitely many weak solutions (with the property that the L^2–norm of their gradient goes to infinity) was proved (see also [63, 129]). Finally, we recall that the nonlocal fractional case, under the Dirichlet boundary condition, was studied in [120]. See also the papers [27] and [93] for related topics.

10.3.2 Ground state solutions

Requiring a suitable behaviour at zero and at infinity on the nonlinearity f, in this subsection we study the existence of a ground state solution for nonlocal periodic problems. Moreover, under additional symmetry assumptions, the existence of infinitely many pairs of weak solutions is achieved.

First of all, we recall the definition of ground state solution. Let E be a real Banach space and $\Phi \in C^1(E;\mathbb{R})$ be a functional. The Fréchet derivative of Φ at z, $\Phi'(z)$, is an element of the dual space E^*, and we will denote $\Phi'(z)$ evaluated at $\varphi \in E$ by $\langle \Phi'(z), \varphi\rangle$.

Suppose $z \neq 0$ is a critical point of Φ, i.e. $\Phi'(z) = 0$. Then, necessarily z is contained in the set

$$\mathcal{N} := \{z \in E\backslash\{0\} : \langle \Phi'(z), z\rangle = 0\}.$$

So $\mathcal{N}$ is a natural constraint for the problem of finding nontrivial (i.e., $z \neq 0$) critical points of Φ. $\mathcal{N}$ is called *Nehari manifold*, though in general it may not be a manifold.

Set

$$c := \inf_{z\in\mathcal{N}} \Phi(z).$$

A critical point $z_0 \in \mathcal{N}$ for the functional Φ such that $\Phi(z_0) = c$ is said to be a *ground state solution* of the equation $\Phi'(z) = 0$.

Szulkin and Weth in [130] give a unified approach to the method of Nehari manifold for functionals which have a local minimum at zero and they provide several examples where this method is applied to the problem of finding ground states and multiple solutions for nonlinear elliptic boundary value problems. Moreover, they also consider a recent generalization of this method to problems where zero is a saddle point of the associated energy functionals.

In particular, they proved the following abstract result (see [130, Theorem 12]):

Theorem 10.13. *Let E be a Hilbert space and suppose that $\Phi : E \to \mathbb{R}$ is given by*

$$\Phi(z) := \frac{1}{2}\|z\|^2 - I(z),$$

where $I : E \to \mathbb{R}$ is such that:

(i) $I'(z) = o(\|z\|)$ *as* $z \to 0$;

(ii) $z \mapsto I'(tz)z/t$ *is strictly increasing for all* $z \neq 0$ *and* $t > 0$;

(iii) $I(tz)/t^2 \to +\infty$ *uniformly for z on weakly compact subsets of* $E\backslash\{0\}$ *as* $t \to +\infty$;

(iv) I' *is completely continuous.*

Then equation $\Phi'(z) = 0$ has a ground state solution. Moreover, if the functional I is even, then this equation has infinitely many pairs of solutions.

Now, let $f : \mathbb{R}^N \times \mathbb{R} \to \mathbb{R}$ be a continuous T–periodic (with respect to $x \in \mathbb{R}^N$) function verifying the following conditions:

(h_1) $f(x,t) = o(t)$ *uniformly in* $x \in (0,T)^N$ *as* $t \to 0$;

(h_2) $t \mapsto f(x,t)/|t|$ *is strictly decreasing on* $(-\infty, 0)$ *and* $(0, +\infty)$;

(h_3) $F(x,t)/t^2 \to -\infty$ *uniformly in* $x \in (0,T)^N$ *as* $|t| \to +\infty$,

where $F(x,t) := \int_0^t f(x,w)dw$.

A direct application of Theorem 10.13 gives the following existence (and multiplicity) result:

Theorem 10.14. *Let $f : \mathbb{R}^N \times \mathbb{R} \to \mathbb{R}$ be a continuous T–periodic (with respect to $x \in \mathbb{R}^N$) function that satisfies hypotheses (h_1)–(h_3). Then, if $\gamma < m^{2s}$, the following problem*

$$\begin{cases} (-\Delta + m^2)^s u = \gamma u + f(x,u) & \text{in } (0,T)^N \\ u(x + Te_i) = u(x) & \text{for all } x \in \mathbb{R}^N \text{ and } i = 1, \dots, N \end{cases},$$

has a ground state solution.

Moreover, if the nonlinear term f is odd in the second variable, then problem (10.14) *admits infinitely many pairs of solutions.*

We notice that the Ambrosetti–Rabinowitz condition, requested in Theorem 10.10, is a superlinear growth assumption on the nonlinearity f. Indeed, integrating (10.3.3) we get that

$$\begin{aligned} &\textit{there exist } a_3, a_4 > 0, \textit{ such that} \\ &F(x,t) \geqslant a_3|t|^\mu - a_4 \textit{ for any } (x,t) \in [0,T]^N \times \mathbb{R}, \end{aligned} \tag{10.3.6}$$

see, for instance, [125, Lemma 4].

As a consequence of (10.3.6) and the fact that $\mu > 2$, we have that

$$\lim_{|t|\to+\infty} \frac{F(x,t)}{t^2} = +\infty \textit{ uniformly for any } x \in [0,T]^N, \tag{10.3.7}$$

which is a superlinear assumption on f at infinity.

Of course there are functions that which increase slower than that and yet satisfy (h_1)–(h_3) of Theorem 10.14 (see also Subsection 10.3.4).

10.3.3 A local minimum result

The methods concerned with the minimization of functionals go under the name of direct methods in the Calculus of Variations, while the ones related to finding critical points of functionals give rise to a branch of nonlinear analysis known as Critical Point Theory. The starting point of the so–called direct methods of the Calculus of Variations is the Weierstrass Theorem (saying that a weakly lower semicontinuous and coercive functional defined on a reflexive Banach space admits a global minimum), as well as in critical points theory the crucial idea is that the existence of critical points is related to the topological properties of the sublevels of the functional, provided some compactness properties are satisfied.

Of course, when using direct minimization, we need that the functional is bounded from below and, in this case, we look for its global minima, which are the most natural critical points. In looking for global minima of a functional, the two relevant notions are the weakly lower semicontinuity and the coercivity, as stated in the Weierstrass Theorem. The coercivity of the functional assures that the minimizing sequence is bounded, while the semicontinuity gives the existence of the minimum for the functional.

In this subsection we consider nonlocal problems depending on a real parameter and we study them by using some abstract methods valid for differentiable functionals. We would emphasize the fact that these results can not be achieved by direct minimization.

More precisely, our aim is to study the existence of solutions for the following nonlocal equation

$$\begin{cases} (-\Delta + m^2)^s u = \mu f(x,u) & \text{in } (0,T)^N \\ u(x+Te_i) = u(x) & \text{for all } x \in \mathbb{R}^N \text{ and } i = 1,\dots,N \end{cases}, \tag{10.3.8}$$

adopting variational techniques. The main tool, in order to prove our main result, stated in Theorem 10.16, is the variational principle contained in [103, Theorem 2.5] that we recall here in the form given below.

Theorem 10.15. *Let E be a reflexive real Banach space, let $\Phi, \Psi : E \to \mathbb{R}$ be two Gâteaux differentiable functionals such that Φ is strongly continuous, sequentially weakly lower semicontinuous and coercive in E and Ψ is sequentially weakly upper semicontinuous in E. Let J_μ be the functional defined as $J_\mu := \Phi - \mu\Psi$, $\mu \in \mathbb{R}$, and for any $r > \inf_E \Phi$ let φ be the function defined as*

$$\varphi(r) := \inf_{u\in\Phi^{-1}\left((-\infty,r)\right)} \frac{\sup\limits_{v\in\Phi^{-1}\left((-\infty,r)\right)} \Psi(v) - \Psi(u)}{r - \Phi(u)} .$$

Then, for any $r > \inf_E \Phi$, and any $\mu \in (0, 1/\varphi(r))$[10.1]*, the restriction of the functional J_μ to $\Phi^{-1}((-\infty, r))$ admits a global minimum, which is a critical point (precisely a local minimum) of J_μ in E.*

We assume that the nonlinearity in (10.3.8) is a continuous T–periodic (with respect to $x \in \mathbb{R}^N$) function $f : \mathbb{R}^N \times \mathbb{R} \to \mathbb{R}$ verifying the following subcritical growth condition:

$$\begin{gathered} \text{there exist } a_1, a_2 \geqslant 0 \text{ and } q \in (2, 2_s^\sharp),\ 2_s^\sharp := 2N/(N-2s), \text{ such that} \\ |f(x,t)| \leqslant a_1 + a_2|t|^{q-1} \text{ for every } x \in \mathbb{R}^N, t \in \mathbb{R}. \end{gathered} \tag{10.3.9}$$

Moreover, we assume the following condition, which is a sort of asymptotic growth condition at zero:

$$\lim_{t\to 0^+} \frac{F(x,t)}{t^2} = +\infty \text{ uniformly in } x \in (0,T)^N. \tag{10.3.10}$$

Here, as usual, F is the primitive of the nonlinearity f with respect to the second variable, i.e.

$$F(x,t) := \int_0^t f(x,w)dw,$$

for every $x \in \mathbb{R}^N$ and any $t \in \mathbb{R}$.

The main result of this subsection can be stated as follows.

Theorem 10.16. *Let $f : \mathbb{R}^N \times \mathbb{R} \to \mathbb{R}$ be a continuous T–periodic (with respect to $x \in \mathbb{R}^N$) verifying (10.3.9). In addition, if $f(x,0) = 0$ for every $x \in \mathbb{R}^N$, assume also (10.3.10).*

Then, there exists $\mu^\star > 0$ such that, for any $\mu \in (0, \mu^\star)$, problem (10.3.8) admits at least one not identically zero weak solution $u_\mu = \mathrm{Tr}(v_\mu) \in \mathbb{H}_T^s$. Also $\mu^\star = +\infty$, provided $q \in (1,2)$.

Moreover,

$$\lim_{\mu\to 0^+} |u_\mu|_{\mathbb{H}_T^s} = 0$$

10.1 Note that, by definition, $\varphi(r) \geqslant 0$ for any $r > \inf_E \Phi$. Here and in the following, if $\varphi(r) = 0$, by $1/\varphi(r)$ we mean $+\infty$, i.e. we set $1/\varphi(r) = +\infty$.

and, for every $r > 0$, the function

$$\mu \mapsto \frac{1}{2}\|v_\mu\|^2_{\mathbb{X}^s_T} - \mu\kappa_s \int_{\partial^0 \mathcal{S}_T} F(x, u_\mu)\, dx,$$

is negative and strictly decreasing in $(0, 1/\varphi^\star(r))$, where the number $\varphi^\star(r)$ is defined by

$$\varphi^\star(r) := \kappa_s \inf_{\|w\|^2_{\mathbb{X}^s_T} < 2r} \frac{\sup\limits_{\|v\|^2_{\mathbb{X}^s_T} < 2r} \int_{\partial^0 \mathcal{S}_T} F(x, \mathrm{Tr}(v))\, dx - \int_{\partial^0 \mathcal{S}_T} F(x, \mathrm{Tr}(w))\, dx}{r - \frac{1}{2}\|w\|^2_{\mathbb{X}^s_T}}.$$

The main novelty of this new framework is that, instead of the usual assumptions on functionals, it requires some hypotheses on the nonlinearity which allow to better understand the existence phenomena in the periodic case. This allows us to enlarge the set of applications of [103] exploiting this abstract methods without the existence of a continuous representative.

We notice that, in general, when $f(\cdot, 0) \neq 0$, problem (10.3.8) admits changing–sign solutions, as it happens if we look at the classical case of the Laplacian.

A simple and direct case of Theorem 10.16 reads as follows.

Corrolary 10.17. *Let $\beta : \mathbb{R}^N \to \mathbb{R}$ be a positive continuous T–periodic (non–constant) map and $f : \mathbb{R} \to \mathbb{R}$ be a continuous and subcritical function. Assume that*

$$\lim_{t \to 0^+} \frac{f(t)}{t} = +\infty. \tag{10.3.11}$$

Then, there exists $\mu^\star > 0$ such that, for any $\mu \in (0, \mu^\star)$, the following problem

$$\begin{cases} (-\Delta + m^2)^s u = \mu\beta(x)f(u) & \text{in } (0, T)^N \\ u(x + Te_i) = u(x) & \text{for all } x \in \mathbb{R}^N \text{ and } i = 1, \dots, N \end{cases}, \tag{10.3.12}$$

admits at least one non–trivial weak solution in $\mathbb{H}^s_T$.

As an application of our main result, we can consider the following model problem

$$\begin{cases} (-\Delta + m^2)^s u = \mu(a(x)|u|^{r-2}u + b(x)|u|^{q-2}u + c(x)) & \text{in } (0, T)^N \\ u(x + Te_i) = u(x) & \text{for } x \in \mathbb{R}^N,\ i = 1, \dots, N \end{cases}. \tag{10.3.13}$$

In this framework, Theorem 10.16 reduces to the following result.

Corrolary 10.18. *Assume that $1 < r < 2 \leqslant q < 2^\sharp_s$ and $a, b, c : \mathbb{R}^N \to \mathbb{R}$ are T–periodic continuous functions. In addition, if $c \equiv 0$ in $(0, T)^N$, assume that* $\inf_{x \in (0,T)^N} a(x) > 0$.

Then, there exists $\mu^\star > 0$ such that for any $\mu \in (0, \mu^\star)$, problem (10.3.13) admits at least one not identically zero weak solution $u_\mu \in \mathbb{H}^s_T$ and

$$\lim_{\mu \to 0^+} |u_\mu|_{\mathbb{H}^s_T} = 0.$$

Also $\mu^\star = +\infty$, *provided* $b \equiv 0$ *in* $(0, T)^N$.

As a final remark, we give an estimate from below for the parameter $\mu^\star$ appearing in Theorem 10.16. Indeed, while when $q \in (1, 2)$ Theorem 10.16 assures that $\mu^\star = +\infty$, the exact value of $\mu^\star$ is not known in the other cases, that is when $q \in [2, 2^\sharp_s)$. We have that

$$\mu^\star := \sup_{r>0} \frac{1}{\varphi^\star(r)} \geqslant \sup_{r>0} \frac{q\kappa_s^{-1}}{\sqrt{2}a_1\sigma_1 q r^{-1/2} + 2^{q/2}a_2\sigma_q^q r^{q/2-1}},$$

where

$$\sigma_t := \frac{1}{\sqrt{\kappa_s}} \sup_{u\in\mathbb{H}^s_T\setminus\{0\}} \frac{|u|_{L^t(0,T)^N}}{|u|_{\mathbb{H}^s_T}} \quad \text{with } t \in \{1, q\}.$$

Then

$$\mu^\star \geqslant \frac{\kappa_s^{-1}}{a_2\sigma_2^2} \quad \text{if} \quad q = 2$$

and

$$\mu^\star \geqslant \frac{q\kappa_s^{-1}}{\sqrt{2}a_1\sigma_1 q r_{\max}^{-1/2} + 2^{q/2}a_2\sigma_q^q r_{\max}^{q/2-1}} \quad \text{if} \quad q \in (2, 2^\sharp_s),$$

where

$$r_{\max} := \frac{1}{2}\left(\frac{a_1\sigma_1 q}{a_2\sigma_q^q(q-2)}\right)^{2/(q-1)},$$

while a_1 and a_2 are as in (10.3.9).

Remark 10.19. We notice that, when $q \in (1, 2)$ the existence of solutions for problem (10.3.8) can be obtained using classical direct methods (see, for instance [129, Chapter I]), but, as it happens with the arguments used along the present paper, we do not know a priori if the solution provided by these classical theorems is the identically zero function or not (of course, we refer to the case $f(\cdot, 0) = 0$, otherwise $v \equiv 0$ does not solve (10.3.8)). Hence, if $f(\cdot, 0) = 0$, also when using standard methods, we need to assume extra conditions on f, in order to prove that the solution of the problem is not the zero function. Finally, we emphasize that, arguing as in [92], combining Theorem 10.15 with the classical Pucci–Serrin result [100, Theorem 4], the existence of at least three weak solutions for problem (10.3.8) can be proved.

Remark 10.20. The statement of Theorem 10.16 remains valid if, instead of condition (10.3.10), we require that

$$\limsup_{t\to 0^+} \frac{F(x,t)}{t^2} = +\infty, \tag{10.3.14}$$

uniformly in $x \in (0, T)^N$.

For instance, Theorem 10.16 can be applied to the following parametric nonlinear problem

$$\begin{cases} (-\Delta + m^2)^s u = \mu\beta(x)h(u) & \text{in } (0,T)^N \\ u(x+Te_i) = u(x) & \text{for all } x \in \mathbb{R}^N \text{ and } i = 1,\dots,N \end{cases}, \tag{10.3.15}$$

where $\beta : \mathbb{R}^N \to \mathbb{R}$ is a positive continuous T–periodic function and

$$h(t) := \begin{cases} 0 & \text{if } t \leqslant 0 \\ \sqrt{t}\left(\frac{1}{2} + \sin\frac{1}{t^2}\right) & \text{if } t > 0 \end{cases}.$$

Indeed, as a direct computation ensures, one has

$$-\infty < \liminf_{t\to 0^+} \frac{\int_0^t h(w)dw}{t^2} < \limsup_{t\to 0^+} \frac{\int_0^t h(w)dw}{t^2} = +\infty.$$

Then, by Theorem 10.16 (where condition (10.3.10) is replaced by(10.3.14)) there exists $\mu^\star > 0$ such that, for any $\mu \in (0,\mu^\star)$, problem (10.3.15) admits at least one not identically zero weak solution $u_\mu \in \mathbb{H}^s_T$. We notice that Theorem 10.10 in Subsection 10.3.1 (as well as Theorem 10.14 in Subsection 10.3.2) cannot be applied to problem (10.3.15) since conditions (10.3.14) and (10.3.2) are mutually exclusive. See [91, Part II, Chapter 7] and [60, 95, 94] for related topics.

If in addition to condition (10.3.11) the function f satisfies also the Ambrosetti–Rabinowitz condition, the existence of at least two solutions can be achieved as the next result shows:

Theorem 10.21. *Let $\beta : \mathbb{R}^N \to \mathbb{R}$ be a positive continuous T–periodic (non–constant) map and $f : \mathbb{R} \to \mathbb{R}$ be a continuous function verifying* (10.3.9) *and* (10.3.11). *Assume that*

there exist $\nu > 2$, $M > 0$ sufficiently large such that

$$0 < \nu F(t) \leqslant tf(t), \tag{10.3.16}$$

for every $|t| \geqslant M$.
Then, for μ sufficiently small, the following problem

$$\begin{cases} (-\Delta + m^2)^s u = \mu\beta(x)f(u) & \text{in } (0,T)^N \\ u(x+Te_i) = u(x) & \text{for all } x \in \mathbb{R}^N \text{ and } i = 1,\dots,N \end{cases}, \tag{10.3.17}$$

admits at least two non–trivial weak solutions in $\mathbb{H}^s_T$.

Proof. Since conditions (10.3.9) and (10.3.11) hold, by Theorem 10.16, for μ sufficiently small, problem (10.3.17) admits at least one non–trivial solution that is the trace of a local minimum (namely $v_{1,\mu}$) of the energy functional

$$\mathcal{J}_\mu(v) := \frac{1}{2}\|v\|^2_{\mathbb{X}^s_T} - \mu\kappa_s \int_{\partial^0 \mathcal{S}_T} \beta(x)F(\mathrm{Tr}(v))dx, \quad \forall\, v \in \mathbb{X}^s_T.$$

In view of condition (10.3.17), arguing in a standard way, it is possible to prove that every Palais–Smale sequence is bounded. Hence, the functional $\mathcal{J}_\mu$ satisfies the (PS) condition. Moreover, as pointed out in Subsection 10.3.2, hypothesis (10.3.16) ensures that

$$\begin{aligned} &\textit{there exist } \gamma_1, \gamma_2 > 0 \textit{ such that} \\ &F(t) \geqslant \gamma_1 |t|^\nu - \gamma_2, \ \textit{for any } t \in \mathbb{R}. \end{aligned} \tag{10.3.18}$$

Now, let $\{\xi_j\}_{j\in\mathbb{N}}$ be a real sequence such that $\lim_{j\to+\infty} \xi_j = +\infty$ and, for every $j \in \mathbb{N}$, let us define

$$w_{\xi_j}(x,y) := \frac{2}{\Gamma(s)} \left(\frac{my}{2}\right)^s K_s(my)\xi_j, \quad \forall (x,y) \in \mathbb{R}^{N+1}_+ \tag{10.3.19}$$

where K_s is the Bessel function of the second kind of order s (see Subsection 10.2.1).

In view of (10.3.19) and (10.3.18), it follows that

$$\mathcal{J}_\mu(w_{\xi_j}) \leqslant \left(\frac{\xi_j^2}{2} m^{2s} T^N - \mu(\gamma_1 \xi_j^\nu - \gamma_2)|\beta|_{L^1(0,T)^N}\right) \kappa_s,$$

bearing in mind that $\mathrm{Tr}(w_{\xi_j}) = \xi_j$, for every $j \in \mathbb{N}$. Clearly, since $\nu > 2$, the above inequality implies that

$$\lim_{j\to+\infty} \mathcal{J}_\mu(w_{\xi_j}) = -\infty.$$

Hence,

$$\liminf_{\|v\|_{\mathbb{X}^s_T}\to+\infty} \mathcal{J}_\mu(v) = -\infty$$

and there exists some $v_0 \in \mathbb{X}^s_T$ such that $\mathcal{J}_\mu(v_0) < \mathcal{J}_\mu(v_{1,\mu})$.

Now, we can assume that $v_{1,\mu}$ is a strict local minimum for $\mathcal{J}_\mu$ in $\mathbb{X}^s_T$. Hence, the Mountain Pass Theorem ensures that there exists a critical point of $\mathcal{J}_\mu$, namely $v_2 \in \mathbb{X}^s_T$, such that $\mathcal{J}_\mu(v_2) > \mathcal{J}_\mu(v_{1,\mu})$.

The proof of Theorem 10.21 is complete. □

10.3.4 A Morse theoretical approach

Let E be a real Banach space, $\mathcal{J} \in C^1(E;\mathbb{R})$, and

$$\mathcal{K} := \{z \in E : \mathcal{J}'(z) = 0\},$$

where E^* denotes the dual space of E endowed by the norm $\|\cdot\|_{E^*}$.

The q–th critical group of $\mathcal{J}$ at an isolated critical point $z \in \mathcal{K}$ with $\mathcal{J}(z) = c$ is defined by

$$C_q(\mathcal{J}, z) := H_q(\mathcal{J}^c \cap U, \mathcal{J}^c \cap U\setminus\{z\}), \quad (q \in \mathbb{N})$$

where

$$\mathcal{J}^c := \{z \in E : \mathcal{J}(z) \leqslant c\},$$

U is any neighborhood of z (containing z as unique critical point), H_* is the *singular relative homology* with coefficients in an Abelian group G. Finally, let ϑ be the trivial homological group.

We say that $z \in \mathcal{K}$ is an homological non–trivial critical point of $\mathcal{J}$ if at least one of its critical groups is non–trivial.

For the sake of completeness, we recall that a C^1–functional $\mathcal{J} : E \to \mathbb{R}$ satisfies the Cerami condition at level $\mu \in \mathbb{R}$, (briefly $(\mathrm{C})_\mu$) if

$(\mathrm{C})_\mu$ *every sequence* $\{z_j\}_{j\in\mathbb{N}}$ *in* E *such that*

$$\mathcal{J}(z_j) \to \mu \quad \text{and} \quad (1 + \|z_j\|)\|\mathcal{J}'(z_j)\|_{E^*} \to 0,$$

as $j \to +\infty$, *possesses a convergent subsequence.*

We say that $\mathcal{J}$ satisfies the Cerami condition (in short (C)) if $(\mathrm{C})_\mu$ holds for every $\mu \in \mathbb{R}$.

Further, a C^1–functional $\mathcal{J} : X \to \mathbb{R}$ satisfies the Palais–Smale condition at level $\mu \in \mathbb{R}$, (briefly $(\mathrm{PS})_\mu$) if

$(\mathrm{PS})_\mu$ *every sequence* $\{z_j\}_{j\in\mathbb{N}}$ *in* E *such that*

$$\mathcal{J}(z_j) \to \mu \quad \text{and} \quad \|\mathcal{J}'(z_j)\|_{E^*} \to 0,$$

as $j \to +\infty$, *possesses a convergent subsequence.*

Hence, we say that $\mathcal{J}$ satisfies the Palais–Smale condition (in short (PS)) if $(\mathrm{PS})_\mu$ holds for every $\mu \in \mathbb{R}$. As it is well–known, the (C) condition is a weak version of the (PS) compactness condition. Finally, if $\mathcal{J}$ satisfied the (PS) condition or the (C) condition, then $\mathcal{J}$ satisfied the deformation condition (see [76]).

Proposition 10.22. *Let* $\mathcal{J} \in C^1(E;\mathbb{R})$ *be bounded from below. Assume that* $\mathcal{J}$ *satisfies the* (PS) *condition. If* $\mathcal{J}$ *has a critical point that is homological non–trivial and is not a minimizer of* $\mathcal{J}$, *then* $\mathcal{J}$ *has at least three critical points.*

See [75, Theorem 2.1].

If $\mathcal{J}$ satisfies the (C) condition and the critical values of $\mathcal{J}$, are bounded from below by some $a < \inf \mathcal{J}(\mathcal{K})$, then the critical groups of $\mathcal{J}$ at infinity were introduced by Bartsch and Li [] as follows

$$C_q(\mathcal{J}, \infty) := H_q(Y, \mathcal{J}^a), \quad (q \in \mathbb{N}). \tag{10.3.20}$$

If $\mathcal{J}$ satisfies the (C) condition, then $\mathcal{J}$ satisfies the deformation lemma, hence the right–hand side of (10.3.20) does not depend on the choice of the constant a.

Proposition 10.23. *Let $\mathcal{J} \in C^1(E;\mathbb{R})$ and assume that condition* (C) *holds. Furthermore, suppose that $\mathcal{J}$ has only finitely many critical points. Then*
(i) *if there is some $q \in \mathbb{N}$ such that $C_q(\mathcal{J},\infty) \neq \vartheta$, then $\mathcal{J}$ has a critical point z with $C_q(\mathcal{J},z) \neq \vartheta$;*
(ii) *let 0 be an isolated critical point of $\mathcal{J}$. If $\mathcal{K} = \{0\}$, then*

$$C_q(\mathcal{J},\infty) = C_q(\mathcal{J},0),$$

for every $q \in \mathbb{N}$.

See [44] for details.

By Proposition 10.23 it follows that if $C_q(\mathcal{J},\infty) \neq C_q(\mathcal{J},0)$ for some $q \in \mathbb{N}$, then $\mathcal{J}$ must have a non–trivial critical point. Therefore, we must compute the critical groups at zero and at infinity.

In what follows we may assume that $\mathcal{J}$ has only finitely many critical points. Since $\mathcal{J}$ satisfies the (C) condition or the (PS) condition, then the critical groups $C_q(\mathcal{J},\infty)$ at infinity make sense.

Proposition 10.24. *Assume that $\mathcal{J}$ has a critical point at zero with $\mathcal{J}(0) = 0$. Suppose that $\mathcal{J}$ has a local linking at zero with respect to $Y = V \oplus W$, with $k := \dim V < +\infty$, that is: there exists $\rho > 0$ such that*
(i) *$\mathcal{J}(z) \leqslant 0$, for every $z \in V$, $\|z\| \leqslant \rho$;*
(ii) *$\mathcal{J}(z) > 0$, for every $z \in W$, $0 < \|z\| \leqslant \rho$.*
Then $C_k(\mathcal{J},0) \neq \vartheta$. Hence, 0 is a homological nontrivial critical point of $\mathcal{J}$.

See [75, Proposition 2.1].

A careful application of Propositions (10.22), (10.23) and (10.24) gives us some existence results for the following problem

$$\begin{cases} (-\Delta + m^2)^{1/2}u = f(x,u) & \text{in } (0,T)^N \\ u(x+Te_i) = u(x) & \text{for all } x \in \mathbb{R}^N \text{ and } i = 1,\dots,N \end{cases}, \tag{10.3.21}$$

where f is a suitable nonlinear term.

More precisely, we assume that $f : \mathbb{R}^N \times \mathbb{R} \to \mathbb{R}$ is T–periodic (with respect to $x \in \mathbb{R}^N$) and continuous function with $f(x,0) = 0$, for every $x \in (0,T)^N$, and

$$F(x,t) := \int_0^t f(x,w)dw,$$

for every $(x,t) \in (0,T)^N \times \mathbb{R}$.

We also use the following conditions:

(k_1) *there exists a constant $c_0 > 0$ such that*

$$|f(x,t)| \leqslant c_0(1+|t|^{q-1}), \quad \text{for any } (x,t) \in (0,T)^N \times \mathbb{R} \tag{10.3.22}$$

where $q \in (1, 2^\sharp)$*, and* $2^\sharp := 2N/(N-1)$*;*

(k_2) *there is some* $r > 0$ *small enough and* $\lambda_k < \lambda < \lambda_{k+1}$ *such that*

$$\lambda_k \frac{t^2}{2} \leqslant F(x, t) \leqslant \lambda \frac{t^2}{2}, \tag{10.3.23}$$

for every $x \in (0, T)^N$ *and* $|t| \leqslant r$*;*

(k_3) *there exist* $\mu > 2,\ M > 0$ *sufficiently large such that*

$$0 < \mu F(x, t) \leqslant t f(x, t), \tag{10.3.24}$$

for every $x \in (0, T)^N$*, and* $|t| \geqslant M$*.*

Here $0 < \lambda_1 < \lambda_2 \leqslant \ldots \leqslant \lambda_k < \lambda_{k+1} \leqslant \ldots$ are the eigenvalues of $(-\Delta + m^2)^{1/2}$ in $\mathbb{H}^{1/2}_T$ (see Section 2 for more details).

Assuming that $f(x, 0) \equiv 0$, then problem (10.3.21) has the zero solution. Therefore, we are interested in the existence of not identically zero solutions. To this goal, it follows by Morse theory that, comparing the critical groups of the energy functional (associated with problem (10.3.21)) at zero and at infinity, one may deduce the existence of non–trivial weak solutions to elliptic equations (see, for instance, [44, 96]).

Our first existence result reads as follows.

Theorem 10.25. *Assume that conditions* (k_1)*–*(k_3) *hold. Then, problem* (10.3.21) *has at least one not identically zero weak solution in* $\mathbb{H}^{1/2}_T$*.*

On the other hand, a lot of works, concerning superlinear elliptic boundary value problems, have been done by using the Ambrosetti–Rabinowitz condition, expressed here by (k_3). As observed in Subsection 10.3.2, the role of this condition is to ensure that every Palais–Smale sequence of the energy functional associated to (10.3.21) is bounded.

However, there are many functions which are superlinear at infinity, but for which that condition fails. For instance, the real function

$$f(x, t) := t\log(1 + |t|), \quad \forall\, (x, t) \in (0, T)^N \times \mathbb{R} \tag{10.3.25}$$

does not satisfy condition (10.3.24).

Therefore, we will use the following hypotheses on the nonlinearity f:

(k_4) $f(x, t)t \geqslant 0$ *for any* $(x, t) \in (0, T)^N \times \mathbb{R}$ *and*

$$\lim_{|t| \to +\infty} \frac{f(x, t)}{t} = +\infty, \tag{10.3.26}$$

uniformly in $(0, T)^N$*;*

(k_5) *there exists* $\vartheta \geqslant 1$ *such that*

$$\vartheta\mathcal{F}(x,t) \geqslant \mathcal{F}(x,\zeta t),$$

for every $(x,t) \in (0,T)^N \times \mathbb{R}$ *and* $\zeta \in [0,1]$, *where we set*

$$\mathcal{F}(x,t) := f(x,t)t - 2F(x,t).$$

It is easy to see that the function (10.3.25) satisfies the conditions (k_4) and (k_5). Condition (k_5) was introduced by Jeanjean in [70] and, in recent years, was often applied to consider the existence of non–trivial solutions for superlinear problems without the Ambrosetti–Rabinowitz condition. See, for instance, the papers [1, 55, 72, 73] and references therein.

We have the following existence property:

Theorem 10.26. *Assume that conditions* (k_1), (k_2), (k_4) *and* (k_5) *are verified. Then, problem* (10.3.21) *has at least one not identically zero weak solution in* $\mathbb{H}_T^{1/2}$.

Condition (k_5) is a global hypothesis on f, hence is not very satisfactory. For this reason, we replace (k_5) with the following monotonicity condition:

(k_6) *there exists* $\nu > 0$ *such that*

$$\frac{f(x,t)}{t} \text{ is increasing in } t \geqslant \nu \text{ and decreasing in } t \leqslant -\nu. \tag{10.3.27}$$

The next theorem holds true:

Theorem 10.27. *Assume that* (k_1), (k_2), (k_4) *and* (k_6) *hold. Then, problem* (10.3.21) *has at least one not identically zero weak solution in* $\mathbb{H}_T^{1/2}$.

Now, we turn to look for a multiplicity result for the problem (10.3.21) (see Section 10.4 for several results on the existence of multiple solutions for nonlocal periodic equations).

Finally, under the following assumption

(k_7) $\limsup_{|t|\to+\infty} \dfrac{F(x,t)}{t^2} < \dfrac{m}{2}$ *uniformly in* $(0,T)^N$,

we have the following result:

Theorem 10.28. *Assume that* (f_1), (f_2) *and* (f_7) *hold. Then, problem* (10.3.21) *has at least two not identically zero weak solutions in* $\mathbb{H}_T^{1/2}$.

Remark 10.29. The main theorems presented here represent a nonlocal counterpart of [74, Theorem 1.1], [55, Theorem 1.1], [73, Theorem 1.2] and [75, Theorem 2.1]. In the treated cases the growth conditions expressed by (10.3.22) and (10.3.23), as well as by (10.3.26) and (10.3.27), are crucial in order to obtain our results. We refer also to [58] where similar results are obtained for nonlocal equations involving the regional fractional Laplacian.

10.3.5 Some localization theorems

In this subsection we study the existence of weak solutions for the following problem

$$\begin{cases} \displaystyle\iint_{\mathcal{S}_T} y^{1-2s}(\nabla v \nabla \varphi + m^2 v\varphi)\, dxdy \\ \qquad\qquad = \lambda\kappa_s \displaystyle\int_{\partial^0 \mathcal{S}_T} f(x, \mathrm{Tr}(v))\mathrm{Tr}(\varphi) dx, \quad \forall\ \varphi \in \mathbb{X}_T^s \\ v \in \mathbb{X}_T^s, \end{cases} \tag{10.3.28}$$

where λ is a positive real parameter. On the contrary of Subsection 10.3.3 we do not assume here any asymptotic condition at zero on the nonlinear term f as requested, for instance, in Theorem 10.16. The main approach is based on the direct methods in the Calculus of Variations (see [91] and [59] where a similar approach has been used in order to study subelliptic problems on Carnot groups).

More precisely, under a suitable subcritical growth condition on the nonlinear term f, we are able to prove the existence of at least one weak solution of problem (10.3.28) provided that λ belongs to a precise bounded interval of positive parameters, as stated here below:

Theorem 10.30. *Let $f : \mathbb{R}^N \times \mathbb{R} \to \mathbb{R}$ be a Carathéodory T-periodic (with respect to $x \in \mathbb{R}^N$) function such that*

$$|f(x, t)| \leqslant \alpha(x) + \beta(x)|t|^p \quad \textit{a.e. in } \mathbb{R}^N \times \mathbb{R}, \tag{10.3.29}$$

where

$$\alpha \in L^{\frac{\gamma 2_s^\sharp}{\gamma 2_s^\sharp - 1}}(0, T)^N \qquad \textit{and} \qquad \beta \in L^{\frac{1}{1-\gamma}}(0, T)^N$$

with $\gamma \in (2/2_s^\sharp, 1)$, $p \in (1, \gamma 2_s^\sharp - 1)$. Furthermore, let

$$0 < \lambda < \frac{(p-1)^{\frac{p-1}{p}}}{p(\kappa_s^{\frac{1}{2}}\kappa_{1,\gamma})^{\frac{p-1}{p}} (\kappa_s^{\frac{1-p}{2}}\kappa_{2,\gamma})^{\frac{p+1}{p}} |\alpha|^{\frac{p-1}{p}}_{L^{\frac{\gamma 2_s^\sharp}{\gamma 2_s^\sharp - 1}}(0,T)^N} |\beta|^{\frac{1}{p}}_{L^{\frac{1}{1-\gamma}}(0,T)^N}}, \tag{10.3.30}$$

where $\kappa_{1,\gamma}$ and $\kappa_{2,\gamma}$ denote the embedding constants of the Sobolev space $\mathbb{H}_T^s$ into $L^{\gamma 2_s^\sharp}(0, T)^N$ and $L^{\frac{p+1}{\gamma}}(0, T)^N$, respectively.

Then, the following nonlocal parametric problem

$$\begin{cases} (-\Delta + m^2)^s u = \lambda f(x, u) & \textit{in } (0, T)^N \\ u(x + Te_i) = u(x) & \textit{for all } x \in \mathbb{R}^N \textit{ and } i = 1, \dots, N \end{cases}, \tag{10.3.31}$$

has a weak solution $u_{0,\lambda} \in \mathbb{H}_T^s$ *and*

$$\|u_{0,\lambda}\|_{\mathbb{H}_T^s} < \left(\lambda p \kappa_{2,\gamma}^{p+1} |\beta|_{L^{\frac{1}{1-\gamma}}(0,T)^N}\right)^{\frac{1}{1-p}}.$$

Let us consider the functional $\mathcal{J}_\lambda : \mathbb{X}_T^s \to \mathbb{R}$ defined by

$$\mathcal{J}_\lambda(v) := \frac{1}{2}\|v\|_{\mathbb{X}_T^s}^2 - \lambda \kappa_s \int_{\partial^0 \mathcal{S}_T} F(x, \mathrm{Tr}(v))dx, \tag{10.3.32}$$

where $\lambda \in \mathbb{R}$ and, as usual, we set $F(x, t) := \int_0^t f(x, w)dw$.

Note that, under our growth condition on f, the functional $\mathcal{J}_\lambda \in C^1(\mathbb{X}_T^s; \mathbb{R})$ and its derivative at $v \in \mathbb{X}_T^s$ is given by

$$\langle \mathcal{J}_\lambda'(v), \varphi\rangle = \iint_{\mathcal{S}_T} y^{1-2s}(\nabla v \nabla \varphi + m^2 v\varphi)dxdy - \lambda\kappa_s \int_{\partial^0 \mathcal{S}_T} f(x, \mathrm{Tr}(v))\mathrm{Tr}(\varphi)dx,$$

for every $\varphi \in \mathbb{X}_T^s$.

Thus, the weak solutions of problem (10.3.31) are exactly the critical points of the energy functional $\mathcal{J}_\lambda$.

Fix $\lambda > 0$ and denote

$$\Phi(v) := \|v\|_{\mathbb{X}_T^s} \quad \text{and} \quad \Psi_\lambda(v) := \lambda\kappa_s \int_{\partial^0 \mathcal{S}_T} F(x, \mathrm{Tr}(v))dx,$$

for every $u \in \mathbb{X}_T^s$.

Note that, thanks to condition (10.3.29), the operator Ψ_λ is well defined and sequentially weakly (upper) continuous. So the operator $\mathcal{J}_\lambda$ is sequentially weakly lower semicontinuous on $\mathbb{X}_T^s$. With the above notations we can prove the next two lemmas, that will be crucial in the sequel.

Lemma 10.31. *Let* $\lambda > 0$ *and suppose that*

$$\limsup_{\varepsilon \to 0^+} \frac{\sup\limits_{v \in \Phi^{-1}([0,\varrho_0])} \Psi_\lambda(v) - \sup\limits_{v \in \Phi^{-1}([0,\varrho_0 - \varepsilon])} \Psi_\lambda(v)}{\varepsilon} < \varrho_0, \tag{10.3.33}$$

for some $\varrho_0 > 0$. *Then,*

$$\inf_{\sigma < \varrho_0} \frac{\sup\limits_{v \in \Phi^{-1}([0,\varrho_0])} \Psi_\lambda(v) - \sup\limits_{v \in \Phi^{-1}([0,\sigma])} \Psi_\lambda(v)}{\varrho_0^2 - \sigma^2} < \frac{1}{2}. \tag{10.3.34}$$

Proof. Firstly, by condition (10.3.33) one has

$$\limsup_{\varepsilon \to 0^+} \frac{\sup\limits_{v \in \Phi^{-1}([0,\varrho_0])} \Psi_\lambda(v) - \sup\limits_{v \in \Phi^{-1}([0,\varrho_0 - \varepsilon])} \Psi_\lambda(v)}{\varrho_0^2 - (\varrho_0 - \varepsilon)^2} < \frac{1}{2}. \tag{10.3.35}$$

Indeed, if $\varepsilon \in (0, \varrho_0)$, one has

$$\frac{\sup\limits_{v\in\Phi^{-1}([0,\varrho_0])} \Psi_\lambda(v) - \sup\limits_{v\in\Phi^{-1}([0,\varrho_0-\varepsilon])} \Psi_\lambda(v)}{\varrho_0^2 - (\varrho_0 - \varepsilon)^2}$$

$$= \frac{\sup\limits_{v\in\Phi^{-1}([0,\varrho_0])} \Psi_\lambda(v) - \sup\limits_{v\in\Phi^{-1}([0,\varrho_0-\varepsilon])} \Psi_\lambda(v)}{\varepsilon} \times$$

$$\times \frac{-\varepsilon/\varrho_0}{\varrho_0\left[\left(1 - \frac{\varepsilon}{\varrho_0}\right)^2 - 1\right]},$$

and

$$\lim_{\varepsilon\to 0^+} \frac{-\varepsilon/\varrho_0}{\varrho_0\left[\left(1 - \frac{\varepsilon}{\varrho_0}\right)^2 - 1\right]} = \frac{1}{2\varrho_0}.$$

Now, by (10.3.35) there exists $\overline{\varepsilon} > 0$ such that

$$\frac{\sup\limits_{v\in\Phi^{-1}([0,\varrho_0])} \Psi_\lambda(v) - \sup\limits_{v\in\Phi^{-1}([0,\varrho_0-\varepsilon])} \Psi_\lambda(v)}{\varrho_0^2 - (\varrho_0 - \varepsilon)^2} < \frac{1}{2},$$

for every $\varepsilon \in]0, \overline{\varepsilon}[$. Setting $\sigma_0 := \varrho_0 - \varepsilon_0$ (with $\varepsilon_0 \in]0, \overline{\varepsilon}[$), it follows that

$$\frac{\sup\limits_{v\in\Phi^{-1}([0,\varrho_0])} \Psi_\lambda(v) - \sup\limits_{v\in\Phi^{-1}([0,\sigma_0])} \Psi_\lambda(v)}{\varrho_0^2 - \sigma_0^2} < \frac{1}{2},$$

and thus inequality (10.3.34) is verified. □

Lemma 10.32. *Let $\lambda > 0$ and suppose that condition* (10.3.34) *holds. Then*

$$\inf_{w\in\Phi^{-1}([0,\varrho_0))} \frac{\sup\limits_{v\in\Phi^{-1}([0,\varrho_0])} \Psi_\lambda(v) - \Psi_\lambda(w)}{\varrho_0^2 - \|w\|^2_{\mathbb{X}^s_T}} < \frac{1}{2}. \tag{10.3.36}$$

Proof. Assumption (10.3.34) yields

$$\sup_{v\in\Phi^{-1}([0,\sigma_0])} \Psi_\lambda(v) > \sup_{v\in\Phi^{-1}([0,\varrho_0])} \Psi_\lambda(v) - \frac{1}{2}(\varrho_0^2 - \sigma_0^2), \tag{10.3.37}$$

for some $0 < \sigma_0 < \varrho_0$. Thanks to the weakly regularity of the functional Ψ_λ, since

$$\sup_{v\in\Phi^{-1}([0,\sigma_0])} \Psi_\lambda(v) = \sup_{\|v\|_{\mathbb{X}^s_T}=\sigma_0} \Psi_\lambda(v),$$

by (10.3.37) there exists $v_0 \in \mathbb{X}^s_T$ with $\|v_0\|_{\mathbb{X}^s_T} = \sigma_0$ such that

$$\Psi_\lambda(v_0) > \sup_{v\in\Phi^{-1}([0,\varrho_0])} \Psi_\lambda(v) - \frac{1}{2}(\varrho_0^2 - \sigma_0^2), \tag{10.3.38}$$

that is

$$\frac{\sup_{v\in\Phi^{-1}([0,\varrho_0])}\Psi_\lambda(v)-\Psi_\lambda(v_0)}{\varrho_0^2-\|v_0\|^2_{\mathbb{X}^s_T}}<\frac{1}{2}, \tag{10.3.39}$$

with $\|v_0\|_{\mathbb{X}^s_T}=\sigma_0$. The proof of Lemma 10.32 is now complete. □

For the proof of Theorem 10.30, first of all, we note that problem (10.3.31) has a variational structure. Indeed, it is the Euler–Lagrange equation of the functional $\mathcal{J}_\lambda$ defined in (10.3.32).

Hence, fix

$$\lambda\in\left(0,\frac{(p-1)^{\frac{p-1}{p}}}{p(\kappa_s^{\frac{1}{2}}\kappa_{1,\gamma})^{\frac{p-1}{p}}(\kappa_s^{\frac{1-p}{2}}\kappa_{2,\gamma})^{\frac{p+1}{p}}|\alpha|^{\frac{p-1}{p}}_{L^{\frac{\gamma2^\sharp_s}{\gamma2^\sharp_s-1}}(0,T)^N}|\beta|^{\frac{1}{p}}_{L^{\frac{1}{1-\gamma}}(0,T)^N}}\right), \tag{10.3.40}$$

and let consider $0<\varepsilon<\varrho$. Setting

$$\Lambda_\lambda(\varepsilon,\varrho):=\frac{\sup_{v\in\Phi^{-1}([0,\varrho])}\Psi_\lambda(v)-\sup_{v\in\Phi^{-1}([0,\varrho-\varepsilon])}\Psi_\lambda(v)}{\varepsilon},$$

one has

$$\Lambda_\lambda(\varepsilon,\varrho)\leqslant\frac{1}{\varepsilon}\left|\sup_{v\in\Phi^{-1}([0,\varrho])}\Psi_\lambda(v)-\sup_{v\in\Phi^{-1}([0,\varrho-\varepsilon])}\Psi_\lambda(v)\right|.$$

Moreover, it easily follows that

$$\Lambda_\lambda(\varepsilon,\varrho)\leqslant\kappa_s\sup_{v\in\Phi^{-1}([0,1])}\int_{(0,T)^N}\left|\int_{(\varrho-\varepsilon)\operatorname{Tr}(v)}^{\varrho\operatorname{Tr}(v)}\lambda\frac{|f(x,t)|}{\varepsilon}dt\right|dx.$$

Moreover, the growth condition (10.3.29) yields

$$\begin{aligned}\sup_{v\in\Phi^{-1}([0,1])}\int_{(0,T)^N}\left|\int_{(\varrho-\varepsilon)\operatorname{Tr}(v)}^{\varrho\operatorname{Tr}(v)}\lambda\frac{|f(x,t)|}{\varepsilon}dt\right|dx&\leqslant\sup_{v\in\Phi^{-1}([0,1])}\int_{(0,T)^N}\lambda\alpha(x)|\operatorname{Tr}(v)|dx\\&+\sup_{v\in\Phi^{-1}([0,1])}\int_{(0,T)^N}\frac{\lambda\beta(x)}{p+1}\left(\frac{\varrho^{p+1}-(\varrho-\varepsilon)^{p+1}}{\varepsilon}\right)|\operatorname{Tr}(v)|^{p+1}dx.\end{aligned}$$

Since the Sobolev space $\mathbb{H}^s_T$ is compactly embedded in $L^q(0,T)^N$, for every $q\in[1,2^\sharp_s)$, bearing in mind that

$$\alpha\in L^{\frac{\gamma2^\sharp_s}{\gamma2^\sharp_s-1}}(0,T)^N\qquad\text{and}\qquad\beta\in L^{\frac{1}{1-\gamma}}(0,T)^N,$$

the above inequality yields

$$\Lambda_\lambda(\varepsilon,\varrho)\leqslant\kappa_s^{\frac{1}{2}}\kappa_{1,\gamma}|\lambda\alpha|_{L^{\frac{\gamma2^\sharp_s}{\gamma2^\sharp_s-1}}(0,T)^N}+\frac{\kappa_s^{\frac{1-p}{2}}\kappa_{2,\gamma}^{p+1}}{p+1}|\lambda\beta|_{L^{\frac{1}{1-\gamma}}(0,T)^N}\left(\frac{\varrho^{p+1}-(\varrho-\varepsilon)^{p+1}}{\varepsilon}\right).$$

Thus, passing to the limsup, as $\varepsilon \to 0^+$, we get

$$\limsup_{\varepsilon\to 0^+} \Lambda_\lambda(\varepsilon,\varrho) < \kappa_s^{\frac{1}{2}} \kappa_{1,\gamma} |\lambda\alpha|_{L^{\frac{\gamma 2_s^\sharp}{\gamma 2_s^\sharp - 1}}(0,T)^N} + \kappa_s^{\frac{1-p}{2}} \kappa_{2,\gamma}^{p+1} |\lambda\beta|_{L^{\frac{1}{1-\gamma}}(0,T)^N} \varrho^p. \tag{10.3.41}$$

Now, consider the real function

$$\varphi_\lambda(\varrho) := \kappa_s^{\frac{1}{2}} \kappa_{1,\gamma} |\lambda\alpha|_{L^{\frac{\gamma 2_s^\sharp}{\gamma 2_s^\sharp - 1}}(0,T)^N} + \kappa_s^{\frac{1-p}{2}} \kappa_{2,\gamma}^{p+1} |\lambda\beta|_{L^{\frac{1}{1-\gamma}}(0,T)^N} \varrho^p - \varrho,$$

for every $\varrho > 0$.

It is easy to see that $\inf_{\varrho>0} \varphi_\lambda(\varrho)$ is attained at

$$\varrho_{0,\lambda} := \kappa_s^{\frac{1}{2}} \left(\lambda p \kappa_{2,\gamma}^{p+1} |\beta|_{L^{\frac{1}{1-\gamma}}(0,T)^N}\right)^{\frac{1}{1-p}}.$$

and, by (10.3.40), one has

$$\inf_{\varrho>0} \varphi_\lambda(\varrho) < 0.$$

Hence, inequality (10.3.41) yields

$$\limsup_{\varepsilon\to 0^+} \Lambda_\lambda(\varepsilon,\varrho_{0,\lambda}) < \varrho_{0,\lambda}.$$

Now, it follows by Lemmas 10.31 and 10.32 that

$$\inf_{w\in\Phi^{-1}([0,\varrho_{0,\lambda}))} \frac{\sup_{v\in\Phi^{-1}([0,\varrho_{0,\lambda}])} \Psi_\lambda(v) - \Psi_\lambda(w)}{\varrho_{0,\lambda}^2 - \|w\|_{\mathbb{X}_T^s}^2} < \frac{1}{2},$$

which implies that there exists $w_\lambda \in \mathbb{X}_T^s$ such that

$$\Psi_\lambda(w) \leqslant \sup_{v\in\Phi^{-1}([0,\varrho_{0,\lambda}])} \Psi_\lambda(v) < \Psi_\lambda(w_\lambda) + \frac{1}{2}(\varrho_{0,\lambda}^2 - \|w_\lambda\|_{\mathbb{X}_T^s}^2),$$

for every $w \in \Phi^{-1}([0,\varrho_{0,\lambda}])$.

Thus

$$\mathcal{J}_\lambda(w_\lambda) := \frac{1}{2}\|w_\lambda\|_{\mathbb{X}_T^s}^2 - \Psi_\lambda(w_\lambda) < \frac{\varrho_{0,\lambda}^2}{2} - \Psi_\lambda(w), \tag{10.3.42}$$

for every $w \in \Phi^{-1}([0,\varrho_{0,\lambda}])$.

Since the energy functional $\mathcal{J}_\lambda$ is sequentially weakly lower semicontinuous, its restriction on $\Phi^{-1}([0,\varrho_{0,\lambda}])$ has a global minimum $v_{0,\lambda} \in \Phi^{-1}([0,\varrho_{0,\lambda}])$.

Note that $v_{0,\lambda}$ belongs to $\Phi^{-1}([0,\varrho_{0,\lambda}))$. Indeed, if $\|v_{0,\lambda}\|_{\mathbb{X}_T^s} = \varrho_{0,\lambda}$, by (10.3.42), one has

$$\mathcal{J}_\lambda(v_{0,\lambda}) = \frac{\varrho_{0,\lambda}^2}{2} - \Psi_\lambda(v_{0,\lambda}) > \mathcal{J}_\lambda(w_\lambda),$$

which is a contradiction.

In conclusion, it follows that $v_{0,\lambda} \in \mathbb{X}_T^s$ is a local minimum for the energy functional $\mathcal{J}_\lambda$ with

$$\|v_{0,\lambda}\|_{\mathbb{X}_T^s} < \varrho_{0,\lambda},$$

and so, a weak solution of problem (10.3.31). This completes the proof.

A simple special case of Theorem 10.30 reads as follows:

Corrolary 10.33. *Let $f : \mathbb{R} \to \mathbb{R}$ be a continuous function such that*

$$\sup_{t\in\mathbb{R}} \frac{|f(t)|}{1+|t|^p} \leqslant \kappa, \tag{10.3.43}$$

where $p \in (1, \gamma 2_s^\sharp - 1)$, with $\gamma \in (2/2_s^\sharp, 1)$. Assume that

$$0 < \kappa < \frac{(p-1)^{\frac{p-1}{p}}}{p(\kappa_s^{\frac{1}{2}}\kappa_{1,\gamma})^{\frac{p-1}{p}}(\kappa_s^{\frac{1-p}{2}}\kappa_{2,\gamma})^{\frac{p+1}{p}}} T^{N\left(\frac{1-\gamma}{p}+\frac{(1-p)(\gamma 2_s^\sharp - 1)}{p\gamma 2_s^\sharp}\right)}, \tag{10.3.44}$$

where $\kappa_{1,\gamma}$ and $\kappa_{2,\gamma}$ denote the embedding constants of the Sobolev space $\mathbb{H}_T^s$ in $L^{\gamma 2_s^\sharp}(0,T)^N$ and $L^{\frac{p+1}{\gamma}}(0,T)^N$, respectively.

Then, the following nonlocal problem

$$\begin{cases} (-\Delta + m^2)^s u = f(u) & \text{in } (0,T)^N \\ u(x+Te_i) = u(x) & \text{for all } x \in \mathbb{R}^N \text{ and } i = 1,\dots,N \end{cases},$$

has a weak solution $u_{0,\kappa} \in \mathbb{H}_T^s$ such that

$$\|u_{0,\kappa}\|_{\mathbb{H}_T^s} < \left(\kappa p \kappa_{2,\gamma}^{p+1} T^{1-\gamma}\right)^{\frac{1}{1-p}}.$$

A direct consequence of Theorem 10.30 in the Euclidean setting has been proved in [16] by exploiting the variational principle obtained by Ricceri in [103].

10.4 Multiple Solutions

The purpose of this section is to study the existence of multiple solutions for nonlocal periodic equations involving a nonlinear term satisfying different growth conditions.

10.4.1 Periodic equations sublinear at infinity

Under the variational viewpoint, we study the existence and non-existence of weak solutions to the following fractional problem

$$\begin{cases} (-\Delta + m^2)^s u = \lambda\beta(x) f(u) & \text{in } (0,T)^N \\ u(x+Te_i) = u(x) & \text{for all } x \in \mathbb{R}^N \text{ and } i = 1,\dots,N \end{cases}, \tag{10.4.1}$$

where λ is a positive real parameter and $\beta \in L^\infty(\mathbb{R}^N)$ is a T–periodic function satisfying

$$\inf_{x\in(0,T)^N} \beta(x) > 0. \tag{10.4.2}$$

In this context, regarding the nonlinear term, we hypothesize that $f : \mathbb{R} \to \mathbb{R}$ is continuous, *superlinear at zero*, i.e.

$$\lim_{t\to 0} \frac{f(t)}{t} = 0, \tag{10.4.3}$$

sublinear at infinity, i.e.

$$\lim_{|t|\to +\infty} \frac{f(t)}{t} = 0, \tag{10.4.4}$$

and such that

$$\sup_{t\in\mathbb{R}} F(t) > 0, \tag{10.4.5}$$

where

$$F(t) := \int_0^t f(w)dw,$$

for any $t \in \mathbb{R}$.

Assumptions (10.4.3) and (10.4.4) are quite standard in presence of subcritical terms; moreover, together with (10.4.5), they assure that the number

$$c_f := \max_{t\in\mathbb{R}\setminus\{0\}} \frac{|f(t)|}{|t|}. \tag{10.4.6}$$

is well–defined and strictly positive. Further, property (10.4.3) is a sublinear growth condition at infinity on the nonlinearity f which complements the classical Ambrosetti–Rabinowitz assumption.

We point out that a similar variational approach we will employ has been extensively used in several contexts, in order to prove multiplicity results of different problems, such as elliptic problems on either bounded or unbounded domains of the Euclidean space (see [69]), elliptic equations involving the Laplace–Beltrami operator on Riemannian manifold (see [67]), and, more recently, elliptic equations on the ball endowed with Funk–type metrics (see [68]). See also [15, 90] for related topics.

Let us consider $\lambda_1 = m^{2s}$ be the first eigenvalue of the linear problem

$$\begin{cases} (-\Delta + m^2)^s u = \lambda u & \text{in } (0,T)^N \\ u(x + Te_i) = u(x) & \text{for all } x \in \mathbb{R}^N \text{ and } i = 1, \dots, N \end{cases}. \tag{10.4.7}$$

The main result of the present paper is an existence theorem for equations driven by the s–power of $-\Delta + m^2$, as stated here below.

Theorem 10.34. *Let $\beta : \mathbb{R}^N \to \mathbb{R}$ be a T–periodic function satisfying* (10.4.2) *and $f : \mathbb{R} \to \mathbb{R}$ be a continuous function verifying* (10.4.3)–(10.4.5). *Then, the following conclusions hold:*

(*i*) *problem* (10.4.1) *admits only the identically zero solution whenever*

$$0 \leqslant \lambda < \frac{m^{2s}}{c_f \|\beta\|_\infty};$$

(*ii*) *there exists* $\lambda^\star > 0$ *such that problem* (10.4.1) *admits at least two distinct and not identically zero weak solutions* $u_{1,\lambda}, u_{2,\lambda} \in \mathbb{H}^s_T$, *provided that* $\lambda > \lambda^\star$.

Furthermore, in the sequel we will give additional information as far as the localization of the parameter $\lambda^\star$ is concerned. More precisely, we show that

$$\lambda^\star \in \left[\frac{m^{2s}}{c_f \|\beta\|_\infty}, \lambda_0\right],$$

where λ_0 is defined in formula (10.4.26) in Remark 10.37.

Theorem 10.34 will be proved by exporting classical variational techniques to the fractional framework. By studying the extended problem to the cylinder with the classical minimization techniques in addition to the Mountain Pass Theorem, we are able to prove the existence of at least two weak solutions whenever the parameter λ is sufficiently large (for instance $\lambda > \lambda_0$).

The Main Theorem: non-existence for small λ

Let us prove the item (*i*) of Theorem 10.34.

Arguing by contradiction, suppose that there exists a weak solution $v_0 \in \mathbb{X}^s_T \setminus \{0\}$ (and let $u_0 := \operatorname{Tr}(v_0) \in \mathbb{H}^s_T$) to problem (10.34), i.e.

$$\iint_{\mathcal{S}_T} y^{1-2s}(\nabla v_0 \nabla \varphi + m^2 v_0 \varphi)\,dx dy = \lambda \kappa_s \int_{\partial^0 \mathcal{S}_T} \beta(x) f(\operatorname{Tr}(v_0)) \operatorname{Tr}(\varphi)\,dx, \qquad (10.4.8)$$

for every $\varphi \in \mathbb{X}^s_T$.

Testing (10.4.8) with $\varphi := v_0$, we have

$$\|v_0\|^2_{\mathbb{X}^s_T} = \lambda \kappa_s \int_{\partial^0 \mathcal{S}_T} \beta(x) f(\operatorname{Tr}(v_0)) \operatorname{Tr}(v_0)\,dx, \qquad (10.4.9)$$

and it follows that

$$\begin{aligned}\int_{\partial^0 \mathcal{S}_T} \beta(x) f(\operatorname{Tr}(v_0)) \operatorname{Tr}(v_0)\,dx &\leqslant \int_{\partial^0 \mathcal{S}_T} \beta(x)\,|f(\operatorname{Tr}(v_0)) \operatorname{Tr}(v_0)|\,dx \\ &\leqslant c_f \|\beta\|_\infty |\operatorname{Tr}(v_0)|^2_{L^2(0,T)^N} \\ &\leqslant \frac{c_f}{\kappa_s m^{2s}} \|\beta\|_\infty \|v_0\|^2_{\mathbb{X}^s_T}. \end{aligned} \qquad (10.4.10)$$

In the last inequality we have used the following fact

$$\lambda_1 = \min_{u \in \mathbb{H}^s_T \setminus \{0\}} \frac{|u|^2_{\mathbb{H}^s_T}}{|u|^2_{L^2(0,T)^N}} \leqslant \frac{|u_0|^2_{\mathbb{H}^s_T}}{|u_0|^2_{L^2(0,T)^N}} \leqslant \frac{|v_0|^2_{\mathbb{X}^s_T}}{\kappa_s |\operatorname{Tr}(v_0)|^2_{L^2(0,T)^N}},$$

and the trace inequality. By (10.4.9), (10.4.10) and the assumption on λ we get

$$\|v_0\|^2_{\mathbb{X}^s_T} \leqslant \lambda \frac{c_f}{m^{2s}} \|\beta\|_\infty \|v_0\|^2_{\mathbb{X}^s_T} < \|v_0\|^2_{\mathbb{X}^s_T},$$

clearly a contradiction.

The Main Theorem: multiplicity

We prove that, under our natural assumptions on the nonlinear term f, weak solutions to problem (10.4.1) below do exist. Our method to determine multiple solutions to (10.4.1), consists in applying classical variational methods to the functional $\mathcal{J}_\lambda$ given by

$$\mathcal{J}_\lambda(v) := \Phi(v) - \lambda\Psi(v),$$

where

$$\Phi(v) := \frac{1}{2}\|v\|^2_{\mathbb{X}^s_T},$$

while

$$\Psi(v) := \kappa_s \int_{\partial^0 \mathcal{S}_T} \beta(x) F(\mathrm{Tr}(v))dx,$$

for every $v \in \mathbb{X}^s_T$. Clearly the functional Φ and Ψ are Fréchet differentiable.

Moreover, the functional $\mathcal{J}_\lambda$ is weakly lower semicontinuous on $\mathbb{X}^s_T$. Indeed, the application

$$v \mapsto \int_{\partial^0 \mathcal{S}_T} \beta(x) F(\mathrm{Tr}(v))dx$$

is continuous in the weak topology of $\mathbb{X}^s_T$.

We prove this simple fact as follows. Let $\{v_j\}_{j\in\mathbb{N}}$ be a sequence in $\mathbb{X}^s_T$ such that $v_j \rightharpoonup v_\infty$ weakly in $\mathbb{X}^s_T$. Then, by using Sobolev embedding results and [29, Theorem IV.9], up to a subsequence, $\{\mathrm{Tr}(v_j)\}_{j\in\mathbb{N}}$ converges to $\mathrm{Tr}(v_\infty)$ strongly in $L^\nu(0,T)^N$ and a.e. in $(0,T)^N$ as $j \to +\infty$, and it is dominated by some function $\kappa_\nu \in L^\nu(0,T)^N$ i.e.

$$|\mathrm{Tr}(v_j)(x)| \leqslant \kappa_\nu(x) \quad \text{a.e. } x \in (0,T)^N, \text{ for any } j \in \mathbb{N} \tag{10.4.11}$$

for any $\nu \in [1, 2^\sharp_s)$.

Due to (10.4.4), there exists $c > 0$ such that

$$|f(t)| \leqslant c(1+|t|), \quad (\forall\, t \in \mathbb{R}). \tag{10.4.12}$$

Then, by the continuity of F and (10.4.12) it follows that

$$F(\mathrm{Tr}(v_j)(x)) \to F(\mathrm{Tr}(v_\infty)(x)) \text{ a.e. } x \in (0,T)^N$$

as $j \to +\infty$ and

$$|F(\mathrm{Tr}(v_j)(x))| \leqslant c\left(|\mathrm{Tr}(v_j)(x)| + \frac{1}{2}|\mathrm{Tr}(v_j)(x)|^2\right) \leqslant c\left(\kappa_1(x) + \frac{1}{2}\kappa_2(x)^2\right)$$

a.e. $x \in (0,T)^N$ and for any $j \in \mathbb{N}$.

Hence, by applying the Lebesgue Dominated Convergence Theorem in $L^1(0,T)^N$, we have that

$$\int_{\partial^0 \mathcal{S}_T} \beta(x) F(\mathrm{Tr}(v_j))\, dx \to \int_{\partial^0 \mathcal{S}_T} \beta(x) F(\mathrm{Tr}(v_\infty))\, dx$$

as $j \to +\infty$, that is the map

$$v \mapsto \int_{\partial^0 \mathcal{S}_T} \beta(x) F(\mathrm{Tr}(v))dx$$

is continuous from $\mathbb{X}_T^s$ with the weak topology to $\mathbb{R}$.

On the other hand, the map

$$v \mapsto \iint\limits_{\mathcal{S}_T} y^{1-2s}(|\nabla v|^2 + m^2 v^2)dxdy$$

is lower semicontinuous in the weak topology of $\mathbb{X}_T^s$.

Hence, the functional $\mathcal{J}_\lambda$ is lower semicontinuous in the weak topology of $\mathbb{X}_T^s$.

SUB–QUADRATICITY OF THE POTENTIAL

Set

$$c_\ell := \max_{u \in \mathbb{H}_T^s \setminus \{0\}} \frac{|u|_{L^\ell(0,T)^N}}{|u|^2_{\mathbb{H}_T^s}},$$

for every $\ell \in [1, 2_s^\sharp)$.

Let us prove that, under the hypotheses (10.4.3) and (10.4.4), one has

$$\lim_{\|v\|_{\mathbb{X}_T^s} \to 0} \frac{\Psi(v)}{\|v\|^2_{\mathbb{X}_T^s}} = 0 \quad \text{and} \quad \lim_{\|v\|_{\mathbb{X}_T^s} \to \infty} \frac{\Psi(v)}{\|v\|^2_{\mathbb{X}_T^s}} = 0. \tag{10.4.13}$$

Fix $\varepsilon > 0$; in view of (10.4.3) and (10.4.4), there exists $\delta_\varepsilon \in (0,1)$ such that

$$|f(t)| \leqslant \frac{\varepsilon}{\|\beta\|_\infty} |t|, \tag{10.4.14}$$

for all $0 < |t| \leqslant \delta_\varepsilon$ and $|t| \geqslant \delta_\varepsilon^{-1}$.

Let us fix $q \in (2, 2_s^\sharp)$. Since the function

$$t \mapsto \frac{|f(t)|}{|t|^{q-1}}$$

is bounded on $[\delta_\varepsilon, \delta_\varepsilon^{-1}]$, for some $m_\varepsilon > 0$ and for every $t \in \mathbb{R}$ one has

$$|f(t)| \leqslant \frac{\varepsilon}{\|\beta\|_\infty} |t| + m_\varepsilon |t|^{q-1}. \tag{10.4.15}$$

As a byproduct, inequality (10.4.15), in addition to the trace inequality, yields

$$|\Psi(v)| \leqslant \kappa_s \int_{\partial^0 \mathcal{S}_T} \beta(x) |F(\mathrm{Tr}(v))| dx$$

$$\begin{aligned}
&\leqslant \kappa_s \int_{\partial^0 \mathbb{S}_T} \beta(x) \left(\frac{\varepsilon}{2\|\beta\|_\infty} |\operatorname{Tr}(v)|^2 + \frac{m_\varepsilon}{q} |\operatorname{Tr}(v)|^q \right) dx \\
&\leqslant \kappa_s \int_{\partial^0 \mathbb{S}_T} \left(\frac{\varepsilon}{2} |\operatorname{Tr}(v)|^2 + \frac{m_\varepsilon}{q} \beta(x) |\operatorname{Tr}(v)|^q \right) dx \\
&\leqslant \kappa_s \frac{\varepsilon}{2} c_2^2 |\operatorname{Tr}(v)|^2_{\mathbb{H}^s_T} + \kappa_s \frac{m_\varepsilon}{q} \|\beta\|_\infty c_q^q |\operatorname{Tr}(v)|^q_{\mathbb{H}^s_T} \\
&\leqslant \frac{\varepsilon}{2} c_2^2 \|v\|^2_{\mathbb{X}^s_T} + \frac{m_\varepsilon}{q \kappa_s^{q/2-1}} \|\beta\|_\infty c_q^q \|v\|^q_{\mathbb{X}^s_T},
\end{aligned}$$

for every $v \in \mathbb{X}^s_T$.

Therefore, for every $v \in \mathbb{X}^s_T \setminus \{0\}$, it follows that

$$0 \leqslant \frac{|\Psi(v)|}{\|v\|^2_{\mathbb{X}^s_T}} \leqslant \frac{\varepsilon}{2} c_2^2 + \frac{m_\varepsilon}{q \kappa_s^{q/2-1}} \|\beta\|_\infty c_q^q \|v\|^{q-2}_{\mathbb{X}^s_T}.$$

Since $q > 2$ and ε is arbitrary, the first limit of (10.4.13) turns out to be zero.

Now, if $r \in (1, 2)$, due to the continuity of f, there also exists a number $M_\varepsilon > 0$ so that

$$\frac{|f(t)|}{|t|^{r-1}} \leqslant M_\varepsilon,$$

for all $t \in [\delta_\varepsilon, \delta_\varepsilon^{-1}]$, where ε and δ_ε are the numbers previously introduced.

The above inequality, together with (10.4.14), yields

$$|f(t)| \leqslant \frac{\varepsilon}{\|\beta\|_\infty} |t| + M_\varepsilon |t|^{r-1}$$

for each $t \in \mathbb{R}$, and hence

$$\begin{aligned}
|\Psi(v)| &\leqslant \kappa_s \int_{\partial^0 \mathbb{S}_T} \beta(x) |F(\operatorname{Tr}(v))| dx \\
&\leqslant \kappa_s \int_{\partial^0 \mathbb{S}_T} \beta(x) \left(\frac{\varepsilon}{2\|\beta\|_\infty} |\operatorname{Tr}(v)|^2 + \frac{M_\varepsilon}{r} |\operatorname{Tr}(v)|^r \right) dx \\
&\leqslant \kappa_s \int_{\partial^0 \mathbb{S}_T} \left(\frac{\varepsilon}{2} |\operatorname{Tr}(v)|^2 + \frac{M_\varepsilon}{r} \beta(x) |\operatorname{Tr}(v)|^r \right) dx \\
&\leqslant \kappa_s \frac{\varepsilon}{2} c_2^2 |\operatorname{Tr}(v)|^2_{\mathbb{H}^s_T} + \kappa_s \frac{M_\varepsilon}{r} \|\beta\|_\infty c_r^r |\operatorname{Tr}(v)|^r_{\mathbb{H}^s_T} \\
&\leqslant \frac{\varepsilon}{2} c_2^2 \|v\|^2_{\mathbb{X}^s_T} + \frac{M_\varepsilon}{r \kappa_s^{r/2-1}} \|\beta\|_\infty c_r^r \|v\|^r_{\mathbb{X}^s_T},
\end{aligned}$$

for each $v \in \mathbb{X}^s_T$.

Therefore, for every $v \in \mathbb{X}^s_T \setminus \{0\}$, it follows that

$$0 \leqslant \frac{|\Psi(v)|}{\|v\|^2_{\mathbb{X}^s_T}} \leqslant \frac{\varepsilon}{2} c_2^2 + \frac{M_\varepsilon}{r \kappa_s^{r/2-1}} \|\beta\|_\infty c_r^r \|v\|^{r-2}_{\mathbb{X}^s_T}. \tag{10.4.16}$$

Since ε can be chosen as small as we wish and $r \in (1, 2)$, taking the limit as $\|v\|_{\mathbb{X}^s_T} \to +\infty$ in (10.4.16), we have proved the second limit of (10.4.13).

The Palais–Smale condition

Lemma 10.35. *Let $f : \mathbb{R} \to \mathbb{R}$ be a continuous function verifying condition* (10.4.4). *Then, for every $\lambda > 0$, the functional $\mathcal{J}_\lambda$ is bounded from below, coercive and satisfies* (PS).

Proof. Fix $\lambda > 0$ and $0 < \varepsilon < 1/(\lambda c_2^2)$. Due to (10.4.16), one has

$$\begin{aligned}\mathcal{J}_\lambda(w) &\geqslant \frac{1}{2}\|v\|^2_{\mathbb{X}^s_T} - \lambda\kappa_s \int_{\partial^0 \mathcal{S}_T} \beta(x)|F(\mathrm{Tr}(v))|dx \\ &\geqslant \frac{1}{2}\|v\|^2_{\mathbb{X}^s_T} - \lambda\frac{\varepsilon c_2^2}{2}\|v\|^2_{\mathbb{X}^s_T} - \lambda\frac{M_\varepsilon}{r\kappa_s^{r/2-1}}\|\beta\|_\infty c_r^r\|v\|^r_{\mathbb{X}^s_T} \\ &= \frac{1}{2}\left(1 - \lambda c_2^2\varepsilon\right)\|v\|^2_{\mathbb{X}^s_T} - \lambda\frac{M_\varepsilon}{r\kappa_s^{r/2-1}}\|\beta\|_\infty c_r^r\|v\|^r_{\mathbb{X}^s_T},\end{aligned}$$

for every $w \in \mathbb{X}^s_T$. Then the functional $\mathcal{J}_\lambda$ is bounded from below and coercive.

Now, let us prove that $\mathcal{J}_\lambda$ satisfies $(\mathrm{PS})_\mu$ for $\mu \in \mathbb{R}$. To this end, let $\{v_j\}_{j\in\mathbb{N}} \subset \mathbb{X}^s_T$ be a Palais–Smale sequence, i.e.

$$\mathcal{J}_\lambda(v_j) \to \mu \quad \text{and} \quad \|\mathcal{J}_\lambda'(v_j)\|_* \to 0,$$

as $j \to +\infty$ where,

$$\|\mathcal{J}_\lambda'(v_j)\|_* := \sup\left\{|\langle \mathcal{J}_\lambda'(v_j), \varphi\rangle| : \varphi \in \mathbb{X}^s_T, \text{ and } \|\varphi\|_{\mathbb{X}^s_T} = 1\right\}.$$

Taking account the coercivity of $\mathcal{J}_\lambda$, the sequence $\{v_j\}_{j\in\mathbb{N}}$ is necessarily bounded in $\mathbb{X}^s_T$. Since $\mathbb{X}^s_T$ is reflexive, we may extract a subsequence, which for simplicity we still denote $\{v_j\}_{j\in\mathbb{N}}$, such that $v_j \rightharpoonup v_\infty$ in $\mathbb{X}^s_T$, i.e.,

$$\iint_{\mathcal{S}_T} y^{1-2s}(\nabla v_j \nabla\varphi + m^2 v_j\varphi)dxdy \to \iint_{\mathcal{S}_T} y^{1-2s}(\nabla v_\infty \nabla\varphi + m^2 v_\infty\varphi)dxdy, \quad (10.4.17)$$

as $j \to +\infty$, for any $\varphi \in \mathbb{X}^s_T$.

Let us prove that $\{v_j\}_{j\in\mathbb{N}}$ strongly converges to $v_\infty \in \mathbb{X}^s_T$. At this purpose, note that

$$\langle \Phi'(v_j), v_j - v_\infty\rangle = \langle \mathcal{J}_\lambda'(v_j), v_j - v_\infty\rangle + \lambda\kappa_s \int_{\partial^0 \mathcal{S}_T} \beta(x) f(\mathrm{Tr}(v_j))\,\mathrm{Tr}(v_j - v_\infty)dx, \quad (10.4.18)$$

where

$$\begin{aligned}\langle \Phi'(v_j), v_j - v_\infty\rangle &= \iint_{\mathcal{S}_T} y^{1-2s}(|\nabla v_j|^2 + m^2 v_j^2)dxdy \\ &\quad - \iint_{\mathcal{S}_T} y^{1-2s}(\nabla v_j \nabla v_\infty + m^2 v_j v_\infty)dxdy.\end{aligned}$$

Since $\|\mathcal{J}_\lambda'(v_j)\|_* \to 0$ and the sequence $\{v_j - v_\infty\}_{j\in\mathbb{N}}$ is bounded in $\mathbb{X}^s_T$, taking account that $|\langle \mathcal{J}_\lambda'(v_j), v_j - v_\infty\rangle| \leqslant \|\mathcal{J}_\lambda'(v_j)\|_*\|v_j - v_\infty\|_{\mathbb{X}^s_T}$, it is easily seen that

$$\langle \mathcal{J}_\lambda'(v_j), v_j - v_\infty\rangle \to 0 \quad (10.4.19)$$

as $j \to +\infty$.

Furthermore, by (10.4.15) and Hölder's inequality one has

$$\begin{aligned}\int_{\partial^0 \mathcal{S}_T} \beta(x)|f(\mathrm{Tr}(v_j))||\,\mathrm{Tr}(v_j - v_\infty)|dx &\leqslant \varepsilon \int_{\partial^0 \mathcal{S}_T} |\,\mathrm{Tr}(v_j)||\,\mathrm{Tr}(v_j - v_\infty)|dx \\ &\quad + m_\varepsilon \|\beta\|_\infty \int_{\partial^0 \mathcal{S}_T} |\,\mathrm{Tr}(v_j)|^{q-1}|\,\mathrm{Tr}(v_j - v_\infty)|dx \\ &\leqslant \varepsilon \left|\mathrm{Tr}(v_j)\right|_{L^2(0,T)^N} \left|\mathrm{Tr}(v_j - v_\infty)\right|_{L^2(0,T)^N} \\ &\quad + m_\varepsilon \|\beta\|_\infty \left|\mathrm{Tr}(v_j)\right|^{q-1}_{L^q(0,T)^N} \left|\mathrm{Tr}(v_j - v_\infty)\right|_{L^q(0,T)^N}.\end{aligned}$$

Since ε is arbitrary and the embedding $\mathrm{Tr}(\mathbb{X}^s_T) \hookrightarrow L^q(0,T)^N$ compact, we obtain

$$\int_{\partial^0 \mathcal{S}_T} \beta(x)|f(\mathrm{Tr}(v_j))||\,\mathrm{Tr}(v_j - v_\infty)|dx \to 0, \tag{10.4.20}$$

as $j \to +\infty$.

Relations (10.4.18), (10.4.19) and (10.4.20) yield

$$\langle \Phi'(v_j), v_j - v_\infty \rangle \to 0, \tag{10.4.21}$$

as $j \to +\infty$ and hence

$$\iint_{\mathcal{S}_T} y^{1-2s}(|\nabla v_j|^2 + m^2 v_j^2)dxdy - \iint_{\mathcal{S}_T} y^{1-2s}(\nabla v_j \nabla v_\infty + m^2 v_j v_\infty)dxdy \to 0, \tag{10.4.22}$$

as $j \to +\infty$.

Thus, by (10.4.22) and (10.4.17) it follows that

$$\lim_{j\to+\infty} \iint_{\mathcal{S}_T} y^{1-2s}(|\nabla v_j|^2 + m^2 v_j^2)dxdy = \iint_{\mathcal{S}_T} y^{1-2s}(|\nabla v_\infty|^2 + m^2 v_\infty^2)dxdy.$$

In conclusion, thanks to [29, Proposition III.30], $v_j \to v_\infty$ in $\mathbb{X}^s_T$ and the proof is complete. □

The following technical lemma will be useful in the proof of our result via minimization procedure.

Lemma 10.36. *Let $f : \mathbb{R} \to \mathbb{R}$ be a continuous function satisfying condition* (10.4.5). *Then, there exists $t_0 \in \mathbb{R}$ such that $\Psi(w_{t_0}) > 0$, where the function w_{t_0} is defined as in* (10.2.7) *with t_0 instead of σ.*

Proof. By condition (10.4.5) there exists $t_0 \in \mathbb{R}$ such that $F(t_0) > 0$. Hence, bearing in mind (10.4.2) and by using (10.2.8) it follows that

$$\Psi(w_{t_0}) := \kappa_s \int_{\partial^0 \mathcal{S}_T} \beta(x)F(\mathrm{Tr}(w_{t_0}))\,dx \geqslant \kappa_s T^N \inf_{x\in(0,T)^N} \beta(x)F(t_0) > 0. \tag{10.4.23}$$

The proof is complete. □

Now, let us prove the item (*ii*) of Theorem 10.34.

First solution via direct minimization

The assumptions on β and (10.4.5) imply that there exists a suitable function $w_{t_0} \in \mathbb{X}_T^s \setminus \{0\}$ so that $\Psi(w_{t_0}) > 0$, and thus the number

$$\lambda^\star := \inf_{\substack{\Psi(v)>0 \\ v \in \mathbb{X}_T^s}} \frac{\Phi(v)}{\Psi(v)} \tag{10.4.24}$$

is well–defined and, in the light of (10.4.13), positive and finite.

Fixing $\lambda > \lambda^\star$ and choosing $v_\lambda^\star \in \mathbb{X}_T^s$ with $\Psi(v_\lambda^\star) > 0$ and

$$\lambda^\star \leqslant \frac{\Phi(v_\lambda^\star)}{\Psi(v_\lambda^\star)} < \lambda,$$

one has

$$c_{1,\lambda} := \inf_{v \in \mathbb{X}_T^s} \mathcal{J}_\lambda(v) \leqslant \mathcal{J}_\lambda(v_\lambda^\star) < 0.$$

Since $\mathcal{J}_\lambda$ is bounded from below and satisfies $(\mathrm{PS})_{c_{1,\lambda}}$, then $c_{1,\lambda}$ is a critical value of $\mathcal{J}_\lambda$, to wit, there exists $v_{1,\lambda} \in \mathbb{X}_T^s \setminus \{0\}$ such that

$$\mathcal{J}_\lambda(v_{1,\lambda}) = c_{1,\lambda} < 0 \quad \text{and} \quad \mathcal{J}_\lambda'(v_{1,\lambda}) = 0.$$

This is the first solution sought for.

Second solution via MPT

The nonlocal analysis that we perform in order to use Mountain Pass Theorem is quite general and may be suitable for other goals too. Our proof will check that the classical geometry of the Mountain Pass Theorem is respected by the nonlocal framework.

Fix $\lambda > \lambda^\star$, where $\lambda^\star$ is defined by (10.4.24), and apply (10.4.15) with $\varepsilon := 1/(2\lambda c_2^2)$. For each $w \in \mathbb{X}_T^s$ one has

$$\begin{aligned}
\mathcal{J}_\lambda(v) &= \frac{1}{2}\|v\|_{\mathbb{X}_T^s}^2 - \lambda\Psi(v) \\
&\geqslant \frac{1}{2}\|v\|_{\mathbb{X}_T^s}^2 - \frac{\lambda}{2}\varepsilon\kappa_s\,|\mathrm{Tr}(v)|_{L^2(0,T)^N}^2 - \kappa_s\frac{\lambda}{q}\|\beta\|_\infty\, m_\lambda\,|\mathrm{Tr}(v)|_{L^q(0,T)^N}^q \\
&\geqslant \frac{1-\lambda\varepsilon c_2^2}{2}\|v\|_{\mathbb{X}_T^s}^2 - \frac{\lambda}{q\kappa_s^{q/2-1}}\|\beta\|_\infty\, m_\lambda c_q^q\,\|v\|_{\mathbb{X}_T^s}^q\,.
\end{aligned}$$

Setting

$$r_\lambda := \min\left\{\|v_\lambda^\star\|_{\mathbb{X}_T^s}, \left(\frac{4q\kappa_s^{q/2-1}}{\lambda\|\beta\|_\infty\, m_\lambda c_q^q}\right)^{1/(q-2)}\right\},$$

due to what seen before, we get

$$\inf_{\|v\|_{\mathbb{X}_T^s}=r_\lambda} \mathcal{J}_\lambda(v) > 0 = \mathcal{J}_\lambda(0) > \mathcal{J}_\lambda(v_\lambda^\star),$$

namely the energy functional possesses the usual mountain pass geometry.

Then, invoking also Lemma 10.35, we can apply the Mountain Pass Theorem to deduce the existence of $v_{2,\lambda} \in \mathbb{X}^s_T$ so that $\mathcal{J}'_\lambda(v_{2,\lambda}) = 0$ and $\mathcal{J}_\lambda(v_{2,\lambda}) = c_{2,\lambda}$, where $c_{2,\lambda}$ has the well-known characterization:

$$c_{2,\lambda} := \inf_{\gamma\in\Gamma} \max_{t\in[0,1]} \mathcal{J}_\lambda(\gamma(t)),$$

where

$$\Gamma := \left\{\gamma \in C^0([0,1]; \mathbb{X}^s_T) : \gamma(0) = 0, \gamma(1) = v^\star_\lambda\right\}.$$

Since

$$c_{2,\lambda} \geqslant \inf_{\|v\|_{\mathbb{X}^s_T}=r_\lambda} \mathcal{J}_\lambda(v) > 0,$$

we have $0 \neq v_{2,\lambda} \neq v_{1,\lambda}$ and the existence of two distinct non-zero weak solutions to (10.4.1) is proved. In conclusion $\mathrm{Tr}(v_{1,\lambda})$ and $\mathrm{Tr}(v_{2,\lambda})$ are two distinct non-zero weak solutions to (10.4.1). The proof is now complete.

Remark 10.37. The proof of Theorem 10.34 gives an exact, but quite involved form of the parameter $\lambda^\star$. In particular we notice that

$$\lambda^\star := \inf_{\substack{\Psi(v)>0\\ v\in\mathbb{X}^s_T}} \frac{\Phi(v)}{\Psi(v)} \geqslant \frac{m^{2s}}{c_f\,\|\beta\|_\infty}. \tag{10.4.25}$$

Indeed, by (10.4.6), one clearly have

$$|f(t)| \leqslant c_f|t|, \quad \forall\, t \in \mathbb{R}.$$

Moreover, since

$$|\,\mathrm{Tr}(v)|_2^2 \leqslant \frac{1}{\kappa_s m^{2s}}\|v\|^2_{\mathbb{X}^s_T}, \quad \forall\, v \in \mathbb{X}^s_T$$

it follows that

$$\begin{aligned}\Psi(v) &\leqslant \kappa_s \int_{(0,T)^N} \beta(x)|F(\mathrm{Tr}(v))|dx\\ &\leqslant c_f\kappa_s \frac{\|\beta\|_\infty}{2}|\,\mathrm{Tr}(v)|^2_{L^2(0,T)^N}\\ &\leqslant c_f\frac{\|\beta\|_\infty}{2m^{2s}}\|v\|^2_{\mathbb{X}^s_T},\end{aligned}$$

for every $v \in \mathbb{X}^s_T$. Hence, inequality (10.4.25) immediately holds. We point out that no information is available concerning the number of solutions of problem (10.4.1) if

$$\lambda \in \left[\frac{m^{2s}}{c_f\,\|\beta\|_\infty}, \lambda^\star\right].$$

Since the expression of $\lambda^\star$ is involved, we give in the sequel an upper estimate of it which can be easily calculated. This fact can be done in terms of the some analytical and geometrical constants.

Due to (10.4.24) one has

$$\lambda^\star \leqslant \frac{\Phi(w_{t_0})}{\Psi(w_{t_0})},$$

where the test function w_{t_0} is defined in Section 10.2, and $t_0 \in \mathbb{R}$ with $F(t_0) > 0$.

More precisely, by using (10.2.9), one has $\lambda^\star \leqslant \lambda_0$, where

$$\lambda_0 := \frac{1}{2}\left(\frac{m^{2s}}{\inf\limits_{x\in(0,T)^N}\beta(x)}\right)\frac{t_0^2}{F(t_0)}. \tag{10.4.26}$$

Now, the conclusions of Theorem 10.34 are valid for every $\lambda > \lambda_0$.

Finally, we point out that, by using [107, Theorem 1] and Remark 10.37, the following result can be proved.

Theorem 10.38. *Let $f : \mathbb{R}^N \times \mathbb{R} \to \mathbb{R}$ be a Carathéodory T–periodic (with respect to $x \in \mathbb{R}^N$) function such that (10.4.3) and (10.4.4) hold and $g : \mathbb{R}^N \times \mathbb{R} \to \mathbb{R}$ be a Carathéodory T–periodic (with respect to $x \in \mathbb{R}^N$) function such that*

$$\begin{aligned}&\text{there exist } a_1, a_2 > 0 \text{ and } q \in (2, 2_s^\sharp),\, 2_s^\sharp := 2N/(N-2s)\,, \text{ such that}\\ &\qquad |g(x,t)| \leqslant a_1 + a_2|t|^{q-1} \text{ for every } x \in (0,T)^N, t \in \mathbb{R}.\end{aligned} \tag{10.4.27}$$

Furthermore, set

$$\lambda^\star := \frac{1}{2}\left(\frac{m^{2s}}{\inf\limits_{x\in(0,T)^N}\beta(x)}\right)\frac{t_0^2}{F(t_0)},$$

where $t_0 \in \mathbb{R}$ is such that $F(t_0) > 0$.

Then, for each compact interval $[a, b] \subset (\lambda^\star, +\infty)$, there exists $\gamma > 0$ with the following property:

for every $\lambda \in [a, b]$, there exists $\delta > 0$ such that, for each $\mu \in [0, \delta]$, the perturbed problem

$$\begin{cases} (-\Delta + m^2)^s u = \beta(x) f(u) + \mu g(x,u) & \text{in } (0,T)^N \\ u(x + Te_i) = u(x) & \text{for all } x \in \mathbb{R}^N \text{ and } i = 1, \dots, N \end{cases},$$

admits at least three weak solutions in $\mathbb{H}_T^s \cap L^\infty(0,T)^N$, whose $\mathbb{H}_T^s$–norms are less than γ.

Remark 10.39. Theorem 10.38 gives a more precise information as the one stated in Theorem 10.34; indeed, it shows that the multiplicity result described below is stable with respect to small periodic subcritical perturbations g. We also note that, a sharp value for the parameter $\lambda^\star$ that appears in Theorem 10.38 is given by:

$$\lambda_\sharp^\star := \frac{1}{2}\left(\frac{m^{2s}}{\inf\limits_{x\in(0,T)^N}\beta(x)}\right)\inf_{\sigma\in S_f}\frac{\sigma^2}{F(\sigma)},$$

where

$$S_f := \left\{\sigma \in \mathbb{R} : \int_0^\sigma f(t)dt > 0\right\}.$$

See Section 10.5 and [89] for related topics.

In the next result, in connection with Theorem 10.38, by adapting the technical approach developed in [56, Theorem 7] to the nonlocal periodic setting, we prove the existence of at least two weak solutions for nonlocal periodic problems under a slight different assumption at infinity on the nonlinear term f and in presence of a sufficiently small periodic perturbation term g.

Theorem 10.40. *Let $f : \mathbb{R}^N \times \mathbb{R} \to \mathbb{R}$ be a Carathéodory T–periodic function (with respect to $x \in \mathbb{R}^N$) such that:*

(j_1) *the following limit holds*

$$\limsup_{|t|\to+\infty} \frac{F(x,t)}{t^2} < \frac{1}{2}m^{2s}$$

uniformly in $x \in (0,T)^N$;

(j_2) *for every $M > 0$, $\sup_{|t|\leqslant M} |F(\cdot,t)| \in L^1(0,T)^N$.*

Furthermore, suppose that there exist positive constants c_0, σ and $d \in \mathbb{R}\setminus\{0\}$ such that:

(j_3) *$|F(x,t)| \leqslant c_0 t^2$ uniformly in $x \in (0,T)^N$ and for every $t \in \mathbb{R}$;*

(j_4) *$F(x,t) \leqslant 0$ for every $t \in (-\sigma,\sigma)$, uniformly in $x \in (0,T)^N$;*

(j_5) *the inequality*

$$\frac{\displaystyle\int_{(0,T)^N} F(x,d)dx}{d^2} > \frac{1}{2}m^{2s}T^N$$

holds.

Finally, let $g : \mathbb{R}^N \times \mathbb{R} \to \mathbb{R}$ be a Carathéodory T–periodic (with respect to $x \in \mathbb{R}^N$) function such that

(j_6) *there exist $\varrho > 1$ and $\alpha > 0$, $\beta \in L^1(0,T)^N$, with $\varrho < 2_s^\sharp$ for which*

$$\left|\int_0^t g(x,w)dw\right| \leqslant \alpha|t|^\varrho + \beta(x), \quad \forall\,(x,t) \in (0,T)^N \times \mathbb{R}.$$

Then, for every $r > 0$, there exists $\lambda_r > 0$ such that, for every $\lambda \in (0,\lambda_r)$, the following problem

$$\begin{cases} \displaystyle\iint_{\mathcal{S}_T} y^{1-2s}(\nabla v \nabla\varphi + m^2 v\varphi)\,dxdy \\ \qquad = \kappa_s \displaystyle\int_{\partial^0\mathcal{S}_T} f(x,\mathrm{Tr}(v))\mathrm{Tr}(\varphi)dx + \lambda\kappa_s \int_{\partial^0\mathcal{S}_T} g(x,\mathrm{Tr}(v))\mathrm{Tr}(\varphi)dx, \quad \forall\,\varphi \in \mathbb{X}_T^s \\ v \in \mathbb{X}_T^s, \end{cases} \tag{10.4.28}$$

has at least two solutions in $\Psi^{-1}((-\infty,r))$, where

$$\Psi(v) := \frac{1}{2}\|v\|_{\mathbb{X}_T^s}^2 - \kappa_s \int_{\partial^0\mathcal{S}_T} F(x,\mathrm{Tr}(v))dx, \quad \forall\, v \in \mathbb{X}_T^s.$$

In order to prove the result, stated in Theorem 10.40, the main tool will be a critical point theorem due to Ricceri [105, Theorem 10]. For the sake of clarity, we recall it here below:

Theorem 10.41. *Let $(E, \|\cdot\|)$ be a uniformly convex and separable real Banach space. Furthermore, let $J, \Phi : E \to \mathbb{R}$ two sequentially weakly lower semicontinuous functionals, with J also strongly continuous.*

For every $z \in E$, put

$$\Psi(z) := \frac{\|z\|^2}{2} + J(z).$$

Assume that Ψ is coercive and has a strict local (not global) minimum, say z_0.

Then, for every $r > \Psi(z_0)$, there exists $\lambda_r^\star > 0$ such that, for each $\lambda \in (0, \lambda_r^\star)$, the functional $\Psi + \lambda\Phi$ has at least two local minima lying in $\Psi^{-1}((-\infty, r))$.

Proof of Theorem 10.40 - For every $v \in \mathbb{X}_T^s$, let us put

$$J(v) := -\kappa_s \int_{\partial^0 \mathcal{S}_T} F(x, \mathrm{Tr}(v))dx,$$

$$\Psi(v) := \frac{1}{2}\|v\|_{\mathbb{X}_T^s}^2 + J(v),$$

and

$$\Phi(v) := -\kappa_s \int_{\partial^0 \mathcal{S}_T} G(x, \mathrm{Tr}(v))dx,$$

where

$$G(x, t) := \int_0^t g(x, w)dw.$$

First of all, we prove that the following facts hold:

(p_1) the functional Ψ is coercive;
(p_2) $\inf_{v \in \mathbb{X}_T^s} \Psi < 0$;
(p_3) Ψ has a strict local minimum at zero.

More precisely, one has:

(p_1) - By (j_1) and (j_2) it follows that for a.e. $x \in (0, T)^N$ and $t \in \mathbb{R}$ one has

$$F(x, t) \leqslant \overline{\lambda} t^2 + k(x), \tag{10.4.29}$$

for some $k \in L^1(0, T)^N$ and $\overline{\lambda} < m^{2s}$. Thus, inequality (10.4.29) yields

$$\int_{\partial^0 \mathcal{S}_T} F(x, \mathrm{Tr}(v))dx < \frac{\overline{\lambda}}{2\kappa_s m^{2s}}\|v\|_{\mathbb{X}_T^s}^2 + |k|_{L^1(0,T)^N}.$$

Hence,

$$\Psi(v) > \frac{1}{2}\left(1 - \frac{\overline{\lambda}}{m^{2s}}\right)\|v\|_{\mathbb{X}_T^s}^2 + \kappa_s^{-1}|k|_{L^1(0,T)^N},$$

for every $v \in \mathbb{X}_T^s$, and $\lim_{\|v\|_{\mathbb{X}_T^s} \to +\infty} \Psi(v) = +\infty$.

(p_2) - Let us define

$$w_d(x,y) := \frac{2d}{\Gamma(s)} \left(\frac{my}{2}\right)^s K_s(my), \quad \forall (x,y) \in \mathbb{R}_+^{N+1}, \tag{10.4.30}$$

where K_s is the Bessel function of the second kind of order s (see Subsection 10.2.1). Hence, $w_d \in \mathbb{X}_T^s$ and by (10.2.9) and (10.2.8) it follows that

$$\Psi(w_d) = \left(\frac{\xi_j^2}{2} m^{2s} T^N - \int_{(0,T)^N} F(x,d)dx\right) \kappa_s.$$

Then, by (j_5), one has $\Psi(w_d) < 0$ and so $\inf_{v \in \mathbb{X}_T^s} \Psi < 0$.

(p_3) - For every $v \in \mathbb{X}_T^s$, let

$$(0,T)_{v,\sigma}^N := \{x \in (0,T)^N : |\operatorname{Tr}(v)(x)| > \sigma\}.$$

If $x \in (0,T)^N \setminus (0,T)_{v,\sigma}^N$, then $|\operatorname{Tr}(v)(x)| \leqslant \sigma$ and, by (j_4), one has $F(x, \operatorname{Tr}(v)(x)) \leqslant 0$. Thus, (j_3) yields

$$\int_{(0,T)^N} F(x, \operatorname{Tr}(v))dx \leqslant c_0 |(0,T)_{v,\sigma}^N|^{1-\frac{2}{2_s^\sharp}} |\operatorname{Tr}(v)|_{2_s^\sharp}^2. \tag{10.4.31}$$

Now, by (10.4.31), we can write

$$\Psi(v) \geqslant \left(\frac{1}{2} - c_0 c_{2_s^\sharp} |(0,T)_{v,\sigma}^N|^{1-\frac{2}{2_s^\sharp}}\right) \|v\|_{\mathbb{X}_T^s}^2,$$

for every $v \in \mathbb{X}_T^s$. Moreover, the right–hand side of the above inequality goes to zero as $\|v\|_{\mathbb{X}_T^s} \to 0$. Indeed, by (10.2.20), it follows that

$$|(0,T)_{v,\sigma}^N| \leqslant \frac{1}{m^{2s}\sqrt{\kappa_s}\sigma^2} \|v\|_{\mathbb{X}_T^s}^2.$$

The conclusion is attained.

Finally, by (j_1) the functional J is sequentially weakly continuous and, by (j_6), Φ is weakly lower semicontinuous. Hence, all the assumptions of Theorem 10.41 are satisfied. Then, for every $r > \Psi(z_0)$, there exists $\lambda_r^\star > 0$ such that, for each $\lambda \in (0, \lambda_r^\star)$, the functional $\Psi + \lambda\Phi$ has at least two local minima lying in $\Psi^{-1}((-\infty, r))$. The proof of Theorem 10.41 is now complete.

10.4.2 On the Ambrosetti–Rabinowitz condition

The aim of this subsection is to prove that under the Ambrosetti–Rabinowitz condition on the nonlinear term f, there exist two (weak) solutions to the problem

$$\begin{cases} (-\Delta + m^2)^s u = \gamma u + \lambda f(x,u) & \text{in } (0,T)^N \\ u(x + Te_i) = u(x) & \text{for all } x \in \mathbb{R}^N \text{ and } i = 1, \dots, N \end{cases}, \tag{10.4.32}$$

where $\gamma < m^{2s}$ and $\lambda > 0$ is sufficiently large. The results presented here are based on the paper [13].

Let us introduce the following functional $\mathcal{J}_\lambda : \mathbb{X}^s_T \to \mathbb{R}$ defined by

$$\mathcal{J}_\lambda(v) := \frac{1}{2\lambda}\left(\|v\|^2_{\mathbb{X}^s_T} - \gamma\kappa_s|\mathrm{Tr}(v)|^2_{L^2(0,T)^N}\right) - \kappa_s\int_{\partial^0\mathcal{S}_T} F(x, \mathrm{Tr}(v))dx, \tag{10.4.33}$$

where $\lambda > 0$ is fixed.

We will prove that $\mathcal{J}_\lambda$ satisfies the assumptions of the following abstract result:

Theorem 10.42. *Let E be a reflexive real Banach space, and let $\Phi, \Psi : E \to \mathbb{R}$ be two continuously Gâteaux differentiable functionals such that Φ is sequentially weakly lower semicontinuous and coercive. Further, assume that Ψ is sequentially weakly continuous. In addition, assume that, for each $\mu > 0$, the functional $J_\mu := \mu\Phi - \Psi$ satisfies the classical compactness Palais-Smale (briefly (PS)) condition. Then, for each $\varrho > \inf_E \Phi$ and each*

$$\mu > \inf_{u\in\Phi^{-1}((-\infty,\varrho))} \frac{\sup_{v\in\Phi^{-1}((-\infty,\varrho))} \Psi(v) - \Psi(u)}{\varrho - \Phi(u)}$$

the following alternative holds: either the functional J_μ has a strict global minimum which lies in $\Phi^{-1}((-\infty,\varrho))$, or J_μ has at least two critical points one of which lies in $\Phi^{-1}((-\infty,\varrho))$.

For a proof of Theorem 10.42, one can see [102]. We refer also to [87, 108, 111] and references therein for recent applications of Ricceri's variational principle [103].

For our purpose, we assume that the right-hand side in equation (10.4.32) is a continuous function $f : \mathbb{R}^N \times \mathbb{R} \to \mathbb{R}$ verifying the following hypotheses:

(f_1) *$f(x,t)$ is T–periodic in $x \in \mathbb{R}^N$, that is $f(x + Te_i, t) = f(x,t)$ for any $x \in \mathbb{R}^N$, $T > 0$, and $i = 1, \ldots, N$;*

(f_2) *there exist $a_1, a_2 > 0$, and $2 < q < 2^\sharp_s := \dfrac{2N}{N-2s}$ such that*

$$|f(x,t)| \leqslant a_1 + a_2|t|^{q-1}$$

for any $x \in \mathbb{R}^N$ and $t \in \mathbb{R}$;

(f_3) *there exist $\alpha > 2$ and $r_0 > 0$ such that*

$$0 < \alpha F(x,t) \leqslant tf(x,t)$$

for $x \in \mathbb{R}^N$ and $|t| \geqslant r_0$, where $F(x,t) := \int_0^t f(x,w)dw$.

In this context, the main result of the present subsection is a multiplicity theorem as stated here below.

Theorem 10.43. *Let $m > 0$ and $\gamma < m^{2s}$. Let $f : \mathbb{R}^N \times \mathbb{R} \to \mathbb{R}$ be a continuous function satisfying the assumptions (f_1)–(f_3).*

Then, for any $\varrho > 0$ and each

$$0 < \lambda < \frac{q\sqrt{\varrho}\left(1 - \frac{\gamma}{m^{2s}}\right)^{q/2}}{2\kappa_s\left(a_1\sigma_1 q\left(1 - \frac{\gamma}{m^{2s}}\right)^{\frac{q-1}{2}} + a_2\sigma_q^q\varrho^{\frac{q-1}{2}}\right)}, \tag{10.4.34}$$

problem (10.4.5) admits at least two weak solutions in $\mathbb{H}_T^s$, one of which lies in

$$\mathbb{S}_\varrho := \left\{u \in \mathbb{H}_T^s : |u|_{\mathbb{H}_T^s} < \sqrt{\frac{\varrho}{\kappa_s\left(1 - \frac{\gamma}{m^{2s}}\right)}}\right\},$$

where

$$\kappa_s := 2^{1-2s}\frac{\Gamma(1-s)}{\Gamma(s)},$$

and

$$\sigma_r := \frac{1}{\sqrt{\kappa_s}} \sup_{u \in \mathbb{H}_T^s \setminus \{0\}} \frac{|u|_{L^r(0,T)^N}}{|u|_{\mathbb{H}_T^s}} \quad \text{with } r \in \{1, q\}.$$

Proof. By using (f_2) and Theorem 10.2, it follows that $\mathcal{J}_\lambda$ is well defined, $\mathcal{J}_\lambda \in C^1(\mathbb{X}_T^s; \mathbb{R})$ and

$$\langle \mathcal{J}_\lambda'(v), \varphi\rangle = \frac{1}{\lambda}\left(\iint_{\mathcal{S}_T} y^{1-2s}(\nabla v \nabla \varphi + m^2 v\varphi)\,dxdy - \gamma\kappa_s \int_{\partial^0 \mathcal{S}_T} \operatorname{Tr}(v)\operatorname{Tr}(\varphi)\,dx\right)$$
$$-\kappa_s \int_{\partial^0 \mathcal{S}_T} f(x, \operatorname{Tr}(v))\operatorname{Tr}(\varphi)\,dx,$$

for every $\varphi \in \mathbb{X}_T^s$.

Let $\varrho > 0$ and set $\mu := 1/(2\lambda)$, with λ as in the statement of Theorem 10.43. We set $E := \mathbb{X}_T^s$, $J_\mu := \mathcal{J}_\lambda$ and

$$\Phi(v) := \|v\|_{\mathbb{X}_T^s}^2 - \gamma\kappa_s|\operatorname{Tr}(v)|_{L^2(0,T)^N}^2 =: \|v\|_e^2,$$

as well as

$$\Psi(v) := \kappa_s \int_{\partial^0 \mathcal{S}_T} F(x, \operatorname{Tr}(v))\,dx,$$

for every $v \in \mathbb{X}_T^s$.

Taking into account that

$$\begin{aligned} m^{2s}|u|_{L^2(0,T)^N}^2 &= m^{2s}\sum_{k\in\mathbb{Z}^N} |c_k|^2 \\ &\leqslant \sum_{k\in\mathbb{Z}^N} (\omega^2|k|^2 + m^2)^s|c_k|^2 = |u|_{\mathbb{H}_T^s}^2, \end{aligned}$$

for any $u = \sum_{k\in\mathbb{Z}^N} c_k \frac{e^{\imath\omega k\cdot x}}{\sqrt{T^N}} \in \mathbb{H}^s_T$, by using Theorem 10.1, it is easily seen that

$$\begin{aligned}\|v\|^2_{\mathbb{X}^s_T} - \gamma\kappa_s|\operatorname{Tr}(v)|^2_{L^2(0,T)^N} &= \|v\|^2_{\mathbb{X}^s_T} - \frac{\gamma}{m^{2s}}\kappa_s m^{2s}|\operatorname{Tr}(v)|^2_{L^2(0,T)^N}\\ &\geqslant \|v\|^2_{\mathbb{X}^s_T} - \frac{\gamma}{m^{2s}}\kappa_s|\operatorname{Tr}(v)|^2_{\mathbb{H}^s_T}\\ &\geqslant \left(1-\frac{\gamma}{m^{2s}}\right)\|v\|^2_{\mathbb{X}^s_T}.\end{aligned}$$

This fact and $0 < \gamma < m^{2s}$ yield

$$\sqrt{1-\frac{\gamma}{m^{2s}}}\|v\|_{\mathbb{X}^s_T} \leqslant \|v\|_e \leqslant \|v\|_{\mathbb{X}^s_T} \tag{10.4.35}$$

for every $v \in \mathbb{X}^s_T$, so $\|\cdot\|_e$ is equivalent to the standard norm $\|\cdot\|_{\mathbb{X}^s_T}$.

Then, Φ is sequentially weakly lower semicontinuous and coercive, and Ψ is sequentially weakly continuous thanks to Theorem 10.2. Now, by using the AR–condition (f_3), it is easily seen that there exists $v_0 \in \mathbb{X}^s_T$ such that

$$J_\mu(tv_0) \to -\infty \tag{10.4.36}$$

as $t \to +\infty$. Hence, the functional J_μ is unbounded from below.

In order to apply our abstract result, we observe that the functional $\mathcal{J}_\lambda$ satisfies the compactness (PS) condition (see [13] for details).

Now, we are in the position to apply Theorem 10.42. We aim to prove that

$$\mu > \chi(\varrho) := \inf_{u\in\mathbb{B}_\varrho} \frac{\kappa_s \sup_{v\in\mathbb{B}_\varrho}\int_{\partial^0\mathcal{S}_T} F(x,\operatorname{Tr}(v))dx - \kappa_s\int_{\partial^0\mathcal{S}_T} F(x,\operatorname{Tr}(u))dx}{\varrho - \|u\|^2_e} \tag{10.4.37}$$

for every $\varrho > 0$, where $\mathbb{B}_\varrho := \{v \in \mathbb{X}^s_T : \|v\|_e < \sqrt{\rho}\}$.

Fix $\varrho > 0$. Since $0 \in \mathbb{B}_\varrho$, it follows that

$$\chi(\varrho) \leqslant \frac{\kappa_s \sup_{v\in\mathbb{B}_\varrho}\int_{\partial^0\mathcal{S}_T} F(x,\operatorname{Tr}(v))dx}{\varrho}. \tag{10.4.38}$$

By using (f_2) we can see that

$$\int_{\partial^0\mathcal{S}_T} F(x,\operatorname{Tr}(v))dx \leqslant a_1|\operatorname{Tr}(v)|_{L^1(0,T)^N} + \frac{a_2}{q}|\operatorname{Tr}(v)|^q_{L^q(0,T)^N}$$

for every $v \in \mathbb{X}^s_T$, so, by (10.4.35), we deduce that

$$\sup_{v\in\mathbb{B}_\varrho}\int_{\partial^0\mathcal{S}_T} F(x,\operatorname{Tr}(v))dx \leqslant \sigma_1\sqrt{\varrho}\frac{a_1}{\sqrt{1-\frac{\gamma}{m^{2s}}}} + \frac{\sigma_q^q a_2}{q\left(1-\frac{\gamma}{m^{2s}}\right)^{q/2}}\varrho^{\frac{q}{2}}.$$

This implies that

$$\frac{\sup\limits_{v\in\mathbb{B}_\varrho}\int_{\partial^0\mathbb{S}_T} F(x,\mathrm{Tr}(v))dx}{\varrho} \leqslant \frac{\sigma_1}{\sqrt{\varrho}}\frac{a_1}{\sqrt{1-\frac{\gamma}{m^{2s}}}} + \frac{\sigma_q^q a_2}{q\left(1-\frac{\gamma}{m^{2s}}\right)^{q/2}}\varrho^{\frac{q}{2}-1}. \tag{10.4.39}$$

Since (10.4.34) holds, conditions (10.4.38) and (10.4.39) immediately yield

$$\chi(\varrho) \leqslant \kappa_s\left[\frac{\sigma_1}{\sqrt{\varrho}}\frac{a_1}{\sqrt{1-\frac{\gamma}{m^{2s}}}} + \frac{\sigma_q^q a_2}{q\left(1-\frac{\gamma}{m^{2s}}\right)^{q/2}}\varrho^{\frac{q}{2}-1}\right] < \frac{1}{2\lambda} =: \mu.$$

Thus, inequality (10.4.37) is proved. Then, in view of Theorem 10.42, problem (10.4.5) admits at least two weak solutions one of which lies in $\mathbb{S}_\varrho$. This completes the proof of Theorem 10.43. □

10.4.3 Three solutions for perturbed equations

In this section we are interested in nonlocal equations depending on multiple parameters. In many mathematical problems deriving from applications the presence of one (or more) parameter is a relevant feature, and the study of how solutions depend on parameters is an important topic.

Most of the results in this direction were obtained through bifurcation theory (for an extensive treatment of this matters we refer to [2, 3] and their bibliography). However, some interesting results can be obtained also by means of variational techniques, or as a combination of the two methods.

In the periodic nonlocal framework, the simplest example we can deal with is given by the fractional Laplacian, according to the following result:

Theorem 10.44. *Let*

$$1 < p < q < \frac{N+2s}{N-2s},$$

and $\beta : \mathbb{R}^N \to \mathbb{R}$ be a positive T–periodic continuous function.

Then, for each $\varepsilon > 0$ small enough, there exists $\lambda_\varepsilon > 0$ such that, for every compact interval $[a, b] \subset (0, \lambda_\varepsilon)$, there exists $\rho > 0$ with the following property:

for every $\lambda \in [a, b]$ and every continuous function $h : \mathbb{R} \to \mathbb{R}$ satisfying

$$\limsup_{|t|\to+\infty}\frac{|h(t)|}{|t|^\alpha} < +\infty,$$

for some

$$\alpha \in \left(0, \frac{N+2s}{N-2s}\right)$$

there exists $\delta > 0$ such that, for every $\nu \in [0, \delta]$, the nonlocal problem

$$\begin{cases} (-\Delta + m^2)^s u = \varepsilon |u|^{p-1}u - \lambda |u|^{q-1}u + \nu\beta(x)h(u) & \text{in } (0, T)^N \\ u(x + Te_i) = u(x) & \text{for all } x \in \mathbb{R}^N \text{ and } i = 1, \dots, N \end{cases},$$

has at least three distinct weak solutions $\{u_j\}_{j=1}^3 \subset \mathbb{H}_T^s$, such that

$$|u_j|_{\mathbb{H}_T^s} < \frac{\rho}{\sqrt{\kappa_S}},$$

for every $j \in \{1, 2, 3\}$.

We give some details concerning the abstract approach that can be used in order to prove Theorem 10.44 and the more general version given in Theorem 10.47. Let E be a non–empty set and $I, \Psi, \Phi : E \to \mathbb{R}$ three given functions. If $\mu > 0$ and $r \in (\inf_E \Phi, \sup_E \Phi)$, we put

$$\beta(\mu I + \Psi, \Phi, r) := \sup_{z \in \Phi^{-1}((r, +\infty))} \frac{\mu I(z) + \Psi(z) - \inf_{\Phi^{-1}((-\infty, r))} (\mu I + \Psi)}{r - \Phi(z)}.$$

When the map $\Psi + \Phi$ is bounded from below, for each

$$r \in (\inf_E \Phi, \sup_E \Phi)$$

such that

$$\inf_{z \in \Phi^{-1}((-\infty, r])} I(z) \leqslant \inf_{z \in \Phi^{-1}(r)} I(z),$$

we put

$$\mu^\star(I, \Psi, \Phi, r) := \inf \left\{ \frac{\Psi(z) - \gamma + r}{\eta_r - I(z)} : z \in E, \Phi(z) < r, I(z) < \eta_r \right\},$$

where

$$\gamma := \inf_E (\Psi(z) + \Phi(z)),$$

and

$$\eta_r := \inf_{z \in \Phi^{-1}(r)} I(z).$$

With the above notations, our abstract tool for proving the main result is [110, Theorem 3] that we recall here for reader convenience.

Theorem 10.45. *Let $(E, \|\cdot\|)$ be a reflexive Banach space; $I : E \to \mathbb{R}$ a sequentially weakly lower semicontinuous, bounded on each bounded subset of E, C^1–functional whose derivative admits a continuous inverse on the topological dual E^*; $\Phi, \Psi : E \to \mathbb{R}$ two C^1–functionals with compact derivative. Assume also that the functional $\Psi + \lambda\Phi$ is bounded from below for all $\lambda > 0$ and that*

$$\liminf_{\|z\| \to +\infty} \frac{\Psi(z)}{I(z)} = -\infty. \tag{10.4.40}$$

Then, for each $r > \sup_S \Phi$, where S is the set of all global minima of I, for each

$$\mu > \max\{0, \mu^\star(I, \Psi, \Phi, r)\},$$

and each compact interval $[a, b] \subset (0, \beta(\mu I + \Psi, \Psi, r))$, there exists a number $\rho > 0$ with the following property:

for every $\lambda \in [a, b]$ and every C^1–functional $\Gamma : E \to \mathbb{R}$ with compact derivative, there exists $\delta > 0$ such that, for each $\nu \in [0, \delta]$, the equation

$$\mu I'(z) + \Psi'(z) + \lambda\Phi'(z) + \nu\Gamma'(z) = 0,$$

has at least three solutions in E whose norm are less than ρ.

We recall that if I is a C^1–functional, the derivative $I' : E \to E^*$ admits a continuous inverse on E^* provided that there exists a continuous operator $h : E^* \to E$ such that

$$h(I'(z)) = z,$$

for every $z \in E$.

Remark 10.46. For completeness, we mention that other results, related to Theorem 10.45, are contained in [106, 107, 108, 109, 110].

We denote by $\mathcal{A}$ the class of all T–periodic (with respect to $x \in \mathbb{R}^N$) continuous (or more generally Carathéodory) functions $f : \mathbb{R}^N \times \mathbb{R} \to \mathbb{R}$ such that

$$\sup_{(x,t)\in\mathbb{R}^N\times\mathbb{R}} \frac{|f(x,t)|}{1 + |t|^{q-1}} < +\infty,$$

for some $q \in [1, 2^\sharp_s)$.

Let $p \in \mathcal{A}$ and set $J_p : \mathbb{X}^s_T \to \mathbb{R}$ the C^1–functional defined by

$$J_P(v) := \kappa_s \int_{\partial^0 \mathcal{S}_T} P(x, \mathrm{Tr}(v))dx,$$

where

$$P(x, t) := \int_0^t p(x, w)\, dw,$$

and whose derivative is given by

$$\langle J_P'(v), \varphi\rangle = \kappa_s \int_\Omega p(x, v)\varphi dx,$$

for every $\varphi \in \mathbb{X}^s_T$.

For each $r > 0$ and each pair of functions $f, g : \mathbb{R}^N \times \mathbb{R} \to \mathbb{R}$ belonging to $\mathcal{A}$ and such that $G - F$ is bounded from below, set

$$\widetilde{\mu}(f, g, r) := 2 \inf\left\{\frac{r - \widetilde{\gamma} - J_f(v)}{\widetilde{\eta}_r - \|v\|^2_{\mathbb{X}^s_T}} : v \in \mathbb{X}^s_T, J_g(v) < r, \|v\|^2_{\mathbb{X}^s_T} < \widetilde{\eta}_r\right\},$$

where

$$G(x,t) := \int_0^t g(x,w)\,dw$$

for every $(x,t) \in \mathbb{R}^N \times \mathbb{R}$,

$$\widetilde{\gamma} := \kappa_S \int_\Omega \inf_{z\in\mathbb{R}} (G(x,z) - F(x,z))\,dx,$$

and

$$\widetilde{\eta}_r := \inf_{u\in J_g^{-1}(r)} \|v\|^2_{\mathbb{X}^s_T}.$$

Finally, for each $\varepsilon \in \left(0, \frac{1}{\max\{0, \widetilde{\mu}(f,g,r)\}}\right)$, denote by $\widetilde{\beta}(\varepsilon, f, g, r)$ the quantity

$$\widetilde{\beta}(\varepsilon,f,g,r) := \sup_{u\in J_g^{-1}((r,+\infty))} \frac{\|v\|^2_{\mathbb{X}^s_T} - 2\varepsilon J_f(v) - \inf_{v\in J_g^{-1}((-\infty,r))} (\|v\|^2_{\mathbb{X}^s_T} - 2\varepsilon J_f(v))}{2(r - J_g(v))}.$$

In order to prove the main result, we will make use of Theorem 10.45.

With the above notations our theorem reads as follows:

Theorem 10.47. *Let $\sigma \in (2, 2^\sharp_s)$ and $f, g : \mathbb{R}^N \times \mathbb{R} \to \mathbb{R}$ be two functions belonging to $\mathcal{A}$ such that*

$$\lim_{t\to+\infty} \frac{\inf_{x\in(0,T)^N} F(x,t)}{t^2} = +\infty, \qquad \limsup_{|t|\to+\infty} \frac{\sup_{x\in(0,T)^N} F(x,t)}{|t|^\sigma} < +\infty$$

and

$$\lim_{|t|\to+\infty} \frac{\inf_{x\in(0,T)^N} G(x,t)}{|t|^\sigma} = +\infty.$$

Then, for each $r > 0$, for each

$$\varepsilon \in \left(0, \frac{1}{\max\{0, \widetilde{\mu}(f,g,r)\}}\right)$$

and for each compact interval $[a,b] \subset (0, \tilde{\beta}(\varepsilon,f,g,r))$, there exists a number $\rho > 0$ with the following property:

for every $\lambda \in [a,b]$ and every function $h \in \mathcal{A}$ there exists $\delta > 0$ such that, for each $\nu \in [0,\delta]$, the problem

$$\begin{cases} (-\Delta + m^2)^s u = \varepsilon f(x,u) - \lambda g(x,u) + \nu h(x,u) & \text{in } (0,T)^N \\ u(x + Te_i) = u(x) & \text{for all } x \in \mathbb{R}^N \text{ and } i = 1,\dots,N \end{cases},$$

has at least three weak solutions whose norms in $\mathbb{H}^s_T$ are less than $\frac{\rho}{\sqrt{\kappa_S}}$.

Remark 10.48. The above result can be viewed as the (periodic) fractional analogous of [110, Theorem 4] in which a Dirichlet problem involving the p–Laplacian operator $\Delta_p u := \mathrm{div}(|\nabla u|^{p-2}\nabla u)$ (where $p > 1$) is studied; see also [91, Part II, Chapter 10] and [86] for some related problems in the nonlocal setting.

Remark 10.49. On the contrary of the results contained in Subsection 10.4.1, in Theorem 10.47 no condition at zero on the nonlinear term f is requested.

We end this subsection studying the existence of at least three non–trivial solutions for the following parametric ($\gamma \in \mathbb{R}$) problem

$$\begin{cases} (-\Delta + m^2)^s u = \gamma u + f(x, u) & \text{in } (0, T)^N \\ u(x + Te_i) = u(x) & \text{for all } x \in \mathbb{R}^N \text{ and } i = 1, \dots, N \end{cases}, \tag{10.4.41}$$

in a suitable left neighborhood of any eigenvalue of $(-\Delta + m^2)^s$.

We assume that $f : \mathbb{R}^N \times \mathbb{R} \to \mathbb{R}$ is a Carathéodory T–periodic (with respect to $x \in \mathbb{R}^N$) function such that:

$$\begin{gathered} \textit{there exist } a_1, a_2 > 0 \textit{ and } q \in (2, 2^\sharp_s), 2^\sharp_s := 2N/(N-2s), \textit{ such that} \\ |f(x,t)| \leqslant a_1 + a_2|t|^{q-1} \textit{ a.e. } x \in (0,T)^N, t \in \mathbb{R}; \end{gathered} \tag{10.4.42}$$

$$\begin{gathered} \textit{there exist two positive constants } a_3 \textit{ and } a_4 \textit{ such that} \\ F(x,t) \geqslant a_3|t|^q - a_4, \quad x \in (0,T)^N, t \in \mathbb{R}; \end{gathered} \tag{10.4.43}$$

$$\lim_{|t| \to 0} \frac{f(x,t)}{|t|} = 0 \textit{ uniformly in } x \in (0,T)^N; \tag{10.4.44}$$

$$0 < qF(x,t) \leqslant tf(x,t) \textit{ a.e. } x \in (0,T)^N, t \in \mathbb{R}\setminus\{0\}, \tag{10.4.45}$$

where q is given in (10.4.42).

The main result about problem (10.4.41) can be stated as follows:

Theorem 10.50. *Assume that $f : \mathbb{R}^N \times \mathbb{R} \to \mathbb{R}$ is a Carathéodory T–periodic (with respect to $x \in \mathbb{R}^N$) function verifying* (10.4.42)–(10.4.45).

Then, for every eigenvalue λ_k of $(-\Delta + m^2)^s$ with periodic boundary data, there exists a left neighborhood $\mathcal{O}_k$ of λ_k such that problem (10.4.41) *admits at least three non–zero weak solutions for all $\gamma \in \mathcal{O}_k$.*

In order to get this result we apply the following abstract critical point theorem (see [77, Theorem 2.10]):

Theorem 10.51 (Sphere–torus linking with mixed type assumptions). *Let H be a Hilbert space and X_1, X_2, X_3 be three subspaces of H such that $H = X_1 \oplus X_2 \oplus X_3$ with $0 < \dim X_i < +\infty$ for $i = 1, 2$. Let $\mathcal{J} : H \to \mathbb{R}$ be a $C^{1,1}$–functional. Let $\rho, \rho', \rho'', \rho_1$ be such that $0 < \rho_1$, $0 \leqslant \rho' < \rho < \rho''$ and*

$$\Delta := \{z \in X_1 \oplus X_2 : \rho' \leqslant \|P_2 z\| \leqslant \rho'', \|P_1 z\| \leqslant \rho_1\} \textit{ and } T := \partial_{X_1 \oplus X_2} \Delta,$$

where $P_i : H \to X_i$ is the orthogonal projection of H onto X_i, $i = 1, 2$, and

$$S_{23}(\rho) := \{z \in X_2 \oplus X_3 : \|z\| = \rho\} \text{ and } B_{23}(\rho) := \{z \in X_2 \oplus X_3 : \|z\| < \rho\}.$$

Assume that

$$a' := \sup \mathcal{J}(T) < \inf \mathcal{J}(S_{23}(\rho)) =: a''.$$

Let a, b be such that $a' < a < a''$, $b > \sup \mathcal{J}(\Delta)$ and

$$\text{the assumption } (\nabla)(\mathcal{J}, X_1 \oplus X_3, a, b) \text{ holds;}$$

$$\text{the Palais–Smale condition holds at any level } c \in [a, b].$$

Then, $\mathcal{J}$ has at least two critical points in $\mathcal{J}^{-1}([a, b])$.

If, furthermore,

$$a_1 < \inf \mathcal{J}(B_{23}(\rho)) > -\infty$$

and the Palais–Smale condition holds at every $c \in [a_1, b]$, then $\mathcal{J}$ has another critical level between a_1 and a'.

One of the main ingredient of Theorem 10.51 in order to get our multiplicity result is the so-called ∇-condition introduced in [77, Definition 2.4]. Let

$$P_C : H \to C$$

be the orthogonal projection of H onto C, where C be a closed subspace of H and let $a, b \in \mathbb{R} \cup \{\pm\infty\}$.

We say that functional $\mathcal{J}$ verifies condition $(\nabla)(\mathcal{J}, C, a, b)$ if there exists $\gamma > 0$ such that

$$\inf\left\{\|P_C \nabla \mathcal{J}(z)\| : a \leqslant \mathcal{J}(z) \leqslant b,\ \operatorname{dist}(z, C) \leqslant \gamma\right\} > 0.$$

Roughly speaking, the condition $(\nabla)(\mathcal{J}, C, a, b)$ requires that $\mathcal{J}$ has no critical points $z \in C$ such that $a \leqslant \mathcal{J}(z) \leqslant b$, with some uniformity.

The proof of Theorem 10.50 is similar to the one given in [83] and [82] (see also [81]).

10.4.4 Periodic equations with bounded primitive

We aim at finding condition on the datum f for which the problem

$$\begin{cases} \displaystyle\iint\limits_{\mathcal{S}_T} (\nabla v \nabla \varphi + m^2 v \varphi)\, dxdy \\ \qquad = \kappa_s \mu \left(\displaystyle\int_{\partial^0 \mathcal{S}_T} \beta(x) F(\operatorname{Tr}(v))\, dx - \lambda \right) \displaystyle\int_{\partial^0 \mathcal{S}_T} \beta(x) f(\operatorname{Tr}(v)) \operatorname{Tr}(\varphi)\, dx \quad \forall\, \varphi \in \mathbb{X}_T^s \\ v \in \mathbb{X}_T^s, \end{cases} \tag{10.4.46}$$

possesses at least three solutions, where

$$F(t) := \int_0^t f(w)\,dw, \quad \forall\, t \in \mathbb{R}$$

moreover $\lambda, \mu \in \mathbb{R}$ and $\beta \in L^\infty(\mathbb{R}^N)$ is a T–periodic function satisfying

$$\inf_{x \in (0,T)^N} \beta(x) > 0.$$

Our variational approach is realizable checking that the associated energy functional $\mathcal{J}_{\lambda,\mu} : \mathbb{X}_T^s \to \mathbb{R}$ defined by

$$\mathcal{J}_{\lambda,\mu}(v) := \frac{1}{2}\|v\|_{\mathbb{X}_T^s}^2 - \kappa_s \frac{\mu}{2}\left(\int_{\partial^0 \mathcal{S}_T} \beta(x) F(\mathrm{Tr}(v))dx - \lambda\right)^2,$$

verifies, thanks to the abstract setting developed in Section 10.2, the assumptions requested by a special case (see Theorem 10.53 below) of a recent and general critical point theorem obtained by Ricceri in [111, Theorem 1.6].

It is worth pointing out that the classical variational approach to attack such problems is not often easy to perform; indeed the presence of the term

$$L(v) := \left(\int_{\partial^0 \mathcal{S}_T} \beta(x) F(\mathrm{Tr}(v))dx - \lambda\right)^2, \quad v \in \mathbb{X}_T^s$$

causes variational methods not to work when applied to these classes of equations.

We rephrase [111, Theorem 1.6] in a slightly different version:

Theorem 10.52. *Let $(E, \|\cdot\|)$ be a separable and reflexive real Banach space, $\eta, J : E \to \mathbb{R}$ be two C^1–functionals with compact derivative such that $\eta(0) = J(0) = 0$. Moreover, assume that J is non–constant and that η is bounded from above. Let $\Phi : E \to \mathbb{R}$ be a sequentially weakly lower semicontinuous and coercive C^1–functional, with $\Phi(0) = 0$ and whose derivative admits a continuous inverse on E^*; $\varphi : [-\operatorname{osc}_E J, \operatorname{osc}_E J] \cap \mathbb{R} \to [0, +\infty)$ a convex C^1 function with $\varphi^{-1}(0) = \{0\}$, where $\operatorname{osc}_E J$ stands for the oscillation of J on E, namely*

$$\operatorname*{osc}_E J := \sup_E J - \inf_E J.$$

Suppose that the number

$$\vartheta := \inf_{z \in J^{-1}(\mathbb{R}\setminus\{0\})} \frac{\Phi(z) - \eta(z)}{\varphi(J(z))},$$

is non–negative and that there exists $\mu > 0$ such that

$$\lim_{\|z\| \to +\infty} (\Phi(z) - \eta(z) - \mu\varphi(J(z) - \lambda)) = +\infty,$$

for every $\lambda \in (\inf_E J, \sup_E J)$.

Then, there exists an open interval $\Lambda \subseteq (\inf_E J, \sup_E J)$ such that, for every $\lambda \in \Lambda$, the equation

$$\Phi'(z) = \eta'(z) + \mu\varphi'(J(z) - \lambda)J'(z), \tag{10.4.47}$$

has at least three solutions in E.

The main novelty obtained the above result is that, in contrast with a large part of the existing literature, the abstract energy functional does not depend on the parameter λ in an affine way.

A key role in Theorem 10.52 is played by the convex function φ: it is responsible for both the nonlocality of equation (10.4.47) and the non–affine dependence on the parameter λ.

Our abstract tool proving the main result is a direct consequence of Theorem 10.52 obtained in [40, Theorem 3.1], where the role of φ is assigned to a parabola:

Theorem 10.53. *Let $(E, \|\cdot\|)$ be a separable and reflexive real Banach space; $J : E \to \mathbb{R}$ be a noncostant C^1–functional with compact derivative such that $J(0) = 0$; $\Psi : E \to [0, +\infty)$ be a sequentially weakly lower semicontinuous and coercive C^1–functional, with $\Psi(0) = 0$ and whose derivative admits a continuous inverse on the topological dual E^*. Set*

$$\widehat{\vartheta} := 2 \inf_{z \in J^{-1}(\mathbb{R}\setminus\{0\})} \frac{\Psi(z)}{|J(z)|^2}.$$

Then, for each $\mu > \widehat{\vartheta}$ satisfying

$$\limsup_{\|z\| \to +\infty} \frac{|J(z)|^2}{\Psi(z)} < \frac{2}{\mu}, \tag{10.4.48}$$

there exists an open interval $\Lambda \subseteq (\inf_E J, \sup_E J)$ such that, for each $\lambda \in \Lambda$, the equation

$$\Psi'(z) = \mu(J(z) - \lambda)J'(z),$$

has at least three solutions.

Some applications of this recent results can be found for instance in [39, 40] and [104] itself. In the last paper, Ricceri has used his abstract result to prove the existence of three weak solutions for a Dirichlet problem involving the Laplace operator in which two nonlocalities appear, both in the divergence term and in the external force.

We denote by $\mathcal{C}$ the class of all continuous function $f : \mathbb{R} \to \mathbb{R}$ such that

$$\sup_{t \in \mathbb{R}} \frac{|f(t)|}{1 + |t|^{q-1}} < +\infty,$$

for some $q \in [1, 2_s^\sharp)$.

With the above notations our result reads as follows.

Theorem 10.54. *Let $f : \mathbb{R} \to \mathbb{R}$ be a not identically zero continuous function belonging to $\mathcal{C}$ such that*

$$\lim_{|t|\to+\infty} \frac{F(t)}{t} = 0. \tag{10.4.49}$$

Then, for each μ satisfying

$$\mu > \frac{1}{\kappa_s} \inf \left\{ \frac{\displaystyle\iint_{\mathcal{S}_T} (|\nabla v|^2 + m^2 v^2)\, dxdy}{\left(\displaystyle\int_{\partial^0 \mathcal{S}_T} \beta(x) F(\mathrm{Tr}(v))dx\right)^2} : v \in \mathbb{X}^s_T, \int_{\partial^0 \mathcal{S}_T} \beta(x) F(\mathrm{Tr}(v))dx \neq 0 \right\},$$

there exists an open interval

$$\Lambda \subseteq \left(T^N \inf_{x\in(0,T)^N} \beta(x) \inf_{t\in\mathbb{R}} F(t), T^N \|\beta\|_\infty \sup_{t\in\mathbb{R}} F(t) \right)$$

such that, for each $\lambda \in \Lambda$, problem (10.4.46) *has at least three solutions.*

Proof. Let us apply Theorem 10.53 by choosing

$$E := \mathbb{X}^s_T, \qquad J(v) := \int_{\partial^0 \mathcal{S}_T} \beta(x) F(\mathrm{Tr}(v))dx, \qquad \Psi(v) := \frac{1}{2} \|v\|^2_{\mathbb{X}^s_T},$$

for every $v \in E$.

Of course Ψ is a weakly lower semicontinuous and coercive C^1–functional with $\Phi(0) = 0$ and its derivative Ψ' is a homeomorphism between the Hilbert space E and its topological dual E^*.

Furthermore, since $f \in \mathcal{C}$ the functional J is a C^1–functional with compact derivative; note that the embedding $j : \mathbb{H}^s_T \hookrightarrow L^q(0, T)^N$ is compact for every $q \in [1, 2^\sharp_s)$. Finally, observe that J is clearly noncostant with $J(0) = 0$.

Now, let us fix μ as in Theorem 10.54. By condition (10.4.49) and the continuity of the potential F, we infer that, for every $\varepsilon > 0$, there exist two positive constants c_1 and c_2 such that

$$J(v) = \int_{\partial^0 \mathcal{S}_T} \beta(x) F(\mathrm{Tr}(v))dx \leqslant c_1 + \varepsilon \|v\|_{\mathbb{X}^s_T},$$

for every $v \in E$. Hence, if $v \neq 0$, we easily get

$$\frac{|J(v)|^2}{\Psi(v)} \leqslant 2 \frac{c_1^2}{\|v\|^2_{\mathbb{X}^s_T}} + 4 \frac{c_1 c_2}{\|v\|_{\mathbb{X}^s_T}} \varepsilon + 2 c_2^2 \varepsilon^2.$$

Thus,

$$\limsup_{\|v\|_{\mathbb{X}^s_T}\to+\infty} \frac{|J(v)|^2}{\Psi(v)} = 0 < \frac{2}{\kappa_s \mu},$$

and so condition (10.4.48) holds as desired. Then, the thesis of Theorem 10.53 follows and the existence of three solutions to problem (10.4.46) is achieved. □

Theorem 10.54 may be seen as a periodic version of the result contained in [85]; see also [91, Part II] for similar multiplicity properties.

10.4.5 Resonant periodic equations

As usual, the classical Ambrosetti–Rabinowitz condition plays a crucial rôle in proving that every Palais–Smale sequence is bounded, as well as the so called Mountain–Pass geometry is satisfied. However, even dealing with different problems than ours, several authors studied different assumptions that still allow to apply min–max procedure in order to assure the existence of critical points. For asymptotically linear problems we show a result that moves along this direction.

Let us denote by $(E, \|\cdot\|)$ a Banach space, Φ a C^1–functional on E, $\Phi^b := \{z \in E : \Phi(z) \leqslant b\}$ the sublevel of Φ corresponding to $b \in \overline{\mathbb{R}} := \mathbb{R} \cup \{\pm\infty\}$ and by

$$K_c := \{z \in E : \ \Phi(z) = c, \ \Phi'(z) = 0\}$$

the set of the critical points of Φ in E at the critical level $c \in \mathbb{R}$.

Let us recall some basic notions of the index theory for an even functional with symmetry group $\mathbb{Z}_2 = \{\mathrm{id}, -\mathrm{id}\}$. Let us set

$$\begin{aligned}\Sigma \ := \ &\{A \subseteq E : A \text{ closed and symmetric w.r.t. the origin,} \\ &\qquad \text{i.e. } -z \in A \text{ if } z \in A\}\end{aligned}$$

and

$$\mathcal{H} := \{h \in C^0(E; E) : h \text{ odd}\}.$$

For $A \in \Sigma$, $A \neq \varnothing$, the genus of A is

$$\gamma(A) \ := \ \inf\{m \in \mathbb{N} : \ \exists \psi \in C^0(A; \mathbb{R}^m \backslash \{0\}) \text{ s.t. } \psi(-z) = -\psi(z), \forall z \in A\}$$

if such an infimum exists, otherwise $\gamma(A) = +\infty$. Assume $\gamma(\varnothing) = 0$.

The index theory $(\Sigma, \mathcal{H}, \gamma)$ related to $\mathbb{Z}_2$ is also called genus (we refer for more details to [129, Section II.5]).

The pseudo–index related to the genus and $S \in \Sigma$ is the triplet $(S, \mathcal{H}^*, \gamma^*)$ such that $\mathcal{H}^*$ is a group of odd homeomorphisms and $\gamma^* : \Sigma \to \mathbb{N} \cup \{+\infty\}$ is the map defined by

$$\gamma^*(A) := \min_{h \in \mathcal{H}^*} \gamma(h(A) \cap S), \quad \forall A \in \Sigma$$

(cf. [26] for more details).

The main result of this subsection, the forthcoming Theorem 10.58, is based on the following abstract result proved in [19, Theorem 2.9]; see also [20, 21] and references therein for related topics.

Theorem 10.55. *Let $a, b, c_0, c_\infty \in \overline{\mathbb{R}}$, $-\infty \leqslant a < c_0 < c_\infty < b \leqslant +\infty$, Φ be an even functional, $(\Sigma, \mathcal{H}, \gamma)$ the genus theory on E, $S \in \Sigma$, $(S, \mathcal{H}^*, \gamma^*)$ the pseudo–index theory related to the genus and S, with*

$$\mathcal{H}^* := \{h \in \mathcal{H} : h \text{ bounded homeomorphism}$$

$$\text{such that } h(z) = z \text{ if } z \notin \Phi^{-1}((a, b))\}.$$

Assume that:

(*i*) *the functional* Φ *satisfies* (PS) *in* (a, b);

(*ii*) $S \subseteq \Phi^{-1}([c_0, +\infty))$;

(*iii*) *there exist* $\tilde{k} \in \mathbb{N}$ *and* $\tilde{A} \in \Sigma$ *such that* $\tilde{A} \subseteq \Phi^{c_\infty}$ *and* $\gamma^*(\tilde{A}) \geqslant \tilde{k}$.

Then, setting $\Sigma_i^* := \{A \in \Sigma : \gamma^*(A) \geqslant i\}$, *the numbers*

$$c_i := \inf_{A \in \Sigma_i^*} \sup_{z \in A} \Phi(z), \quad \forall i \in \{1, \ldots, \tilde{k}\}, \tag{10.4.50}$$

are critical values for Φ *and*

$$c_0 \leqslant c_1 \leqslant \ldots \leqslant c_{\tilde{k}} \leqslant c_\infty.$$

Furthermore, if $c = c_i = \ldots = c_{i+r}$, *with* $i \geqslant 1$ *and* $i + r \leqslant \tilde{k}$, *then* $\gamma(K_c) \geqslant r + 1$.

Remark 10.56. In the applications, a lower bound for the pseudo–index of a suitable $\tilde{A}$ as in (*iii*) of Theorem 10.55 is needed: considering the genus theory $(\Sigma, \mathcal{H}, \gamma)$ on E and V, W two closed subspaces of E, if

$$\dim V < +\infty \quad \text{and} \quad \operatorname{codim} W < +\infty,$$

then

$$\gamma(V \cap h(\partial B \cap W)) \geqslant \dim V - \operatorname{codim} W$$

for every bounded $h \in \mathcal{H}$ and every open bounded symmetric neighbourhood B of 0 in E (cf. [19, Theorem A.2]). The latter result is then employed to solve nonlocal periodic equations patterned after the main problems in [22, 23].

From now on we will suppose that $f : \mathbb{R}^N \times \mathbb{R} \to \mathbb{R}$ is a Carathéodory function satisfying the following assumptions:

(h_1) $f(x, t)$ *is* T*–periodic in* $x \in \mathbb{R}^N$, *that is* $f(x + Te_i, t) = f(x, t)$ *for any* $x \in \mathbb{R}^N$ $(T > 0)$, *for every* $i = 1, \ldots, N$ *and*

$$\sup_{|t| \leqslant a} |f(\cdot, t)| \in L^\infty(0, T)^N \text{ for any } a > 0;$$

(h_2) *there exist*

$$\lim_{|t| \to +\infty} \frac{f(x, t)}{t} = 0 \tag{10.4.51}$$

and

$$\lim_{t \to 0} \frac{f(x, t)}{t} = \lambda_0 \in \mathbb{R} \tag{10.4.52}$$

uniformly with respect to a.e. $x \in \mathbb{R}^N$.

Further, let us denote respectively by $\sigma((-\Delta + m^2)^s)$ and by

$$0 < \lambda_1 < \lambda_2 \leqslant \ldots \leqslant \lambda_k \leqslant \ldots$$

the spectrum and the non–decreasing, diverging sequence of eigenvalues of the following problem

$$\begin{cases} (-\Delta + m^2)^s u = \lambda u & \text{in } (0, T)^N \\ u(x + Te_i) = u(x) & \text{for all } x \in \mathbb{R}^N \text{ and } i = 1, \ldots, N \end{cases},$$

repeated according to their multiplicity (see Section 10.2 for details).

Under the above assumptions on f it is possible to prove some existence and multiplicity results for asymptotically linear T–periodic fractional problems of the form

$$\begin{cases} (-\Delta + m^2)^s u = \lambda_\infty u + f(x, u) & \text{in } (0, T)^N \\ u(x + Te_i) = u(x) & \text{for all } x \in \mathbb{R}^N \text{ and } i = 1, \ldots, N \end{cases}. \tag{10.4.53}$$

By using the classical Saddle Point Theorem due to Rabinowitz [101], the following existence result has been achieved.

Theorem 10.57. *Suppose that $f : \mathbb{R}^N \times \mathbb{R} \to \mathbb{R}$ is a Carathéodory function satisfying (h_1) and (h_2). Then, problem* (10.4.53) *has at least a weak solution in $\mathbb{H}^s_T$, provided that $\lambda_\infty \notin \sigma((-\Delta + m^2)^s)$.*

When the nonlinear term f is symmetric the pseudo–index theory ensures that the next result holds.

Theorem 10.58. *Suppose that $f : \mathbb{R}^N \times \mathbb{R} \to \mathbb{R}$ is a Carathéodory function satisfying conditions (h_1) and (h_2). Further, let us assume that $f(x, \cdot)$ is odd for a.e. $x \in \mathbb{R}^N$ and verifies the following condition*
(C_{λ_∞}) there exist $h, k \in \mathbb{N}$, with $k \geqslant h$, such that

$$\lambda_0 + \lambda_\infty < \lambda_h \leqslant \lambda_k < \lambda_\infty.$$

Then, problem (10.4.53) *has at least $k - h + 1$ distinct pairs of non–trivial weak solutions in $\mathbb{H}^s_T$, provided that $\lambda_\infty \notin \sigma((-\Delta + m^2)^s)$.*

See [85] for a complete and detailed proof.

Remark 10.59. We point out that, combining the proof of our main results and those of [21, Theorem 3.1], Theorem 10.58 holds if we require that

$$\lambda_\infty < \lambda_h \leqslant \lambda_k < \lambda_0 + \lambda_\infty,$$

instead of condition (C_{λ_∞}).

We emphasize that further difficulties arise in the so-called "resonant case", that is $\lambda_\infty \in \sigma((-\Delta + m^2)^s)$: indeed, the resonance affects both the compactness property and the geometry of the Euler–Lagrange functional arising in a suitable variational approach (cf., e.g., [19] and references therein for the classical elliptic case). We will consider this interesting case via some further investigations. Finally, we also point out that our results should be viewed as a periodic nonlocal version of Proposition 1.1 and Theorem 1.2 of [22] (see also [23]).

10.5 Critical and Supercritical Nonlinearities

Although in the current literature there are a lot of papers concerning subcritical and critical fractional Laplacian problems, only few of them deal with nonlinearities having supercritical growth (see, for instance, [54, 115, 132]).

Moreover, to the best of our knowledge, neither critical nor supercritical periodic fractional equations have been investigated until now due to several difficulties that naturally appear in treating such problems. We present here a first result in this direction, concerning the existence of multiple periodic solutions for a class of nonlocal equations with critical and supercritical growth (see the recent paper [12] for a detailed proof and additional comments and remarks).

By combining the Moser iteration scheme in the nonlocal framework with an abstract multiplicity result valid for differentiable functionals we show that the problem under consideration admits at least three periodic solutions with the property that their Sobolev norms are bounded by a suitable constant.

If E is a real Banach space, we denote by $\mathcal{W}_E$ the class of all functionals $\Phi : E \to \mathbb{R}$ possessing the following property:

if $\{z_j\}_{j\in\mathbb{N}}$ is a sequence in E converging weakly to $z \in E$ and

$$\liminf_{j\to+\infty} \Phi(z_j) \leqslant \Phi(z),$$

then $\{z_j\}_{j\in\mathbb{N}}$ has a subsequence converging strongly to z.

With the above notation, [107, Theorem 1] reads as follows:

Theorem 10.60. *Let E be a separable and reflexive real Banach space; $I \subset \mathbb{R}$ be an interval; $\Phi : X \to \mathbb{R}$ be a sequentially weakly lower semicontinuous C^1-functional belonging to $\mathcal{W}_E$, bounded on each bounded subset of E and whose derivative admits a continuous inverse on the dual E^* of E; $J : E \to \mathbb{R}$ be a C^1-functional with compact derivative.*

Assume that, for each $\lambda \in I$, the functional $\Phi - \lambda J$ is coercive and has a strict local, not global minimimum, say $\overline{z}_\lambda$. Then, for each compact interval $[a, b] \subseteq I$ for which

$$\sup_{\lambda\in[a,b]} (\Phi(\overline{z}_\lambda) - \lambda J(\overline{z}_\lambda)) < +\infty,$$

there exists $\gamma > 0$ with the following property:

for every $\lambda \in [a, b]$ and every C^1–functional $\Psi : E \to \mathbb{R}$ with compact derivative, there exists $\delta_0 > 0$ such that, for each $\mu \in [0, \delta_0]$, the equation

$$\Phi'(z) = \lambda J'(z) + \mu\Psi'(z)$$

has at least three solutions in E whose norms are less than γ.

10.5.1 A periodic critical equation

Let $\beta \in L^\infty(\mathbb{R}^N)$ be a T–periodic function satisfying

$$\inf_{x\in(0,T)^N} \beta(x) > 0.$$

A special case of our main result reads as follows.

Theorem 10.61. *Assume that $f : \mathbb{R} \to \mathbb{R}$ is a continuous function such that*

(f_1') $\lim\limits_{t\to 0} \dfrac{f(t)}{t} = \lim\limits_{|t|\to+\infty} \dfrac{f(t)}{t} = 0;$

(f_2') $\sup\limits_{t\in\mathbb{R}} F(t) > 0$, *where* $F(t) := \int_0^t f(w)dw.$

Furthermore, set

$$\lambda^\star := \frac{1}{2}\left(\frac{m^{2s}}{\inf\limits_{x\in(0,T)^N} \beta(x)}\right)\frac{t_0^2}{F(t_0)},$$

where $t_0 \in \mathbb{R}$ is such that $F(t_0) > 0$.

Then, for each compact interval $[a, b] \subset (\lambda^\star, +\infty)$, there exists $\gamma > 0$ with the following property:

for every $\lambda \in [a, b]$, there exists $\delta > 0$ such that, for each $\mu \in [0, \delta]$, the critical problem

$$\begin{cases} [(-\Delta + m^2)^s]u = \lambda\beta(x)f(u) + \mu|u|^{2_s^\sharp-2}u & \text{in } (0, T)^N \\ u(x + Te_i) = u(x) & \text{for all } x \in \mathbb{R}^N, i = 1, \dots, N \end{cases}, \tag{10.5.1}$$

admits at least three weak solutions in $\mathbb{H}_T^s \cap L^\infty(0, T)^N$, whose $\mathbb{H}_T^s$–norms are less than γ.

As a byproduct of Theorem 10.61 we immediately have the following multiplicity result valid for unperturbed periodic problems.

Corrolary 10.62. *Assume that $f : \mathbb{R} \to \mathbb{R}$ is a continuous function satisfying conditions (f_1') and (f_2'). Then, for each compact interval $[a, b] \subset (\lambda^\star, +\infty)$, there exists $\gamma > 0$ such that, for every $\lambda \in [a, b]$, the following problem*

$$\begin{cases} [(-\Delta + m^2)^s]u = \lambda\beta(x)f(u) & \textit{in } (0, T)^N \\ u(x + Te_i) = u(x) & \textit{for all } x \in \mathbb{R}^N, i = 1, \dots, N \end{cases}, \tag{10.5.2}$$

admits at least three weak solutions in $\mathbb{H}_T^s \cap L^\infty(0, T)^N$, whose $\mathbb{H}_T^s$–norms are less than γ.

We also notice that Theorem 10.34 in Section 10.4 can be achieved as a direct consequence of Theorem 10.60.

10.5.2 Supercritical periodic problems

We end this note by presenting the existence and multiplicity of periodic solutions for the following class of perturbed problems

$$\begin{cases} [(-\Delta + m^2)^s]u = \lambda f(x, u) + \mu g(x, u) & \text{in } (0, T)^N \\ u(x + Te_i) = u(x) & \text{for all } x \in \mathbb{R}^N, i = 1, \dots, N \end{cases}. \tag{10.5.3}$$

Let us assume that $f : \mathbb{R}^N \times \mathbb{R} \to \mathbb{R}$ is a Carathéodory T–periodic (with respect to $x \in \mathbb{R}^N$) function satisfying the following hypotheses:

(f_1) $\displaystyle\lim_{|t|\to+\infty} \frac{f(x, t)}{t} = 0$ *uniformly in $x \in (0, T)^N$;*

(f_2) $\displaystyle\lim_{t\to 0} \frac{f(x, t)}{t} = 0$ *uniformly in $x \in (0, T)^N$;*

(f_3) *there is $t_0 \in \mathbb{R}$ such that*

$$\int_{(0,T)^N} F(x, t_0)dx > 0,$$

where

$$F(x, t) := \int_0^t f(x, w)\, dw;$$

(f_4) *for any $M > 0$, the function*

$$\sup_{|t|\leqslant M} |f(\cdot, t)| \in L^\infty(0, T)^N.$$

We also suppose that $g : \mathbb{R}^N \times \mathbb{R} \to \mathbb{R}$ is a Carathéodory T–periodic (with respect to $x \in \mathbb{R}^N$) such that

(g_1) *there exist $c_0 > 0$ and $p \geqslant 2_s^\sharp$ such that*

$$|g(x, t)| \leqslant c_0(1 + |t|^{p-1})$$

for a.e. $x \in \mathbb{R}^N$, $t \in \mathbb{R}$.

With the above notations our main result can be stated as follows:

Theorem 10.63. *Let* $f : \mathbb{R}^N \times \mathbb{R} \to \mathbb{R}$ *be a Carathéodory* T*−periodic (with respect to* $x \in \mathbb{R}^N$*) function such that* (f_1)*–*(f_4) *hold and* $g : \mathbb{R}^N \times \mathbb{R} \to \mathbb{R}$ *be a Carathéodory* T*−periodic (with respect to* $x \in \mathbb{R}^N$*) function for which* (g_1) *is verified.*

Then, it follows that

$$\vartheta := 2 \sup \left\{ \frac{\kappa_s}{\|v\|^2_{\mathbb{X}^s_T}} \int_{(0,T)^N} F(x, \operatorname{Tr}(v))dx : \|v\|_{\mathbb{X}^s_T} \neq 0 \right\} > 0,$$

and, for each compact interval $[a, b] \subset (1/\vartheta, +\infty)$*, there exists* $\gamma > 0$ *with the following property: for every* $\lambda \in [a, b]$*, there exists* $\delta > 0$ *such that, for each* $\mu \in [0, \delta]$*, problem* (10.4.5) *admits at least three weak solutions in* $\mathbb{H}^s_T \cap L^\infty(0, T)^N$*, whose* $\mathbb{H}^s_T$*−norms are less than* γ*.*

In order to prove Theorem 10.63 we are led to consider the following problem

$$\begin{cases} -\operatorname{div}(y^{1-2s}\nabla v) + m^2 y^{1-2s} v = 0 & \text{in } \mathcal{S}_T \\ v_{|\{x_i=0\}} = v_{|\{x_i=T\}} & \text{on } \partial_L \mathcal{S}_T \\ \partial^{1-2s}_\nu v = \kappa_s[\lambda f(x, v) + \mu g(x, v)] & \text{on } \partial^0 \mathcal{S}_T \end{cases}, \tag{10.5.4}$$

where

$$\partial^{1-2s}_\nu v(x, 0) := -\lim_{y\to 0^+} y^{1-2s}\frac{\partial v}{\partial y}(x, y)$$

is the conormal exterior derivative of v.

Problem (10.5.4) is (super)critical, so we cannot use directly variational techniques in order to obtain existence and multiplicity results for the initial problem (10.5.3).

First of all we consider the case

$$g(x, t) := |t|^{p-2}t,$$

for every $x \in (0, T)^N$ and $t \in \mathbb{R}$. Therefore, we adapt a truncation argument due to Chabrowski and Yang [43]: precisely, we consider the following auxiliary truncated problem

$$\begin{cases} -\operatorname{div}(y^{1-2s}\nabla v) + m^2 y^{1-2s} v = 0 & \text{in } \mathcal{S}_T \\ v_{|\{x_i=0\}} = v_{|\{x_i=T\}} & \text{on } \partial_L \mathcal{S}_T \\ \partial^{1-2s}_\nu v = \kappa_s\left[\lambda f(x, v) + \mu g_\kappa(x, v)\right] & \text{on } \partial^0 \mathcal{S}_T \end{cases}, \tag{10.5.5}$$

where

$$g_\kappa(x, t) := \begin{cases} g(x, t) & \text{if } |t| \leqslant \kappa \\ \kappa^{p-q}|t|^{q-2}t & \text{if } |t| > \kappa, \end{cases} \tag{10.5.6}$$

and κ is a suitable constant and $q \in (2, 2^\sharp_s)$.

Since the nonlinearity g_κ which appears in (10.5.5) is a superlinear function with subcritical growth, thanks to the results contained in Section 10.2, we can apply Theorem 10.60 to establish a multiplicity result for problem (10.5.5).

Now, we notice that if $v \in \mathbb{X}^s_T$ is a weak solution of problem (10.5.5) whose trace $u := \mathrm{Tr}(v)$ is bounded in $L^\infty(0,T)^N$, with

$$|u|_{L^\infty(0,T)^N} \leqslant \kappa, \tag{10.5.7}$$

then, thanks to the definition of g_κ, the function $v \in \mathbb{X}^s_T$ is a weak solution of (10.5.4) and $u := \mathrm{Tr}(v) \in \mathbb{H}^s_T$ weakly solves (10.5.3).

A direct computation ensures that, if κ is sufficiently large, every solution obtained by using Theorem 10.60 satisfies (10.5.7) and consequently solves problem (10.5.3).

Now, suppose that g verifies the (super)critical growth (g_1) and set

$$g_\kappa(x,t) := \begin{cases} g(x,t) & \text{if } |t| \leqslant \kappa \\ \min\{g(x,t), c_0(1+\kappa^{p-q}|t|^{q-2}t)\} & \text{if } |t| > \kappa \end{cases},$$

where $q \in (2, 2^\sharp_s)$. Since

$$|g_\kappa(x,t)| \leqslant c_0(1+\kappa^{p-q}|t|^{q-1}) \text{ for a.e. } x \in \mathbb{R}^N, \forall\, t \in \mathbb{R}, \tag{10.5.8}$$

we can easily adapt the previous iteration arguments by using (10.5.8) instead of (10.5.6).

Remark 10.64. Exploiting the arguments introduced in Section 10.2 it easy to see that the statements of Theorem 10.63 are valid for each compact interval $[a,b]$ contained in

$$\left(\frac{m^{2s}T^N}{2}\frac{t_0^2}{\displaystyle\int_{(0,T)^N} F(x,t_0)dx}, +\infty\right).$$

Remark 10.65. In [57, Theorem 1.1] the authors studied perturbed elliptic equations proving that the method of [133] can be adapted to arbitrary perturbations not necessarily governed by a polynomial growth. Their proof employs a variant of the Moser iteration argument inspired by [80, Theorem C]; see [57, Lemma 2.3]. An analogous approach can be adapted to our nonlocal periodic framework obtaining the existence of at least three weak solutions for problem (10.5.3) in presence of an arbitrary T–periodic Carathéodory term g such that:

for any $M > 0$, the function

$$\sup_{|t|\leqslant M} |g(\cdot,t)| \in L^\infty(0,T)^N.$$

This fact is due to the peculiar nature of the conclusion of the abstract Ricceri's result. Indeed, the interplay between [107, Theorem 1] with the Moser–type iteration scheme turns out to be successful also in this case thanks to the preliminary properties presented in Section 10.2.

Remark 10.66. To the best of our knowledge Theorem 10.63 (as well as its consequences) is the first result that has appeared in the literature concerning the existence of multiple solutions for (super)critical fractional periodic equations. For completeness, we mention here the papers [61, 78, 121, 122, 124, 126] in connection with the celebrated Brezis–Nirenberg problem, where the authors studied, under the Dirichlet boundary condition, nonlocal elliptic equations involving the fractional Laplacian operator and a nonlinear term with critical growth.

Acknowledgment: This work was realized within the auspices of the INdAM–GNAMPA Project 2016 *problemi variazionali su varietà riemanniane e gruppi di Carnot*, by the DiSBeF Research Project 2015 *Fenomeni non–locali: modelli e applicazioni*, by the DiSPeA Research Project 2016 *Implementazione e testing di modelli di fonti energetiche ambientali per reti di sensori senza fili autoalimentate* and by the PRIN 2015 Research Project *Variational methods, with applications to problems in mathematical physics and geometry*.

Bibliography

[1] C.O. Alves, and S.B. Liu, *On superlinear $p(x)$–Laplacian equations in $\mathbb{R}^N$*, Nonlinear Anal. TMA **73** (2010), 2566–2579.

[2] A. Ambrosetti and A. Malchiodi, Nonlinear Analysis and Semilinear Elliptic problems, Cambridge Stud. Adv. Math. **104** Cambridge: *Cambridge Univ. Press* (2007).

[3] A. Ambrosetti and G. Prodi, A primer of nonlinear analysis, Cambridge Studies in Advanced Mathematics 34. Cambridge: *Cambridge University Press* (1993).

[4] A. Ambrosetti and P. Rabinowitz, *Dual variational methods in critical point theory and applications*, J. Funct. Anal. **14** (1973), 349–381.

[5] V. Ambrosio, *Variational methods for a pseudo–relativistic Schrödinger equation*, (PhD Thesis) DOI: 10.6092/unina/fedoa/10261.

[6] V. Ambrosio, *Periodic solutions for a pseudo–relativistic Schrödinger equation*, Nonlinear Anal. TMA **120** (2015), 262–284.

[7] V. Ambrosio, *Periodic solutions for the nonlocal operator pseudo–relativistic $(-\Delta + m^2)^s - m^{2s}$ with $m \geqslant 0$*, Topol. Methods Nonlinear Anal. (2016), DOI: 10.12775/tmna.2016.063, Preprint. ArXiv: 1510.05808.

[8] V. Ambrosio, *Ground states for superlinear fractional Schrödinger equations in $\mathbb{R}^N$*, Ann. Acad. Sci. Fenn. Math. **41** (2016), 745–756.

[9] V. Ambrosio, *Periodic solutions for a superlinear fractional problems without the Ambrosetti–Rabinowitz condition*, Discrete and Continuous Dinamical Systems - Series A (in press).
[10] V. Ambrosio, *Infinitely many periodic solutions for a fractional problem under perturbation*, JEPE Journal of Elliptic and Parabolic Equations (in press).
[11] V. Ambrosio, *On the existence of periodic solutions for a fractional Schrödinger equation*, Proc. Amer. Math. Soc. (in press).
[12] V. Ambrosio, J. Mawhin, and G. Molica Bisci, *Supercritical periodic nonlocal equations*, submitted for publication (2016).
[13] V. Ambrosio and G. Molica Bisci, *Periodic solutions for nonlocal fractional equations*, Commun. Pure Appl. Anal. **16** (2017), no. 1, DOI: 10.3934/cpaa.2017.
[14] V. Ambrosio and G. Molica Bisci, *Periodic solutions for a fractional asymptotically linear problem*, Proc. Edinb. Math. Soc. Sect. A (in press).
[15] V. Ambrosio, G. Molica Bisci, and D. Repovš, *Nonlinear equations involving the square root of the Laplacian*, submitted for publication (2016).
[16] G. Anello and G. Cordaro, *An existence and localization theorem for the solutions of a Dirichlet problem*, Ann. Pol. Math. **83** (2004), 107–112.
[17] A. Bahri and H. Berestycki, *A perturbation method in critical point theory and applications*, Trans. Amer. Math. Soc. **267** (1981), 1–32.
[18] A. Bahri and P.L. Lions, *Morse index of some min-max critical points I. Application to multiplicity results*, Comm. Pure Appl. Math. **41**, no. 8 (1988), 1027–1037.
[19] P. Bartolo, V. Benci, and D. Fortunato, *Abstract critical point theorems and applications to some nonlinear problems with "strong" resonance at infinity*, Nonlinear Anal. TMA **7** (1983), 981–1012.
[20] R. Bartolo, A.M. Candela, and A. Salvatore, *p–Laplacian problems with nonlinearities interacting with the spectrum*, Nonlinear Differ. Equ. Appl. **20** (2013), 1701–1721.
[21] R. Bartolo, A.M. Candela, and A. Salvatore, *Perturbed asymptotically linear problems*, Ann. Mat. Pura Appl. **193** (2014), 89–101.
[22] R. Bartolo and G. Molica Bisci, *A pseudo–index approach to fractional equations*, Expo. Math. **33** (2015), 502–516.
[23] R. Bartolo and G. Molica Bisci, *Asymptotically linear fractional p–Laplacian equations*, Ann. Mat. Pura Appl. **196** (2017), 427–442.
[24] T. Bartsch, and S.J. Li, *Critical point theory for asymptotically quadratic functionals and applications to problems with resonance*, Nonlinear Anal. TMA **28** (1997), 419–441.
[25] B. Barrios, E. Colorado, A. de Pablo and U. Sánchez, *On some critical problems for the fractional Laplacian operator*, J. Differential Equations **252** (2012), 6133–6162.
[26] V. Benci, *On the critical point theory for indefinite functionals in the presence of symmetries*, Trans. Amer. Math. Soc. **274** (1982), 533–572.
[27] Z. Binlin, G. Molica Bisci, and R. Servadei, *Superlinear nonlocal fractional problems with infinitely many solutions*, Nonlinearity **28** (2015), 2247–2264.

[28] C. Brändle, E. Colorado, A. de Pablo, and U. Sánchez, *A concave–convex elliptic problem involving the fractional Laplacian*, Proc. Roy. Soc. Edinburgh Sect. A **143** (2013), 39–71.

[29] H. Brézis, Analyse fonctionelle. Théorie et applications, *Masson*, Paris (1983).

[30] H. Brézis and L. Nirenberg, *Positive solutions of nonlinear elliptic equations involving critical Sobolev exponents*, Comm. Pure Appl. Math. **36** (1983), 437–477.

[31] X. Cabré and Y. Sire, *Nonlinear Equations for fractional Laplacians I: Regularity, maximum principles, and Hamiltonian estimates*, Ann. Inst. H. Poincaré Anal. Non Linéaire **31** (2014), 23–53.

[32] X. Cabré and J. Tan, *Positive solutions of nonlinear problems involving the square root of the Laplacian*, Adv. Math., **224**, no. 5 (2010), 2052–2093.

[33] L. Caffarelli, *Nonlocal equations, drifts and games*, Nonlinear Partial Differential Equations, Abel Symposia **7** (2012), 37–52.

[34] L. Caffarelli, J.M. Roquejoffre, and Y. Sire, *Variational problems with free boundaries for the fractional laplacian*, J. Eur. Math. Soc. **12** (2010), 1151–1179.

[35] L. Caffarelli, S. Salsa, and L. Silvestre, *Regularity estimates for the solution and the free boundary of the obstacle problem for the fractional Laplacian*, Invent. Math. **171** (2008), 425–461.

[36] L. Caffarelli and L. Silvestre, *An extension problem related to the fractional Laplacian*, Comm. Partial Differential Equations **32** (2007), 1245–1260.

[37] L. Caffarelli and L. Silvestre, *Regularity theory for fully nonlinear integro–differential equations*, Comm. Pure Appl. Math. **62** (2009), 597–638.

[38] L. Caffarelli and L. Silvestre, *Regularity results for nonlocal equations by approximation*, Arch. Ration. Mech. Anal. **200** (2011), 59–88.

[39] F. Cammaroto and F. Faraci, *Multiple solutions for some Dirichlet problems with nonlocal terms*, Ann. Polon. Math. **105** (2012), 31–42.

[40] F. Cammaroto and L. Vilasi, *Three solutions for some Dirichlet and Neumann nonlocal problems*, Appl. Anal. **92** (2013), 1717–1730.

[41] A. Capella, *Solutions of a pure critical exponent problem involving the half–Laplacian in annular-shaped domains*, Commun. Pure Appl. Anal. **10** (2011), 1645–1662.

[42] R. Carmona, W.C. Masters, and B. Simon, *Relativistic Schrödinger operators: Asymptotic behavior of the eigenfunctions*, J. Func. Anal. **91** (1990), 117–142.

[43] J. Chabrowski and J. Yang, *Existence theorems for elliptic equations involving supercritical Sobolev exponent*, Adv. Differential Equations **2** (1997), 231–256.

[44] K.C. Chang, Infinite dimensional Morse theory and multiple solution problems, *Birkhäuser*, Boston, 1993.

[45] S. Cingolani and S. Secchi, *Ground states for the pseudo–relativistic Hartree equation with external potential*, Proc. Roy. Soc. Edinburgh Sect. A **145** (2015), 73–90.

[46] V. Coti Zelati and M. Nolasco, *Existence of ground state for nonlinear, pseudorelativistic Schröedinger equations*, Rend. Lincei Mat. Appl. **22** (2011), 51–72.

[47] M. Dabkowski, *Eventual regularity of the solutions to the supercritical dissipative quasi–geostrophic equation*, Geom. Funct. Anal. **21** (2011), 1–13.
[48] A. Di Castro, T. Kuusi, and G. Palatucci, *Local behavior of fractional p–minimizers*, Ann. Inst. H. Poincaré Anal. Non Linéaire **267** (2015), 1807–1836.
[49] A. Di Castro, T. Kuusi, and G. Palatucci, *Nonlocal Harnack inequalities*, J. Funct. Anal. **267** (2014), 1807–1836.
[50] E. Di Nezza, G. Palatucci, and E. Valdinoci, *Hitchhiker's guide to the fractional Sobolev spaces*, Bull. Sci. Math. **136** (2012), 521–573.
[51] G. Dinca, P. Jebelean, and J. Mawhin, *A result of Ambrosetti-Rabinowitz type for p–Laplacian*, Qualitative problems for Differential Equations and Control Theory, World Scientific Publishing, New Jersey (1995), 231–242.
[52] G. Dinca, P. Jebelean, and D. Motreanu, *Existence and approximation for a general class of differential inclusions*, Houston J.Math. **28** (2002), 193–215.
[53] A. Erdélyi, W. Magnus, F. Oberhettinger, and F. Tricomi, Higher Trascendental Functions, vol. 1,2 *McGraw-Hill*, New York (1953).
[54] M.M. Fall and T. Weth, *Nonexistence results for a class of fractional elliptic boundary value problems*, J. Funct. Anal. **263** (2012), 2205–2227.
[55] F. Fang, and S.B. Liu, *Nontrivial solutions of superlinear p–Laplacian equations*, J. Math. Anal. Appl. **351** (2009) 138–146.
[56] F. Faraci, *Multiple solutions for two nonlinear problems involving the p–Laplacian*, Nonlinear Anal. TMA **63** (2005), 1017–1029.
[57] F. Faraci and L. Zhao, *Bounded multiple solutions for p–Laplacian problems with arbirary perturbations*, J. Aust. Math. Soc. **99** (2015), 175–185.
[58] M. Ferrara, G. Molica Bisci, and B. Zhang, *Existence of weak solutions for non–local fractional problems via Morse theory*, Discrete Contin. Dyn. Syst. B **19** (2014), 2493–2499.
[59] M. Ferrara, G. Molica Bisci, and D. Repovš, *Nonlinear Elliptic Equations on Carnot groups*, Rev. R. Acad. Cienc. Exactas Fis. Nat. Ser. A Math. RACSAM **111**, n. 3 (2017), 707–718.
[60] A. Fiscella and P. Pucci, *On certain nonlocal Hardy–Sobolev critical elliptic Dirichlet problems*, Adv. Differential Equations **21** (2016), 571–599.
[61] A. Fiscella, G. Molica Bisci, and R. Servadei, *Bifurcation and multiplicity results for critical nonlocal fractional problems*, Bull. Sci. Math. **140** (2016), 14–35.
[62] J. Fröhlich, B. Jonsson, G. Lars, and E. Lenzmann, *Boson stars as solitary waves*, Comm. Math. Phys. **274** (2007), 1–30.
[63] O. Kavian, Introduction à la théorie des points critiques et applications aux problèmes elliptiques, *Mathématiques & Applications*, Springer–Verlag, Paris (1993).
[64] A. Kiselev, F. Nazarov and A. Volberg, *Global well–posedness for the critical 2D dissipative quasi-geostrophic equation*, Invent. Math. **167** (2007), 445–453.
[65] T. Kuusi, G. Mingione, and Y. Sire, *Nonlocal equations with measure data*, Comm. Math. Phys. **337** (2015), 1317–1368.

[66] T. Kuusi, G. Mingione, and Y. Sire, *Nonlocal self-improving properties*, Anal. PDE **8** (2015), 57–114.

[67] A. Kristály and V. Rădulescu, *Sublinear eigenvalue problems on compact Riemannian manifolds with applications in Emden–Fowler equations*, Studia Math. **191** (2009), no. 3, 237–246.

[68] A. Kristály and I.J. Rudas, *Elliptic problems on the ball endowed with Funk-type metrics*, Nonlinear Anal. TMA **119** (2015), 199–208.

[69] A. Kristály and Cs. Varga, *Multiple solutions for a degenerate elliptic equation involving sublinear terms at infinity*, J. Math. Anal. Appl. **352** (2009), 139–148.

[70] L. Jeanjean, *On the existence of bounded Palais–Smale sequences and application to a Landesman-Lazer type problem set on* $\mathbb{R}^N$, Proc. Roy. Soc. Edinburgh **129** (1999), 787–809.

[71] E.H. Lieb and M. Loss, Analysis, *American Mathematical Society*, Providence, RI, 2001.

[72] S.B. Liu, *On ground states of superlinear p–Laplacian equations in* $\mathbb{R}^N$, J. Math. Anal. Appl. **361** (2010), 48–58.

[73] S.B. Liu, *On superlinear problems without Ambrosetti–Rabinowitz condition*, Nonlinear Anal. TMA **73** (2010), 788–795.

[74] S.B. Liu, *Existence of solutions to a superlinear p–Laplacian equation*, Electron. J. Differential Equations **66** (2001), 1–6.

[75] J.Q. Liu and J.B. Su, *Remarks on multiple nontrivial solutions for quasi–linear resonant problems*, J. Math. Anal. Appl. **258** (2001), 209–222.

[76] Q.S. Liu and J.B. Su, *Existence and multiplicity results for Dirichlet problems with p–Laplacian*, J. Math. Anal. Appl. **281** (2003), 587–601.

[77] A. Marino and C. Saccon, *Some variational theorems of mixed type and elliptic problems with jumping nonlinearities*, Ann. Scuola Norm. Sup. Pisa Cl. Sci. (4) **25** (1997), 631–665.

[78] J. Mawhin and G. Molica Bisci, *A Brezis–Nirenberg type result for a non–local fractional operator*, J. Lond. Math. Soc. (2) **95** (2017), 73–93.

[79] J. Mawhin and M. Willem, Critical point theory and Hamiltonian systems, Applied Mathematical Sciences, *Springer–Verlag*, New York **74** (1989).

[80] S. Miyajima, D. Motreanu, and M. Tanaka, *Multiple existence results of solutions for the Neumann problems via super– and sub–solutions*, J. Funct. Anal. **262** (2012), 1921–1953.

[81] D. Mugnai, *Multiplicity of critical points in presence of a linking: application to a superlinear boundary value problem*, NoDEA Nonlinear Differential Equations Appl., **11** (2004), 379–391.

[82] D. Mugnai, G. Molica Bisci, and R. Servadei, *On multiple solutions for nonlocal fractional problems via* ∇*-theorems*, Differential Integral Equations **30** (2017), 641–666.

[83] D. Mugnai and D. Pagliardini, *Existence and multiplicity results for the fractional Laplacian in bounded domains*, Adv. Calc. Var., to appear, DOI: 10.1515/acv-2015-0032.

[84] G. Molica Bisci, *Sequence of weak solutions for fractional equations*, Math. Res. Lett. **21** (2014), 241–253.

[85] G. Molica Bisci, *Fractional equations with bounded primitive*, Appl. Math. Lett. **27** (2014), 53–58.

[86] G. Molica Bisci and B.A. Pansera, *Three weak solutions for nonlocal fractional equations*, Adv. Nonlinear Stud. **14** (2014), 619–629.

[87] G. Molica Bisci and V. Rădulescu, *A characterization for elliptic problems on fractal sets*, Proc. Amer. Math. Soc. **143** (2015), 2959–2968.

[88] G. Molica Bisci and V. Rădulescu, *A sharp eigenvalue theorem for fractional elliptic equations*, Israel J. Math. **219** (2017), 331–351.

[89] G. Molica Bisci and V. Rădulescu, *Ground state solutions of scalar field fractional Schrödinger equations*, Calc. Var. Partial Differential Equations **54** (2015), 2985–3008.

[90] G. Molica Bisci and V. Rădulescu, *Multiplicity results for elliptic fractional equations with subcritical term*, NoDEA Nonlinear Differential Equations Appl. **22** (2015), 721–739.

[91] G. Molica Bisci, V. Rădulescu, and R. Servadei, Variational Methods for Nonlocal Fractional problems. With a Foreword by Jean Mawhin, Encyclopedia of Mathematics and its Applications, *Cambridge University Press*, **162** Cambridge, 2016.

[92] G. Molica Bisci, D. Repovš, and L. Vilasi, *Multiple solutions of nonlinear equations involving the square root of the Laplacian*, Appl. Anal **96** (2017), 1483–1496.

[93] G. Molica Bisci, D. Repovš, and R. Servadei, *Nontrivial solutions of superlinear nonlocal problems*, Forum Math. **28** (2016), 1095–1110.

[94] G. Molica Bisci and R. Servadei, *Lower semicontinuity of functionals of fractional type and applications to nonlocal equations with critical Sobolev exponent*, Adv. Differential Equations **20** (2015), 635–660.

[95] G. Molica Bisci and R. Servadei, *A bifurcation result for nonlocal fractional equations*, Anal. Appl. **13** (2015), 371–394.

[96] K. Perera, R.P. Agarwal, and D. O'Regan, Morse theoretic aspects of p–Laplacian type operators, Mathematical Surveys and Monographs, 161. *American Mathematical Society*, Providence, RI (2010), xx+141 pp.

[97] P. Pucci and V. Rădulescu, *The impact of the mountain pass theory in nonlinear analysis: a mathematical survey*, Boll. Unione Mat. Ital. Series IX, No. 3 (2010), 543–584.

[98] P. Pucci and S. Saldi, *Critical stationary Kirchhoff equations in* $\mathbb{R}^N$ *involving nonlocal operators*, Rev. Mat. Iberoam. **32** (2016), 1–22.

[99] P. Pucci and S. Saldi, *Multiple solutions for an eigenvalue problem involving nonlocal elliptic* p*–Laplacian operators*, in Geometric Methods in PDE's - Springer

INdAM Series - Vol. 11, G. Citti, M. Manfredini, D. Morbidelli, S. Polidoro, F. Uguzzoni Eds., pages 16.
[100] P. Pucci and J. Serrin, *Extensions of the mountain pass theorem*, J. Funct. Anal. **59** (1984), no. 2, 185–210.
[101] P.H. Rabinowitz, Minimax methods in critical point theory with applications to differential equations, *CBMS Reg. Conf. Ser. Math.* **65** *American Mathematical Society*, Providence, RI (1986).
[102] B. Ricceri, *On a classical existence theorem for nonlinear elliptic equations*, in *Esperimental, constructive and nonlinar analysis*, M. Théra ed., CMS Conf. Proc. **27**, Canad. Math. Soc. (2000), 275–278.
[103] B. Ricceri, *A general variational principle and some of its applications*, J. Comput. Appl. Math., Special Issue on Fixed point theory with applications in Nonlinear Analysis **113** (2000), 401–410.
[104] B. Ricceri, *On a three critical points theorem*, Arch. Math. (Basel) **75** (2000), 220–226.
[105] B. Ricceri, *Sublevel sets and global minima of coercive functionals and local minima of their perturbation*, J. Nonlinear Convex Anal. **5** (2004), 157–168.
[106] B. Ricceri, *A three critical points theorem revisited*, Nonlinear Anal. TMA **70** (2009), 3084–3089.
[107] B. Ricceri, *A further three critical points theorem*, Nonlinear Anal. TMA **71** (2009), 4151–4157.
[108] B. Ricceri, *Nonlinear eigenvalue problems*, in: D.Y. Gao, D. Motreanu (Eds.), Handbook of Nonconvex Analysis and Applications, International Press (2010), 543–595.
[109] B. Ricceri, *A multiplicity result for nonlocal problems involving nonlinearities with bounded primitive*, Stud. Univ. Babęs-Bolyai, Math. **55** (2010), 107–114.
[110] B. Ricceri, *A further refinement of a three critical points theorem*, Nonlinear Anal. TMA **74** (2011), 7446–7454.
[111] B. Ricceri, *A new existence and localization theorem for Dirichlet problem*, Dynam. Systems Appl. **22** (2013), 317–324.
[112] F. Riesz and B. Szökefalvi-Nagy, *Lecons d'analyse fonctionnelle*, Académie des Sciences de Hongrie, Akadémiai Kiadó, Budapest (1952), viii+449 pp.
[113] L. Roncal and P.R. Stinga, *Fractional Laplacian on the torus*, Commun. Contemp. Math. **18** (2016), no. 3, 1550033, 26 pp.
[114] L. Roncal and P.R. Stinga, *Transference of fractional Laplacian regularity*, in Special functions, partial differential equations, and harmonic analysis, 203–212, Springer Proc. Math. Stat., **108**, Springer, Cham. (2014).
[115] X. Ros-Oton and J. Serra, *Nonexistence results for nonlocal equations with critical and supercritical nonlinearities*, Comm. Partial Differential Equations 40 (2015), 115–133.
[116] M. Ryznar, *Estimate of Green function for relativistic α−stable processes*, Potential Anal. **17** (2002), 1–23.

[117] S. Salsa, *The problems of the obstacle in lower dimension and for the fractional Laplacian*, Regularity estimates for nonlinear elliptic and parabolic problems, 153–244, Lecture Notes in Math., 2045, *Springer*, Heidelberg (2012).

[118] S. Secchi, *Ground state solutions for nonlinear fractional Schrödinger equations in* $\mathbb{R}^N$, J. Math. Phys. **54**, 031501 (2013).

[119] S. Secchi, *Perturbation results for some nonlinear equations involving fractional operators*, Differ. Equ. Appl. **5** (2013), 221–236.

[120] R. Servadei, *Infinitely many solutions for fractional Laplace equations with subcritical nonlinearity*, Contemporary Mathematics **595** (2013), 317–340.

[121] R. Servadei, *The Yamabe equation in a non-local setting*, Adv. Nonlinear Anal. **2** (2013), 235–270.

[122] R. Servadei, *A critical fractional Laplace equation in the resonant case*, Topol. Methods Nonlinear Anal. **43** (2014), 251–267.

[123] R. Servadei and E. Valdinoci, *Mountain Pass solutions for non–local elliptic operators*, J. Math. Anal. Appl. **389** (2012), 887–898.

[124] R. Servadei and E. Valdinoci, *A Brezis–Nirenberg result for non–local critical equations in low dimension*, Commun. Pure Appl. Anal. **12** (2013) 2445–2464.

[125] R. Servadei and E. Valdinoci, *Variational methods for non–local operators of elliptic type*, Discrete Contin. Dyn. Syst. **33** (2013), 2105–2137.

[126] R. Servadei and E. Valdinoci, *The Brezis–Nirenberg result for the fractional Laplacian*, Trans. Amer. Math. Soc. **367** (2015), 67–102.

[127] R. Stinga and J.L. Torrea, *Extension problem and Harnack's inequality for some fractional operators*, Comm. Partial Diff. Eq. (2000).

[128] M. Struwe, *Infinitely many critical points for functionals which are not even and applications to superlinear boundary value problems*, Manuscripta Math. **32** (1980), 335–364.

[129] M. Struwe, Variational methods, Applications to nonlinear partial differential equations and Hamiltonian systems, *Ergebnisse der Mathematik und ihrer Grenzgebiete*, 3, *Springer Verlag*, Berlin–Heidelberg (1990).

[130] A. Szulkin and T. Weth, *The method of Nehari manifold*, in Handbook of Nonconvex Analysis and Applications, edited by D.Y. Gao and D. Montreanu (International Press, Boston, 2010), pp. 597–632.

[131] K. Yosida, *Functional Analysis*, Die Grundlehren der Mathematischen Wissenschaften, Band 123 Academic Press, Inc., New York; Springer-Verlag, Berlin (1965), xi+458 pp.

[132] J. Zhang and X. Liu, *Three solutions for a fractional elliptic problems with critical and supercritical growth*, Acta Math. Sci. Ser. B Engl. Ed. **36** (2016), 1819–1831.

[133] L. Zhao and P. Zhao, *The existence of solutions for p-Laplacian problems with critical and supercritical growth*, Rocky Mountain J. Math. **44** (2014), 1383–1397.

Stefania Patrizi

Change of Scales for Crystal Dislocation Dynamics

Abstract: We present some recent results obtained in [12, 13, 19, 26, 28, 29, 35, 36, 37, 38] for equations of evolutionary type, run by fractional and possibly anisotropic fractional operators. The models considered arise natural in crystallography, in which the solution of the equation has the physical meaning of the atom dislocation inside the crystal structure. Since different scales come into play in such description, different models have been adopted in order to deal with phenomena at atomic, microscopic, mesoscopic and macroscopic scale. We show that, looking at the asymptotic states of the solutions of equations modeling the dynamics of dislocations at a given scale, one can deduce the model for the motion of dislocations at a larger scale.

11.1 Introduction

Dislocations are line defects in crystals. Their typical length is of the order of $10^{-6}m$ and their thickness of order of $10^{-9}m$. When the material is submitted to shear stress, these lines can move in the crystallographic planes and their dynamics is one of the main explanation of the plastic behavior of metals. Dislocations can be described at several scales by different models:

(a) atomic scale (Frenkel-Kontorova model),
(b) microscopic scale (Peierls-Nabarro model),
(c) mesoscopic scale (Discrete Dislocation Dynamics),
(d) macroscopic scale (elasto-visco-plasticity with density of dislocations).

We refer the reader to the book of Hirth and Lothe [23] for a detailed introduction to dislocations. In physics and mechanics, it is a great challenge to try to predict macroscopic elasto-visco-plasticity properties of materials (like metals), based on microscopic properties (like dislocations). The classical Frenkel-Kontorova model describes a chain of classical particles evolving in a one dimensional space, coupled with their neighbors and subjected to a periodic potential, see the book of Braun and Kivshar [6] for a detailed presentation of the model. The Peierls-Nabarro model has been originally introduced as a variational (stationary) model (see [32, 23]), in which the microscopic effects are described by a partial differential equation involving a fractional and possibly anisotropic operator of order 1 of elliptic type. The asymptotics of the sta-

Stefania Patrizi, Department Of Mathematics, University of Texas at Austin, 2515 Speedway, Austin TX 78712, United States, E-mail: spatrizi@math.utexas.edu

https://doi.org/10.1515/9783110571561-013

tionary model have been characterized in a number of mathematical papers within the framework of Γ-convergence, see, e.g., the works by Garroni, Leoni, Müller and collaborators [17, 18, 9, 8, 2] and references therein also for related models. In the face cubic structured (FCC) observed in many metals and alloys, dislocations move at low temperature on the slip plane.The dynamics of dislocations at microscopic scale is then described by the evolutive version of the Peierls-Nabarro model, see for instance [11, 30].

Several changes of scale exist in the literature. In dimension 1, the passage from the Frenkel-Kontorova model to Peierls-Nabarro model (*from (a) to (b)*) has been performed in [14]. The evolutive Peierls-Nabarro equation in dimension 1 where the fractional operator is the half-Laplacian, has been considered by González and Monneau [19]. The equation models the case of parallel straight edge dislocation lines in the same slip plane. In this setting, after a suitable section of a three-dimensional crystal with a transverse plane, dislocation lines can be identified with points lying in the same line. We will refer to them as *particles* (though no "material" particle is really involved). In [19], looking at the sharp interface limit of the phase transitions of the Peierls-Nabarro model, the authors identify a dynamics of particles that corresponds to the classical discrete dislocation dynamics (*from (b) to (c)*). The results of [19] have been extended to the fractional Laplacian of order $2s \in (0, 2)$ by Dipierro, Figalli, Palatucci and Valdinoci [13, 12]. The evolutive, generalized ($s \in (0, 1)$) Peierls-Nabarro equation in dimension 1 has been considered again by the author and E. Valdinoci in [35]. Here, differently from the existing literature, dislocation particles are allowed to have different orientations. This produces a new phenomenon: *collision of particles*. Indeed particles with opposite orientations have the tendency to attract each other. The main difficulty here is that when a collision occurs the mesoscopic scale model becomes singular and we loose information about the dynamics of the particles after the collision. We have been able to overcome this difficulty in [37, 38], where we describe the dynamics of dislocations for times bigger than the collision time. We want to mention the papers [5, 1], for related results about the dynamics of dislocations for short times.

For $N \geqslant 2$, the large scale limit of a single phase transition described by Peierls-Nabarro shows that the line tension effect is the much stronger term. The limit model for $s \geqslant \frac{1}{2}$ appears to be the mean curvature motion as proven by Imbert and Souganidis [26]. For $s < \frac{1}{2}$, only partial but significant results have been obtained in [26], suggesting that in this case the front moves by fractional mean curvature. The results of [26] can be seen as the non-local counterpart of those obtained for the Allen-Cahn equation (see [10]). The double limit behavior of the Peierls-Nabarro model in dimensions greater than 1 has been also stressed in [39]. Here the authors show that the Peierls-Nabarro energy in a bounded domain of $\mathbb{R}^N$, $N \geqslant 2$, Γ-converges to the classical minimal surface functional when $s \in \left[\frac{1}{2}, 1\right)$, while it Γ-converges to the non-local minimal surface functional when $s \in \left(0, \frac{1}{2}\right)$.

The passage from the Discrete Dislocation Dynamics model to the Dislocation Density model in dimension 1 has been performed in [15] *(from (c) to (d))*.

In [29] we have investigated the large scale limit of the evolutive Peierls-Nabarro model in any dimension N, in the case of a large number of phase transitions (i.e. of dislocations), recovering at the limit a model with evolution of dislocation densities. In other words, a direct passage from the microscopic scale (Peierls-Nabarro model) to the macroscopic scale (elasto-visco-plasticity with density of dislocations), has been performed *(from (b) to (d))*. From a mathematical point of view, this is an homogenization problem for an evolutive equation run by a fractional and possibly anisotropic operator of order 1 of elliptic type, usually called Lévy operator. The homogenized limit equation can be interpreted as the plastic flow rule in a model for macroscopic crystal plasticity. In [28] we have been able to explicitly characterize the macroscopic equation in the case of parallel straight edge dislocation lines in the same slip plane with the same Burgers' vectors, moving with self-interactions. This result recovers the so called Orowan's law. In the Physics literature this was proposed by Head in [22]. The results of [29, 28] have been extended to equations with anisotropic fractional operators of any order $2s \in (0, 2)$ in [36]. The scaling of the system and the results obtained are different according to the fractional parameter s. Namely, when $s > \frac{1}{2}$ the effective Hamiltonian "localizes" and it only depends on a first order differential operator. Conversely, when $s < \frac{1}{2}$, the non-local features are predominant and the effective Hamiltonian involves the fractional operator of order s.

The aim of the present paper is to describe and explain some recent results for problems involving non-linear and non-local partial differential equations with application to the theory of crystal dislocations. Due to their mathematical interest and in view of the concrete applications in physical models, these problems have been extensively studied in the recent literature. The results presented here, in particular, are contained in [12, 13, 19, 26, 28, 29, 35, 36, 37, 38]. Proofs are quite technical, so we have decided not to include them in the paper. We prefer instead to give heuristic explanations of the theorems we are going to state. We want to mention the paper [21] for a presentation of results prior to [12, 13, 28, 29, 35, 36, 37, 38], related to passages of scales in dimension 1, from the atomic one to the macroscopic one.

Organization of the paper

The paper is organized as follows. We first recall the classical Peierls-Nabarro model, in Section 11.2. In Section 11.3 we describe the passage from the Peierls-Nabarro model to the Discrete Dislocation Dynamics model. The results here presented are contained in [12, 13, 19, 35, 37, 38]. In Section 11.4 we review the homogenization of the Peierls-Nabarro model which represents the passage from the microscopic to the macroscopic scale, and it is studied in [28, 29, 36]. In Section 11.5, we give a heuristic explanation of some of the results proven in [26], which can be seen as the analogous of the results of Section 11.3 in dimension greater than 1. Finally, some open problems are presented in Section 11.6.

11.2 The Peierls-Nabarro Model

The Peierls-Nabarro model is a phase field model for dislocation dynamics incorporating atomic features into continuum framework. In a phase field approach, the dislocations are represented by transition of a continuous field. We briefly review the model. There are two basic types of dislocations: the edge dislocation and the screw dislocation. In both cases the motion of a dislocation is a result of shear stress, but for a screw dislocation, the defect line movement is perpendicular to direction of the stress and the atom displacement, while for an edge dislocations it is parallel, see [23] for more details. As an example, consider an edge dislocation in a crystal with simple cubic lattice. In a Cartesian system of coordinates $x_1x_2x_3$, we assume that the dislocation is located in the slip plane x_1x_2 (where the dislocation can move) and that the Burgers' vector (i.e. a fixed vector associated to the dislocation) is in the direction of the x_1 axis. We write this Burgers' vector as be_1 for a real b. The disregistry of the upper half crystal $\{x_3 > 0\}$ relative to the lower half $\{x_3 < 0\}$ in the direction of the Burgers' vector is $u(x_1, x_2)$, where u is a phase parameter between 0 and b. Then the dislocation loop can be for instance localized by the level set $u = b/2$. For a closed loop, we expect to have $u \simeq b$ inside the loop and $u \simeq 0$ far outside the loop. In the Peierls-Nabarro model, the total energy is given by

$$\mathcal{E} = \mathcal{E}^{el} + \mathcal{E}^{mis}. \tag{11.2.1}$$

In (11.2.1), $\mathcal{E}^{el}$ is the elastic energy induced by the dislocation, and $\mathcal{E}^{mis}$ is the so called *misfit energy* due to the nonlinear atomic interaction across the slip plane,

$$\mathcal{E}^{mis}(u) = \int_{\mathbb{R}^2} W(u(x))\,dx \quad \text{with} \quad x = (x_1, x_2),$$

where $W(u)$ is the interplanar potential. In the classical Peierls-Nabarro model [34, 31], $W(u)$ is approximated by the sinusoidal potential

$$W(u) = \frac{\mu b^2}{4\pi^2 d}\left(1 - \cos\left(\frac{2\pi u}{b}\right)\right),$$

where d is the lattice spacing perpendicular to the slip plane.

The elastic energy $\mathcal{E}^{el}$ induced by the dislocation is (for $X = (x, x_3)$ with $x = (x_1, x_2)$)

$$\mathcal{E}^{el}(u, U) = \frac{1}{2}\int_{\mathbb{R}^3} e : \Lambda : e\,dX \quad \text{with} \quad e = e(U) - u(x)\delta_0(x_3)e^0,$$

and

$$\begin{cases} e(U) = \frac{1}{2}\left(\nabla U + (\nabla U)^T\right) \\ e^0 = \frac{1}{2}\left(e_1 \otimes e_3 + e_3 \otimes e_1\right) \end{cases}$$

where $U : \mathbb{R}^3 \to \mathbb{R}^3$ is the displacement and $\Lambda = \{\Lambda_{ijkl}\}$ are the elastic coefficients. Given the field u, we minimize the energy $\mathcal{E}^{el}(u, U)$ with respect to the displacement U and define

$$\mathcal{E}^{el}(u) = \inf_U \mathcal{E}^{el}(u, U).$$

Following the proof of Proposition 6.1 (iii) in [3], we can see that (at least formally)

$$\mathcal{E}^{el}(u) = -\frac{1}{2}\int_{\mathbb{R}^2} (c_0 \star u)u$$

where c_0 is a certain kernel. In the case of isotropic elasticity, we have

$$\Lambda_{ijkl} = \lambda\delta_{ij}\delta_{kl} + \mu\left(\delta_{ik}\delta_{jl} + \delta_{il}\delta_{jk}\right)$$

where λ, μ are the Lamé coefficients. Then the kernel c_0 can be written (see Proposition 6.2 in [3], translated in our framework):

$$c_0(x) = \frac{\mu}{4\pi}\left(\partial_{22}\frac{1}{|x|} + \omega\partial_{11}\frac{1}{|x|}\right) \quad \text{with} \quad \omega = \frac{1}{1-\nu} \quad \text{and} \quad \nu = \frac{\lambda}{2(\lambda+\mu)}$$

where $\nu \in (-1, 1/2)$ is called the Poisson ratio.

The equilibrium configuration of the edge dislocation is obtained by minimizing the total energy with respect to u, under the constraint that far from the dislocation core, the function u tends to 0 in one half plane and to b in the other half plane. In particular, the phase transition u is then solution of the following equation

$$\mathbb{I}[u] = W'(u) \quad \text{on } \mathbb{R}^N, \tag{11.2.2}$$

where

$$\mathbb{I}[u] = \text{PV}\, c_0 \star u = \text{PV}\int_{\mathbb{R}^N} \frac{u(x+y) - u(x)}{|y|^{N+2s}}\, g\left(\frac{y}{|y|}\right) dy, \tag{11.2.3}$$

where PV stands for principal value, $s = \frac{1}{2}$, $N = 2$ and $g(z_1, z_2) = \frac{\mu}{4\pi}\left((2\omega - 1)z_1^2 + (2-\omega)z_2^2\right)$. This operator is known as anisotropic (if $g \not\equiv$ constant) Lévy operator (of order 2s=1). If $\omega = 1$ and $\mu = 2$, then $\mathcal{I}_{\frac{1}{2}} = -(-\Delta)^{\frac{1}{2}}$ (isotropic case). In that special case, we recall that the solution u of (11.2.2) satisfies $u(x) = \tilde{u}(x, 0)$ where $\tilde{u}(X)$ is the solution of (see [27, 19])

$$\begin{cases} \Delta\tilde{u} = 0 & \text{in} \quad \{x_3 > 0\} \\ \dfrac{\partial\tilde{u}}{\partial x_3} = W'(\tilde{u}) & \text{on} \quad \{x_3 = 0\}. \end{cases}$$

Moreover, we have in particular an explicit solution for $b = 1$, $d = 2$ (with $W'(\tilde{u}) = \frac{1}{2\pi}\sin(2\pi\tilde{u})$)

$$\tilde{u}(X) = \frac{1}{2} + \frac{1}{\pi}\arctan\left(\frac{x_1}{x_3+1}\right). \tag{11.2.4}$$

In the original Peierls-Nabarro model, the dislocation line is assumed to be straight, say perpendicular to the x_1 axis. In this simplified case, the displacement is a function of only one variable and the phase field function satisfies (11.2.2) with $N = 1$. It is easy to check that we can recover the explicit solution found in Nabarro [31] by rescaling (11.2.4):

$$\begin{cases} u(x) = \dfrac{b}{2} + \dfrac{b}{\pi}\arctan\left(\dfrac{2(1-\nu)x_1}{d}\right) & \text{(edge dislocation)} \\ u(x) = \dfrac{b}{2} + \dfrac{b}{\pi}\arctan\left(\dfrac{2x_2}{d}\right) & \text{(screw dislocation).} \end{cases}$$

In a more general model, one can consider a potential W satisfying

$$\begin{aligned} &(i)\;\; W(v+b) = W(v) \quad \text{for all } v \in \mathbb{R}; \\ &(ii)\;\; W(b\mathbb{Z}) = 0 < W(a) \quad \text{for all } a \in \mathbb{R}\setminus b\mathbb{Z}. \end{aligned} \tag{11.2.5}$$

The periodicity of W reflects the periodicity of the crystal, while the minimum property is consistent with the fact that the perfect crystal is assumed to minimize the energy. We can assume wlog that $b = 1$. Then the 1-D phase transition is solution to:

$$\begin{cases} \mathbb{I}[u] = W'(u) & \text{in } \mathbb{R} \\ u' > 0 & \text{in } \mathbb{R} \\ \lim\limits_{x\to-\infty} u(x) = 0, \quad \lim\limits_{x\to+\infty} u(x) = b, \quad u(0) = \dfrac{1}{2}. \end{cases} \tag{11.2.6}$$

The existence of a unique solution of (11.2.6), when $\mathbb{I} = -(-\Delta)^s$ and under an additional non degeneracy assumption on the second derivative of the potential, has been proven independently by Palatucci, Savin and Valdinoci in [33] and by Cabré and Y. Sire in [7] for any $s \in (0,1)$. Asymptotic estimates for u and u' are given in [33]. Finer estimates on u are shown in [13] and [12] respectively when $s \in \left[\frac{1}{2}, 1\right)$ and $s \in \left(0, \frac{1}{2}\right)$. We collect these results in the following lemma

Lemma 11.2.1. *Assume $g \equiv 1$ and*

$$\begin{cases} W \in C^{3,\alpha}(\mathbb{R}) & \text{for some } 0 < \alpha < 1 \\ W(v+1) = W(v) & \text{for any } v \in \mathbb{R} \\ W = 0 & \text{on } \mathbb{Z} \\ W > 0 & \text{on } \mathbb{R}\setminus\mathbb{Z} \\ W'' > 0 & \text{on } \mathbb{Z}. \end{cases} \tag{11.2.7}$$

Then there exists a unique solution $u \in C^{2,\alpha}(\mathbb{R})$ of (11.2.6). Moreover, there exists a constant $C > 0$ and $\kappa > 2s$ (only depending on s) such that

$$\left|u(x) - H(x) + \frac{1}{2sW''(0)}\frac{x}{|x|^{2s+1}}\right| \leqslant \frac{C}{|x|^{\kappa}}, \quad \textit{for } |x| \geqslant 1, \tag{11.2.8}$$

and

$$|u'(x)| \leqslant \frac{C}{|x|^{1+2s}} \quad \textit{for } |x| \geqslant 1, \tag{11.2.9}$$

where H is the Heaviside function.

According to [13], the constant κ in (11.2.8) can be chosen to be optimal equal to $1+2s$. Remark that when $g \equiv 1$, then $-\mathbb{I} = C_{N,s}^{-1}(-\Delta)^s$, for a suitable constant $C_{N,s}$ depending on N and s.

In the face cubic structured (FCC) observed in many metals and alloys, dislocations move at low temperature on the slip plane. A collection of dislocations curves all contained in a single slip plane x_1x_2, and moving in a landscape with periodic obstacles (that can be for instance precipitates in the material) are represented by a single phase parameter $u(t, x_1, x_2)$ defined on the plane x_1x_2. The dynamics of dislocations is then described by the evolutive version of the Peierls-Nabarro model (see for instance [30] and [11]):

$$\partial_t u = \mathbb{I}[u(t,\cdot)] - W'(u) + \sigma_{13}^{\text{obst}}(t,x) \quad \text{in} \quad \mathbb{R}^+ \times \mathbb{R}^N \tag{11.2.10}$$

with the physical dimension $N = 2$. In the model, the component $\sigma_{13}^{\text{obst}}$ of the stress (evaluated on the slip plane) has been introduced to take into account the shear stress not created by the dislocations themselves. This shear stress is created by the presence of the obstacles and the possible external applied stress on the material.

11.3 From the Peierls-Nabarro Model to the Discrete Dislocation Dynamics Model

The evolutive Peierls-Nabarro model (11.2.10) in dimension $N = 1$, describes at microscopic scale the dynamics of a collection of *parallel* and *straight* edge dislocations all lying in the same slip plane x_1x_2. Suppose that the dislocation lines are perpendicular to the x_1 axis, then, after a section of a three-dimensional crystal with the plane x_1x_3, they can be identified with points lying in the x_1 axis. We will refer to them as *particles*. We want to identify at a larger scale, the mesoscopic one, an evolution model for the Discrete Dislocation Dynamics. In the entire section we will assume that $g \equiv 1$, that is $\mathbb{I}$ is, up to a multiplicative constant, the operator $-(-\Delta)^s$. Assume in addition that the exterior stress in (11.2.10) has the following form $\sigma_{13}^{\text{obst}} = \varepsilon^{2s}\sigma(\varepsilon^{1+2s}t, \varepsilon x)$. We perform the following rescaling of the solution u of (11.2.10)

$$v_\varepsilon(t,x) = u\left(\frac{t}{\varepsilon^{1+2s}}, \frac{x}{\varepsilon}\right),$$

where $\varepsilon > 0$ is a small parameter representing the ratio between microscopic scale and the mesoscopic scale. Then, the function $v_\varepsilon(t,x)$ solves

$$\begin{cases} (v_\varepsilon)_t = \dfrac{1}{\varepsilon}\left(\mathbb{I}[v_\varepsilon] - \dfrac{1}{\varepsilon^{2s}}W'(v_\varepsilon) + \sigma(t,x)\right) & \text{in } (0,+\infty)\times\mathbb{R} \\ v_\varepsilon(0,\cdot) = v_\varepsilon^0 & \text{on } \mathbb{R}, \end{cases} \tag{11.3.1}$$

for a suitable initial condition to be chosen. Assume that the potential W satisfies (11.2.7). We suppose in addition that σ satisfies

$$\begin{cases} \sigma \in BUC([0,+\infty)\times\mathbb{R}) \quad \text{and for some } M > 0 \text{ and } \alpha \in (s,1) \\ \|\sigma_x\|_{L^\infty([0,+\infty)\times\mathbb{R})} + \|\sigma_t\|_{L^\infty([0,+\infty)\times\mathbb{R})} \leqslant M \\ |\sigma_x(t,x+h) - \sigma_x(t,x)| \leqslant M|h|^\alpha, \quad \text{for every } x, h \in \mathbb{R} \text{ and } t \in [0,+\infty). \end{cases} \tag{11.3.2}$$

Given $x_1^0 < x_2^0 < \dots < x_N^0$, we say that the function $u\left(\frac{x-x_i^0}{\varepsilon}\right)$, where u is the solution of (11.2.6), is a transition layer centered at x_i^0 and positively oriented. Similarly, we say that the function $u\left(\frac{x_i^0-x}{\varepsilon}\right) - 1$ is a transition layer centered at x_i^0 and negatively oriented. Then the positively oriented transition layer connects the "rest states" 0 and 1, while the negatively oriented one connects 0 with -1. Remark that, since the equation in (11.2.6) is invariant by translation, the potential W is 1-periodic and $\mathbb{I}[u(-\cdot)](x) = \mathbb{I}[u(\cdot)](-x)$, we have that $u\left(\zeta\frac{x-x^0}{\varepsilon}\right) - k$ is solution of the same equation for any $x^0 \in \mathbb{R}$, $k \in \mathbb{Z}$ and $\zeta \in \{-1,1\}$. The positively (resp., negatively) oriented transition layer identifies a dislocation particle located at the position x_i^0 with Burgers' vector e_1 (resp., $-e_1$). We consider as initial condition in (11.3.1) the state obtained by superposing N copies of the transition layer, centered at $x_1^0, \dots, x_N^0$, $N-K$ of them positively oriented and the remaining K negatively oriented, that is

$$v_\varepsilon^0(x) = \frac{\varepsilon^{2s}}{W''(0)}\sigma(0,x) + \sum_{i=1}^N u\left(\zeta_i\frac{x-x_i^0}{\varepsilon}\right) - K, \tag{11.3.3}$$

where $\zeta_1, \dots, \zeta_N \in \{-1,1\}$, $\sum_{i=1}^N (\zeta_i)^- = K$, $0 \leqslant K \leqslant N$. Here, we denote by $(\zeta)^-$ the function defined by: $(\zeta)^- = 0$ if $\zeta \geqslant 0$ and $(\zeta)^- = \zeta$ if $\zeta < 0$. The first term on the right-hand side of (11.3.3) takes into account the influence of the exterior stress σ. The initial condition (11.3.3) models an initial configuration in which there are N parallel and straight edge dislocation lines, all lying in the same slip plane, x_1x_2, $N-K$ of them with the same Burgers' vector e_1, the remaining K with Burgers' vector $-e_1$.

Let us introduce the solution $(x_i(t))_{i=1,\dots,N}$ to the system

$$\begin{cases} \dot{x}_i = \gamma \left(\sum_{j \neq i} \zeta_i \zeta_j \frac{x_i - x_j}{2s|x_i - x_j|^{1+2s}} - \zeta_i \sigma(t, x_i) \right) & \text{in } (0, T_c) \\ x_i(0) = x_i^0, \end{cases} \tag{11.3.4}$$

where $\gamma := \left(\int_{\mathbb{R}} (u'(x))^2 dx \right)^{-1}$, with u solution of (11.2.6). The physical properties of the singular potential of this ODE's system depend on the orientation of the dislocation function at the transition points. Namely, if the particles x_i and x_{i+1} have the same orientation, i.e., $\zeta_i \zeta_{i+1} = 1$, then the potential induces a repulsion between them. Conversely, when they have opposite orientation, i.e., $\zeta_i \zeta_{i+1} = -1$, then the potential becomes attractive, and the two particles may collide in a finite time T_c. Therefore, $(0, T_c)$ is the maximal interval where the system (11.3.4) is well defined. In formulas, in the collision case we have that $x_i(t) \neq x_{i+1}(t)$ for any $t \in [0, T_c)$ and any $i = 1, \dots, N$, with

$$\lim_{t \to T_c^-} x_{i_0}(t) = \lim_{t \to T_c^-} x_{i_0+1}(t), \tag{11.3.5}$$

for some $i_0 \in \{1, \dots, N\}$. Estimates of the collision time in the case of particles with alternating orientations and in the case in which two consecutive particles with opposite orientation are sufficiently close at the initial time, are given in [35].

We are now ready, to describe the asymptotic behavior of the dislocation function v_ε. For small ε, the solution v_ε of (11.3.1)-(11.3.3) approaches a piecewise constant function. The plateaus of this asymptotic limit correspond to the periodic sites induced by the crystalline structure, but its jump points evolve in time. Roughly speaking, one can imagine that these points behave like a particle system driven by the system of ordinary differential equations (11.3.4). System (11.3.4) *can be interpreted as a mesoscopic model for the Discrete Dislocation Dynamics.*

Let us state the result precisely. First, we recall that the (upper and lower) semi-continuous envelopes of a function v are defined as

$$v^*(t, x) := \limsup_{(t',x') \to (t,x)} v(t', x')$$

and

$$v_*(t, x) := \liminf_{(t',x') \to (t,x)} v(t', x').$$

Theorem 11.3.1 (Theorem 1.1, [35]). *Assume that* (11.2.7), (11.3.2) *and* (11.3.3) *hold, and let*

$$v_0(t, x) = \sum_{i=1}^{N} H(\zeta_i (x - x_i(t))) - K, \tag{11.3.6}$$

where H is the Heaviside function and $(x_i(t))_{i=1,\dots,N}$ *is the solution to* (11.3.4)*. Then, for every* $\varepsilon > 0$ *there exists a unique solution* v_ε *to* (11.3.1)*. Furthermore, as* $\varepsilon \to 0^+$*, the* *solution v_ε exhibits the following asymptotic behavior:*

$$\limsup_{\substack{(t',x')\to(t,x)\\ \varepsilon\to 0+}} v_\varepsilon(t',x') \leqslant (v_0)^*(t,x) \tag{11.3.7}$$

and

$$\liminf_{\substack{(t',x')\to(t,x)\\ \varepsilon\to 0+}} v_\varepsilon(t',x') \geqslant (v_0)_*(t,x), \tag{11.3.8}$$

for any $(t,x)\in[0,T_c)\times\mathbb{R}$.

Theorem 11.3.1 was already proven in [19, 13, 12] respectively in the case $s=\frac{1}{2}$, $s\in\left(\frac{1}{2},1\right)$ and $s\in\left(0,\frac{1}{2}\right)$ for a collection of dislocation lines all with the same orientation, i.e, $K=0$. In this particular case the particles points have the tendency of repel each other, so collisions do not occur, i.e, $T_c=+\infty$.

Let us now give an heuristic explanation of the result of Theorem 11.3.1.

11.3.1 Heuristics of the dynamics

This subsection is contained in [35]. We think that it could be useful to understand the heuristic derivation of (11.3.4) in the simpler setting of two particles with different orientations (i.e. $N=2$ and $K=1$).

For this, let u be the solution of (11.2.6). Let us introduce the notation

$$u_{\varepsilon,1}(t,x) := u\left(\frac{x-x_1(t)}{\varepsilon}\right),\quad u_{\varepsilon,2}(t,x) := u\left(\frac{x_2(t)-x}{\varepsilon}\right)-1,$$

and with a slight abuse of notation

$$u'_{\varepsilon,1}(t,x) := u'\left(\frac{x-x_1(t)}{\varepsilon}\right),\quad u'_{\varepsilon,2}(t,x) := u'\left(\frac{x_2(t)-x}{\varepsilon}\right).$$

Let us consider the following ansatz for v_ε

$$v_\varepsilon(t,x) \simeq u_{\varepsilon,1}(t,x)+u_{\varepsilon,2}(t,x) = u\left(\frac{x-x_1(t)}{\varepsilon}\right)+u\left(\frac{x_2(t)-x}{\varepsilon}\right)-1.$$

Then, we compute

$$\begin{aligned}(v_\varepsilon)_t &= -u'\left(\frac{x-x_1(t)}{\varepsilon}\right)\frac{\dot{x}_1(t)}{\varepsilon}+u'\left(\frac{x_2(t)-x}{\varepsilon}\right)\frac{\dot{x}_2(t)}{\varepsilon}\\ &= -u'_{\varepsilon,1}(t,x)\frac{\dot{x}_1(t)}{\varepsilon}+u'_{\varepsilon,2}(t,x)\frac{\dot{x}_2(t)}{\varepsilon},\end{aligned}$$

and using the equation (11.2.6) and the periodicity of W

$$
\begin{aligned}
\mathbb{I}v_\varepsilon(t,x) &= \frac{1}{\varepsilon^{2s}}\mathbb{I}u\left(\frac{x-x_1(t)}{\varepsilon}\right) + \frac{1}{\varepsilon^{2s}}\mathbb{I}u\left(\frac{x_2(t)-x}{\varepsilon}\right)\\
&= \frac{1}{\varepsilon^{2s}}W'\left(u\left(\frac{x-x_1(t)}{\varepsilon}\right)\right) + \frac{1}{\varepsilon^{2s}}W'\left(u\left(\frac{x_2(t)-x}{\varepsilon}\right)\right)\\
&= \frac{1}{\varepsilon^{2s}}W'(u_{\varepsilon,1}(t,x)) + \frac{1}{\varepsilon^{2s}}W'(u_{\varepsilon,2}(t,x)).
\end{aligned}
$$

By inserting into (11.3.1), we obtain

$$
-u'_{\varepsilon,1}\frac{\dot{x}_1}{\varepsilon} + u'_{\varepsilon,2}\frac{\dot{x}_2}{\varepsilon} = \frac{1}{\varepsilon^{2s+1}}\Big(W'(u_{\varepsilon,1}) + W'(u_{\varepsilon,2}) - W'(u_{\varepsilon,1}+u_{\varepsilon,2})\Big) + \frac{\sigma}{\varepsilon}. \tag{11.3.9}
$$

Now we make some observations on the asymptotics of the potential W. First of all, we notice that the periodicity of W and the asymptotic behavior of u imply

$$
\int_{\mathbb{R}} W'(u(x))u'(x)dx = \int_{\mathbb{R}} \frac{d}{dx}W(u(x))dx = W(1) - W(0) = 0, \tag{11.3.10}
$$

and similarly

$$
\int_{\mathbb{R}} W''(u(x))u'(x)dx = 0. \tag{11.3.11}
$$

Next, we use estimate (11.2.8) and make a Taylor expansion of W' at 0 to compute for $x \neq x_2$

$$
\begin{aligned}
W'\left(u\left(\frac{x_2-x}{\varepsilon}\right)\right) &\simeq W'\left(H\left(\frac{x_2-x}{\varepsilon}\right) + \frac{\varepsilon^{2s}(x-x_2)}{2sW''(0)|x-x_2|^{1+2s}}\right)\\
&= W'\left(\frac{\varepsilon^{2s}(x-x_2)}{2sW''(0)|x-x_2|^{1+2s}}\right)\\
&\simeq W''(0)\frac{\varepsilon^{2s}(x-x_2)}{2sW''(0)|x-x_2|^{1+2s}}\\
&= \frac{\varepsilon^{2s}(x-x_2)}{2s|x-x_2|^{1+2s}}.
\end{aligned}
$$

So, we use the substitution $y = (x - x_1)/\varepsilon$ to see that

$$
\begin{aligned}
\frac{1}{\varepsilon}\int_{\mathbb{R}} W'(u_{\varepsilon,2}(t,x))u'_{\varepsilon,1}(t,x)dx &\simeq \frac{1}{\varepsilon}\int_{\mathbb{R}} \frac{\varepsilon^{2s}(x-x_2)}{2s|x-x_2|^{1+2s}}u'\left(\frac{x-x_1}{\varepsilon}\right)dx\\
&= \int_{\mathbb{R}} \frac{\varepsilon^{2s}(\varepsilon y + x_1 - x_2)}{2s|\varepsilon y + x_1 - x_2|^{1+2s}}u'(y)dy\\
&\simeq \frac{\varepsilon^{2s}(x_1-x_2)}{2s|x_1-x_2|^{1+2s}}\int_{\mathbb{R}} u'(y)dy\\
&= \frac{\varepsilon^{2s}(x_1-x_2)}{2s|x_1-x_2|^{1+2s}},
\end{aligned}
$$

if $x_1 \neq x_2$. Hence

$$
\frac{1}{\varepsilon^{2s+1}}\int_{\mathbb{R}} W'(u_{\varepsilon,2}(t,x))u'_{\varepsilon,1}(t,x)dx \simeq \frac{x_1-x_2}{2s|x_1-x_2|^{1+2s}}, \tag{11.3.12}
$$

if $x_1 \neq x_2$. We use again the substitution $y = (x - x_1)/\varepsilon$, (11.3.10) and (11.3.11) to get

$$\begin{aligned}
&\frac{1}{\varepsilon}\int_{\mathbb{R}} W'(u_{\varepsilon,1}(t,x) + u_{\varepsilon,2}(t,x))u'_{\varepsilon,1}(t,x)dx \\
&\simeq \frac{1}{\varepsilon}\int_{\mathbb{R}} W'\left(u\left(\frac{x-x_1}{\varepsilon}\right) + H(x) + \frac{\varepsilon^{2s}(x-x_2)}{2sW''(0)|x-x_2|^{1+2s}}\right)u'\left(\frac{x-x_1}{\varepsilon}\right)dx \\
&= \int_{\mathbb{R}} W'\left(u(y) + \frac{\varepsilon^{2s}(\varepsilon y + x_1 - x_2)}{2sW''(0)|\varepsilon y + x_1 - x_2|^{1+2s}}\right)u'(y)dy \\
&\simeq \int_{\mathbb{R}} W'(u(y))u'(y)dy + \int_{\mathbb{R}} W''(u(y))\frac{\varepsilon^{2s}(\varepsilon y + x_1 - x_2)}{2sW''(0)|\varepsilon y + x_1 - x_2|^{1+2s}}u'(y)dy \\
&\simeq \frac{\varepsilon^{2s}(x_1 - x_2)}{2sW''(0)|x_1 - x_2|^{1+2s}}\int_{\mathbb{R}} W''(u(y))u'(y)dy \\
&= 0.
\end{aligned}$$

We deduce

$$\frac{1}{\varepsilon^{1+2s}}\int_{\mathbb{R}} W'(u_{\varepsilon,1}(t,x) + u_{\varepsilon,2}(t,x))u'_{\varepsilon,1}(t,x)dx \simeq 0. \tag{11.3.13}$$

Moreover, we have

$$\begin{aligned}
\frac{1}{\varepsilon}\int_{\mathbb{R}} \sigma(t,x)u'_{\varepsilon,1}(t,x)dx &= \int_{\mathbb{R}} \sigma(t,\varepsilon y + x_1)u'(y)dy \\
&\simeq \sigma(t,x_1)\int_{\mathbb{R}} u'(y)dy \\
&= \sigma(t,x_1).
\end{aligned} \tag{11.3.14}$$

Finally

$$\frac{1}{\varepsilon}\int_{\mathbb{R}} (u'_{\varepsilon,1}(t,x))^2 dx = \int_{\mathbb{R}} (u'(y))^2 dy = \gamma^{-1}, \tag{11.3.15}$$

and using (11.2.9)

$$\begin{aligned}
\frac{1}{\varepsilon}\int_{\mathbb{R}} u'_{\varepsilon,1}(t,x)u'_{\varepsilon,2}(t,x)dx &\simeq \frac{1}{\varepsilon}\int_{\mathbb{R}} u'\left(\frac{x-x_1}{\varepsilon}\right)\frac{\varepsilon^{1+2s}}{|x-x_2|^{1+2s}}dx \\
&= \int_{\mathbb{R}} u'(y)\frac{\varepsilon^{1+2s}}{|\varepsilon y + x_1 - x_2|^{1+2s}}dy \\
&\simeq \frac{\varepsilon^{1+2s}}{|x_1 - x_2|^{1+2s}}\int_{\mathbb{R}} u'(y)dy \\
&\simeq 0,
\end{aligned} \tag{11.3.16}$$

if $x_1 \neq x_2$. Now we multiply (11.3.9) by $u'_{\varepsilon,1}(t,x)$, we integrate on $\mathbb{R}$ and we use (11.3.10), (11.3.12), (11.3.13), (11.3.14), (11.3.15) and (11.3.16), to get

$$-\gamma^{-1}\dot{x}_1 = \frac{x_1 - x_2}{2s|x_1 - x_2|^{1+2s}} + \sigma(t,x_1).$$

A similar equation is obtained if we multiply (11.3.9) by $u'_{\varepsilon,2}(t,x)$ and integrate on $\mathbb{R}$. Therefore we get the system

$$
\begin{cases} \dot{x}_1 = -\gamma \dfrac{x_1 - x_2}{2s|x_1 - x_2|^{1+2s}} - \gamma\sigma(t, x_1) \\ \dot{x}_2 = -\gamma \dfrac{x_2 - x_1}{2s|x_2 - x_1|^{1+2s}} + \gamma\sigma(t, x_2), \end{cases} \tag{11.3.17}
$$

which is (11.3.4) with $N = 2$ and $K = 1$. This is a heuristic justification of the link between the partial differential equation in (11.3.1) and the system of ordinary differential equations in (11.3.4).

11.3.2 Dislocation dynamics after the collision time

System (11.3.4), describing at mesoscopic scale the dynamics of dislocation lines, becomes singular at the collision time $t = T_c$, and so no information about the dynamics of dislocations for times bigger than T_c can be inferred from it. To overcome this difficulty the idea consists in looking instead to the solution of the PDE (11.3.1) for small but fixed ε. Indeed such a function v_ε is well defined for any positive time. For simplicity here we present only the cases with two and three particles. We refer to [38] for the general case of N particles. Roughly speaking, in case of two particles, the collision of the two particles "annihilate" all the dynamics, nothing more is left and the system relaxes to the trivial equilibrium.

The case of three particles is, on the other hand, different from the case of two particles, since the steady state associated with the case of three particles is the heteroclinic orbit (and not the trivial function as in the case of two particles). In the case of three particles, one has that two particles "annihilate" each other, but the third particle "survives", and this produces a jump in the dislocation function – indeed, as explained in the previous sections, these "purely mathematical" particles correspond to an excursion of the dislocation, from two equilibria, which is modeled by the standard transition layer in (11.2.6). The precise results, which are proven in [37], are stated in the next two subsections.

11.3.3 The case of two transition layers

Given $x_1^0 < x_2^0$ let us consider as initial condition in (11.3.1)

$$
v_\varepsilon^0(x) = \frac{\varepsilon^{2s}}{W''(0)}\sigma(0, x) + u\left(\frac{x - x_1^0}{\varepsilon}\right) + u\left(\frac{x_2^0 - x}{\varepsilon}\right) - 1, \tag{11.3.18}
$$

where u is the solution of (11.2.6).

In general, it may happen that $T_c = +\infty$, i.e. no collision occurs. On the other hand, it can be shown that when either the external stress is small or the particles are initially close to collision, then $T_c < +\infty$. More precisely, in [36] we proved that if the following condition is satisfied

$$\text{either } \sigma \leqslant 0 \quad \text{or} \quad x_2^0 - x_1^0 < \left(\frac{1}{2s\|\sigma\|_\infty}\right)^{\frac{1}{2s}},$$

then the collision time T_c is finite.

In the setting of finite collision time, the dislocation function v_ε, after a time T_ε, which is only slightly larger than the collision time T_c, becomes small with ε. Indeed, the following two theorems are proven in [37].

Theorem 11.3.2 (Theorem 1.1, [37]). *Assume that* (11.2.7), (11.3.2) *hold and* $T_c < +\infty$. *Let* v_ε *be the solution of* (11.3.1)- (11.3.18). *Then there exists* $\varepsilon_0 > 0$ *such that for any* $\varepsilon < \varepsilon_0$ *there exist* T_ε, $\varrho_\varepsilon > 0$ *such that*

$$T_\varepsilon = T_c + o(1), \quad \varrho_\varepsilon = o(1) \quad \text{as } \varepsilon \to 0$$

and

$$v_\varepsilon(T_\varepsilon, x) \leqslant \varrho_\varepsilon \quad \text{for any } x \in \mathbb{R}. \tag{11.3.19}$$

The result above can be made precise by saying that, if the system is not subject to any external stress, then the dislocation function v_ε decays in time exponentially fast. More precisely, we have:

Theorem 11.3.3 (Theorem 1.2, [37]). *Assume that* (11.2.7), (11.3.2) *hold and that* $\sigma \equiv 0$. *Let* v_ε *be the solution of* (11.3.1)- (11.3.18). *Then there exist* $\varepsilon_0 > 0$ *and* $c > 0$ *such that for any* $\varepsilon < \varepsilon_0$ *we have*

$$|v_\varepsilon(t,x)| \leqslant \varrho_\varepsilon e^{c\frac{T_\varepsilon - t}{\varepsilon^{2s+1}}}, \quad \text{for any } x \in \mathbb{R} \text{ and } t \geqslant T_\varepsilon, \tag{11.3.20}$$

where T_ε *and* ϱ_ε *are given in Theorem 11.3.2.*

The evolution of the two particle system and of the associated dislocation function, as obtained in Theorems 11.3.2 and 11.3.3, is described in Figure 11.1.

11.3.4 The case of three transition layers

Next, we consider the case in which the initial condition in (11.3.1) is a superposition of three transition layers with different orientation. Precisely, let $\zeta_1 = 1$, $\zeta_2 = -1$, $\zeta_3 = 1$. Given $x_1^0 < x_2^0 < x_3^0$, let us consider as initial condition in (11.3.1)

$$v_\varepsilon^0(x) = \frac{\varepsilon^{2s}}{W''(0)}\sigma(0,x) + \sum_{i=1}^{3} u\left(\zeta_i \frac{x - x_i^0}{\varepsilon}\right) - 1, \tag{11.3.21}$$

where u is the solution of (11.2.6).

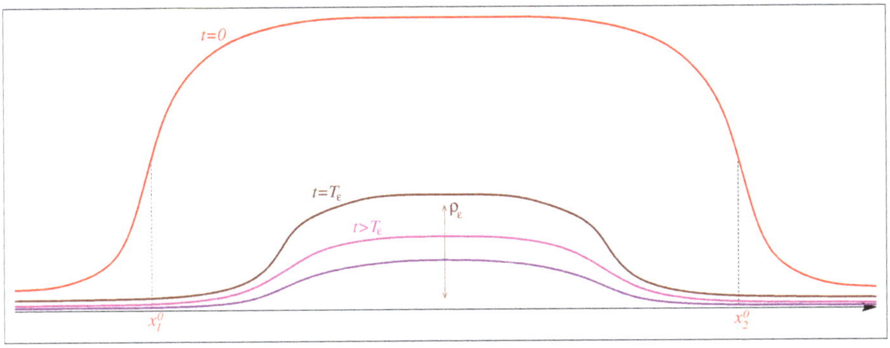

Fig. 11.1: (Figure 1 in [37]) Evolution of the dislocation function in case of two particles.

Theorem 11.3.4 (Theorem 1.3, [37]). *Assume that* (11.2.5), (11.3.2), *hold and* $T_c < +\infty$. *Let* v_ε *be the solution of* (11.3.1)-(11.3.21). *Then there exists* $\varepsilon_0 > 0$ *such that for any* $\varepsilon < \varepsilon_0$ *there exist* T_ε^1, T_ε^2, $\varrho_\varepsilon > 0$ *and* y_ε, z_ε *such that*

$$T_\varepsilon^1,\ T_\varepsilon^2 = T_c + o(1), \quad \varrho_\varepsilon = o(1) \quad as\ \varepsilon \to 0,$$

$$|z_\varepsilon - y_\varepsilon| = o(1) \quad as\ \varepsilon \to 0$$

and for any $x \in \mathbb{R}$

$$v_\varepsilon(T_\varepsilon^1, x) \leqslant u\left(\frac{x - y_\varepsilon}{\varepsilon}\right) + \varrho_\varepsilon \tag{11.3.22}$$

and

$$v_\varepsilon(T_\varepsilon^2, x) \geqslant u\left(\frac{x - z_\varepsilon}{\varepsilon}\right) - \varrho_\varepsilon, \tag{11.3.23}$$

where u *is the solution of* (11.2.6).

Next result is the analogue of Theorem 11.3.3 in the three particle setting. Roughly speaking, it says that, after a small transition time after the collision, the dislocation function relaxes towards the standard layer solution exponentially fast. The formal statement is the following:

Theorem 11.3.5 (Theorem 1.4, [37]). *Assume that* (11.2.5), (11.3.2), *hold and that* $\sigma \equiv 0$. *Let* v_ε *be the solution of* (11.3.1)-(11.3.21). *Then there exist* $\varepsilon_0 > 0$ *and* $\mu > 0$ *such that for any* $\varepsilon < \varepsilon_0$ *there exists* $K_\varepsilon = o(1)$ *as* $\varepsilon \to 0$ *such that*

$$v_\varepsilon(t, x) \leqslant u\left(\frac{x - y_\varepsilon + K_\varepsilon \varrho_\varepsilon \left(1 - e^{-\frac{\mu(t - T_\varepsilon^1)}{\varepsilon^{2s+1}}}\right)}{\varepsilon}\right) + \varrho_\varepsilon e^{-\frac{\mu(t - T_\varepsilon^1)}{\varepsilon^{2s+1}}},$$

$$for\ any\ x \in \mathbb{R}\ and\ t \geqslant T_\varepsilon^1, \tag{11.3.24}$$

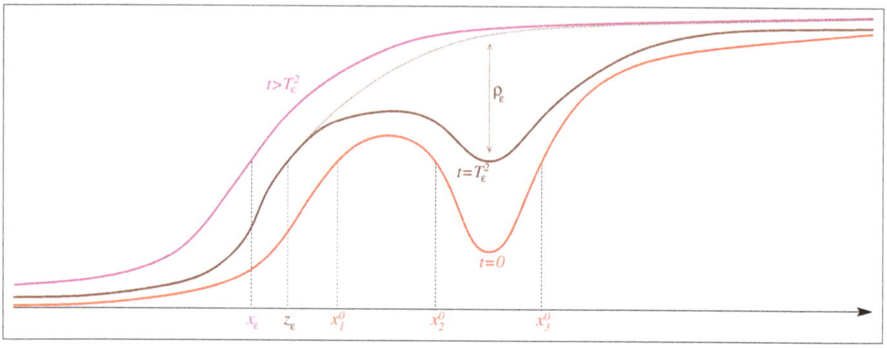

Fig. 11.2: (Figure 2 in [37]) Evolution of the dislocation function in case of three particles.

$$v_\varepsilon(t,x) \geqslant u\left(\frac{x - z_\varepsilon - K_\varepsilon \varrho_\varepsilon\left(1 - e^{-\frac{\mu(t-T^2_\varepsilon)}{\varepsilon^{2s+1}}}\right)}{\varepsilon}\right) - \varrho_\varepsilon e^{-\frac{\mu(t-T^2_\varepsilon)}{\varepsilon^{2s+1}}},$$

for any $x \in \mathbb{R}$ and $t \geqslant T^2_\varepsilon$ (11.3.25)

where T^1_ε, T^2_ε, ϱ_ε, y_ε and z_ε are given in Theorem 11.3.4 and u is the solution of (11.2.6).

Corrolary 11.1 (Corollary 1.5, [37]). *Under the assumptions of Theorem 11.3.5, there exists $\varepsilon_0 > 0$ such that for any $\varepsilon < \varepsilon_0$, there exist a sequence $t_k \to +\infty$ as $k \to +\infty$, and a point $x_\varepsilon \in \mathbb{R}$ with*

$$y_\varepsilon - K_\varepsilon \varrho_\varepsilon < x_\varepsilon < z_\varepsilon + K_\varepsilon \varrho_\varepsilon, \tag{11.3.26}$$

such that

$$v_\varepsilon(t_k, x) \to u\left(\frac{x - x_\varepsilon}{\varepsilon}\right) \quad \text{as } k \to +\infty, \tag{11.3.27}$$

where y_ε, z_ε, K_ε and ϱ_ε are given in Theorem 11.3.4 and u is the solution of (11.2.6).

The results of Theorems 11.3.4 and 11.3.5 and Corollary 11.1 are represented in Figure 11.2, where we sketched the evolution of the dislocation function and of the associated particle system in the case of three particles with alternate orientations.

It is worth to point out that the case of three particles provides structurally richer phenomena than the case of two particles. Indeed, in the case of three particles we have two different types of collision: simple and triple. The simple collision occurs when only two particles collide at time T_c, i.e., either

$$x_1(T_c) = x_2(T_c) \quad \text{and} \quad x_3(T_c) > x_2(T_c),$$

or

$$x_2(T_c) = x_3(T_c) \quad \text{and} \quad x_1(T_c) < x_2(T_c).$$

In the triple collision case, the three particles collide together and simultaneously, i.e.

$$x_1(T_c) = x_2(T_c) = x_3(T_c).$$

In [35], we proved that if $\sigma \equiv 0$, then for any choice of the initial condition (x_1^0, x_2^0, x_3^0) we have a collision in a finite time. Moreover a triple collision is possible if and only if

$$x_2^0 - x_1^0 = x_3^0 - x_2^0.$$

11.4 From the Peierls-Nabarro Model to the Dislocation Density Model

Consider the evolutive Peierls-Nabarro model in any dimension (11.2.10), where $\mathcal{I}_s$ is the anisotropic Lévy operator of order 2s, defined in (11.2.3). Let us first consider the case $s = \frac{1}{2}$, which is studied in [29, 28].

We want to identify at *macroscopic scale* an evolution model for the dynamics of a density of dislocations. We consider the following rescaling

$$u^\varepsilon(t, x) = \varepsilon u\left(\frac{t}{\varepsilon}, \frac{x}{\varepsilon}\right),$$

where ε is the ratio between the typical length scale for dislocation (of the order of the micrometer) and the typical macroscopic length scale in mechanics (millimeter or centimeter). Moreover, assuming suitable initial data

$$u(0, x) = \frac{1}{\varepsilon} u_0(\varepsilon x) \quad \text{on } \mathbb{R}^N, \tag{11.4.1}$$

(where u_0 is a regular bounded function), we see that the function u^ε is solution of

$$\begin{cases} \partial_t u^\varepsilon = \mathcal{I}_1[u^\varepsilon(t, \cdot)] - W'\left(\frac{u^\varepsilon}{\varepsilon}\right) + \sharp\left(\frac{t}{\varepsilon}, \frac{x}{\varepsilon}\right) & \text{in } \mathbb{R}^+ \times \mathbb{R}^N \\ u^\varepsilon(0, x) = u_0(x) & \text{on } \mathbb{R}^N. \end{cases} \tag{11.4.2}$$

This indicates that at the limit $\varepsilon \to 0$, we will recover a model for the dynamics of (renormalized) densities of dislocations. For $N = 2$, (11.2.10) with initial condition (11.4.1) models, at microscopic scale, the dynamics of a collections of edge dislocation lines moving in the same slip plane, with same Burgers' vectors, such that the number of dislocations is of the order of $1/\varepsilon$ per unit of macroscopic scale.

Here, we assume that the function g in (11.2.3) satisfies

(H1) $g \in C(\mathbf{S}^{N-1})$, $g > 0$, g even.

On the functions W, σ and u_0 we assume:

(H2) $W \in C^{1,1}(\mathbb{R})$ and $W(v + 1) = W(v)$ for any $v \in \mathbb{R}$;

(H3) $\sigma \in C^{0,1}(\mathbb{R}^+ \times \mathbb{R}^N)$ and $\sharp(t+1,x) = \sharp(t,x)$, $\sigma(t,x+k) = \sigma(t,x)$ for any $k \in \mathbb{Z}^N$ and $(t,x) \in \mathbb{R}^+ \times \mathbb{R}^N$;
(H4) $u_0 \in W^{2,\infty}(\mathbb{R}^N)$.

Identifying the limit solution of the function u_ε, when $\varepsilon \to 0$, means solving an homogenization problem. In homogenization, both the limit function and the equation satisfied by it are unknown of the problem. In [29], we show that the limit u^0 of u^ε as $\varepsilon \to 0$ exists and is the unique solution of the homogenized problem

$$\begin{cases} \partial_t u = \overline{H}(\nabla_x u, \mathcal{I}_1[u(t,\cdot)]) & \text{in} \quad \mathbb{R}^+ \times \mathbb{R}^N \\ u(0,x) = u_0(x) & \text{on} \quad \mathbb{R}^N, \end{cases} \tag{11.4.3}$$

for some continuous function $\overline{H}$ usually called *effective Hamiltonian*. The function u^0 will be interpreted later as a macroscopic plastic strain satisfying the macroscopic plastic flow rule (11.4.3). Moreover $\mathcal{I}_s[u^0]$ will be the stress created by the macroscopic density of dislocations.

As usual in periodic homogenization, the limit equation is determined by a *cell problem*. In our case, such a problem is for any $p \in \mathbb{R}^N$ and $L \in \mathbb{R}$ the following:

$$\begin{cases} \lambda + \partial_\tau v = \mathcal{I}_1[v(\tau,\cdot)] + L - W'(v + \lambda\tau + p\cdot y) + \sharp(\tau,y) & \text{in} \quad \mathbb{R}^+ \times \mathbb{R}^N \\ v(0,y) = 0 & \text{on} \quad \mathbb{R}^N, \end{cases} \tag{11.4.4}$$

where $\lambda = \lambda(p,L)$ is the unique number for which there exists a solution v of (11.4.4) which is bounded on $\mathbb{R}^+ \times \mathbb{R}^N$. In order to solve (11.4.4), we show for any $p \in \mathbb{R}^N$ and $L \in \mathbb{R}$ the existence of a unique solution of

$$\begin{cases} \partial_\tau w = \mathcal{I}_1[w(\tau,\cdot)] + L - W'(w + p\cdot y) + \sharp(\tau,y) & \text{in} \quad \mathbb{R}^+ \times \mathbb{R}^N \\ w(0,y) = 0 & \text{on} \quad \mathbb{R}^N, \end{cases} \tag{11.4.5}$$

and we look for some $\lambda \in \mathbb{R}$ for which $w - \lambda\tau$ is bounded. Precisely we have:

Theorem 11.4.1 (Theorem 1.1, [29]). *Assume (H1)-(H4). For $L \in \mathbb{R}$ and $p \in \mathbb{R}^N$, there exists a unique viscosity solution $w \in C_b(\mathbb{R}^+ \times \mathbb{R}^N)$ of (11.4.5) and there exists a unique $\lambda \in \mathbb{R}$ such that w satisfies: $\frac{w(\tau,y)}{\tau}$ converges towards λ as $\tau \to +\infty$, locally uniformly in y. The real number λ is denoted by $\overline{H}(p,L)$. The function $\overline{H}(p,L)$ is continuous on $\mathbb{R}^N \times \mathbb{R}$ and non-decreasing in L.*

In Theorem 11.4.1, we denoted by $C_b(\mathbb{R}^+ \times \mathbb{R}^N)$ the set of continuous functions on $\mathbb{R}^+ \times \mathbb{R}^N$ which are bounded on $(0,T) \times \mathbb{R}^N$ for any $T > 0$. The non-local equation (11.4.2) is related to the local equation

$$\begin{cases} \partial_t u^\varepsilon = F\left(\frac{x}{\varepsilon}, \frac{u^\varepsilon}{\varepsilon}, \nabla u^\varepsilon\right) & \text{in} \quad \mathbb{R}^+ \times \mathbb{R}^N \\ u^\varepsilon(0,x) = u_0(x) & \text{on} \quad \mathbb{R}^N, \end{cases}$$

that was studied in [24] under the assumption that $F(x,u,p)$ is periodic in (x,u) and coercive in p. As in the local case, the presence of the term $\frac{u^\varepsilon}{\varepsilon}$ in (11.4.2) does not allow to use directly the bounded solution of (11.4.4), usually called a corrector. Indeed, a corrector in dimension $N+1$ needs to be introduced. Nevertheless, we have the following convergence result:

Theorem 11.4.2 (Theorem 1.2, [29]). *Assume (H1)-(H4). The solution u^ε of* (11.4.2) *converges towards the solution u^0 of* (11.4.3) *locally uniformly in (t,x), where $\overline{H}$ is defined in Theorem 11.4.1.*

11.4.1 Viscosity solutions for non-local operators

The classical notion of viscosity solution can be adapted for a quite general class of equations involving non-local operators, which includes equations (11.4.2) and (11.4.3), see for instance [4]. For equation (11.4.3), the property of the effective Hamiltonian $\overline{H}(p,L)$ to be non-decreasing with respect to L is a sort of ellipticity condition, which allows to define a well-posed notion of viscosity solution. The definition of viscosity solution for equations involving the Lévy operator $\mathfrak{I}_s$, comes from this simple observation: for a smooth function φ, one has that, for any $r>0$,

$$\begin{aligned}&\mathrm{PV}\int_{|z|\leqslant r}(\varphi(x+z)-\varphi(x))\frac{1}{|z|^{N+2s}}g\left(\frac{z}{|z|}\right)dz\\ =&\mathrm{PV}\int_{|z|\leqslant r}(\varphi(x+z)-\varphi(x)-\nabla\varphi(x)\cdot z)\frac{1}{|z|^{N+2s}}g\left(\frac{z}{|z|}\right)dz,\end{aligned}$$

as

$$\mathrm{PV}\int_{|z|\leqslant r}\nabla(\varphi(x)\cdot z)\frac{1}{|z|^{N+2s}}g\left(\frac{z}{|z|}\right)dz=\lim_{\delta\to 0+}\int_{\delta<|z|\leqslant r}\nabla(\varphi(x)\cdot z)\frac{1}{|z|^{N+2s}}g\left(\frac{z}{|z|}\right)dz=0,$$

being the integrand an odd function. Now, if φ is sufficiently regular, then the following integrand is convergent,

$$\int_{|z|\leqslant r}(\varphi(x+z)-\varphi(x)-\nabla\varphi(x)\cdot z)\frac{1}{|z|^{N+2s}}g\left(\frac{z}{|z|}\right)dz.$$

On the other hand, if φ is bounded, the following integral is convergent too

$$\int_{|z|>r}(\varphi(x+z)-\varphi(x))\frac{1}{|z|^{N+2s}}g\left(\frac{z}{|z|}\right)dz.$$

Taking into account this simple remark, a well-posed definition of viscosity solution for a general non-local equation with associated initial condition can be given. Consider

$$\begin{cases}u_t=F(t,x,u,Du,\mathfrak{I}_s[u]) & \text{in}\quad \mathbb{R}^+\times\mathbb{R}^N\\ u(0,x)=u_0(x) & \text{on}\quad \mathbb{R}^N,\end{cases}\tag{11.4.6}$$

where $F(t, x, u, p, L)$ is continuous and non-decreasing in L. Set

$$\mathcal{I}_s^{1,r}[\varphi, x] := \int_{|z|\leqslant r} (\varphi(x+z) - \varphi(x) - \nabla\varphi(x)\cdot z)\frac{1}{|z|^{N+2s}} g\left(\frac{z}{|z|}\right) dz,$$

$$\mathcal{I}_s^{2,r}[\varphi, x] := \int_{|z|> r} (\varphi(x+z) - \varphi(x))\frac{1}{|z|^{N+2s}} g\left(\frac{z}{|z|}\right) dz.$$

Denote by $USC_b(\mathbb{R}^+ \times \mathbb{R}^N)$ (resp., $LSC_b(\mathbb{R}^+ \times \mathbb{R}^N)$) the set of upper (resp., lower) semicontinuous functions on $\mathbb{R}^+ \times \mathbb{R}^N$ which are bounded on $(0, T) \times \mathbb{R}^N$ for any $T > 0$.

Definition 11.4.3 (r-viscosity solution). *A function $u \in USC_b(\mathbb{R}^+ \times \mathbb{R}^N)$ (resp., $u \in LSC_b(\mathbb{R}^+ \times \mathbb{R}^N)$) is a r-viscosity subsolution (resp., supersolution) of* (11.4.6) *if $u(0, x) \leqslant (u_0)^*(x)$ (resp., $u(0, x) \geqslant (u_0)_*(x)$) and for any $(t_0, x_0) \in \mathbb{R}^+ \times \mathbb{R}^N$, any $\tau \in (0, t_0)$ and any test function $\varphi \in C^2(\mathbb{R}^+ \times \mathbb{R}^N)$ such that $u - \varphi$ attains a local maximum (resp., minimum) at the point (t_0, x_0) on $Q_{(\tau,r)}(t_0, x_0)$, then we have*

$$\partial_t\varphi(t_0, x_0) - F(t_0, x_0, u(t_0, x_0), \nabla_x\varphi(t_0, x_0), \mathcal{I}_s^{1,r}[\varphi(t_0, \cdot), x_0] + \mathcal{I}_s^{2,r}[u(t_0, \cdot), x_0]) \leqslant 0$$

(resp., $\geqslant 0$).

A function $u \in C_b(\mathbb{R}^+ \times \mathbb{R}^N)$ is a r-viscosity solution of (11.4.6) *if it is a r-viscosity sub and supersolution of* (11.4.6).

It is classical that the maximum in the above definition can be supposed to be global. We have also the following property, see e.g., [4]:

Proposition 11.4.4 (Equivalence of the definitions). *Assume $F(t, x, u, p, L)$ continuous and non-decreasing in L. Let $r > 0$ and $r' > 0$. A function $u \in USC_b(\mathbb{R}^+ \times \mathbb{R}^N)$ (resp., $u \in LSC_b(\mathbb{R}^+ \times \mathbb{R}^N)$) is a r-viscosity subsolution (resp., supersolution) of* (11.4.6) *if and only if it is a r'-viscosity subsolution (resp., supersolution) of* (11.4.6).

The non-decreasing property of $F(t, x, u, p, L)$ with respect to L is crucial to prove comparison principles between viscosity sub and supersolutions. Comparison principles for viscosity sub and supersolutions of non-local equations including (11.4.2) are proven for instance in [20]. The comparison principle for (11.4.3) has been proven in [25]. Existence of viscosity solutions for non-local equations for which the comparison principle holds, follows by using the Perron's method, after providing a suitable sub and supersolution.

11.4.2 Mechanical interpretation of the homogenization

Let us briefly explain the meaning of the homogenization result. In the macroscopic model, the function $u^0(t, x)$ can be interpreted as the plastic strain (localized in the slip plane $\{x_3 = 0\}$). Then the three-dimensional displacement $U(t, X)$ is obtained as a minimizer of the elastic energy

$$U(t, \cdot) = \arg\min_{\tilde{U}} \mathcal{E}^{el}(u^0(t, \cdot), \tilde{U})$$

and the stress is

$$\sigma = \Lambda : e \quad \text{with} \quad e = e(U) - u^0(t, x)\delta_0(x_3)e^0.$$

Then the resolved shear stress is

$$\mathcal{I}_1[u^0] = \sigma_{13}^{\text{obst}}.$$

The homogenized equation (11.4.3), i.e.,

$$\partial_t u^0 = \overline{H}(\nabla_x u^0, \mathcal{I}_1[u^0(t, \cdot)])$$

which is the evolution equation for u^0, can be interpreted as the plastic flow rule in a model for macroscopic crystal plasticity. This is the law giving the plastic strain velocity $\partial_t u^0$ as a function of the resolved shear stress $\sigma_{13}^{\text{obst}}$ and the dislocation density ∇u^0.

The typical example of such a plastic flow rule is the Orowan's law:

$$\overline{H}(p, L) \simeq |p|L.$$

This is also the law that we recover in dimension $N = 1$ in paper [29] in the case where there are no obstacles (i.e., $\sigma_{13}^{\text{obst}} \equiv 0$) and for small stress L and small density $|p|$.

11.4.3 The Orowan's law

The limit equation of an homogenization problem is defined through a cell problem, but its explicit expression is usually unknown. In [28] we are able to explicitly characterize the effettive Hamiltonian $\overline{H}(p, L)$ defined in Theorem 11.4.1 for small values of p and L, in the case $\mathcal{I}_1 = -(-\Delta)^{\frac{1}{2}}$, $N = 1$ and $\sigma \equiv 0$. In this setting, equation (11.4.2) models the dynamics of parallel straight edge dislocation lines in the same slip plane with the same Burgers' vector, moving with self-interactions. In other words equation (11.4.2) simply describes the motion of dislocations by relaxation of the total energy (elastic + misfit). In [28] we study the behavior of $\overline{H}(p, L)$ for small p and L, and in this regime we recover Orowan's law, which claims that

$$\overline{H}(p, L) \simeq \gamma|p|L \tag{11.4.7}$$

for some constant of proportionality $\gamma > 0$. The precise result is stated in the following

Theorem 11.4.5 (Theorem 1.2, [28]). *Assume* $g = \frac{1}{\pi}$, $N = 1$, $\sigma \equiv 0$, $W \in C^{4,\alpha}$, *for some* $0 < \alpha < 1$, *and* (11.2.7). *Let* p_0, $L_0 \in \mathbb{R}$. *Then the function* $\overline{H}$ *defined in Theorem 11.4.1 satisfies*

$$\frac{\overline{H}(\delta p_0, \delta L_0)}{\delta^2} \to \gamma |p_0| L_0 \quad \text{as } \delta \to 0^+ \quad \text{with} \quad \gamma = \left(\int_{\mathbb{R}} (u')^2\right)^{-1}, \tag{11.4.8}$$

where u *is the solution of* (11.2.6).

11.4.4 Heuristic for the proof of Orowan's law

Define $u(\tau, y) := w(\tau, y) + py$, where w is the corrector solution of (11.4.5). Then, u satisfies

$$\begin{cases} \partial_\tau u = L + \mathcal{I}_1[u(\tau, \cdot)] - W'(u) & \text{in} \quad \mathbb{R}^+ \times \mathbb{R} \\ u(0, y) = py & \text{on} \quad \mathbb{R}. \end{cases} \tag{11.4.9}$$

Moreover, by Theorem 11.4.1, we have that

$$u(\tau, y) \sim py + \lambda\tau + \text{bounded},$$

where $\lambda = \overline{H}(p, L)$. The idea underlying the proof of Orowan's law is related to a fine asymptotics of equation (11.4.9). From Theorem 11.3.1 we know that if u solves (11.4.9) with $L = \delta L_0$, i.e.

$$\partial_\tau u = \delta L_0 + \mathcal{I}_1[u(\tau, \cdot)] - W'(u) \tag{11.4.10}$$

for a choice of initial data with a finite number of indices i:

$$u(0, y) = \frac{\delta L_0}{W''(0)} + \sum_{x_i^0 \geq 0} \varphi\left(y - \frac{x_i^0}{\delta}\right) + \sum_{x_i^0 < 0} \left(\varphi\left(y - \frac{x_i^0}{\delta}\right) - 1\right)$$

then

$$u^\delta(t, x) := u\left(\frac{t}{\delta^2}, \frac{x}{\delta}\right) \to u^0(t, x) = \sum_{x_i^0 \geq 0} H(x - x_i(t)) + \sum_{x_i^0 < 0} (H(x - x_i(t)) - 1) \quad \text{as } \delta \to 0$$

where H is the Heaviside function and with the dynamics

$$\begin{cases} \dfrac{dx_i}{dt} = \gamma\left(-L_0 + \dfrac{1}{\pi}\displaystyle\sum_{j \neq i} \frac{1}{x_i - x_j}\right) \\ \\ x_i(0) = x_i^0. \end{cases} \tag{11.4.11}$$

System (11.4.11) is the (rescaled as here we have $g = 1/\pi$ instead than $g = 1$) system (11.3.4), with $s = \frac{1}{2}$ and $\zeta_i = 1$ for any i. Moreover for the choice $p = \delta p_0$ with $p_0 > 0$ and $x_i^0 = i/p_0$ that we extend formally for all $i \in \mathbb{Z}$, we see (at least formally) that

$$|u(0, y) - \delta p_0 y| \leq C_\delta.$$

This suggests also that the infinite sum in (11.4.11) should vanish (by antisymmetry) and then the mean velocity should be

$$\frac{dx_i}{dt} \simeq -\gamma L_0$$

i.e., after scaling back

$$u(\tau, y) \simeq \delta p_0(y - c_1\tau) + \text{bounded}$$

with the velocity

$$c_1 = \frac{d(x_i/\delta)}{d(t/\delta^2)} \simeq -\gamma L_0 \delta$$

i.e.,

$$u(\tau, y) \simeq \delta p_0 y + \lambda\tau + \text{bounded} \quad \text{with} \quad \lambda \simeq \delta^2 \gamma p_0 L_0.$$

We deduce that we should have

$$\frac{u(\tau, y)}{\tau} \to \lambda \simeq \delta^2 \gamma p_0 L_0 \quad \text{as} \quad \tau \to +\infty.$$

We see that this $\lambda = \overline{H}(\delta p_0, \delta L_0)$ is exactly the one we expect asymptotically in Theorem 11.4.5 when $p_0 > 0$.

11.4.5 Homogenization and Orowan's law for anisotropic fractional operators of any order

The results of [28, 29] have been generalized in [36] to Lévy operators of any order $2s$, with $s \in (0, 1)$. In [36], for $s > \frac{1}{2}$ we considered the following homogenization problem:

$$\begin{cases} \partial_t u^\varepsilon = \varepsilon^{2s-1} \mathcal{I}_s[u^\varepsilon(t, \cdot)] - W'\left(\frac{u^\varepsilon}{\varepsilon}\right) + \sharp\left(\frac{t}{\varepsilon}, \frac{x}{\varepsilon}\right) & \text{in} \quad \mathbb{R}^+ \times \mathbb{R}^N \\ u^\varepsilon(0, x) = u_0(x) & \text{on} \quad \mathbb{R}^N, \end{cases} \tag{11.4.12}$$

and for $s < \frac{1}{2}$:

$$\begin{cases} \partial_t u^\varepsilon = \mathcal{I}_s[u^\varepsilon(t, \cdot)] - W'\left(\frac{u^\varepsilon}{\varepsilon^{2s}}\right) + \sharp\left(\frac{t}{\varepsilon^{2s}}, \frac{x}{\varepsilon}\right) & \text{in} \quad \mathbb{R}^+ \times \mathbb{R}^N \\ u^\varepsilon(0, x) = u_0(x) & \text{on} \quad \mathbb{R}^N. \end{cases} \tag{11.4.13}$$

Remark that the scalings for $s > \frac{1}{2}$ and $s < \frac{1}{2}$ are different. They formally coincide when $s = \frac{1}{2}$. We proved that the solution u^ε of (11.4.12) converges as $\varepsilon \to 0$ to the solution u^0 of the homogenized problem

$$\begin{cases} \partial_t u = \overline{H}_1(\nabla_x u) & \text{in} \quad \mathbb{R}^+ \times \mathbb{R}^N \\ u(0, x) = u_0(x) & \text{on} \quad \mathbb{R}^N, \end{cases} \tag{11.4.14}$$

with an effective Hamiltonian $\overline{H}_1$ which does not depend on $\mathcal{I}_s$ anymore, while the solution u^ε of (11.4.13) converges as $\varepsilon \to 0$ to u^0 solution of the following

$$\begin{cases} \partial_t u = \overline{H}_2(\mathcal{I}_s[u]) & \text{in} \quad \mathbb{R}^+ \times \mathbb{R}^N \\ u(0,x) = u_0(x) & \text{on} \quad \mathbb{R}^N, \end{cases} \tag{11.4.15}$$

with an effective Hamiltonian $\overline{H}_2$ not depending on the gradient. That is, roughly speaking, for any $s \in (0,1)$, the effective Hamiltonian is an operator of order $\min\{2s, 1\}$, which reveals the stronger non-local effects present in the case $s < \frac{1}{2}$. As before, the functions $\overline{H}_1$ and $\overline{H}_2$ are determined by the following cell problem:

$$\begin{cases} \partial_\tau w = \mathcal{I}_s[w(\tau,\cdot)] + L - W'(w + p\cdot y) + \sharp(\tau, y) & \text{in} \quad \mathbb{R}^+ \times \mathbb{R}^N \\ w(0,y) = 0 & \text{on} \quad \mathbb{R}^N, \end{cases} \tag{11.4.16}$$

and we look for some λ such that $w - \lambda\tau$ is bounded. As in the case $s = \frac{1}{2}$, we proved the following ergodic result.

Theorem 11.4.6 (Theorem 1.1 [36]). *Assume (H1)-(H4). For $L \in \mathbb{R}$ and $p \in \mathbb{R}^N$, there exists a unique viscosity solution $w \in C_b(\mathbb{R}^+ \times \mathbb{R}^N)$ of (11.4.5) and there exists a unique $\lambda \in \mathbb{R}$ such that w satisfies:*

$$\frac{w(\tau,y)}{\tau} \text{ converges towards } \lambda \text{ as } \tau \to +\infty, \text{ locally uniformly in } y.$$

The real number λ is denoted by $\overline{H}(p,L)$. The function $\overline{H}(p,L)$ is continuous on $\mathbb{R}^N \times \mathbb{R}$ and non-decreasing in L.

Once the cell problem was solved, we could prove the following convergence results:

Theorem 11.4.7 (Theorem 1.2 [36]). *Assume (H1)-(H4). The solution u^ε of (11.4.12) converges towards the solution u^0 of (11.4.14) locally uniformly in (t,x), where*

$$\overline{H}_1(p) := \overline{H}(p, 0)$$

and $\overline{H}(p,L)$ is defined in Theorem 11.4.6.

Theorem 11.4.8 (Theorem 1.3 [36]). *Assume (H1)-(H4). The solution u^ε of (11.4.13) converges towards the solution u^0 of (11.4.15) locally uniformly in (t,x), where*

$$\overline{H}_2(L) := \overline{H}(0, L)$$

and $\overline{H}(p,L)$ is defined in Theorem 11.4.6.

We point out that the effective Hamiltonians $\overline{H}_1$ and $\overline{H}_2$ represent the speed of propagation of the dislocation dynamics according to (11.4.14) and (11.4.15). In particular, due to Theorems 11.4.7 and 11.4.8, such speed only depends on the slope of the dislocation in the weakly non-local setting $s > \frac{1}{2}$ and only on an operator of order s of the dislocation in the strongly non-local setting $s < \frac{1}{2}$.

Finally, when $N = 1$, $\mathcal{I}_s = -(-\Delta)^s$ and $\sigma \equiv 0$, and we make the further assumptions (11.2.7) on the potential W, we proved the following extension of the Orowan's law:

Theorem 11.4.9 (Theorem 1.4, [36]). *Assume* $\mathcal{I}_s = -(-\Delta)^s$, $N = 1$, $\sigma \equiv 0$, $W \in C^{4,\alpha}$, *for some* $0 < \alpha < 1$, (11.2.7), *and* W *even when* $s \in \left(0, \frac{1}{2}\right)$. *Let* p_0, $L_0 \in \mathbb{R}$ *with* $p_0 \neq 0$. *Then the function* $\overline{H}$ *defined in Theorem 11.4.6 satisfies*

$$\frac{\overline{H}(\delta p_0, \delta^{2s} L_0)}{\delta^{1+2s}} \to \gamma |p_0| L_0 \quad \text{as } \delta \to 0^+ \quad \text{with} \quad \gamma = \left(\int_{\mathbb{R}} (u')^2\right)^{-1}, \tag{11.4.17}$$

where u *is solution of* (11.2.6).

11.5 Non-local Allen-Cahn Equation

Imbert and Souganidis [26] have considered the following rescaled in time and space version of the evolutive Peierl-Nabarro model in dimension $N \geqslant 2$: for $t > 0$ and $x \in \mathbb{R}^N$,

$$\partial_t u + \frac{1}{\varepsilon \eta_\varepsilon}\left\{-\varepsilon^{2s}\mathcal{I}_s[u^\varepsilon] + W'(u^\varepsilon)\right\} = 0 \tag{11.5.1}$$

where $\mathcal{I}_s$ is the Lévy operator of order $2s \in (0, 2)$, introduced in (11.2.3), W' is a bistable nonlinearity and the parameter η_ε depends on s and it is defined as follows:

$$\eta_\varepsilon = \begin{cases} \varepsilon & \text{if } s > \frac{1}{2}, \\ \varepsilon|\log(\varepsilon)| & \text{if } s = \frac{1}{2}, \\ \varepsilon^{2s} & \text{if } s < \frac{1}{2}. \end{cases} \tag{11.5.2}$$

For our purposes, we assume that W satisfies (11.2.7). Let q be the phase transition function, solution to: for $e \in \mathbb{S}^{N-1}$,

$$\begin{cases} \mathbb{I}^e[q] = W'(q), & \text{in } \mathbb{R} \\ q' > 0 & \\ \lim_{\xi\to-\infty} q(\xi) = 0, \quad \lim_{\xi\to+\infty} q(\xi) = 1, & \end{cases} \tag{11.5.3}$$

where

$$\mathcal{I}_s^e[q](\xi) := \mathrm{PV}\int_{\mathbb{R}^N} (q(\xi + e\cdot z) - q(\xi))J(z)dz$$

and

$$J(z) := g\left(\frac{z}{|z|}\right)\frac{1}{|z|^{N+2s}}. \tag{11.5.4}$$

When $g \equiv c_{N,s}$, then $\mathcal{J}_s^e$ is actually independent of e and a solution is providing by $q(\xi) = u(\xi)$, where u is solution of (11.2.6). In [26] it is proven that when $s \geqslant \frac{1}{2}$, the solution u_ε of the diffusion-reaction equation (11.4.2) with initial datum

$$u_\varepsilon^0(x) = q\left(\frac{d_0(x)}{\varepsilon}, Dd_0(x)\right), \tag{11.5.5}$$

where d_0 is the signed distance function to the boundary of a smooth set Ω_0, can only have, as $\varepsilon \to 0$, two limits: the stable equilibria of the bistable non-linearity W' in $[0, 1]$, i.e., 0 and 1. The resulting interface, $\partial\Omega_t$, evolves by anisotropic mean curvature. For $s < \frac{1}{2}$ only partial but significant results have been obtained. In the theory of crystal dislocations when $N = 2$, $\partial\Omega_t$ represents the dislocation line at time t moving on the slip plane x_1x_2. In the one dimensional space, moving interfaces are points. Their dynamics is then described by the system of ODE's (11.3.4). In this section, we want to give a heuristic proof of the results contained in [26]. Assume for simplicity $g \equiv c_{N,s}$, then the phase transition q is independent of the direction e. In this setting we have the following Ansatz for u_ε:

$$u_\varepsilon(t, x) \sim q\left(\frac{d(t, x)}{\varepsilon}\right)$$

where $d(t, x)$ is the signed distance function from the front that propagates starting from the initial configuration Ω_0. Close to the front, the distance function d is smooth in x and $|Dd| = 1$, which implies in particular that $D^2dDd = 0$. Inserting the derivatives of the Ansatz into the equation (11.5.1), multiplying by ε and using equation (11.5.3), we get (close to the front):

$$\begin{aligned}\dot{q}\left(\frac{d(t,x)}{\varepsilon}\right)\partial_t d(t,x) &= \frac{1}{\eta_\varepsilon}\left\{\varepsilon^{2s}\mathcal{J}_s\left[q\left(\frac{d(t,\cdot)}{\varepsilon}\right)\right](x) - W'\left(q\left(\frac{d(t,x)}{\varepsilon}\right)\right)\right\}\\ &= \frac{1}{\eta_\varepsilon}\left\{\mathcal{J}_s\left[q\left(\frac{d(t,\varepsilon\cdot)}{\varepsilon}\right)\right]\left(\frac{x}{\varepsilon}\right) - \mathcal{J}_s^e[q]\left(\frac{d(t,x)}{\varepsilon}\right)\right\}.\end{aligned} \tag{11.5.6}$$

Let us introduce the notation $\xi = \frac{d(t,x)}{\varepsilon}$, $y = \frac{x}{\varepsilon}$ and $e = Dd(t, x)$. Then, we can write the right-hand side of the previous equation as follows

$$\begin{aligned}\mathcal{J}_s\left[q\left(\frac{d(t,\varepsilon\cdot)}{\varepsilon}\right)\right](y) - \mathcal{J}_s^e[q](\xi) &= \mathrm{PV}\int_{\mathbb{R}^N}\left[q\left(\frac{d(t,x+\varepsilon z)}{\varepsilon}\right) - q(\xi + e\cdot z)\right]J(z)dz\\ &= \mathrm{PV}\int_{\mathbb{R}^N}[q(\xi + e\cdot z + \varepsilon W_\varepsilon(t,x,z)) - q(\xi + e\cdot z)]J(z)dz,\end{aligned}$$

where $W_\varepsilon(t, x, z) = \frac{1}{\varepsilon^2}[d(t, x+\varepsilon z) - d(t, x) - \varepsilon Dd(t, x)\cdot z]$. Notice that W_ε is bounded in ε if d is $C^{1,1}$ with respect to the space variable, in a neighborhood of x. Now, if the front is smooth, for (t, x) close to the front, we can assume that the slow variables (t, x) and the fast variable ξ are independent. Therefore, multiplying equation (11.5.6) by $\dot{q}(\xi)$ and integrating in ξ, we get

$$\gamma^{-1}\partial_t d(t, x) - \frac{1}{\eta_\varepsilon}\overline{a}(t, x, e) = 0, \tag{11.5.7}$$

where

$$\gamma^{-1} := \int_{\mathbb{R}} (\dot{q})^2(\xi)d\xi, \tag{11.5.8}$$

$$\overline{a}_\varepsilon(t, x, e) = \int_{\mathbb{R}} \dot{q}(\xi)a_\varepsilon(t, x, \xi, e)d\xi, \tag{11.5.9}$$

and

$$a_\varepsilon(t, x, \xi, e) = \mathrm{PV}\int_{\mathbb{R}^N} [q(\xi + e \cdot z + \varepsilon W_\varepsilon(t, x, z)) - q(\xi + e \cdot z)] J(z)dz.$$

From Lemma 4 in [26], we know that when $s \geqslant \frac{1}{2}$, there is a matrix A depending on s and N (but independent of e in the isotropic case), such that, as $\varepsilon \to 0$, $\frac{1}{\eta_\varepsilon}\overline{a}_\varepsilon(t, x, e) \to \mathrm{tr}(AD^2d(t, x))$. Passing to the limit as $\varepsilon \to 0$ in (11.5.7), we find the following equation for d

$$\partial_t d(t, x) = \gamma \mathrm{tr}(AD^2 d) = \gamma \mathrm{tr}\left(\left(I - \frac{Dd \otimes Dd}{|Dd|^2}\right) AD^2 d\right), \tag{11.5.10}$$

since $D^2dDd = 0$. The mean curvature equation just obtained gives the propagation law of the front. When $s < \frac{1}{2}$, the quantity $\frac{1}{\eta_\varepsilon}\overline{a}(t, x, e)$ converges as $\varepsilon \to 0$ to a fractional mean curvature operator, as proven in Lemma 10 of [26]. So in this case one would get (11.5.10) with the local mean curvature operator replaced by a fractional one.

Let us now state the precise result. To simplify the presentation, we consider the isotropic case. For the anisotropic case we refer to [26].

Theorem 11.5.1 (Theorem 1, [26]). *Let J be given by (11.5.4) with $g \equiv c_{N,s}$ and $s \in \left[\frac{1}{2}, 1\right)$. Let u^ε be the unique solution of (11.5.1) with initial datum (11.5.5), where $q(x, e) = q(x) = u(x)$ is the solution of (11.2.6) and d_0 is the signed distance function to the boundary of a smooth set Ω_0. Then, there exists a symmetric matrix $A \in \mathcal{S}(n)$ depending on q, s and N, such that if u is the unique (generalized flow) solution of the geometric equation (11.5.10) with initial condition $u(0, x) = d_0(x)$, where γ is defined by (11.5.8), the function u^ε satisfies, for $t > 0$, $x \in \mathbb{R}^N$,*

$$\begin{cases} u_\varepsilon(t, x) \to 1 & \text{in } \{u(t, x) > 0\} \\ u_\varepsilon(t, x) \to 0 & \text{in } \{u(t, x) < 0\} \end{cases} \quad \text{as } \varepsilon \to 0.$$

Moreover both limits are local uniform.

11.6 Some Open Problems

A first open problem, to the best of our knowledge, is the extension of Theorem 11.5.1 to the case $s \in \left(0, \frac{1}{2}\right)$. Indeed, in [26] it is proven that, for a smooth function d, such that $|\nabla d| = 1$, then

$$\frac{1}{\eta_\varepsilon}\overline{a}_\varepsilon(t, x, e) \to k[d](x)$$

where k is a fractional operator and $\overline{a}_\varepsilon$ is defined in (11.5.9). This suggests that the front moves according the following fractional mean-curvature equation:

$$\partial_t d(t,x) = \gamma k[d(t,\cdot)](x)|Dd|$$

with μ and k depending on the gradient variable in the anisotropic case. The proof of Theorem 11.5.1 relies on the construction of barriers, i.e., sub and supersolution of (11.5.1), which are suitable correction of the Ansatz. In the case $s < \frac{1}{2}$, the authors are not able to construct a barrier far from the front. Indeed, in the case $s < \frac{1}{2}$ the contributions from far from the front are not negligible. This is somehow expected, see for instance [33].

A further interesting problem in this direction, with important applications in the theory of crystal dislocations, consists in extending the result of Section 11.3 to higher dimensions, i.e., proving analogous results to those presented in Section 11.5 in the case of two or more dislocation lines. To provide a concrete example, suppose that there are two closed dislocation lines in the slip plane x_1x_2. This situation can be modeled by equation (11.5.1), with associated initial condition

$$u_\varepsilon^0(x) = q\left(\frac{d_0^1(x)}{\varepsilon}, Dd_0^1(x)\right) + q\left(\frac{d_0^2(x)}{\varepsilon}, Dd_0^2(x)\right), \tag{11.6.1}$$

where d_0^i is the signed distance function to the boundary of a smooth set Ω_0^i, $i = 1, 2$, $\Omega_0^1 \subset \Omega_0^2$ and the two sets are at positive distance one from each other. Here q is solution of (11.5.3). Existence of such layer solutions needs to be proven as well. In the isotropic case, solutions of (11.5.3) are given by $q(x) = u(x - x_0)$, for any $x_0 \in \mathbb{R}$, where u is the unique solution of (11.2.6).

The solution u_ε of (11.5.1)-(11.6.1), will convergence as $\varepsilon \to 0$ to the following stable equilibria of W': 0,1 and 2. Let $\Omega_1(t)$ and $\Omega_2(t)$ be the resulting interphases at time t starting from the initial configurations Ω_0^1 and Ω_0^2 respectively. We expect a double behavior for the motion of these fronts. Indeed, we think that in the case $s \geqslant \frac{1}{2}$, the fronts move by local mean curvature and the interaction between the two fronts is negligible at a first level of approximation, i.e., the mean curvature motion is predominant. On the other hand, when $s < \frac{1}{2}$, we expect that the interaction is not negligible anymore and that the equations for the motion of the fronts involve a fractional mean curvature operator and a local term taking into account the interaction of the fronts.

If the expected results hold true, this would suggest to investigate a higher order of asymptotics for the solution of (11.5.1)-(11.6.1) to gather how dislocation lines interact, having in mind that the case $s = \frac{1}{2}$ has a physical interest in view of its applications to the theory of crystal dislocations.

Finally, one could also consider dislocation lines with different orientations. In this case, the initial datum, with two of them would be

$$u_\varepsilon^0(x) = q\left(\frac{d_0^1(x)}{\varepsilon}, Dd_0^1(x)\right) + q\left(-\frac{d_0^2(x)}{\varepsilon}, -Dd_0^2(x)\right).$$

In this situation, we expect some sort of collision in finite time.

Bibliography

[1] R. Alicandro, L. De Luca, A. Garroni and M. Ponsiglione, Dynamics of discrete screw dislocations on glide directions, *Journal of the Mechanics and Physics of Solids*, **92** (2016), 87-104.

[2] R. Alicandro, L. De Luca, A. Garroni and M. Ponsiglione, Metastability and dynamics of discrete topological singularities in two dimensions: a Γ-convergence approach, *Arch. Rational Mech. Anal.*, **214** (2014), no. 1, pp 269-330.

[3] O. Alvarez, P. Hoch, Y. Le Bouar and R. Monneau, Dislocation dynamics: short-time existence and uniqueness of the solution, *Arch. Ration. Mech. Anal.*, **181** (2006), no. 3, 449-504.

[4] S. Awatif, Équations d'Hamilton-Jacobi du premier ordre avec termes intégro-différentiels. I. Unicité des solutions de viscosité. II. Existence de solutions de viscosité,*Comm. Partial Differential Equations*, **16** (1991), no. 6-7, 1057-1093.

[5] T. Blass, I. Fonseca, G. Leoni and M. Morandotti, Dynamics for systems of screw dislocations, *SIAM J. Appl. Math.* **75** (2015), no. 2, 393-419.

[6] O. M. Braun and Y. S. Kivshar, The Frenkel-Kontorova Model, Concepts, Methods and Applications, *Springer-Verlag*, (2004).

[7] X. Cabré and Y. Sire, Nonlinear equations for fractional Laplacians II: existence, uniqueness, and qualitative properties of solutions, *Trans. Amer. Math. Soc.*, **367** (2015) no. 2, 911-941.

[8] P. Cermelli and G. Leoni, Renormalized energy and forces on dislocations, *SIAM J. Math. Anal.*, **37** (2005), no. 4, 1131-1160.

[9] S. Conti, A. Garroni and S. Müller, Dislocation microstructures and strain-gradient plasticity with one active slip plane, *J. Mech. Phys. Solids*, to appear.

[10] X. Chen, Generation and propagation of interfaces for reaction-diffusion equations, *J. Differential Equations*, **96** (1992), no. 1, 116-141.

[11] C. Denoual, Dynamic dislocation modeling by combining Peierls Nabarro and Galerkin methods, *Phys. Rev. B*, **70** (2004), 024106.

[12] S. Dipierro, A. Figalli and E. Valdinoci, Strongly nonlocal dislocation dynamics in crystals, *Commun. Partial Differ. Equations* **39** (2014) no. 12, 2351-2387.

[13] S. Dipierro, G. Palatucci and E. Valdinoci, Dislocation dynamics in crystals: a macroscopic theory in a fractional Laplace setting, *Comm. Math. Phys.*, **333** (2015) no. 2, 1061-1105.

[14] A. Z. Fino, El Hajj, H. Ibrahim and R. Monneau, The Peierls-Nabarro model as a limit of a Frenkel-Kontorova model, *J. Differential Equations*, **252** (2012), no. 1, 258-293.

[15] N. Forcadel, C. Imbert and R. Monneau, Homogenization of dislocation dynamics and some particle systems with two-body interactions. *Discrete and Continuous Dynamical Systems - A*, **23** (2009), no.3, 785 - 826.

[16] N. Forcadel, C. Imbert and R. Monneau, Homogenization of fully overdamped Frenkel-Kontorova models, *Journal of Differential Equations*, **246** (2009), no. 1, 1057-1097.

[17] A. Garroni, G. Leoni and M. Ponsiglione, Gradient theory for plasticity via homogenization of discrete dislocations, *J. Eur. Math. Soc.*, **12** (2010), no. 5, 1231-1266.

[18] A. Garroni and S. Müller, A variational model for dislocations in the line tension limit, *Arch. Ration. Mech. Anal.*, **181** (2006), 535-578.

[19] M. González and R. Monneau, Slow motion of particle systems as a limit of a reaction-diffusion equation with half-Laplacian in dimension one, *Discrete Contin. Dyn. Syst.*, **32** (2012), no. 4, 1255-1286.

[20] E. R. Jakobsen and K. H. Karlsen, Continuous dependence estimates for viscosity solutions of integro-PDEs. *J. Differential Equations*, **212** (2005), 278-318.

[21] A. El Hajj, H. Ibrahim and R. Monneau, Dislocation dynamics: from microscopic models to macroscopic crystal plasticity, *Continuum Mechanics and Thermodynamics*, **21** (2009), no. 2, 109-123.

[22] A. K. Head, Dislocation group dynamics III. Similarity solutions of the continuum approximation, *Phil. Magazine*, **26**, (1972), 65-72.

[23] J. R. Hirth and L. Lothe, Theory of dislocations, Second Edition. Malabar, Florida: Krieger, 1992.

[24] C. Imbert and R. Monneau, Homogenization of first order equations with u/ε-periodic Hamiltonians. Part I: local equations, *Archive for Rational Mechanics and Analysis*, **187** (2008), no. 1, 49-89.

[25] C. Imbert, R. Monneau and E. Rouy, Homogenization of first order equations with u/ε-periodic Hamiltonians. Part II: application to dislocations dynamics, *Communications in Partial Differential Equations*, **33** (2008), no. 1-3, 479-516.

[26] C. Imbert and P. E. Souganidis, Phasefield theory for fractional diffusion-reaction equations and applications, *preprint*.

[27] N.S. Landkof, Foundations of Modern Potential Theory, Springer-Verlag, 1972.

[28] R. Monneau and S. Patrizi, Derivation of the Orowan's law from the Peierls-Nabarro model, *Comm. Partial Differential Equations*, **37** (2012), no. 10, 1887-1911.

[29] R. Monneau and S. Patrizi, Homogenization of the Peierls-Nabarro model for dislocation dynamics, *J. Differential Equations*, **253** (2012), no. 7, 2064-2015.

[30] A.B. Movchan, R. Bullough and J.R. Willis, Stability of a dislocation: discrete model, *Eur. J. Appl. Math.* **9** (1998), 373-396.

[31] F.R.N. Nabarro, Dislocations in a simple cubic lattice, *Proc. Phys. Soc.*, **59** (1947), 256-272.

[32] F. R. N. Nabarro, Fifty-year study of the Peierls-Nabarro stress. *Mat. Sci. Eng. A*, **234-236** (1997), 67-76.

[33] G. Palatucci, O. Savin and E. Valdinoci, Local and global minimizers for a variational energy involving a fractional norm. *Ann. Mat. Pura Appl.*, (4) **192** (2013), no. 4, 673-718.
[34] R. Peierls, The size of a dislocation, *Proc. Phys. Soc.*, **52**(1940), 34-37.
[35] S. Patrizi and E. Valdinoci, Crystal dislocations with different orientations and collisions, *Arch. Rational Mech. Anal.*, **217** (2015), 231-261.
[36] S. Patrizi and E. Valdinoci, Homogenization and Orowan's law for anisotropic fractional operators of any order, *Nonlinear Analysis: Theory, Methods and Applications*, **119** (2015), 3-36.
[37] S. Patrizi and E. Valdinoci, Relaxation times for atom dislocations in crystals, *Calc. Var. Partial Differential Equations*, **55** (2016) no. 3, 1-44.
[38] S. Patrizi and E. Valdinoci, Long time behavior for crystal dislocation dynamics, *submitted.*
[39] O. Savin and E. Valdinoci, Γ-convergence for nonlocal phase transitions. *Ann. Inst. H. Poincaré Anal. Non Linéaire*, **29** (2012), no. 4, 479-500.

Kolibri 360 GmbH
Ferdinand-Jühlke-Straße 7
99095 Erfurt, Deutschland
produktsicherheit@kolibri360.de